AF685836

Proceedings of the Conference on

Geotechnical Practice In Offshore Engineering

Edited by Stephen G. Wright

Sponsored by the Geotechnical Engineering Division of the American Society of Civil Engineers

University of Texas at Austin
Austin, Texas
April 27-29, 1983

Published by the
American Society of Civil Engineers
345 East 47th Street
New York, New York 10017-2398

The Society is not responsible for any statements made or opinions expressed in its publications.

Copyright © 1983 by the American Society of Civil Engineers,
All Rights Reserved.
Library of Congress Catalog Card No.: 83-70397
ISBN 0-87262-360-2
Manufactured in the United States of America.

FOREWORD

This Specialty Conference, one of a group designed to further the 1985 International Conference on Soil Mechanics and Foundation Engineering to be held in the United States, includes papers on geotechnical engineering concerned primarily with site specific studies, case histories, and research topics.

Robert M. Semple wrote recently, "The size of offshore foundations precludes load testing, virtually ensures significant inhomogeneity of ground conditions, and results in complex soil-structure interaction. Environmental loading introduces variable repeated and dynamic forces." Semple goes on to say that analytical techniques have been introduced that are valuable to the designer but that the validity of the techniques is unknown (*Ground Engineering,* November, 1982). These views coincide with those of the Task Committee. Strong efforts went into making sure that the conference itself and these Proceedings contribute to the aims of providing for presentations on the state-of-the-art, bringing recent advances to the attention of the profession, and indicating needs and means for additional research.

Appreciation is expressed to the Local Committee: R. E. Olson, Budget and Finance; K. H. Stokoe, II and Howard Hendrickson, Arrangements and Publicity; D. E. Daniel, Program and Entertainment; S. G. Wright, Publication; J. M. Roesset, International Visitors; and Eva Lee Reese, Ouida Wright, and Judy Roesset, Spouses Program. Special thanks go to the reviewers who worked under a strict time schedule: J. M. E. Audibert, F. Bryant, R. D. Darragh, E. Doyle, W. Dunlap, J. Hooper, W. N. Houston, L. M. Kraft, Jr., C. C. Ladd, J. D. Murff, I. Noorany, J. Pelletier, J. Roesset, R. L. Schiffman, G. Thompson, and B. Watt.

The Conference would not have been possible without the willingness of distinguished engineers to submit papers and to take part in the program. The authors are recognized by the papers that are included, and others are shown in the brief listing of the program that follows:

WEDNESDAY

MAIN SESSION, LBJ Auditorium

Moderator: Lymon C. Reese, The University of Texas at Austin, TX
Welcome, Earnest F. Gloyna, Dean, College of Engineering

Overview of Offshore Practice
Bramlette McClelland, McClelland Engineers, Inc., Houston, TX

Geotechnical Investigation for Offshore Sites
Jaap de Ruiter, Fugro Consultants International B.V., The Netherlands

TOPICAL SESSION

SOIL PROPERTIES
Moderator: Ronald F. Scott
California Institute of Technology, Pasadena, CA

TESTS ON MODEL PILES
Moderator: Robert E. Smith
ARCO Oil and Gas Company, Dallas, TX

SITE INVESTIGATION
Moderator: Wayne Ingram
Seafloor Engineers, Houston, TX

DESIGN OF PILES FOR AXIAL LOAD
Moderator: Robert D. Darragh
Dames & Moore, San Francisco, CA

THURSDAY

MAIN SESSION, LBJ Auditorium

Moderator: F. Patrick Dunn, Shell Oil Company, Houston, TX

Design of Single Piles,
Hudson Matlock, Ertec, Inc., Long Beach, CA

Group Action in Offshore Piles
Michael W. O'Neill, The University of Houston, Houston, TX

TOPICAL SESSION

DESIGN OF PILES FOR LATERAL LOAD
Moderator: John A. Focht, Jr.
McClelland Engineers, Inc., Houston, TX

IN-SITU TESTING
Moderator: James K. Mitchell
The University of California at Berkeley, Berkeley, CA

GRAVITY STRUCTURES AND ANCHORS
Moderator: Kaare Hoeg
Norwegian Geotechnical Institute, Oslo, Norway

PILE INSTALLATION
Moderator: Griff Lee
J. Ray McDermott Company, New Orleans, LA

FRIDAY

MAIN SESSION, LBJ Auditorium

Moderator: H. Ray Brannon, Jr.
EXXON Production Research Company, Houston, TX

Design for Dynamic Loadings
Robert G. Bea
PMB, Offshore Division, San Francisco, CA

Instrumentation in the Offshore Environment
Elmo DiBiagio, Norwegian Geotechnical Institute, Oslo, Norway

Landslide at Port of Nice
H. Bolton Seed, The University of California at Berkeley, Berkeley, CA

PANEL DISCUSSIONS: *Future Directions*

Moderator: Bramlette McClelland
McClelland Engineers, Inc., Houston, TX

Panel Members:
Robert G. Bea, PMB, Offshore Division, San Francisco, CA

John A. Focht, Jr. McClelland Engineers, Inc., Houston, TX

Kaare Hoeg, Norwegian Geotechnical Institute, Oslo, Norway

Griff Lee, J. Ray McDermott Company, New Orleans, LA

J. D. Murff, EXXON Production Research Company, Houston, TX

Stephen G. Wright, The University of Texas at Austin, Austin, TX

W. J. Rigden, British Petroleum, London, England

Francois Baguelin, Laboratoire Central des Ponts et Chaussees, Bouguenais, France

Jaap de Ruiter, Fugro Consultants International B.V., The Netherlands

OVERVIEW OF CONFERENCE
H. Bolton Seed, The University of California at Berkeley, Berkeley, CA

Others making significant contributions include: Dr. Earnest F. Gloyna, Dean of the College of Engineering, The University of Texas at Austin, for his support of the Conference and contribution to the opening program; B. McClelland, J. de Ruiter, H. Matlock, M. W. O'Neill, R. G. Bea, E. DiBiagio, H. B. Seed, J. A. Focht, Jr., K. Hoeg, G. Lee, J. D. Murff, S. G. Wright, W. J. Rigden, and F. Baguelin, for their contributions in the main sessions of the Conference. Also, special thanks go to the following for agreeing to serve as moderators of sessions: H. R. Brannon, Jr., R. D. Darragh, F. P. Dunn, J. A. Focht, Jr., K. Hoeg, W. Ingram, G. Lee, B. McClelland, J. K. Mitchell, R. F. Scott, and R. E. Smith.

The Task Committee, acting for the Geotechnical Engineering Division, ASCE, includes: W. R. Cox, E. Doyle, W. Ingram, H. Matlock, H. W. Olsen, L. C. Reese (Chairman), and H. B. Seed.

L. C. Reese
February 15, 1983

CONTENTS

STATE-OF-THE-ART PAPERS

SUBMITTED PAPERS

SOIL PROPERTIES

SITE INVESTIGATIONS

IN-SITU TESTING

MARINE GEOTECHNICAL INVESTIGATIONS, A MATURE TECHNOLOGY

By J. de Ruiter, M.ASCE[1] and A.F. Richards, F.ASCE[2]

INTRODUCTION

Offshore site investigations found their origin in the shallower and relatively friendly waters of the Gulf of Mexico. The piled foundations of the small jacket structures could be designed and installed on the basis of fairly simple soil investigations, consisting usually of testing percussion samples taken from open boreholes. This technique proved to be inadequate later for the deeper waters, hostile environments and more difficult soil profiles in other offshore areas of the world, including the North Sea, where major developments of the petroleum industry have taken place since the early 1970's. In the meantime, huge production platforms have been installed in various parts of the world requiring careful and precise geotechnical investigations and design studies for safe and economic foundations.

These circumstances have led to vast and rapid improvements in technology. It can safely be said that at present the quality of geotechnical data collection offshore can rival and may even exceed any such investigation onshore. The big difference remains the cost of the investigations, which is many times greater offshore than onshore. An inevitable consequence is that in many instances offshore investigations are still inferior, because the available budgets are insufficient.

Obtaining good quality soil samples is even more difficult offshore than onshore. New techniques were primarily aimed therefore at determination of soil properties by in situ testing. Cone penetrometer testing is now a standard part of many important soil surveys, while a range of other devices, such as the pressuremeter and the in situ vane can be applied if necessary. For unusual soil conditions, e.g. calcareous sediments, special techniques have been developed that can be deployed with the basic systems used during a standard soil investigation program. In the meantime, techniques for obtaining undisturbed samples and for rock coring also have been improved.

Thus a mature technology has come into existence with a degree of sophistication that makes it possible to meet the demands of an offshore industry that continues its move to greater water depths with an increase in size and complexity of the structures. This technology is an intricate combination of vessels, drilling equipment, down-hole carrier tools and sensors in which the human factor, i.e. the skill, experience and dedication of the entire crew, remains often decisive

[1]Director of Fugro Consultants International B.V., The Netherlands.
[2]Senior Consultant, Fugro B.V., The Netherlands.

for the quality of the results. An important part has become the sample testing and data interpretation on board the survey vessel to have an immediate evaluation of the adequacy of the data. The cost of a second investigation is usually prohibitive, so that lack of precision and completeness of the results would be detrimental.

The present technology was clearly developed in response to the demands of the petroleum industry. A salient aspect therein is that full scale load tests offshore are practically impossible; consequently, precise geotechnical data and advanced engineering analyses are essential. The maturity of the technology now available is demonstrated by its use for other purposes than foundation studies for offshore oil and gas developments, and its influence on geotechnical studies for land projects.

This paper places the geotechnical data collection offshore in the context of a general site investigation program. Subsequently, it presents an overview of the current state of the art in marine geotechnical investigations to an extent relevant to indicate a mature technology. Only brief references are made to equipment that is now considered standard and on which ample information has been published by various authors in the past. Tools and techniques that are relatively novel are discussed more extensively. This paper deals specifically with investigations in water depths of over 30 meters, where land techniques are no longer applicable. Finally, it touches upon some applications for purposes that are not related to the offshore petroleum industry and takes a look at expected future developments. Further details of offshore site investigations have been summarized recently (22).

GENERAL APPROACH

A typical field program comprises two principal components: the geophysical survey and the geotechnical investigation. These two phases are usually carried out with different vessels and at different times, with the geophysical part preceding the geotechnical.

The program is generally defined by the question whether the investigation is considered site specific or structure specific. In the first instance, the location of the site is definite and the soil conditions may influence the foundation design and even the basic structure type. In the latter case, the type of structure has already been selected and some latitude in the location of the site should be allowed to ensure that suitable soil conditions exist.

A complete offshore soils study can be divided in a number of distinct activities. These are summarized in Table 1. Some of these activities require clarification and elaboration and they are discussed further in the following sections. Other factors that need to be considered for site selection are discussed in detail by reference 30.

GEOPHYSICAL AND BATHYMETRIC SURVEY

Bathymetry and shallow seismic profiling should be performed early enough so that the results can be studied for a proper planning of the subsequent geotechnical investigations. The survey should cover a large enough area to obtain a reliable picture of the seabed conditions at and in the vicinity of the proposed structure sites. The operation is

Table 1 SEQUENCE OF ACTIVITIES FOR A PROGRAM OF MARINE GEOTECHNICAL STUDIES

1. Definition of the site or sites to be investigated
2. Examination of available soil data at the site or in the vicinity of the site
3. Preliminary definition of the required scope of the investigations as a function of the type of the proposed structure and of the anticipated soil and environmental conditions
4. Selection and possible installation of a positioning system, which allows an accurate determination of the location of survey lines, boreholes etc. relative to the position of the structure that will be placed at a later date
5. Performance of a bathymetric and shallow geophysical survey to determine seabed topography and subbottom stratification
6. Definition of the final program of geotechnical investigations, taking into account the conditions and variability in soil profile derived from the geophysical survey, and the requirements of the proposed structure and site specific factors, e.g. seismicity and slope stability.
7. Performance of the geotechnical investigations, comprising sampling and often in situ testing and including the on board testing of samples for a preliminary evaluation of the data
8. Further testing of samples in an onshore laboratory and complete interpretation of all data to establish a final soil profile and the strength and deformation characteristics of the materials encountered
9. Foundation design, including bearing capacity and load deformation behavior and also covering installation feasibility
10. Monitoring the behavior of the structure during placement on the seabed and subsequently the installation of the foundation to check the design assumptions

carried out usually with an array of instruments that include an echo sounder, side scan sonar and seismic reflection tools, such as a high resolution boomer, a deep penetration sparker or a 3.5 kHz system. It may further include seabed sampling by means of grab samples, vibrocorers or gravity corers. Apart from the seabed topography, the results will reveal subbottom features such as buried channels and bedrock. In addition, the records should be studied with respect to possible signs of gas in the soils and evidence for seabed instability. All these phenomena give a first indication of the feasibility of the site/structure combination being contemplated.

Conventional echo sounders are often inadequate for water depth determination because they are accurate only to about 2% of the water depth. Precision depth recorders coupled with at-site calibration of the mean sound velocity using a velocimeter yield a theoretical accuracy of 0.01%, but in practice nearer to 0.1%. Normally this is adequate for siting structures offshore.

Two new types of echo sounders have become available: (1) dual-frequency echo sounders, which permit concurrent surface and very high resolution subbottom profiling and (2) the 'Seabeam' system that automatically computes and draws an actual bathymetric map over a swath of terrain beneath the vessel in almost any water depth. The latter system is not yet in use for geotechnical site selection, but it has applicability, particularly for surveying on the continental slope and in the ocean basins.

Maps of surficial physiographic features, and particularly for delineating the extent of seafloor surface instability, are obtained from side-scan sonar records. Three types of side-scan sonars (Table 2) have become available in the past several years that permit the observation of an entire continental margin (shelf to abyssal plain) in one record (Fig. 1) at one end of the spectrum (see Ref. 35, for example) to highly detailed, rectified small-scale maps at the other end (Fig. 2 and Ref. 24). This permits offshore engineers for the first time to be selective in the area covered and the detail desired.

Table 2 SIDE-SCAN SONARS*

Range	Examples	Approximate Maximum Water Depth, Km	Approximate Slant Range, Km
Long	GLORIA	7.5	70
Intermediate	SEA MARC	6	6
Short	Digitally-acquired scale-corrected type	1 - 6	0.5

* After 28

Seismic reflection techniques that are conventionally used for site investigations have not greatly changed in recent years; however, introduction of deeply towed instruments has improved the quality of the results. Increased resolution can also be obtained by digital signal processing by computer.

It should be emphasized that the geophysical survey provides invaluable data about the stratigraphy and therefore of the degree of uniformity of the soil conditions in the area considered for development. Because of the uninterrupted horizontal continuity of the survey, a picture of the stratigraphy can be obtained that is often superior to geophysical profiling on land. A number of strata are usually encountered that are further identified during the geotechnical investigations. The combination of geophysical and geotechnical data from a multi-line survey produces a three-dimensional picture of the

FIG. 1— A GLORIA sonogram of the continetal slope west of France showing submarine canyons extending transverse to the slope (16). Horizontal scale about 20 nautical miles.

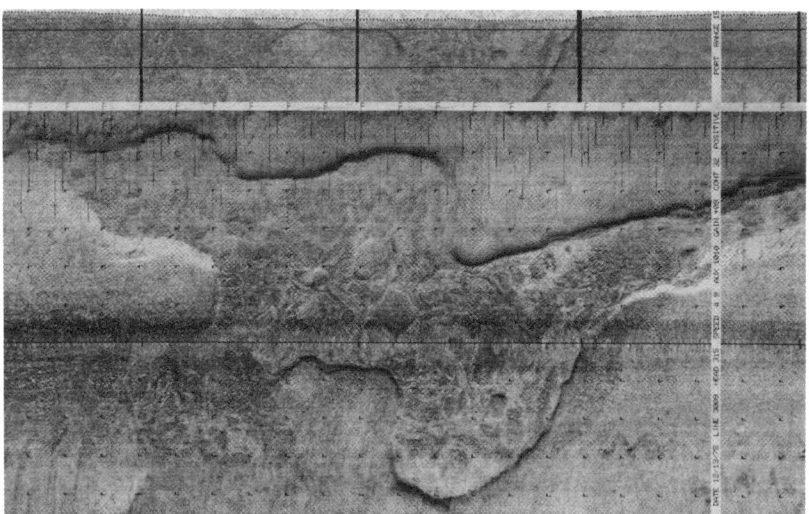

FIG. 2— Part of a submarine slide off British Columbia (24). Swath width 300m.

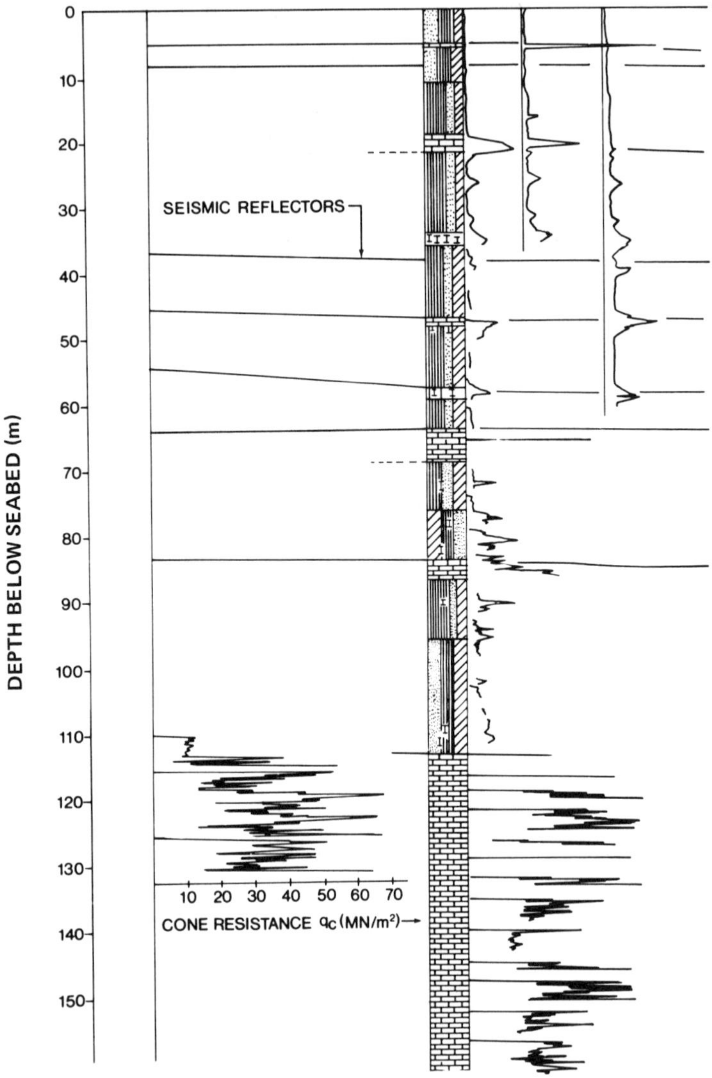

FIG. 3– Typical CPT's and seismic reflections in calcareous sediments (17).

soil conditions. It allows an assessment of the degree of variability of the conditions and a precise identification of the soil strata, which will be critical in the foundation design.

Figure 3 shows geophysical and geotechnical data obtained in the Northwest Shelf of Australia in water depths of about 125 meters (17). There is remarkable agreement between the peaks in the cone resistance and the reflectors in the geophysical profile. The soil conditions in this area show an unusual uniformity over great distances, making it possible to reduce the number of boreholes to the minimum necessary to take a sufficient number of samples and perform a required program of in situ testing.

GEOTECHNICAL INVESTIGATION VESSELS AND DRILLING SYSTEMS

The selection of vessel and drill rig for the geotechnical investigation is controlled by a number of considerations of which the more important ones are: (1) sea conditions, including water depth, expected sea state and seabed condition in connection with anchoring; (2) program of investigations as determined by the proposed structure and the anticipated soil conditions; and (3) the availability and cost of the vessel and drilling equipment.

Standard North Sea practice is that vessel and drill rig are provided by the vessel owner as a single package. The ships have a large central moonpool of about 4 x 4 m to enable the easy lowering of heavy pieces of equipment to the seabed. Tools for sampling and in situ testing are provided by the geotechnical consultant. These have been developed to match the vessel capabilities and are compatible with the 5 inch diameter API drill pipe used as a standard.

North Sea vessels traditionally have employed a spread of 4 to 6 anchors to maintain position. This system is adequate to water depths of about 300 m. The vessels are self anchoring with the assistance of a bow thruster, although in greater water depths the help of a tug boat may be required. Some vessels with dynamic positioning (DP) are entering the market currently and they have no practical limitation with respect to water depth. They have an additional advantage in that they can operate close to existing structures, pipelines and seabed installations. DP vessels are larger than the anchored vessels and therefore more stable, which has a positive influence on their rate of geotechnical production; however, they demand a higher day rate.

Drilling is performed with an open mud system. Drilling mud is circulated through the drill pipe and returned through the annular space between the drill pipe and the wall of the hole. It stabilizes the hole and transports the drill cuttings. Mud and cuttings are spilled over the seafloor, i.e. no recirculation occurs. The drill pipe remains in the hole during the entire drilling and testing program. Tools for sampling and in situ testing are lowered through the drill pipe, while the drilling process is interrupted. Normal practice on North Sea vessels is to rotate the drill pipe with a power swivel that is located on top of the drill string (Fig. 4). This system has some advantages over the power tongs used previously: (1) the tools for sampling and testing can be passed through the power swivel after opening a lid at the top, which allows introduction of those tools at any penetration depth without disengaging the drill string, and (2) the integrated system of mud circulation and drill pipe rotation allows a

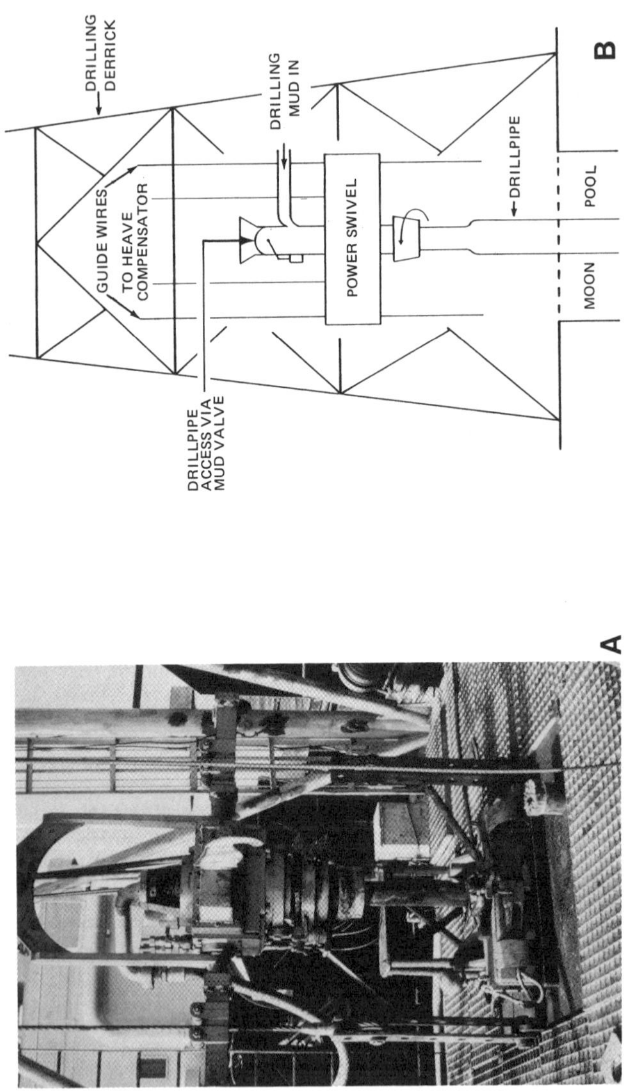

FIG. 4— Power swivel
A photograph on board drilling vessel
B schematic diagram of system

higher production rate in the range of 2.5 to 5 m of penetration per hour with full sampling and in situ testing.

The drilling system of a North Sea vessel also comprises a hydraulically activated motion compensator to overcome the problems associated with the heave of the vessel. It minimizes the variations in bit pressure and reduces the disturbance of the soil underneath the drill bit below the bottom of the hole.

Extensive use currently is made by North Sea vessels of drill string control units which serve at the same time as re-entry frames such as the Fugro 'Seaclam' (Fig. 5) and the McClelland 'Stingray' (32). These units are lowered to the seabed through the moonpool prior to or simultaneously with the drill pipe. They contain a clamping device that holds the drill string in position during sampling and in situ testing. Thus the available reaction for push sampling and cone penetration can be increased to about 100 kN. A utility guide frame can slide up and down the rigging lines providing a re-entry facility for the drill string in the hole. The seabed control units sit on the seabed and permit a fixed reference point to be established on the rigging above the drill floor. This has enhanced the accuracy of the depth determination during drilling and testing. In general, these units have improved the quality and the speed of geotechnical investigations at sea.

The equipment on board further includes the facilities for mixing and pumping drilling mud, the hydraulic and electric power supplies for the sampling and testing tools and the read-out and recording sets to monitor the tests. In addition the ship should provide space for a soil testing laboratory and for workshops for repair and maintenance of the equipment. The whole set-up should ensure a degree of self support that allows completion of a program of investigations without interruptions while waiting for assistance from shore.

The North Sea vessels offer an ideal combination of facilities for a complete geotechnical investigation, including a wide range of in situ tests. For this reason they also are used regularly for marine investigations in other areas of the world, provided that the scope of work and the financial budget make it possible to accept their high cost of mobilization and demobilization. North Sea vessels have been used for soil surveys around the world from India to the West Coast of the U.S.A. and from Japan to Argentina. They are furthermore attractive because their advanced equipment and experienced crews ensure a fast rate of production together with a consistent quality of results.

An alternative to the special drilling vessels are the supply boats with a detachable drill rig, used traditionally in the Gulf of Mexico (23). The most popular drill rigs are the Failing machines, which can be installed quickly on the vessel when it is being prepared for the investigation. These boats have a central moonpool with a diameter of 1.0 m or less and are equipped with a four point anchoring system. After the investigation the drill rig is removed and the vessel returns to its normal operations. The Failing rigs are quite suitable for conventional drilling operations with an open mud system and for percussion sampling. The smaller types customarily used on supply boats cannot handle a seabed control unit and may have limitations for in situ testing and high quality sampling.

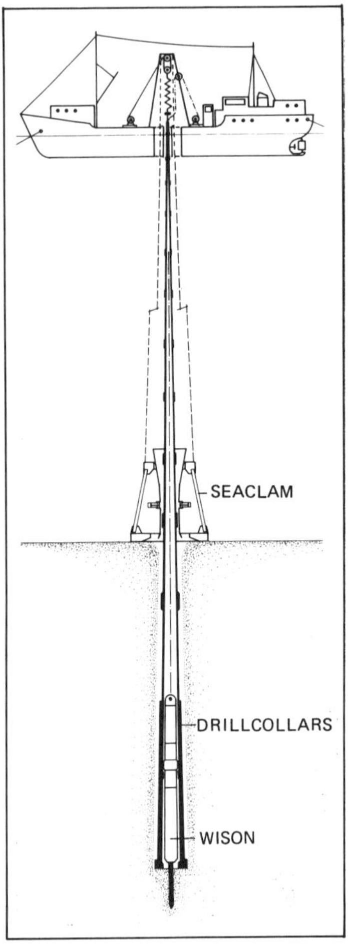

↑ FIG. 6– Wirth offshore drilling rig

←FIG. 5– Seaclam (after Fugro)

To overcome the shortcomings of the supply boat with Failing rig operation, the authors' company has modified a Wirth drilling machine (Fig. 6) for marine geotechnical investigations. It is a skid mounted rig with the full capabilities of the rigs installed on North Sea vessels. It is equipped with a power swivel and motion compensation system. It uses the standard 5 inch diameter API drill pipe and has a hoisting capacity to drill to a depth of 600 m below the sea surface. It can be mounted on a variety of vessels and barges, either over a moonpool or cantilevered over the side or over the stern. It can handle a seabed control unit and perform a full scale investigation including in situ testing. The rig itself and the ancillary equipment, including mud tanks and pumps, have ISO container dimensions permitting easy and expedient shipment on standard container vessels to any location.

MEASUREMENT OF SOIL AND ROCK PROPERTIES

An ever increasing variety of improved samplers and sensors that are mated to drilling and remote seabed systems has been developed in recent years. This development tends to follow closely the actual and perceived needs of different segments of industry. In particular, specialized sensors are becoming available that are used in conjunction with the basic cone penetrometer test (CPT). The literature describing and reviewing recent developments is extensive. References 8 and 9 have summarized the use and interpretation of the cone penetrometer and some of the other ancillary sensors that may be used with this tool. Reference 5 contains a useful summary of general in situ test methods. Reference 27 has reviewed the extent and performance of marine in situ equipment. Table 3 indicates the variety of deployment systems that have been used offshore.

In the 1970's, hydraulically operated, wireline push samplers were developed and extensively used for offshore site investigations using stainless steel, thin-walled shelby tubes 75 mm in diameter. In the early 1980's, these samplers are being superseded by hydraulically operated, wireline piston samplers, which substantially improve sample quality in soft soils when the drill string has been stabilized to prevent vertical movement during the drive stroke. The Deep Sea Drilling Project at the University of California, San Diego, for a number of years has utilized a hydraulically operated, wireline piston sampler to raise samples of quite good quality from the ocean basin seafloor.

A relatively new development is the pressurized core barrel. Although the Deep Sea Drilling Project for several years has used a pressurized sampler for geochemical purposes, this wireline tool takes a homogenized sample that is remolded. Several companies have developed downhole pressure core barrels to collect soil samples for residual fluid analysis in petroleum reservoir studies, but at present the large diameter of the sampler precludes its use with 5 inch drill rods and the sample quality is understood to be poor for geotechnical purposes. A prototype pressurized core barrel for geotechnical sampling has been developed and successfully used in continental shelf water depths in the Gulf of Mexico by personnel at Texas A & M University (7). This sampler is reported to take cores of reasonably good quality that have been tested in a hyperbaric chamber by divers trained in operating

Table 3 TYPES OF IN SITU TESTING EQUIPMENT DEPLOYMENT SYSTEMS*

Deployment System	Notes
Wire-line, Carrier Tool, & Sensor Carrier tool contains jack Entire drillstring jacked by seabed jack	Both subtypes require drilling rig, drillstring, carrier tool, and sensor.
Seabed Unit Unmanned Test rods jacked or motor driven	
Test rods attached to vessel	Water depth limited
Test rods integral with unit, which is fully remotely operated	Water depth limited
Manned Jack operated by free divers	Divers use SCUBA
Jack and divers in a diving bell	Divers in a 1-atmosphere chamber
Vessel Jack	Test rods jacked from vessel, shallow water only
Submersible Unit	Jacked or motor-driven test rods and sensor attached in-board or outboard of sub-mersible
Free-fall Penetrator Free-fall near seabed	Non-expendable unit cable-attached to vessel
Total free-fall through water column	Unit may be fully or partly expendable and the signal may be transmitted over a wire or acoustically through the water column

* After 27

geotechnical laboratory equipment. The tool is not yet suitable for routine use by the offshore industry.

Sampling at sea frequently requires almost immediate knowledge of soil properties, resulting in a number of laboratory tests that are often made on board the survey vessel. In addition to measures of index properties, water content, wet density of the soil, and simple estimations of shear strength from the unconfined compression test, Torvane, pocket penetrometer, etc., triaxial testing can also be undertaken. The authors' company uses a modified Pilcon Engineering Tricon machine to perform automated triaxial tests at sea. On board testing offers the additional advantage that loss of quality of the samples due to time effects and transportation is minimized.

Geotechnical samplers for rock offshore may be grouped into four types: (1) the marine wireline coring system utilizing the Christensen core barrel, which is most suitable for very competent rock; (2) the two types of "piggyback" sytems, one of which uses the standard Longyear coring equipment and the other the Diamec 250 core drill, either of which are positioned on top of casing and thus become very similar to onshore methods; (3) diver-operated, shallow water coring systems; and (4) remotely operated, deep water systems that are self-contained or attached to a submersible, which are not commercially used at present. The Diamec 250 core drill, used on a project for the authors' company, is a high speed, light-weight drilling rig, having highly-automated, drill-rod handling, which cuts a 35-mm diameter sample. It appears to have the capability of sampling softer materials, such as calcareous rocks, than can be readily handled by the Longyear system. An interesting varient of type 2 is the use of a diving bell containing divers, who recently operated rotary drilling equipment to obtain rock cores form the seafloor between Nova Scotia and Newfoundland (34).

The piezocone and the pressuremeter are two sensors that are rapidly gaining widespread use offshore. The pressuremeter has been in use offshore for a decade (2). Within the past few years, second generation pressuremeters are becoming available and used offshore, particularly in Europe but also elsewhere in the world. Equipment for use offshore by wireline and other methods is readily available from British, Canadian, French, and Japanese companies. Comparison tests currently are in progress at the Norwegian Geotechnical Institute to evaluate whether self-boring pressuremeters have a significant advantage over the pushed-in type for offshore soil investigations (33).

The Marchetti (21) dilatometer is another type of stress-strain sensor somewhat similar to the pressuremeter. It has been used in shallow waters offshore in the U.S.A. and Canada. The advantages and disadvantages of this sensor are currently being evaluated as the sensor begins to be deployed offshore.

The piezocone (Fig. 7) was first used offshore by the authors' company in 1981 (36) and is being rapidly adopted for use with or without a friction sleeve. It is important to realize that the pore pressure sensed does not represent in situ ambient pore pressure and that this pressure may consist of a combination of pore-gas and pore-fluid pressure in many offshore soils. While the piezocone yields very useful stratigraphic information (Fig. 8), which is particularly helpful as an aid to soil classification, there are

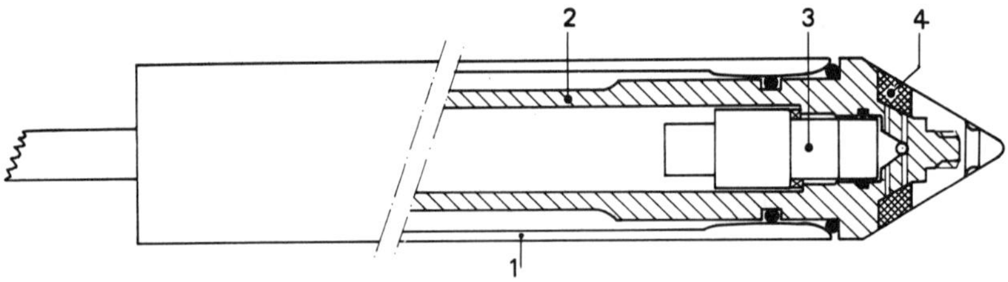

FIG. 7— Fugro Piezocone

1 Mantle
2 Cone loadcell
3 Pressure transducer
4 Filter

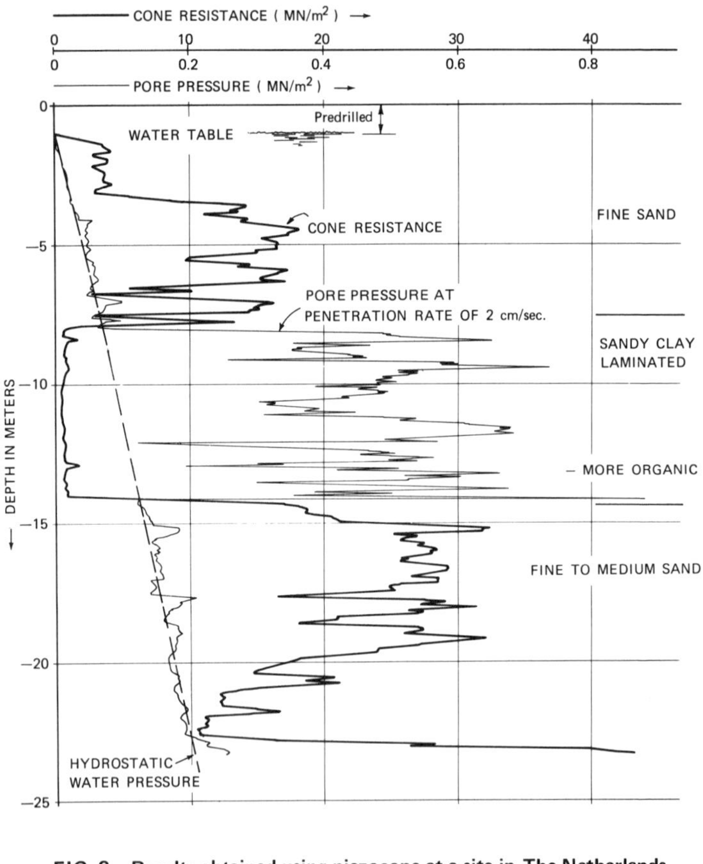

FIG. 8— Results obtained using piezocone at a site in The Netherlands.

problems in how to calibrate the piezometer part of the sensor for a more quantitative interpretation of the dynamic pore pressure. Reference 5 has summarized some of the problems. Research on the theory and practice of the use of the piezocone currently is being conducted at a number of universities.

In situ testing also may involve special tests. Examples include performing axial load tests on a model pile installed in the bottom of a borehole using a modified CPT carrier tool (17); measuring the peak and residual values of grout-soil frictional resistance on grouted pile sections by means of a seabed jacking system (14); and determing the critical mud pressure required to hydraulically fracture the soil adjacent to the bottom of the borehole using a seabed jack, downhole and/or on-deck pumps and a drill string anchor (36).

GEOTECHNICAL MONITORING

Monitoring during and after installation of the structure has become standard practice for the larger platforms or in cases where special problems are anticipated. The most extensive instrumentation has been applied to the gravity structures in the North Sea (11). Measurement is made of the load/penetration behavior during the placing of the structure on the seabed while the skirts penetrate to their design depth. It has been found that the penetration resistance can be predicted accurately on the basis of cone penetration test results (20). After full installation of the structure, a series of measurements are made to observe the development of pore pressures and to follow the long term settlements under operational conditions.

The most common type of geotechnical monitoring offshore are measurements taken during the driving of piles for jacket structures. These measurements can be divided in two types: (1) measurement of the impact velocity of the ram, i.e. of the energy delivered by the driving hammer, and (2) measurement of strain and acceleration at the pile head, which constitutes the so called dynamic pile test.

The foundation design study includes an assessment of the drivability of the piles to ensure that the design penetrations can be achieved with the available hammers. Premature refusal leads to expensive and time consuming remedial measures, which must be avoided. Measurement of the impact velocity of the ram provides a check on the hammer performance, which should correspond with the specifications of the manufacturer. In addition, the observations provide immediate insight whether changes in blow count reflect changes in soil resistance or wether they are connected with variations in hammer efficiency. Records indicate that during the driving of one pile the efficiency can fluctuate from 60 to over 90 % of the rated energy.

The dynamic pile test provides the strain-time and acceleration-time relationships at the head of a pile during a hammer blow. From these measurements a number of data can be derived by means of a wave equation type analysis (4,25): (1) energy losses in pile cap and cushion block by a comparison with the energy delivered by the hammer as determined from ram velocity; (2) the soil resistance during driving and its distribution along the length of the pile, which provides a check on the soil parameters assumed in the design and the drivability study; (3) estimate of the static bearing capacity through correlations between dynamic and static resistances; and (4) integrity of the pile that may be affected by local overstressing.

The static capacity of the piles can best be assessed from a redrive test. The time elapsed since the original drive should be long enough to have a major portion of the ultimate set up on the pile. This period may vary from half an hour in sandy soils to several weeks in normally consolidated clays. On the average, a period of a few hours to some two days should be adequate. The first few blows of the redrive are critical, as the resistance drops back to the residual value when the displacement of the pile increases.

The dynamic measurements can be analyzed on site for an immediate interpretation of the driving behavior and assessment of the pile capacity. Software has been developed for the small computer, which is part of the standard set of equipment mobilized for a program of dynamic pile testing. The data are stored on magnetic tape, so that a more complete analysis can be performed in the office later. Figure 9 shows the set up of a dynamic pile test and a typical result is shown in figure 10.

It is customary to monitor the hammer performance continually by measuring the ram velocity. The dynamic pile test is performed on a small number of representative piles, e.g. one in each group. There may be a variation in the quality of the signals and it is important therefore to record a relatively large number of blows per pile. The most reliable records are selected for complete analysis.

Monitoring during installation is an important part of a program of offshore geotechical studies. Full scale load tests, customary for onshore projects, are virtually impossible under offshore conditions. The data obtained from the monitoring systems are the only check therefore on the accuracy of the soil parameters and the assumptions made during the design. In this connection, it is obvious that driven piles offer an advantage over drilled and grouted piles. The driving of a pile is in fact a large scale dynamic penetration test, which can be analyzed satisfactorily with currently available techniques. With drilled and grouted piles, no information can be obtained during installation that would increase the level of confidence in the design.

NON-PETROLEUM RELATED INVESTIGATIONS

Geotechnical investigations offshore have been predominantly related to the petroleum industry. As would be expected in a mature technology, the same type of investigations also has been steadily increasing in a number of other existing or potential industries, often with financial support from governmental agencies, as well as by government or university engineers. Four examples have been selected to indicate the wide range of these activities: (1) ocean thermal energy conversion (OTEC), (2) polymetallic nodule mining, (3) possible subseabed high-level radioactive waste disposal, and (4) the Deep Sea Drilling Project.

OTEC is a method of extracting energy utilizing the thermal difference between warm surface sea water and water more than about 20°C colder occurring at a depth of approximately 1000 m below the surface. Although the concept was proven more than 50 years ago (6), it has only been seriously considered since the early 1970's. A 30kW$_e$ commercial plant already is operating on the Pacific island of Nauru. The geotechncial concerns involve anchoring floating 40 to 200 MW$_e$ OTEC plants in deep water or attaching the long, and large diameter, cold water pipe to the seabed if the plants are located on the shore.

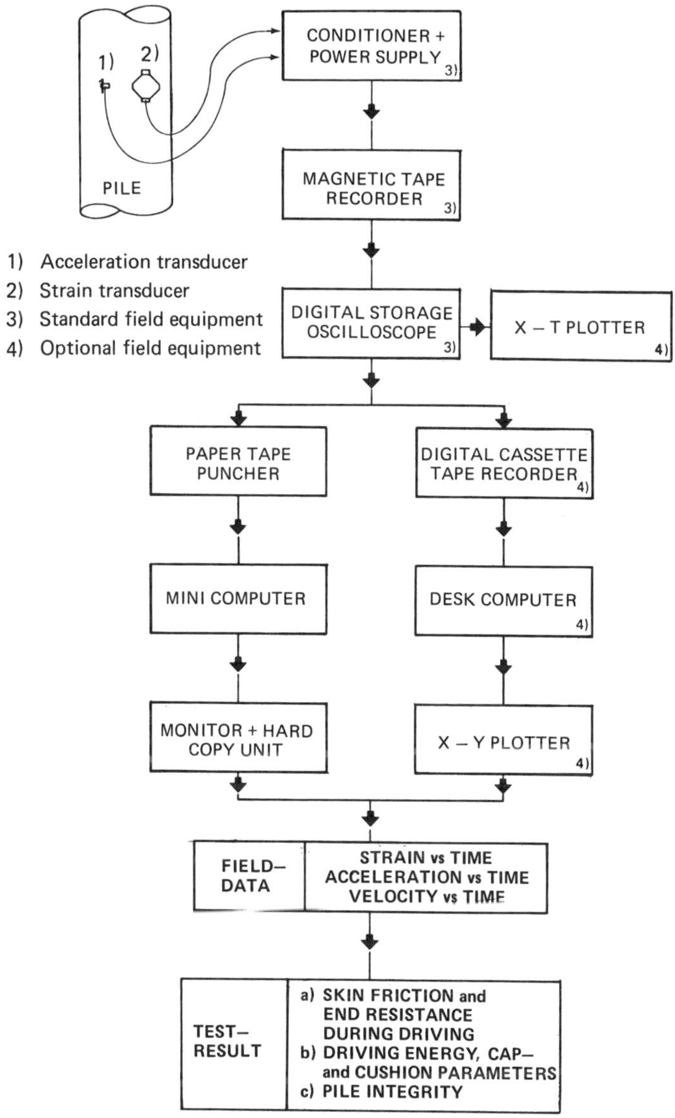

FIG. 9— Dynamic pile testing set - up

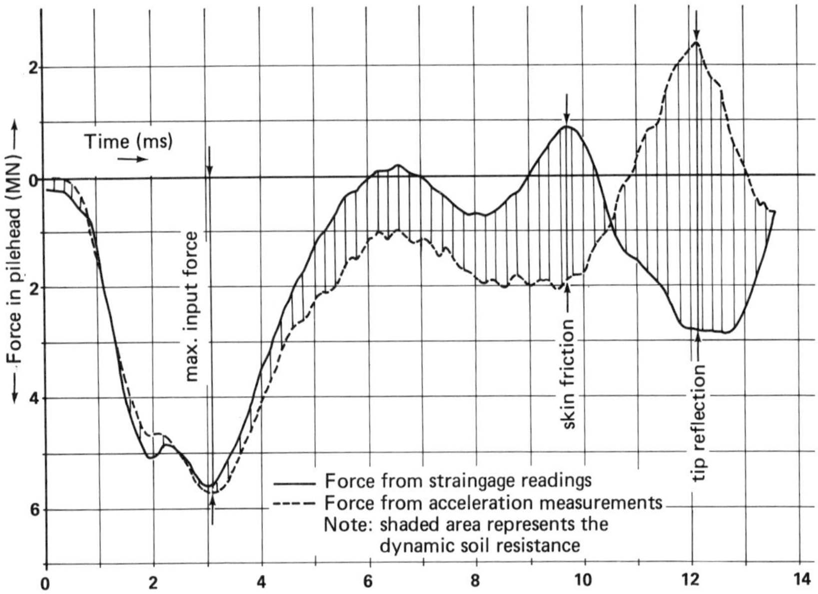

FIG. 10— Example D.P.T.— force versus time

Anchoring concepts for floating plants center generally around gravity or embedment anchors and drilled and grouted piles. It has been estimated that the costs for very large gravity anchors are similar to drilling and grouting (37). At the site of the proposed 10-40 MW_e OTEC plant on the island of Hawaii, the soil cover is thin to non-existant and the seabed slope may exceed 70° where the basalt has no soil cover. A conceptional design to anchor the cold water pipe to the seabed proposes using explosive embedment anchors. The site investigation, which has not yet occurred, promises a number of interesting technical challenges.

Mining the ocean floor in water depths exceeding 4 to 5 km for polymetallic (nickel, copper, cobalt, and manganese) nodules is a newly developing industry that shows considerable future potential. Site investigations have been made using cameras and television, as well as raising soil samples by wireline gravity corers and by free-fall corers and in situ measuring devices. The geotechnical properties of the soils from the primary mining area between Hawaii and Revillagigedo Archipelago of Mexico have been summarized (26,28). They are sufficiently well known so that determination of the trafficability of nodule-harvesting sleds or self-propelled seabed crawlers no longer appears to be a geotechnical problem, which major conclusion was based on prototype mining system tests at sea in the late 1970's and announced during a workshop held several years ago. As a consequence, this fledgling industry may be considered already to be mature from a geotechnical standpoint. Some of the soils in the primary nodule region are fascinating. They may be lightly mineralized and have a high shear strength of 15-30 kPa, a water content exceeding 350 % by dry weight, and exhibit brittleness.

Scientists and engineers in a number of countries are in the process of assessing the feasibility of disposal of high-level radioactive wastes in the pelagic clay soils of the abyssal plains (13,1). Reference 15 has a particularly good overview on the geological properties of the soils. Geotechnical investigations in The Netherlands, U.K., and U.S.A. are presently concerned with soil properties, including shear strength, thermal properties, permeability and creep behavior; penetrator development and deployment; centrifuge modelling of penetrator penetration of clay soils; and developing in situ geotechnical testing and monitoring methods both at or near the seabed surface and at penetration depths of 100 m or more in water depths of about 6 km.

The Deep Sea Drilling Project (DSDP) has had an extensive global program of soil sampling for scientific purposes for about fifteen years. Marine geotechnical investigations have been conducted at irregular intervals until about 1980. In 1981, a South Atlantic Ocean boring, sampled using the DSDP hydraulic piston corer, was dedicated for geotechnical investigations. The entire 196 m long core was removed from the Glomar Challenger in unopened 1.5 m long sections and protected from shock during transit to a shore laboratory. A geotechnical consortium of academic and governmental engineers was formed to undertake routine and specialized geotechnical testing of the core. A subsequent boring from the North Pacific Ocean was studied in 1982. An open invitation exists for geotechnical personnel from academia, government, and industry from any country in the world to participate in the activities of the consortium in future years. Following the recommendation of a Joint Oceanographic Institutes Deep

Earth Sampling (JOIDES) Advisory Committe, a downhole piezocone tool has been proposed by the authors' company for field testing by the Glomar Challenger in 1983.

It is likely that tools and techniques developed outside of the petroleum industry may in time be utilized for geotechnical investigations within the industry. It is also likely that the number of non-petroleum marine geotechnical activities will increase in the near future and that the complexity of their geotechnical investigations will also increase.

FUTURE DEVELOPMENTS

Economics continues to be the driving force for site investigations and the bulk of the business continues to be with the petroleum companies. Some new developments, however, are controlled more by environmental concerns, sociology and politics than by economics. In various countries, governmental agencies are increasingly aware of the need to protect the environment, and environmental impact statements are more often becoming an integral part of site investigations throughout the world. Governmental agencies are also concerned with the use of the environment; for example, in the disposal of all manner of wastes in or on the seafloor, thus leading to new technologies. There follows a few examples of expected future developments. The selection presented is not intended to be complete.

Significant improvements in methods of soil characterization when the vessel is moving are predicted. Multiple-frequency echo sounders; increasingly sophisticated stacked and computer-massaged digital data, together with pattern classifiers; and mini 'Seabeam' computerized mapping systems are likely in the future. Soils may be geotechnically characterized by geophysical methods, particularly with regard to measures of in situ bulk density and water content. The relationship of geophysical, geotechnical, and acoustic properties by reference 10 is but one example. Preliminary results using an expendable doppler penetrator (3) indicate that shear strength can be estimated to within 30 % of values measured in the laboratory on cores from the approximate location and to about 50 % of probable values in areas where no strength data from cores exist. Such a system is capable of operating in water depths to 6 km. A number of papers on penetrators, recently published in the 1981 Annual Conference Record of the Marine Technology Society, give further details. Penetrators are a particularly significant development because when the deacceleration records can be more exactly related to in situ shear strength the method is particularly attractive because of the potential for relatively deep penetration without requiring a survey vessel to stop.

Once a site has been selected, a number of innovative developments appear on the horizon for obtaining extensive information at low unit cost. Multiple sensors, perhaps combined with a new type of side-wall sampler, may be able to measure most physical and geotechnical parameters in situ. There is no technical reason why downhole tools or remote seabed probe systems cannot be used in almost any water depth. In the 1960's simple prototypes, having very shallow penetration capability, wire-telemetered shear strength data from vane and cone penetrometer seabed probes and bulk density data from a nuclear transmission densitometer probe in water depths of nearly 4 km (31,12).

Methods of data gathering and analysis soon are expected to benefit from the systems approach, in which analog to digital converters, perhaps operating with telemetry and microprocessor controlled, feed appropriately-massaged data from the sensor directly into a computer-plotter. There is no reason why one cannot have a final graph or table ready for a company report or for publication essentially in real time. Another trend on the horizon is to use the interactive graphic computer to store, to verify, to correct raw data gathered from a number of different sources and to present these data in a variety of three-dimensional displays.

Analytical methods are also rapidly improving. It is now possible to obtain ambient pore pressures in situ directly measured by total- or differential-pressure piezometers, estimated from piezocones inserted into the seabed at a constant speed, or measured by piezocones held in position long enough for pore-pressures generated by driving to fully or partially dissipate. It is also possible to measure in situ bulk density directly using nuclear back-scatter densitometers coupled with a CPT (19). The obvious next step is to combine a densitometer with the piezocone, which will enable the direct in situ measurement of the key effective stress parameters. Electrical resistivity can also be used for the measurement of in situ bulk density, but this method requires the concurrent measurement of pore-water salinity in water depths of less than the shelf-break (about 120 m). Although both parameters have been measured offshore (18), the apparatus is complicated and appears not easily adapted for commercial use requiring deep penetration data on a routine basis. In addition, with the measurement of in situ stress-strain relationships and K_0, the deformation behavior of the soil can be much better defined.

Most of the U.S. national needs for future developments in marine geotechnics with regard to soil properties and behavior, environmental hazards and soil-structure problems are as valid today as five years ago (29). Reference 33 identified similar needs in the U.K. specific to offshore foundation design that could also be considered indicators of future developments. It is likely that within a few years very significant advances in the knowledge of in situ stress, compressibility and deformation characteristics of offshore soils will enable foundation designs to be based more on fundamental soil properties than empirical relationships. This, in turn, should lead to a better defined factor of safety and lower costs for siting structures on the seafloor. It is also believed that there will be an increasing flow of geotechnical information from offhore projects to onshore projects, further indicators of a mature technology in offshore site investigations.

ACKNOWLEDGMENTS

The authors wish to express their appreciation for the assistance of N.J. Withers and various other people in the Fugro organisation in the preparation of the paper and for the willingness of their company to make the required information available. Dr. D. Prior, Louisiana State University, and Dr. N. Kenyon, Institute of Oceanographic Sciences, kindly made available photographs used as illustrations in this paper.

REFERENCES

1. Anderson, D.R., "Seventh International NEA/Seabed Working Group Meeting, La Jolla, California, March 15-19, 1982," Sandia National Laboratories Report SAND82-0460, 1982, 221 pp.
2. Baguelin, F., and Le Tirant, P., "The Pressuremeter Method Onshore and Offshore Applications," Symposium on the Pressuremeter and its Marine Applications, Editions Technip, Paris, 1982, 5-21 pp.
3. Beard, R.M., "Expendable Doppler Penetrometer: A Performance Evaluation," U.S. Navy Civil Engineering Laboratory Tech. Rept. R855, 1977.
4. Beringen, F.L., Van Hooydonk, W.R. and Schaap, L., "Dynamic Pile Testing, an Aid in Analysing Driving Behaviour," Preprints Seminar on Application of Stress-wave Theory on Piles, Stockholm, 1980, 22 pp.Campanella, R.G., and Robertson, P.K., "State of the Art in
5. In-situ Testing of Soils, Developments Since 1978," Conference on Updating Subsurface Sampling of Soils and Rocks and Their In-situ Testing, Engineering Foundation, New York, in press.
6. Claude, G., "Power from the Tropical Seas," Mechanical Engineering, Vol. 52, 1930, pp. 1039-1044.
7. Denk, E.W., Dunlap, W.A., Bryant, W.R., Milberger, L.J., and Whelan, T.J., III, "A Pressurized Core Barrel for Sampling Gas-charged Marine Sediments," Proceedings of the Offshore Technology Conference, Vol. 1, Paper 4120, 1981, pp. 43-52.
8. De Ruiter, J., "Current Penetrometer Practice," Cone Penetration Testing and Experience, ASCE, New York, 1981, pp. 1-48.
9. De Ruiter, J., "The Static Cone Penetration Test State-of-the-Art-Report," Penetration Testing, Vol. 2, Proceedings Second European Conference on Penetration Testing, A.A. Balkema, Rotterdam, 1982, pp. 389-405.
10. Dunsiger, A.D., Chari, T.R., Fader, G.B., Peters, G.R., Simpkin, P.G., and Zielinski, A., "Ocean Sediments-a Study Relating Geophysical, Geotechnical, and Acoustic Properties," Canadian Geotechnical Journal, Vol. 18, 1981, pp. 492-501.
11. Eide, O., Andersen, K.H., and Lunne T., "Observed Foundation Behaviour of Concrete Gravity Platforms Installed in the North Sea 1973-1978," Proceedings BOSS Conference, London, 1979, pp. 435-456.
12. Hirst, T.J., Burton, B.S., Richards, A.F., Perlow, M., Jr., and Van Sciver, W.J., "Improved In Situ Gamma Transmission Densitometer for Marine Sediments," Ocean Engineering, Vol. 3, 1975, pp. 17-27.
13. Hollister, C.D., Anderson, D.R., and Heath, G.R., "Subseabed Disposal of Nuclear Wastes," Science, Vol. 213, 1981, pp. 1321-1326.
14. Horsnell, M.R., and Toolan, F.E., "The Analysis of the Piled Foundations of Moored and Tethered Offshore Structures," Proceedings of the 2nd International Conference on Numerical Methods in Offshore Piling," University of Texas, Austin, 1982, pp. 597-615.
15. Institute of Oceanographic Sciences, "Oceanography Related to Deep Sea Waste Disposal," Rept. No. 77, 1978.

16. Kenyon, N.H., Belderson, R.H., and Stride, A.H., "Channels, Canyons and Slump Folds on the Continental Slope Between South-West Ireland and Spain," Oceanologica Acta, Vol. 1, 1978, pp. 369-380.
17. King, R.W., Van Hooydonk, W.R., Kolk, H.J., and Windle, D., "Geotechnical Investigations of Calcareous Soils of the North West Shelf, Australia, Proceedings of the Offshore Technology Conference, Vol. 1, Paper No. 3772, 1980, pp. 303-313.
18. Koning, H.L., "On an Explaination of Marine Flow Slides," Marine Slides and Other Mass Movements, Plenum Press, London, 1982, pp. 83-94.
19. Ledoux, J.L., Menard, J., and Soulard, P., "The Penetro-Gamma-densimeter," Penetration Testing, Vol. 2, Proceedings Second European Conference on Penetration Testing, A.A. Balkema, Rotterdam, 1982, pp. 679-681.
20. Lunne, T., and St. John, H.D., "The Use of Cone Penetrometer Tests to Compute Penetration Resistance of Steel Skirts Underneath North Sea Gravity Platforms," Proceedings 7th European Conference on Soil Mechanics and Foundation Engineering, Brighton, England, Vol. 2, 1979, pp. 233-238.
21. Marchetti, S., "In-situ Tests by Flat Dilatometer," Journal of the Geotechnical Engineering Division, ASCE, Vol. 106, No. GT 3, 1980, pp. 299-321.
22. Marsland, A., and Windle, D., "Developments in Offshore Site Investigation," Oceanlogy International Exhibition and Conference Papers, Vol. 1, Paper 2.7, 1982, 38 pp.
23. McClelland, B., "Techniques Used in Soil Sampling at Sea," Offshore, Vol. 32, 1972, pp. 51-57.
24. Prior, D.B., Bornhold, B.D., Coleman, J.M., and Bryant, W.R., "Morphology of a Submarine Slide, Kitimat Arm, British Columbia," Geology, Vol. 10, 1982, pp. 588-592.
25. Rausche, F., Moses, F., and Goble, G., "Soil Resistance Predictions from Pile Dynamics," Journal of the Soil Mechanics and Foundation Division, ASCE, Vol. 98, No. SM 9, 1972, pp. 917-937.
26. Richards, A.F., 1978, "Marine Geotechnical Data for North Pacific Sediments," U.S. Geological Survey Open-file Report 78-231.
27. Richards, A.F., "Review of Marine In Situ Geotechnical Testing Equipment," Oceanology International Exhibition and Conference Papers, Vol. 1, Paper 2.1, 1982. 10 pp.
28. Richards, A.F., and Chaney, R.C., "Present and Future Geotechnical Research Needs in Deep Ocean Mining," Marine Mining, Vol. 2, 1981, pp. 315-337.
29. Richards, A.F., and Matlock, H., "Present and Recommended U.S. Government Research in Seafloor Engineering," NOAA Office of Ocean Engineering, U.S. Government Printing Office, Washington, 1978.
30. Richards, A.F., Ling, S.C., and Gerwick, B.C., Jr., "Site Selection for Offshore Facilities," Ocean Engineering, Vol. 3, 1976, pp. 189-206.
31. Richards, A.F., McDonald, V.J., Olson, R.E., and Keller, G.H., "In-place Measurement of Deep-sea Soil Shear Strength," Symposium on Underwater Soil Sampling, Testing, and Construction Control, ASTM STP 501, 1972, pp. 55-68.

32. Semple, R.M., and Johnston, J.W., "Performance of 'Stingray' in Soil Sampling and In-situ Testing," Proceedings International Conference on Offshore Site Investigation, Graham and Trotman, London, 1979, pp. 169-182.
33. Semple, R.M., St. John, H.D., and Toolan, F.E., "Statement of Research Needs in Offshore Foundation Design," Society for Underwater Technology, Geotechnics Committee, 1982.
34. Smits, A.P., "Mission II Now Equipped for all Seabed Investigations," Preprints of the Second Canadian Conference on Marine Geotechnical Engineering, Halifax, Nova Scotia, 1982, 3 pp.
35. Stride, A.H., "Geotechnical Importance of Gloria Survey of the European Continental Slope," Oceanology International Exhibition and Conference Papers, Vol. 1, Paper 2.5, 1982, 6 pp.
36. Toolan, F.E., "Recent Improvements in Soil Investigation Techniques," Conference on Design and Construction of Offshore Structures, Institution of Civil Engineers, London, in press.
37. Vadus, J.R., "OTEC Ocean Engineering Technology Development," Oceans 82 Conference Record, IEEE, 1982, pp. 547-559.

GROUP ACTION IN OFFSHORE PILES

Michael W. O'Neill,[1] M.A.S.C.E.

INTRODUCTION

Piles have long been used in groups for the support of onshore structures. The application of grouped-pile foundation systems to the offshore regime is a relatively recent innovation, necessitated by severe environmental loadings coupled with the impracticality of installing individual piles to resist these loadings in many stratigraphic situations. Grouped piling systems have been constructed in Cook Inlet, Alaska, to reduce ice loads, and in the North Sea, where it is difficult to obtain large penetrations (103). Present and future concepts for deep water structures, including conventional, guyed tower (7, 99), and tension leg structures, also envision the intense use of piles in groups. Typical offshore pile group applications are illustrated in Fig. 1.

This paper summarizes the state of the art in group action, incorporating recent experimental data and analytical techniques. It does not attempt to provide definitive solutions but rather to define problems and, where possible, to suggest bounds on their effects.

Extensive use is made of model pile data and of data for full-scale groups of onshore piles. For reasons to be discussed subsequently, the model data must be assumed to be qualitative. The latter data are assumed to be quantitative and applicable offshore due to the absence of data concerning offshore groups. However, differences in pile size and length, geometric arrangements (typically circular offshore and typically rectangular onshore), and in loading conditions between the offshore and onshore environments must be borne in mind when applying the information contained herein.

PRINCIPAL DESIGN PROBLEMS

Wright (197) defines two fundamentally different classes of pile group problems. In the first class individual piles are widely spaced, and the responses to loading of the piles are dictated by their overall geometry in the structural system, their individual stiffnesses, the details of pile-structure connections, the flexibility of the superstructure, and the structural loads. This class, which is not the primary concern of this paper, can be solved without resorting to "group action" principles.

In the second class the design considerations are identical to the first, except that the piles are situated closely enough to each other

---

[1] Assoc. Prof. of Civil Engineering, University of Houston, Houston, Texas.

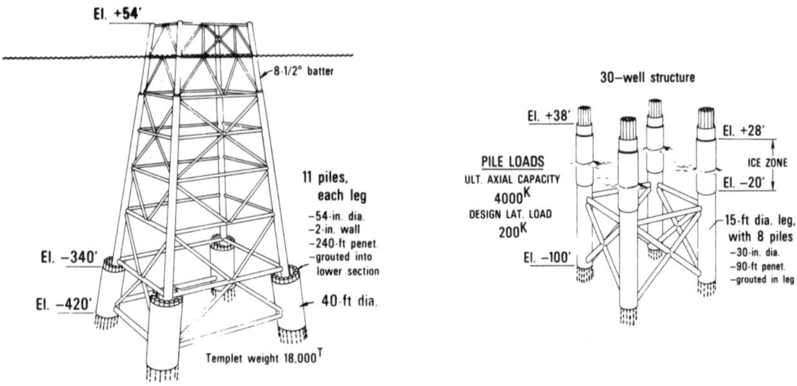

b. Cook Inlet (103)

a. North Sea (103)

Fig. 1. Offshore Pile Groups.

that the individual responses of the piles are influenced by the presence of and loadings upon neighboring piles. Piles interact with each other through the supporting soil itself in such a way that group action, or "pile-soil-pile interaction," must be considered.

The primary special concerns of the designer of foundations in which group action exists may be classified as:

1. Evaluating the efficiency of the group, where efficiency (η) is defined by

$$\eta = \frac{Qg}{(n)Qs} , \tag{1}$$

where n = number of piles, Qg is the capacity of the group, and Qs is the capacity of a reference pile that is identical to a group pile but is isolated from the group. Strictly speaking, Eq. 1 applies either to vertical or lateral loading; however, it is not normally applied to the lateral loading case, where deformations and/or pile stresses are of more concern than efficiency.

2. Evaluating the increased pile-head flexibilities that grouped piles possess in comparison with isolated piles. Under vertical loading the term settlement ratio (SR) is often used to quantify group flexibility:

$$SR = \frac{\delta g}{\delta s} , \tag{2}$$

where δg = settlement of a pile group under n times the load on an isolated pile undergoing a settlement δs. Settlement (or uplift) ratio can be measured experimentally, or it can be computed by any of the several mathematical models described later. Settlement ratio varies with magnitude of load. Conventionally, the load magnitutde considered is one-half the failure load of the isolated pile.

In laterally loaded groups, other means, described subsequently, are used to describe flexibility differences between isolated and group piles.

3. Estimating the effect of a cap (or closed-bottom template) in contact with the soil on Items 1 and 2.

4. Determining loads in the individual piles and assuring that pile stresses are not excessive. While this problem is no different in principle from that for a non-grouped pile foundation, pile-soil-pile interaction considerations require modified approaches to pile load and stress evaluation.

5. Optimizing the geometric arrangement of piles. For example, in a guyed tower foundation of the type depicted in Fig. 1, the group must be large enough to provide resistance to torques applied from the superstructure, but it must be small enough to prevent excessive movements due to tower tilting (99). The geometry of small groups can be optimized approximately by trial and error, considering pile-soil-pile interaction. Theoretically perfect optimization may be prevented by superstructural constraints, such as leg and brace locations that prevent driving piles in certain positions within the group.

Formal optimization procedures are necessary for large pile groups. Such a procedure has been developed recently for semi-rectangular arrays of piles beneath massive structures (74).

Structural Interaction Models (SIM's)

When piles are widely spaced and pile-soil-pile interaction can be neglected, pile loads and settlement can be determined by using SIM's, some of which crudely represent the piles and/or the superstructure as linear or rigid two-dimensional bodies and which can be solved by hand or with a small computer program (76, 85, 189). More advanced SIM's permit consideration of nonlinear soil support, three-dimensional geometry and loading, and superstructure flexibility (5, 6, 8, 22, 24, 26, 46, 62, 73, 111, 117, 157, 158, 162, 163). Many SIM's, which treat all piles as isolated piles (95), are relatively simple and are often useful in preliminary phases of design, even where pile-soil-pile interaction must eventually be considered.

Pile-Soil-Pile Interaction Models (PSPM's)

The nature of pile-soil-pile interaction is both complex and poorly understood. Simply put, it consists of two components: (a) alteration of soil stress states, densities, and perhaps grain size distributions, when piles are installed in close proximity to each other, in a manner different from those alterations that occur with the installation of single piles ("installation effects"), and (b) superposition of strains and alteration of failure zones (71, 115, 118, 185) in the soil mass due to simultaneous loading of two or more neighboring piles ("mechanical effects"). These two components are interdependent. Installation effects include increases or decreases in lateral effective stress, retarded reconsolidation rates, reduction of residual stresses (36, 126), or excess soil heave (70) that can produce substantial uplift forces on piles and/or become expressed as negative skin friction. These effects are influenced strongly by pile spacing and construction techniques, including the means of installing the piles (e.g., boring and grouting versus driving), the rate and order in which the piles are installed, and the permeability, shear strength, compressibility and initial state of effective stress of the soil. They may tend either to increase or decrease efficiency and have a substantial effect on group flexibility and on patterns of distribution of loads to the piles. Mechanical effects vary according to the distribution of stiffness in the soil profile and the distribution of loads to the piles.

Most existing PSPM's are directly concerned with only that portion of the mechanical component of interaction that considers strain superposition in the soil mass, neglecting both installation effects and failure zone alteration. As a consequence, considerable judgment is required to design and analyze closely spaced groups.

Test Data

The sources of information that the engineer must rely on to develop judgment are largely experimental. Model tests comprise the majority of available experimental data. They are a convenient and inexpensive means of assessing relative parametric effects, such as spacing, pile penetration, and group geometry, but should be used with caution in developing quantified design criteria (132). The scaling problems associated with model tests are primarily, but not exclusively, associated with incorrect modeling of effective soil stresses. For example, a dense sand may dilate successively as small model piles are installed, while *in situ*, effective stresses may be sufficient to affect failure of sand particles and thereby prevent or retard dilation (193). Model test efficiencies may therefore be less than unity, whereas in full-scale the efficiencies would exceed unity. On the other hand, $\eta > 1$ may result from increases in lateral effective stresses in the soil surrounding model group piles due to the insertion of neighboring piles (185), but under the high *in situ* pressure states found around long offshore piles, the lateral effective stress increase can be accompanied by a decrease in soil friction angle if the Mohr-Coulomb envelope is nonlinear (193). The net effect here may be field efficiencies less than those in models. Intermediate scale modelling at 1 g (49) and centrifuge modelling (148, 149), including attempts to reproduce installation effects (84) are presently being studied and appear to show promise in reproducing correct soil response.

Full-scale testing currently provides the only true picture of pile group response. Unfortunately, "full-scale" test results are still subject to scaling problems, generally are of lower quality than laboratory model tests, and are not suitable for parameter studies.

Before describing design-related capabilities of PSPM's futher, results of both model and field tests of pile groups will be summarized.

GROUP ACTION: EXPERIMENTAL PERSPECTIVE

Efficiency

Sand. Figure 2 summarizes results of several 1 g model compression tests on vertically loaded rectangular, freestanding groups of vertical piles in uniform sand. The tests include only piles that were inserted into the soil, that had l/d (length-to-diameter) ratios of 11-30, and that were "rough" (e.g., wood or unpolished metal). The figure is separated into two parts to provide an indication of the effect on η of the number of piles installed. The trends for compressive loading appear as follows: (a) η in loose sands always exceeds unity, with highest values occurring at spacing-to-diameter ratios, s/d, of about 2, and generally higher η occurring with increasing numbers of piles. Block failure (i.e., simultaneous failure of the piles and mass of soil within the pile group, 38) affects η in 4-pile groups only below s/d $\simeq$ 1.5 and in 9-16-pile groups only below s/d $\simeq$ 2.0. (b) η in dense sands may be either greater or less than unity, although the trend is toward $\eta > 1$ in groups of all sizes for $2 < s/d < 4$. Efficiency of less than unity is probably a result of dilatency and would not generally be expected in the field for other than bored or partially jetted piles, although theoretical interference studies (115) suggest η slightly below 1 at $s/d > 4$. The variation in η among tests on geometrically similar groups may be due to differences in order and method of installation and in sand mineralogy and gradation.

Little data concerning uplift efficiency exist. The results of two 4-pile group uplift studies in loose sand are included in Fig. 2. The results indicate sharp reductions in η at all s/d; however, these tests were conducted by molding sand around the piles and undoubtedly represent lower bounds to field conditions for driven piles. More representative η values can possibly be obtained by applying theoretical frictional models of Nishida (118) or Brinch Hansen (23), which yield η only slightly less than unity for uplift loading (30).

The applicability of the data in Fig. 2 to typical offshore group geometry is illustrated at the bottom of Fig. 2b, in tests by Tejchman (179), which showed that efficiencies in small groups with perimeter piles only were 10-20% lower than those for square matrix groups.

Full-scale evidence of efficiency in deposits of sand is summarized in Fig. 3 (128), in which those tests conducted both with and without the pile cap in contact with the soil are shown. The parameters used in the figure, pile penetration (p), pile spacing (s), average group width (B), and mean pile diameter (d), represent the largest set of common parameters available from the test reports and not necessarily the most appropriate variables for correlation. The non-dimensional groupings are those that provided the best data correlations. The trends that emerge weakly from full-scale tests on freestanding piles are: (a) the observation from models that efficiency in compression always exceeds one unless the piles are bored or partially jetted is verified, and (b) peak efficiency is achieved for groups having B/d ratios of 2-7. Sixteen or fewer piles were present in all groups tested.

The high efficiency in sand is apparently due to increased lateral stresses against the piles (185) and/or the tendency for sands with

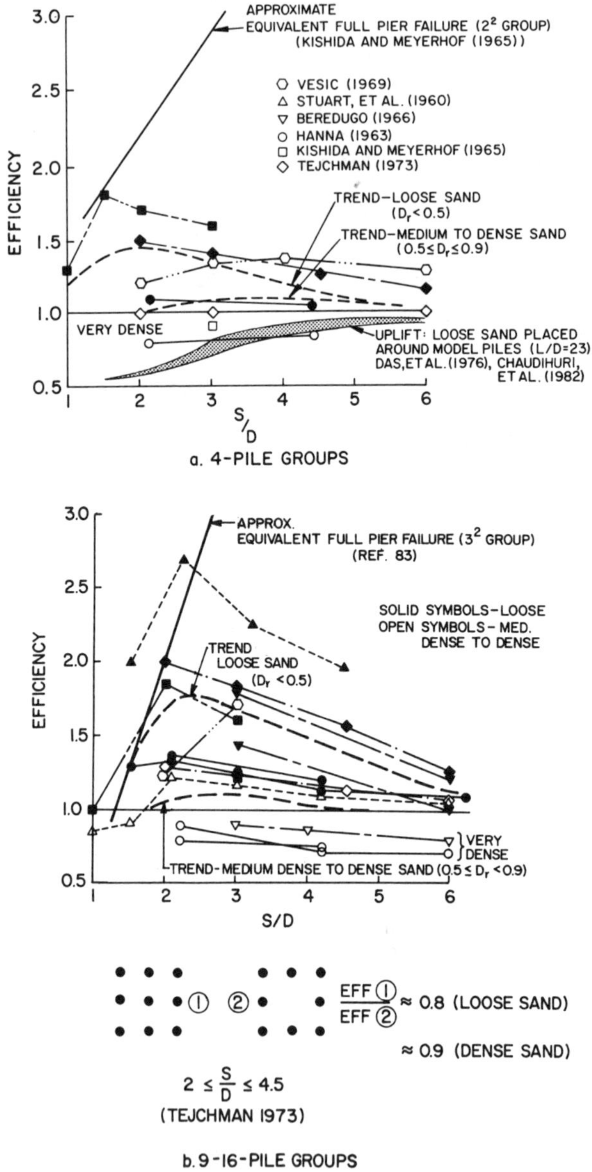

Fig. 2. Vertical Efficiencies for Model Groups in Sand.

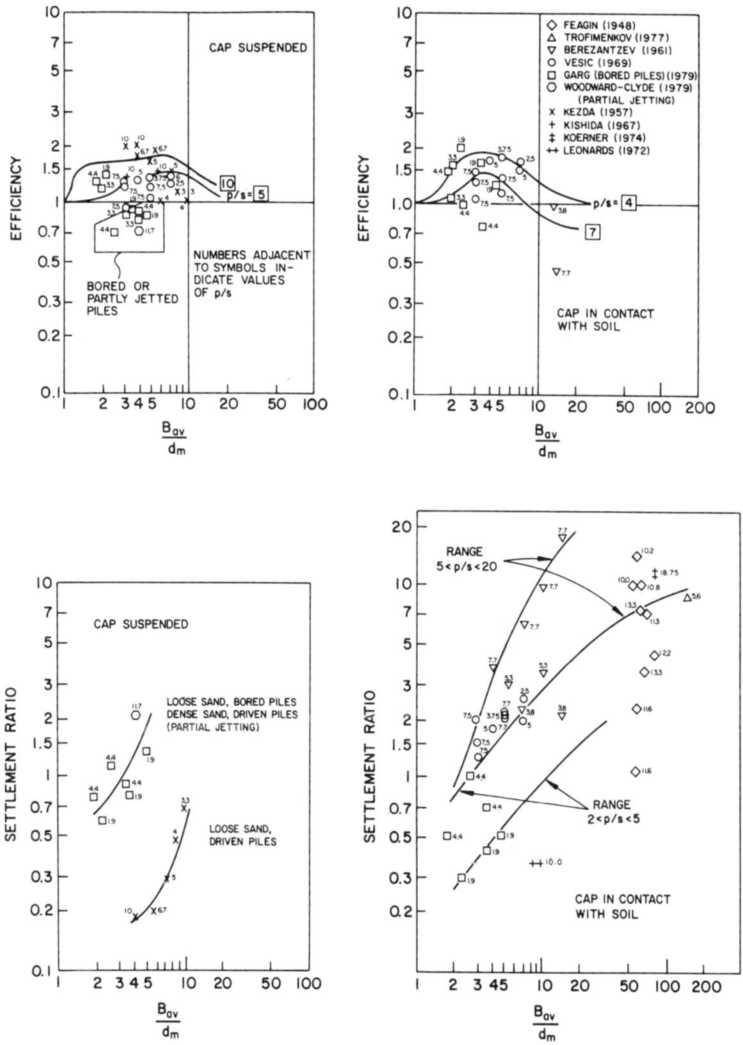

Fig. 3. Full-Scale Groups in Sand (128).

friction angles less than about 40 deg to density during driving (82). It is manifested in higher side resistance and generally unchanged tip capacity (185). Conventional practice generally assigns a conservative bound of $\eta = 1$ for driven piles and 0.67 for bored piles (105, 184).

Clay. Model test results on freestanding friction pile groups in clay and full-scale test results on both freestanding and cap-in-contact groups in clay are shown in Figs. 4 and 5 (128), respectively. Model tests in clay, in contrast to those in sand, always yield $\eta < 1$ and a distinct trend toward block failure in square groups at s/d values of less than about 2. The classical tests of Whitaker are highlighted in Fig. 4, in which block failure (evidenced by the sharp break in the curves) occurred at progressively larger relative spacings with increasing numbers of piles, but at spacings slightly less than the theoretical block failure spacings. Most studies in Fig. 4 showed a slight increase in η with increasing l/d. η is also higher in stiff clays than in soft clays (12) (0.89 vs. 0.69 at s/d = 2 in 5 × 5 group).

Efficiency in Figs. 4 and 5 is < 1 at spacings beyond theoretical block failure spacing. It is possible to explain this effect partially by considering that piles fail progressively. However, progressive failure has been observed to reduce efficiency by less than 5% in overconsolidated clay (125). The relative magnitudes of pore pressure generation and rates of dissipation around driven isolated piles and group piles of the displacement type probably have a stronger influence. Relevant field test data are summarized in Figs. 6 and 7. Figure 6 shows initial excess pore pressure Δu_i normalized by vertical preconsolidation pressure σ'_{vm} (160) vs. radial distance from pile centerline, x, ÷ pile radius, R. The several groups considered in Fig. 6 contain 2 to 230 piles at spacings of 2.5 to 6 d. The magnitude of initial excess pore pressure generated in the immediate vicinity of a pile wall appears much more highly dependent on soil characteristics than on the proximity of piles in the group. Pore pressure generation near isolated piles in normally consolidated clay and sensitive, overconsolidated clay follows reasonably well the trend expressed by the theoretical equation in Fig. 6c, using typical normally consolidated clay parameters. (α_f is the Henkel pore pressure parameter associated with deviatoric stress.) Initial excess pore pressures are considerably lower in the insensitive, overconsolidated clays, which also contained secondary structure networks.

Within the zone of plastic remolding for a single pile, determined in the derivation of the equation on Fig. 6c, relatively small superposition of pore pressures due to driving adjacent piles in normally consolidated or sensitive, overconsolidated clay is evidenced by the relative positions of the group and isolated pile trend lines. Beyond the plastic zone (6-8 diameters from a given pile), superposition becomes more significant, as evidenced by the relatively larger values of Δu around group piles in comparison with those around isolated piles (91). Figure 6 represents data for piles installed over a finite period of time, usually several days. During installation, volume change occurs as pore pressures begin to dissipate (126) so that rates of installation approaching simultaneous insertion should reveal relatively higher group pore pressures.

While initial excess pore pressures in the immediate vicinity of group piles are not significantly larger than those around isolated

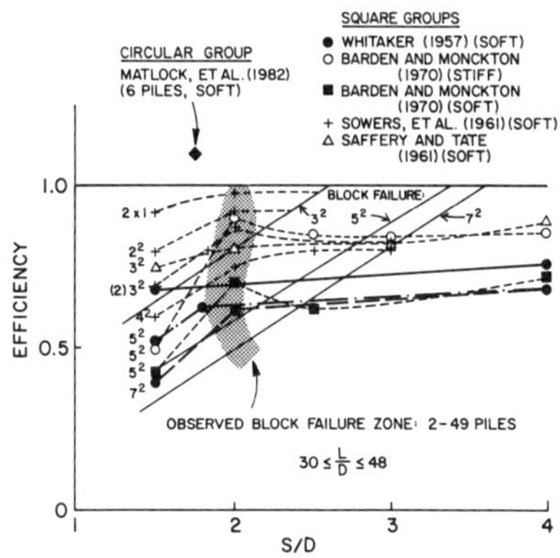

Fig. 4. Vertical Efficiencies for Model Groups in Clay.

piles, a significantly smaller initial horizontal pressure gradient exists around group piles. Figure 7 summarizes measured rates of excess pore pressure dissipation near single piles and typical group piles for the field tests described in Fig. 6. A direct time scale, rather than a nondimensional time factor scale, has been used due to the lack of consolidation data for many of the tests, so only relative trends are observable. Nonetheless, it is clear that dissipation rates around group piles are much slower than around single piles. For example, 80% of the initial excess pore pressure was dissipated around the single piles within about five days, while 50-60 days were required in the median group for the same level of dissipation. Additional indications of the rate of set-up in small-diameter piles in slightly sensitive clays have been provided by Flaate in tests on individual piles within an 8-pile group (59). Capacities of interior piles were found to increase in direct proportion to the logarithm of time up to about 1000 days after installation. This rather slow set-up rate is probably an extreme and was possibly influenced somewhat by pore pressure regeneration during the several tests to failure during the test period.

Figure 7 is based on data from relatively small-diameter displacement piles (generally less than 0.5 m). The relative pore pressure dissipation rates around single and group piles should be approximately correct for larger diameters, but absolute rates in similar soils would be inversely proportional to the square of R. Absolute rates for single piles are calculable from several theories (e.g., 155, 172).

Excess pore pressure generation and dissipation probably has a less significant effect on efficiency in non-displacement piles.

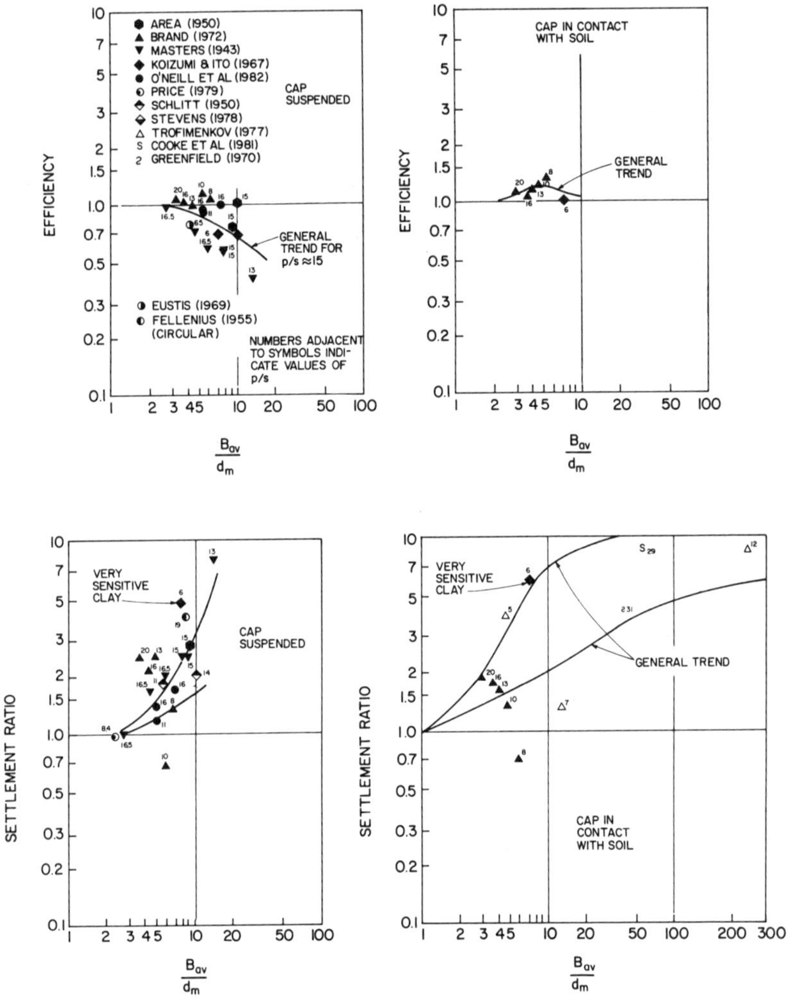

Fig. 5. Non-End-Bearing Full-Scale Groups in Clay (128).

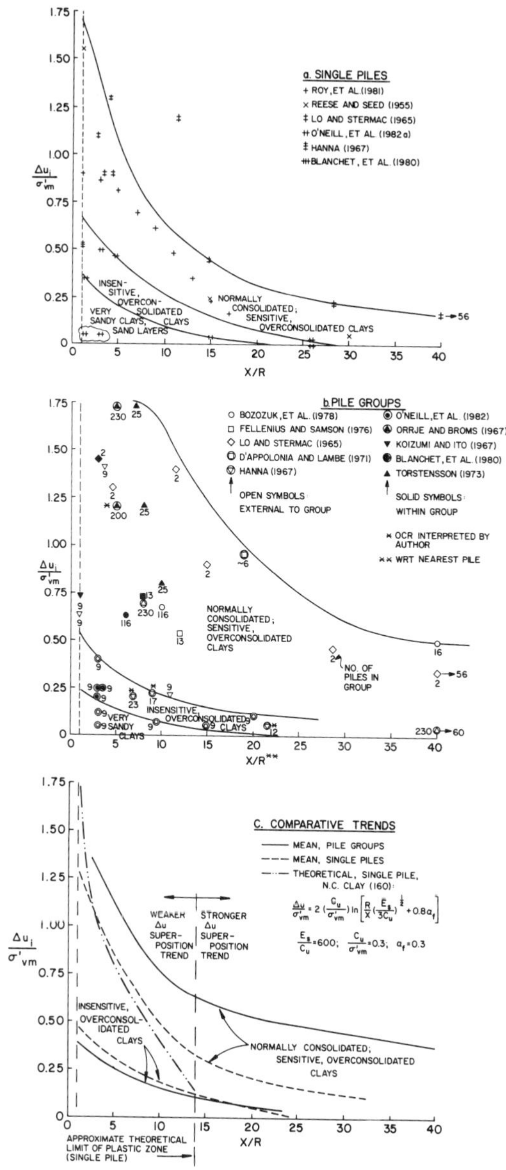

Fig. 6. Pore Water Pressure Generation.

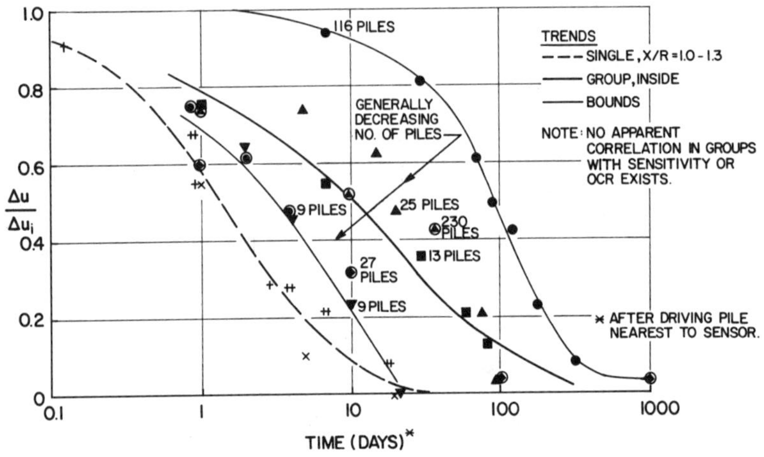

Fig. 7. Pore Water Pressure Dissipation Rates: Full-Scale Single Piles and Groups in Clay.

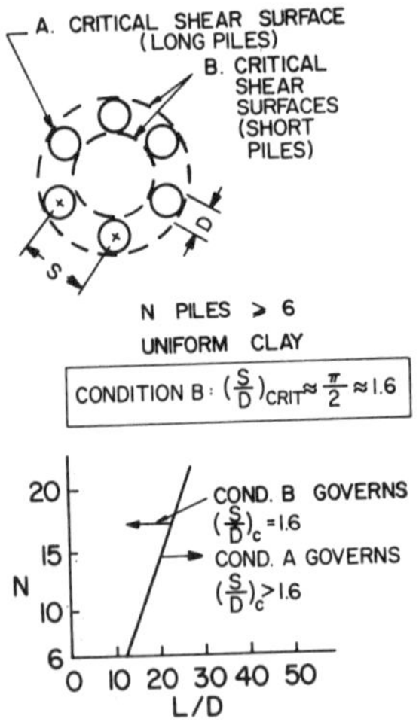

Fig. 8. Circular Group.

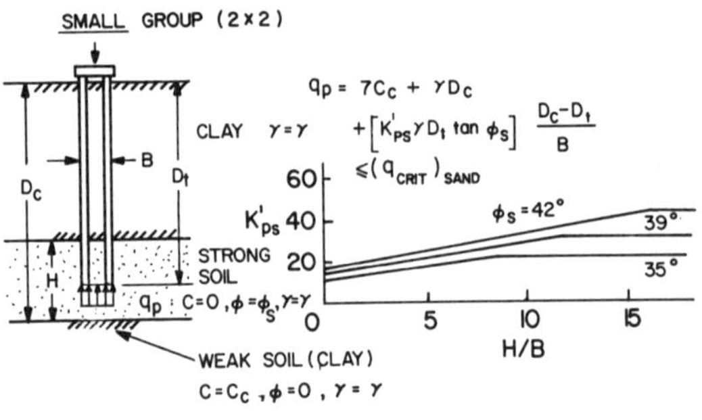

Fig. 9. Tip Punching in Small Groups (109).

However, at least one case study (39) has reported very high pore pressure generation in H-pile groups where the piles were driven through silt layers confined vertically between less permeable clays.

To return to the original argument, reference (isolated) piles and pile groups tested at approximately equal times after installation (as was the case for most of the model and full-scale tests in Figs. 4 and 5) are not likely to yield group efficiencies equal to unity at any reasonable spacing. This fact may be of practical significance if design loads are expected to be applied prior to the end of the reconsolidation period.

Few test results on circular groups in clay have been published, but one well-documented test (96) suggests that efficiences may be near unity at s/d down to about 1.75. Figure 8 examines the conditions of pile spacing in uniform clay under which block failure governs in a circular group. For relatively short piles, a double shear surface can form, resulting in a theoretical critical s/d = 1.6 (Cond. B). Otherwise (Cond. A), conventional perimeter shear considerations govern.

Layered Soils. In clay-over-sand layered systems, according to model studies on displacement piles (108), $\eta = 1$ in 2×2 groups with s/d = 3 for penetrations of the sand layer of 20 d. Where a second clay layer exists beneath the sand (109), however, pile tip (and therefore total) efficiencies are reduced, because groups tend to exhibit punch-through failure at smaller penetrations of the sand than do individual piles. A procedure suggested in Ref. 109 for the calculation of tip capacity of a group is given in Fig. 9. (The term q-crit refers to the ultimate value of unit end bearing normally assigned to a single pile [103]).

Cap (Template) Contact. Several experimental studies have been concerned with the effect of cap-soil contact on efficiency in compression. Model studies in sands indicate that η increases with increasing B(av)/L (2), partly because the vertical effective stress produced in the soil by the cap increases the capacity of the individual piles:

$$(Q_g)\ \text{cap in contact}/(Q_g)\ \text{cap suspended} \simeq 5.5\ (L/B\ (av))^{-1.5}. \quad (3)$$

Other model studies (185) suggest that the added capacity produced by the pressure of the cap can be computed by adding the capacity of the portion of the cap outside the perimeter of the group (computed by simple bearing capacity theory) to that of the freestanding group. Although the general trend in Fig. 3 suggests minor effects of cap contact, in the two test series in which tests were performed with both freestanding piles and with cap in contact, η was enhanced considerably by cap contact.

In clay, cap contact also enhances efficiency, as suggested in Fig. 5. Figure 10 indicates that the cap pressure tends to force a block failure in clay, even at spacings at which a freestanding group would fail as individual piles. Settlements at cap-pile system failure are very large, however, and may not be tolerable by the superstructure.

In laterally loaded groups, contact between the cap base and the soil has been found to have little effect on deformations and pile stresses if piles are battered against the load, but such contact has

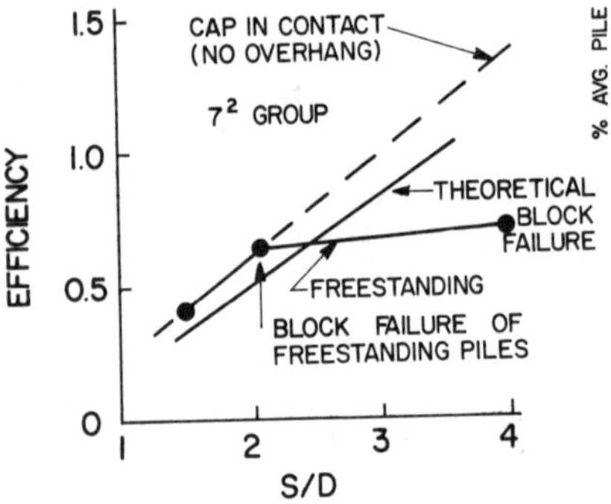

Fig. 10. Cap Effect (191).

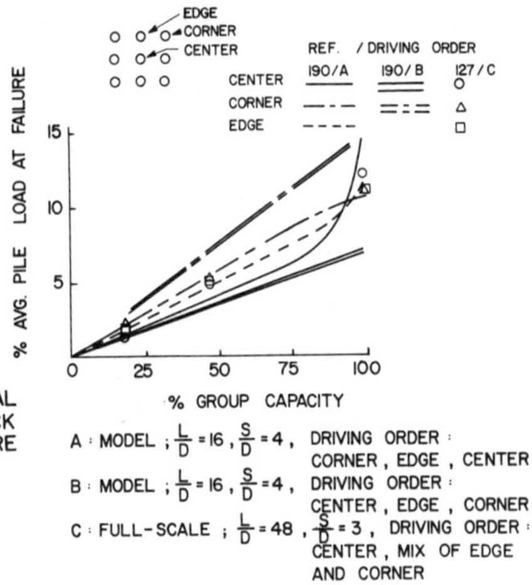

Fig. 11. Clay: Driving Order and Load Level Effects (127,190).

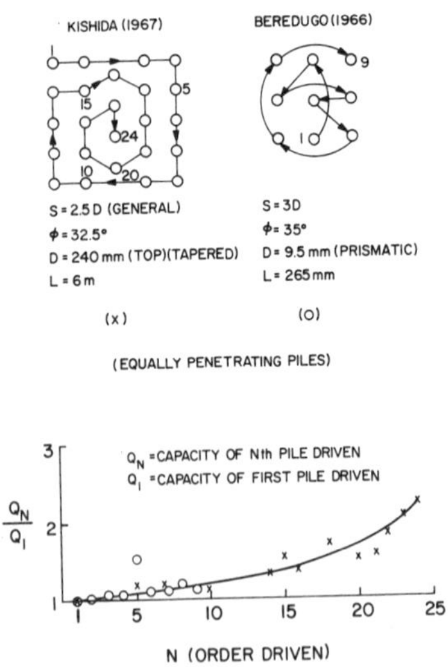

Fig. 12. Sand: Driving Order Effects (15,82).

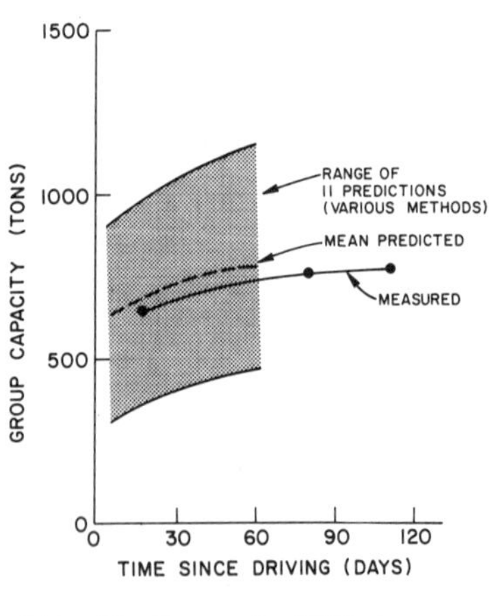

Fig. 13. Range in Efficiency Predictions (188).

been found to reduce effectively maximum bending moments and cap deformation by a factor of about 2 at working loads when the piles are vertical (81).

Inclined and Eccentric Loads. Non-concentric vertical loading has a small effect on η in both clays (125, 161) and sands (83), as long as the eccentricity is less than about 10%. Subjecting vertical piles in sand to inclined loadings apparently decreases single pile and group capacities in about equal amounts and so has little effect on η (106, 107).

Efficiency Formulas. In view of the preceding observations concerning the effects on efficiency of time, soil characteristics (consistency, initial density, stress history, permeability), and cap contact, and subsequent discussions of the effects of driving order and pile capacity variability it is not reasonable to apply general efficiency formulas in designs that consider only the plan geometry of a group (55, 112, 168). Formulas developed for a specific condition (e.g., tapered piles in soft clay soils, 88) and based on actual test data may be appropriate.

Flexibility

Axial. Statically loaded groups are usually more flexible than isolated piles in either the vertical or lateral mode of loading. Additional mechanically induced settlements in vertically loaded groups, with respect to isolated piles, are due to (a) increased elastic compressions of the piles due to load transfer occurring at lower elevations (37, 127), and (b) increased distortion or compression of the soil beneath the pile tips due to stress overlaps. The latter effect is more pronounced in friction piles, while the former is more pronounced in end-bearing piles. Since effect (b) produces generally larger absolute deformations, friction pile groups tend to have higher SR's than end bearing groups. SR in friction pile groups may tend to decrease with load level, as evidence exists that stress increments in the soil above the pile tips, and thereby effects due to (a), decrease with load level (13).

Several investigators have developed empirical equations for SR for square groups in sand (104, 171, 186):

$$SR = (s/d)(5 - s/3d)/(1 + 1/r)^2 \; ; \qquad (4)$$

$$SR = (4B + 9)^2/(B + 12)^2 \; ; \qquad (5)$$

$$SR = (B/d)^{0.5} \; ; \qquad (6)$$

where B = group width in feet and r = number of rows of piles.

Figure 3 suggests that penetration-to-spacing ratio and method of installation also may influence SR, and that pile groups in loose sands and bored piles with high end bearing in medium dense sands can develop SR's < 1. (Leonards [90] reports SR << 1 for driven piles in dense sand overlying very dense gravel. See Fig. 3. Both the single reference pile and the group were tipped a few feet above the sand-gravel interface, and it is likely that this particular layering produced the low SR.) Eqs. 4-6 are in general agreement with Fig. 3 (driven piles, cap in contact) for p/s > 5.

In clay, Fig. 5, only limited instances of SR < 1 exist. Figure 5 describes SR at end of construction in groups where 2.5 < s/d < 3.5. At that stage SR < 10 even for groups consisting of several hundred piles. The tendency for SR to approach an apparent constant with increasing numbers of piles probably reflects the influence of stiffer strata at some depth below the pile tips which limit the additional deflections due to Effect (b), above. Figure 5 also suggests that Eq. 6 may provide a reasonable general approximation for SR in clay for freestanding and cap-in-contact groups in which p/s ≧ 15.

Under long-term loading more creep and consolidation may occur beneath a group than beneath a single pile, so that SR's tend to increase with time over those shown in Fig. 5. Traditional equivalent pier methods (25, 102) or elastic methods involving superposition of vertical effective stresses (64, 65) can be used to predict long-term settlement. Observations of groups in sensitive clay suggest that the depth of load transfer is best assessed at the level of the pile tips (66). One observation of a pile group under long-term loading in soft, slightly overconsolidated marine clay suggests the use of $E = 20\sigma'_v$ for long-term settlement computations using elastic methods, where E is soil Young's modulus and σ'_v is vertical effective stress.

Little data exist with respect to group flexibility in uplift. It has been suggested that groups may be more flexible in monotonic uplift loading than in compression because of release of residual stresses in the individual piles (124).

Lateral. Lateral flexibility is significantly influenced by the degree of fixity between the pile head and the superstructure (97). Therefore, consideration of lateral group action is usually more properly made through means other than single-pile-to-pile-group pile-head lateral movement ratios (LMR's). One simple way of expressing increased group flexibility at working loads in sands or clays with depthwise increasing stiffness is to assume that $E = n_h z$, where z is depth and n_h is a constant of proportionality influenced by soil density and pile deflection. Table 1 summarizes several model and small full-scale test results in terms of n_h effective for pile top displacements in the range of 0.03-0.1 d. Model tests in dense sand by Williams (193) and in loose sand by Davisson and Salley (43) yield relatively constant expressions for relative n_h for small, rectangular groups under both static and cyclic loading where spacings are less than 6d. Studies by Prakash (42, 146, 169) have shown that wider spacings or provision of zero rotational restraint of the pile heads at spacing of 4d (or more) result in relative n_h values of about 1.

Small groups in clay with spacings of 5d or greater likewise tend to yield relative n_h's of about 1. Closer spacings result in relative n_h values as low as 0.5, as shown in Table 1.

Where reference and group piles have similar pile-head fixities, direct comparison of head displacements are possible, as in Refs. 98 and 178, and these comparisons are shown in Table 1. The data from Ref. 98 are particularly valuable for the design of circular offshore groups, as they show the effects of pile spacing in a circular group and further indicate that cyclic and static LMR's are about equal. The latter observation indicates that lateral cyclic degradation in groups in clay is primarily a single pile problem and that no additional significant degradation occurs due to group action. Model studies in the axial mode suggest a similar effect (96).

TABLE 1. RELATIVE FLEXIBILITY OF LATERALLY LOADED VERTICAL PILE GROUPS IN SAND AND CLAY (1psf=47.9Pa, 1K=4.45 kN).

$\frac{N_h^g}{N_h^s}$ (SOILS CHARACTERIZED BY $E_s = N_h Z$)	$\frac{\delta_h^g}{\delta_h^s}$ = LMR	D_r (SANDS)	C_u (psf) (CLAYS)	GROUP GEOMETRY	COMMENTS	SOURCE OF DATA
$0.5\,\frac{M}{N}^{-0.5}\left(\frac{S}{D}\right)^{0.75}$	-	0.94	-	$1.5D \leq S \leq 6D$ UP TO 3 x 3. M ROWS PARALLEL TO LOAD, N ROWS NORMAL TO LOAD	STATIC AND CYCLIC (MODEL)	193
$0.12\left(\frac{S}{D}\right)^{0.75}$	-	LOOSE	-	$3D \leq S \leq 4D$ 9-TO 100-PILE RECTANGULAR GROUPS	STATIC AND CYCLIC (MODEL)	43
1.0	-	"	-	$S \geq 8D$ IN DIRECTION OF LOAD $S \geq 2.5D$ NORMAL TO LOAD. UP TO 3 x 3	STATIC AND CYCLIC (MODEL)	42, 146
1.0	-	0.80	-	S = 4D 2 x 2 FREE-HEADED	STATIC AND CYCLIC (MODEL)	169
≈1.0	-	-	SOFT	S = 6D 2 x2 TO 3 x3	STATIC (MODEL)	147
≈1.0	-	-	1000-2000	S = 5D 2 x 3 (2 ROWS PARALLEL TO LOAD)	STATIC (FULL-SCALE)	80
0.5-0.8	-	-	"	" (S = 3.5D)	"	80
0.5	-	-	700	S = 3D 1 x 2	STATIC (FULL-SCALE)	93
-	1.5 1.9	-	300	5-PILE CIRCULAR S = 3.3D 2K/PILE	STATIC CYCLIC (FULL-SCALE)	98
	3.2 3.3			10-PILE CIRCULAR S = 1.75D 2K/PILE	STATIC CYCLIC (FULL-SCALE)	98
-	SEE NOTE	0.77	-	VARIOUS SQUARE GROUPS	STATIC (MODEL)	178

NOTE

$$LMR = 1 - 5\left[1 - (0.6 - 0.25f)\left(\frac{S}{D}\right)^{(0.3+0.2f)}\right]\left[1 - M^{-0.22}N^{-0.09}\right] \leq 1$$

(f = PILE HEAD MOMENT/MOMENT FOR PERFECT FIXITY)

Combined loading may alter the preceding effects. For example, Swain (177) reports that lateral deflections in a widely spaced group of piles supporting a model offshore structure resulted in elevated side resistance (stiffening of the piles axially).

It is likely that the preceding comments are applicable to laterally loaded groups containing batter piles. However, the effective spacing in such groups should probably be established at a depth of perhaps 0.5 $(EI/n_h)^{0.2}$ (one-half the characteristic depth of the pile). The presence of vertical loads has been reported to provide a slight stiffening effect when the lateral load is applied in the direction of the horizontal component of batter (54) or when the vertical load prevents pile-head rotation (14). Feagin (52) reported that initial flexibility in groups of closely spaced, full-scale, rigidly capped groups of tapered, vertical piles in medium-dense sand actually decreased with increasing numbers of piles. It is not clear whether the densification afforded by driving tapered piles contributed to soil stiffening or whether the overall moment resistance offered by tension and compression with increasing width of the base of the group caused the stiffening. However, Stevens, et al. (175), noted a similar effect in small groups of up to 12 tapered piles, especially when piles were driven in the soil surrounding the test group, which appears to support the first hypothesis. Piles were partially jetted in both tests.

Distribution of Loads to Piles

Concentric Axial Loading. Distributions of axial loads are influenced by geometric position of the piles in the group, the rigidity of the soil (especially below the pile tips), the rigidity of the cap (superstructure interface) and whether the cap bears on the soil, order of pile installation, variability of the soil across the group, pile alignment, and magnitude of applied load.

The effects of geometry and driving order in model and full-scale 3 × 3 freestanding groups in clay are demonstrated in Fig. 11, in which all piles had $L/d = 48$. Driving Order B produced much larger variations than Driving Order A in the model group, possibly because B resulted in higher undissipated pore pressures in the group's interior at the time of testing. In the full-scale test, conducted in over-consolidated slickensided clay (92), the driving order was similar to B, but significant reductions in induced pore pressures occurred during and immediately after installation, so that the load distribution pattern resembled A. That resemblance extends to a "crossover" near failure, in which the center pile became more heavily loaded than the exterior piles. This effect in the full-scale test was brought about largely by progressive failure (corner piles first), followed by relaxation (125). Evidence exists (87) that the pattern for B may also apply to any driving order if the soil is very sensitive, even after full pore pressure dissipation.

The effects of geometric position and driving order for axially loaded groups in sand are illustrated in Fig. 12, where an "outside-in" driving order (Kishida) produced higher capacities near the center of the group. A tendency toward development of higher capacities in the center of a group has also been observed when piles were installed simultaneously (185). The second driving pattern in Fig. 12 (Beredugo) shows a similar dependence on driving order but also implies that an interior pile (No. 5) will carry more load even if driven early in the sequence.

Cap contact can affect load distribution to some degree. This subject is considered during the description of mathematical models.

Lateral Loading. Full-scale tests on freestanding circular groups of up to ten vertical piles in soft clay (s/d ≧ 1.75) (98) revealed little dependence of pile-head shears and moments on geometric position or driving order. Tests on a full-scale rectangular group in sand (s/d ≃ 3) (75), on the other hand, indicated that pile-head shears and moments in piles or the leading row exceeded those on trailing (fourth) row by about 80%.

Random Effects. Evangelista, et al. (51), have suggested that random vertical pile capacity variations within a concentric, vertically loaded group may be characterized by a standard deviation of 15% of the mean capacity at a typical onshore site. Such a variation may produce induced shears, moments and torques. Other evidence from full-scale tests (36, 127) suggests that capacities of individual piles in identical geometric positions within small groups can vary by ±10%, with flexibility variations of about twice that amount. Such variations probably reflect soil property variations, as well as variations in driving procedures.

Effects of minor inadvertent misalignment of H-piles in a real full-scale test group subjected to combined loading and containing both vertical and batter piles (80) likely resulted in differences of 16-23% in pile-head moments and shears and 2-70% errors in axial thrusts compared to the same loads for ideal geometry (123).

PREDICTIVE CAPABILITIES

Considering the complexity of the variables and their interactions described in the preceding section, it should be apparent that our present (1983) ability to model the behavior of pile groups is not especially good. This fact is illustrated in Fig. 13 (188), which summarizes predictions of vertical capacity made by eleven deep foundations authorities using various predictive methods for the full-scale tests reported in Ref. 127. While the average of the predictions was good, the range of predictions was ±50%.

Vesić (187) suggested that our present PSPM's were inappropriate and proposed that a new, comprehensive model be developed that incorporated the effects of installation, soil compressibility, and mechanical pile-soil-pile interaction. Unfortunately, no such model has yet been formulated. Even if one is produced, it will likely remain a research tool until appropriate parametric studies can be done and the model simplified for design use. The designer in the meantime must use available mathematical models which do not have the direct capability of reflecting the phenomena described in the preceding section. Intelligent interpretation of the model outputs may yield reasonably good predictions, but the ranges in Fig. 13 should be borne in mind when selecting factors of safety.

MODELING GROUP ACTION WITH PSPM's

Categories of PSPM's

PSPM's can be used to estimate group flexibilities and load distributions but are not, in their present usage, appropriate for computing η. PSPM's can be categorized into five classes, each of which has advantages and disadvantages with respect to the others:

Finite Element Model. The finite element model has been used by a number of investigators to investigate mechanical interaction (45, 130, 153, 198) and can be feasible in practice for highly stratified soils. It provides the most nearly mathematically correct means of modeling mechanical interactions, since it includes the effects of the presence of the piles themselves directly in the soil "continuum." But it is tedious to use, and, considering the usual lack of availability of high quality soil data for offshore sites and the model's present inability to model installation effects, it probably must remain a research tool for the near term. The finite element model will not be pursued further.

Continuum Model. The continuum model is generally predicated on the use of Mindlin's three-dimensional elasticity equations (110) to relate soil reactions at one discrete zone on a pile to displacement produced through the soil (modeled as an elastic medium) at all other similar zones on all piles. Numerous variations of the continuum model have been developed that permit modeling of soil layering, pile-soil slip, and three-dimensional geometry.

Continuum models are either of the 2-pile interaction type, which consider pile-head flexibility adjustments based on superposition of the effect of a generic loaded pile on the response of any neighboring pile (134, 136, 139, 140, 141, 143, 144), or the direct solution type. Figures 14 and 15 demonstrate how 2-pile interaction factors, denoted α, can be used to analyze a free-headed pile group (Fig. 14) or a fixed-headed group (Fig. 15). The piles are all identical in each of the groups in these two figures, and torsional loads are absent. Interactions among piles and loading conditions have been expressed through modified flexibility matrix equations. A similar approach has been used to develop flexibility matrices for circular groups under general loading (154). The procedure, which can be executed on a micro-computer, requires knowledge of the flexibilities at the head of a single pile under axial and lateral unit forces and moments. These flexibilities can be obtained from single pile theoretical solutions (e.g., 95, 101, 135, 137) or from load tests. Approximate evaluations of the 2-pile interaction factors are given in Table 2. These factors presumably can be applied to situations in which local soil yield occurs, provided the flexibilities of the single piles are adjusted for those effects.

The 2-pile interaction method fails to account properly for the trend for redistribution of axial loads toward the pile tips with increasing numbers of piles in a group and so may tend to overpredict SR's by perhaps 20-40% for groups of 10-15 piles (152). Direct continuum models (boundary element models) take this effect into account because they affect complete simultaneous solutions for pile-soil-pile interaction between all piles in a group (9, 11, 27, 29, 116, 133). However, direct solutions are inconvenient in design because they require the use of a large computer program (11).

3-PILE, VERTICAL GROUP (FREE-HEADED)

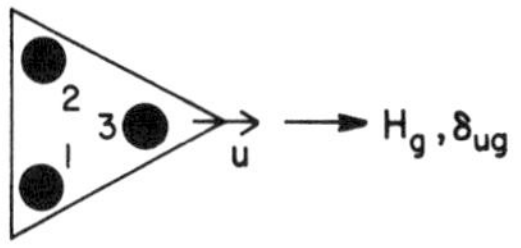

(V_g = VERTICAL LOAD ON GROUP; δ_{vg} = VERTICAL DEFLECTION OF GROUP)

VERTICAL INTERACTION FLEXIBILITY EQUATION:

$$\delta'_v \begin{bmatrix} 1 & \alpha_v^{12} & \alpha_v^{13} & 1 \\ \alpha_v^{21} & 1 & \alpha_v^{23} & 1 \\ \alpha_v^{31} & \alpha_v^{32} & 1 & 1 \\ 1 & 1 & 1 & 0 \end{bmatrix} \begin{bmatrix} V_1 \\ V_2 \\ V_3 \\ \dfrac{\delta_{vg}}{\delta'_v} \end{bmatrix} = \begin{bmatrix} 0 \\ 0 \\ 0 \\ V_g \delta'_v \end{bmatrix}$$

HORIZONTAL INTERACTION FLEXIBILITY EQUATION:
(IN DIRECTION OF LOAD)

$$\delta'_u \begin{bmatrix} 1 & \alpha_{uH}^{12} & \alpha_{uH}^{13} & 1 \\ \alpha_{uH}^{21} & 1 & \alpha_{uH}^{23} & 1 \\ \alpha_{uH}^{31} & \alpha_{uH}^{32} & 1 & 1 \\ 1 & 1 & 1 & 0 \end{bmatrix} \begin{bmatrix} H_1 \\ H_2 \\ H_3 \\ \dfrac{\delta_{ug}}{\delta'_u} \end{bmatrix} = \begin{bmatrix} 0 \\ 0 \\ 0 \\ H_g \delta'_u \end{bmatrix}$$

- H = SHEAR, V = THRUST
- δ'_v = VERTICAL DEFLECTION OF SINGLE PILE UNDER UNIT LOAD
- δ'_u = HORIZONTAL DEFLECTION OF SINGLE PILE UNDER UNIT LOAD
- LATERAL LOADING IS AT GROUND LINE
- α_{ij}^{kl} = INTERACTION FACTOR BETWEEN SUBSCRIPTED MODES (u=HORIZONTAL DEFLECTION; H=HORIZONTAL LOAD) BETWEEN SUPERSCRIPTED PILES.

Fig. 14. Example Flexibility Equation: Free-Headed Pile Group.

3-PILE, VERTICAL GROUP (FIXED-HEADED)

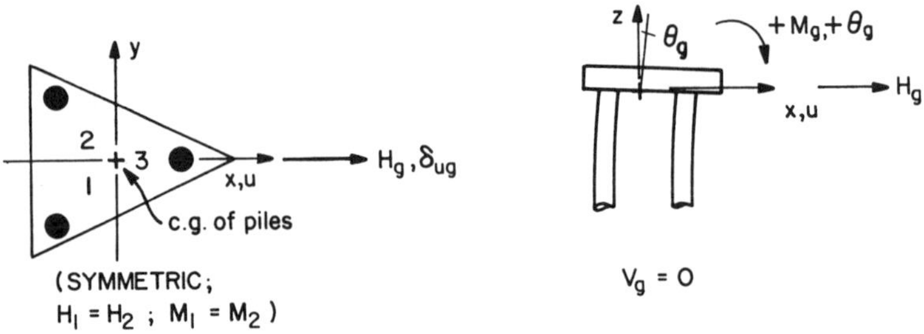

HORIZONTAL INTERACTION FLEXIBILITY EQUATION :

$$\begin{bmatrix} \delta^1_{uH}(1+\alpha^{12}_{uH}) & \delta^1_{uM}(1+\alpha^{12}_{uM}) & \delta^1_{uH}\,\alpha^{13}_{uH} & \delta^1_{uM}\,\alpha^{13}_{uM} & 1 & 0 \\ \theta^1_{\theta H}(1+\alpha^{12}_{\theta H}) & \theta^1_{\theta M}(1+\alpha^{12}_{\theta M}) & \theta^1_{\theta H}\,\alpha^{13}_{\theta H} & \theta^1_{\theta M}\,\alpha^{13}_{\theta M} & 0 & 1 \\ \delta^1_{uH}(2\alpha^{31}_{uH}) & \delta^1_{uM}(2\alpha^{31}_{uM}) & \delta^1_{uH} & \delta^1_{uM} & 1 & 0 \\ \theta^1_{\theta H}(2\alpha^{31}_{\theta H}) & \theta^1_{\theta M}(2\alpha^{31}_{\theta M}) & \theta^1_{\theta H} & \theta^1_{\theta M} & 0 & 1 \\ 2 & 0 & 1 & 0 & 0 & 0 \\ 0 & 2 & 0 & 1 & 0 & Q \end{bmatrix} \begin{bmatrix} H_1 \\ M_1 \\ H_3 \\ M_3 \\ -\delta_{ug} \\ -\theta_g \end{bmatrix} = \begin{bmatrix} 0 \\ 0 \\ 0 \\ 0 \\ H_g \\ M_g \end{bmatrix}$$

δ^1_{uH} = horizontal deflection of free-headed pile under unit force

δ^1_{uM} = horizontal deflection of free-headed pile under unit moment

$\theta^1_{\theta H}$ = rotation of free-headed pile under unit force

$\theta^1_{\theta M}$ = rotation of free-headed pile under unit moment

Q = push - pull moment due to axial reactions

$= -(\delta^1_v)^{-1} \sum_{i=1}^{3} X_i^2$ (no interaction in vertical mode assumed)

Fig. 15. Example Flexibility Equation:
Fixed-Headed Laterally Loaded Group.

TABLE 2. GROUP INTERACTION FACTORS FOR FLOATING PILES [RANDOLPH AND POULOS (1982)]

MODE OF INTERACTION	EQUATION
VERTICAL (α_v)	$\alpha_v = [0.5 \ln(\frac{L}{s})] / \ln(\frac{L}{D\rho})$ ($\lambda = 500$)
LATERAL DEFLECTION, FIXED HEAD (α_{uf})	$\alpha_{uf} = 0.6\rho_c (\frac{E_p^*}{G_c})^{0.143} (\frac{R}{S})(1+\cos^2\beta) \leq 0.33$ $= 1-2/[27\,\alpha_{uf}\text{ (above)}]^{0.5} \geq 0.33$
LATERAL DEFLECTION DUE TO SHEAR (α_{uH})	$\alpha_{uH} \approx 0.7\ \alpha_{uf}$
ROTATION DUE TO MOMENT ($\alpha_{\theta M}$)	$\alpha_{\theta M} \approx \alpha_{uH}^3$
LATERAL DEFLECTION DUE TO MOMENT OR ROTATION DUE TO SHEAR ($\alpha_{uM} = \alpha_{\theta H}$)	$\alpha_{uM} = \alpha_{\theta H} \approx \alpha_{uH}^2$

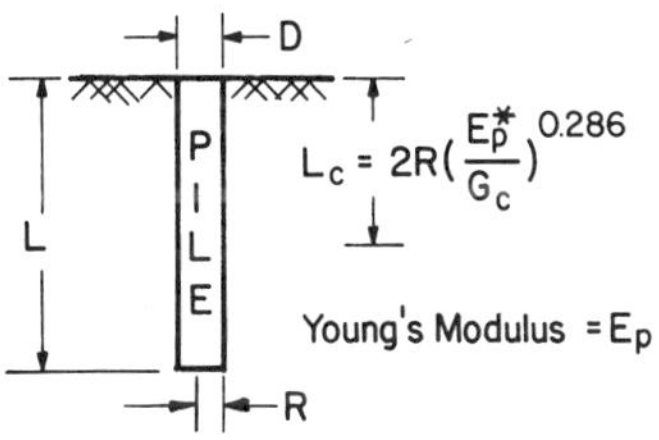

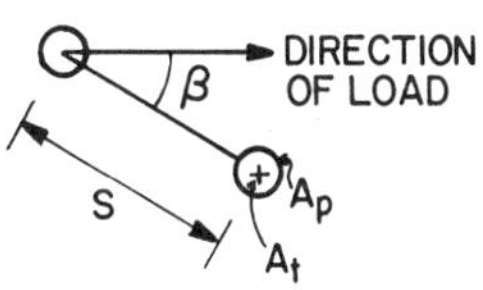

$\rho = \dfrac{G(\text{avg. over } L)}{G(\text{at depth} = L)}$

$G_c = G(\text{avg. over } L_c)$

$\rho_c = \dfrac{G(\text{at depth} = L_c/4)}{G(\text{at depth } L_c)}$

E_p = YOUNG'S MODULUS OF PILE

G = SOIL SHEAR MODULUS × 1.35

$\lambda = \dfrac{E_p(A_p/A_t)}{G(\text{depth} = L_c)}$

$E_p^* = (E_p I_p)/(\pi r_o^4/4)$

(LATERAL FACTORS ARE FOR RELATIVELY FLEXIBLE PILES, $L \geq 20R$)

Further, since the designer does not use interaction factors, the relative effects of one pile on another are not as readily apparent, making geometric rearrangements more cumbersome.

Practical considerations associated with computing group flexibilities using the 2-pile interaction model are illustrated in Figs. 16 and 17. Figure 16 demonstrates the importance of proper quantification of the soil stiffness profile. Figure 17 compares computed axial interaction factors with those measured in full-sized groups in overconsolidated clay profiles and suggests that a Gibson (uniformly increasing modulus) profile is much more appropriate than the assumption of a constant modulus. Absurdities such as introduction of negative loads in interior piles also appear in the 2-pile interaction model if the soil is represented as a uniform elastic halfspace.

Modified Continuum Model. This model is actually a set of linear models that simplify the general continuum solution without apparent significant loss of accuracy (113, 119, 151, 156, 165). These models integrate and superimpose soil strains produced by pile loading by assuming that axial or lateral interactions occur only in horizontal planes. That is, a plane strain or Winkler assumption is made regarding coupling of piles through the soil rather than using the three-dimensional solutions used in continuum models. This class of models can approximate a general soil profile, and simple expressions have been developed for use in the design office (119, 156). In view of the physical complexities of pile group action described earlier, models more complex than this or the 2-pile continuum model are probably not justified for evaluating group flexibility in design at present, except perhaps for geometrically unique groups.

Modified Unit Load Transfer Model. The use of unit load transfer (p-y and f-z) curves for the analysis of nonlinear behavior of single piles (95) has been prevalent for over two decades. Reference 19 suggests that circular groups of laterally loaded piles can be modeled by developing p-y curves for a group considered as a single pile whose diameter is the width of the group and distributing the computed p-values equally among the individual piles in the group. Appropriate adjustments must be made for gapping phenomena under cyclic loading (i.e., no group action occurs to the depth of the gap zone around any individual pile), and the sum of the p-values assigned to the piles, above, must not exceed the sum of the p-values obtained from consideration of the piles acting individually. This scheme requires all piles in the group to respond equally to load and hence would not be appropriate, for example, for the analysis of the rectangular group in sand described earlier (175). Once the modified p-y curves are obtained, group analysis can be conducted using a single pile analysis method.

Hybrid Model. This class of models is intended for one-of-a-kind, nonlinear analyses and is generally not well-adapted to parametric solutions. It assumes that all nonlinear soil behavior occurs in the near field around the individual piles in the group and all pile-soil-pile interaction is linear and predictable through Mindlin's equations (34, 60, 68, 69, 89, 122). Some versions of the model are applicable to three-dimensional groups and to flexible superstructures. Nonlinear behavior around the individual piles is modeled by inputting p-y, f-z, and/or unit torque-twist (48) curves that are valid for isolated piles

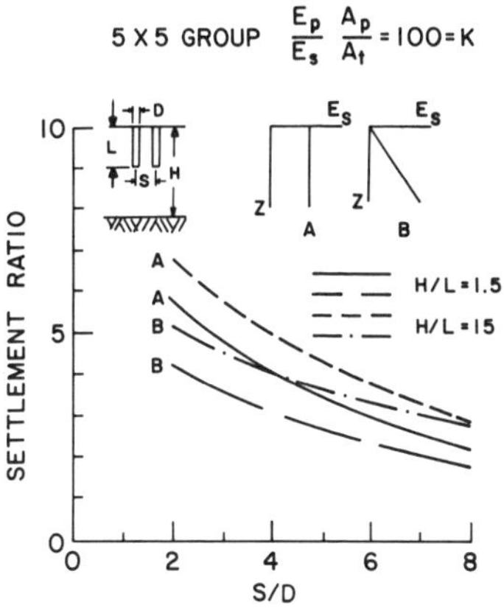

Fig. 16. Effect of Soil Profile (140).

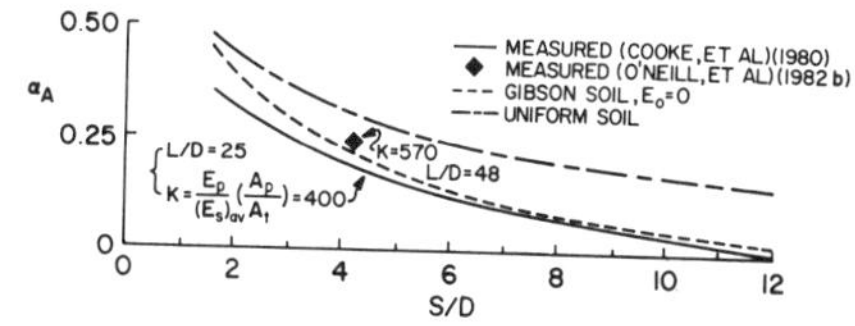

Fig. 17. Measured and Computed Interaction Factors.

or that are empirically modified for installation effects. These curves are then modified internally by stretching or condensing them along the displacement axes by appropriate consideration of displacements produced by reaction forces between the piles and the soil (122, 124). Torque-twist curves are normally not modified.

The hybrid model algorithm has the advantage that separate stiffness relationships are used for individual pile action and for pile-soil-pile interaction in a group, thereby theoretically permitting the independent adjustment of the two relationships in a model calibration process. This feature is not inherent in continuum models, which use the same elastic soil properties to model individual pile action and pile-soil-pile interaction.

The original hybrid model (60), which is still commonly used in design, used an algorithm that multiplied the y-values on all p-y curves along a given pile by an equal value instead of adjusting the p-y curves according to the magnitude of superimposed deformation in the soil at a given elevation. This approximation makes it difficult to obtain correct pile-head deformation and pile moments with a common set of input parameters (61, 197). Later contributions (122) circumvented this problem.

None of the models described should be expected to provide generally accurate predictions of redistributions of loads among piles in a group due to pile-soil-pile interaction because none accounts for installation effects. Reference 124 documents this assertion for two continuum and hybrid model algorithms.

Cap Contact Effects

Several continuum and finite element algorithms can model cap-soil-pile interaction effects in addition to pile-soil-pile interaction effects (28, 130, 192). Analytical studies (28, 130) show that integral cap-soil contact reduces slightly (perhaps by 10 percent) the SR value and significantly influences distribution of load among piles. These effects become less pronounced with increasing relative pile-soil stiffness. For example, for a 3 × 3 × 3-d-spaced group in a uniform elastic soil (130), all piles carry about 9% of a concentric vertical group load, while a cap of plan dimensions of the pile system carries about 22% of the load if K (Fig. 17) is 200 (stiff soil). If K = 800 (soft soil), the loads are 11%, 9% and 7% at the corner, edge, and center piles, respectively, and 11% at the cap. Confirmation of these general effects have come principally through carefully controlled model tests (e.g., 192), although further experimental studies are needed in this area. If the cap-soil contact is disturbed, considerably different trends are likely.

Calibration of Models

Despite the failure of mathematical models to reproduce installation effects, they can still be of service to the designer in predicting group flexibility and in giving insight into differences in load transfer patterns between isolated and group piles provided they are calibrated to real behavior. Table 3 summarizes several attempts to calibrate some of the models against measured flexibilities of pile heads in primarily statically loaded full-sized groups. The parameters calibrated are elastic moduli, through E/c ratios in clay, and primarily through n_h for laterally loaded groups in sand. In the hybrid model calibrations, the unit load transfer curves were generally those that modeled isolated pile behavior at the test site correctly. No general indications are available concerning the appropriateness of these correlations for computing distribution of loads to piles and load transfer. The correlations are probably appropriate for flexibility modeling for cyclic loading, provided cyclic degradation effects are included (e.g., in the unit load transfer curves in the hybrid model).

The hybrid model generally requires higher soil moduli than the continuum model because the moduli that are input for the latter model must represent any decrease in stiffness in the soil in the near field around a pile in addition to pile-soil-pile effects, which occur at lower average soil strain levels.

OTHER CONSIDERATIONS

Downdrag

Negative skin friction can be produced in groups of displacement piles during reconsolidation of disturbed soil (57) and possibly by rapid infilling of subsidence bowls in which offshore platforms are situated (18). Negative skin friction can reduce the effective confining pressure in the stratum in which the pile tips are situated for conditions of undrained (rapid) loading to failure. Procedures for

TABLE 3. CALIBRATED SOIL MODULI FOR VERTICAL GROUPS OF DRIVEN PILES FOR SEVERAL PILE GROUP MODELS.

MODE	SOIL TYPE	MODEL (SPECIFIC ALGORITHM)	DEFINITIVE CORRELATION AND/OR OBSERVATIONS	SOURCE (BASIS OF CORRELATION)
VERTICAL	CLAY	2-PILE CONTINUUM (DEFPIG)	$\frac{E_s}{C_U(AV.)} = 3600\left(\frac{A_p}{A_g}\right)^{0.4}\left(\frac{C_U(TIP)}{C_U(AV.)}\right)^{0.25}$ (UNIFORM SOILS) $\frac{E_s}{C_U(TIP)} = 2300\left[\cos(1000\frac{A_p}{A_g})\right]^{-3}\left(\frac{L}{D}\right)^{-0.25}$ (VARIABLE SOILS)	O'NEILL AND HA (1982) (FULL-SCALE)
		HYBRID (PILGP1)	$\frac{E_s}{C_U(AV.)} = 8100\left(\frac{A_p}{A_s}\right)^{0.5}\left[\frac{C_U(TIP)}{C_U(AV.)}\right]^{0.5}$ (UNIFORM SOILS) $\frac{E_s}{C_U(TIP)} = 4800\left(\frac{A_p}{A_s}\right)^{0.4}\left[\frac{L}{D}\right]^{0.2}$ (VARIABLE SOILS)	
		DIRECT CONTINUUM (PGROUP)	$\frac{E_s}{C_U}$ = 450, WITH CAP IN CONTACT WITH SOIL	BUTTERFIELD AND GHOSH (1980) (MODELS)
		2-PILE CONTINUUM (UNSPECIFIED)	E_s = 6000 PSI (C_U > 1 ATM) $\frac{E_s}{C_U}$ = 300–400 ($C_U \leq$ 1 ATM)	POULOS (1972) (MODELS)
		2-PILE CONTINUUM (UNSPECIFIED)	$E_s' = 20\sigma_v'$ (LONG-TERM SETTLEMENT)	BLANCHET, ET AL. (1980) (FULL-SCALE)
VERTICAL	SAND	HYBRID (PILGP1)	$E_s = E_s$ (CROSS HOLE TEST) TO 50% OF FAILURE $E_s = E_s$ (PRESSUREMETER, RECOMPRESSION) AT FAILURE	AGGARWAL (1980) (FULL-SCALE)
		2-PILE CONTINUUM (UNSPECIFIED)	E_s = 275–550 ATM. ($D_r \leq 0.4$) E_s = 550–680 ATM. ($0.4 < D_r \leq 0.6$) E_s = 680–1100 ATM. ($D_r > 0.6$)	POULOS (1972) (MODELS)
LATERAL	CLAY	HYBRID (GP3A)	$E_s = 750C_U$ (TO Z = 10D) (BATTER-PILE GROUP)	O'NEILL, ET AL. (1979) (FULL-SCALE)
		HYBRID	E_s = 500–1000 C_U TO 10% OF FAILURE = 200 AT 50% OF FAILURE	TOOLAN AND HORSNELL (1980) (RECOMMENDED)
		HYBRID (PILGP2)	$E_s = 500\ C_U$ (CYLCIC LOADING)	CHEN (1982) (FULL-SCALE)
		DIRECT CONTINUUM (PGROUP)	$\frac{E_s}{C_U}$ = 100–300 UP TO 50% OF FAILURE (HIGHER END OF RANGE APPLIES FOR STIFF CLAYS)	BANERJEE AND DAVIES (1980) (MODELS)
		2-PILE CONTINUUM (DEFPIG)	$\frac{E_s}{C_U}$ = 300, LOW LOADS (LINEAR LOAD-DEFL.) = 190, MODERATE LOADS (DEFL. = 10% OF SINGLE PILE DIAMETER)	POULOS AND RANDOLPH (1981) (FULL-SCALE)
		MOD. CONTINUUM (PIGLET)	$\frac{E_s}{C_U}$ = 320, LOW LOADS = 220, MODERATE LOADS	
		HYBRID (FOCHT-KOCH)	$\frac{E_s}{C_U}$ = 100, BASED ON MATCHES OF MOMENT, NOT DEFLECTION. HYBRID RESULTS APPEARS NOT NON-LINEAR ENOUGH.	FOCHT (1982) (FULL-SCALE)
		2-PILE CONTINUUM (UNSPECIFIED)	$\frac{E_s}{C_U}$ = 180–400, FOR MODEL PERMITTING LOCAL YIELD	POULOS (1975) (FULL-SCALE, MODELS)
LATERAL	SAND	MOD. CONTINUUM (PIGLET)	$N_h = 1270\gamma_w$ (MEDIUM DENSE SAND)	POULOS AND RANDOLPH (1981) (FULL-SCALE)
		2-PILE CONTINUUM (DEFPIG)	$N_h = 2240\gamma_w$ (MEDIUM DENSE SAND)	
		DIRECT CONTINUUM (PGROUP)	$N_h = 1100\gamma_w$ (MEDIUM DENSE SAND)	BANERJEE AND DAVIES (1980) (FULL-SCALE)
		2-PILE CONTINUUM (UNSPECIFIED)	$N_h = 550\gamma_w$ (LOOSE SAND) $N_h = 1630\gamma_w$ (MEDIUM DENSE SAND) $N_h = 3460\gamma_w$ (DENSE SAND)	WRIGHT (1982) (AFTER POULOS)
		HYBRID (GP3A)	USE E_s = TWICE THE VALUE OBTAINED FROM LABORATORY TESTS TO REPLICATE DEFLECTION. GP3A PREDICTS LOCATION OF MAX. MOMENTS TOO DEEP	HOLLOWAY, ET AL. (1982) (FULL-SCALE)

NOTES: A_p = CROSS-SECTIONAL AREA OF PILE MATERIAL WITHIN GROUP.
A_s = CROSS-SECTIONAL AREA OF SOIL INCLUDED WITHIN GROUP.
A_g = TOTAL CROSS-SECTIONAL AREA OF GROUP, INCLUDING PILES.
γ_w = UNIT WEIGHT OF WATER.
ATM. = ATMOSPHERES OF PRESSURE.

calculating the magnitude of this effect are given in Refs. 167 and 199.

Only a few experimental studies have been reported on the effects of group action for downdrag loading. One model study (44) indicates no group action in clay for small square grops of 3-6 d spacing (i.e., total downdrag on the group is independent of pile spacing). A second model study (167) suggests major group action in a 3 × 3 group with $2 < s/d < 2.5$ (i.e., total downdrag less than sum of downdrags on individual piles tested as isolated piles).

Meyerhof (105) suggests that negative skin friction in a group be computed from the smaller of (a) the sum of 0.67 times the maximum drained, remolded shear strength integrated over the individual pile perimeters in the drag zone and (b) the weight of soil between the piles plus the integral of the undrained cohesion around the perimeter of the group in the drag zone.

Soil Moving Laterally Against Piles

Group action can also be anticipated in mudslide areas, where the soil moves laterally with respect to the piles. It arises from the necessity for the soil to flow through the constricted spaces between the piles. Ito, et al. (77, 100), have proposed and verified experimentally theoretical formulations for the soil force against a row of equally spaced piles. Typically, increases in lateral soil loading forces of about 50% are achieved over those for isolated piles in very soft clay for a very long row of piles where $s/d = 2$. Load values against isolated piles are usually obtained by using p-values from p-y curves in the sliding zone and offset p-y curves in the shear drag zone below the depth of sliding or, alternatively, by the method described in Ref. 138.

Scour

Scour influences lateral load-deformation behavior and bending moment diagrams for piles. In model groups in sand local scour occurs around the individual piles, but complex water-soil-pile interactions also apparently produce general ripple patterns in the direction of surface wave propagation (32), so that some allowances should be made for group geometry in establishing scour depths.

Dynamic Loading

For loadings at frequencies sufficient to make pile and soil inertial forces significant (seismic and vibrating machinery loadings), group response may be either "softer" or "stiffer" than group response due to static or very low frequency cyclic wave loadings.

Early design models suggested that group action be considered by computing pile-head impedance functions (stiffness and damping, either exact frequency dependent or approximate frequency independent) for the individual piles and applying static 2-pile interaction factors or empirical adjustments to the results to simulate group action (4, 162, 170, 174). This approach has been found experimentally to yield relatively good results at and below the fundamental frequency of a limited number of groups (47, 121, 183). Reference 183, however, describes model tests in which continued lateral vibration at resonance increased group displacements in sands by fourfold, demonstrating time-dependent, in addition to force-dependent, nonlinearities.

Recent finite element and continuum mathematical model studies have shown that phase effects between vibrating piles and incident soil body waves produced by neighboring piles can invalidate the approach just described, especially at high frequencies (33, 120, 166 194, 195). Kaynia and Kausel (78) have recently produced limited design charts for linear, dynamic loading of pile groups that consider this effect and that may be used in a manner analogous to the 2-pile interaction continuum method for statically loaded groups.

CONCLUSIONS

1. Pile group design requires a prediction of both installation and mechanical effects. The former are presently considered only empirically by interpreting results of full-scale and model tests, which this paper has attempted to summarize. Empirical installation effects have not been systematically quantified, but ranges of certain components of these effects have been given.
2. No comprehensive mathematical models are available to compute efficiency. Present design practice uses η = 1 for driven piles in sand. In insensitive clay $\eta \approx 1$ if block failure does not occur and if sufficient time has elapsed between installation and first application of the design load to permit excess pore pressure dissipation (Fig. 7).
3. Several classes of mathematical models are available to compute group flexibility and distribution of loads to piles. Calibrations for group flexibility determination have been given in Table 3.
4. Computed load distributions may be significantly in error unless effects of order of driving, soil variation across the group, pore pressure dissipation rates, and similar effects are empirically included in the computations.
5. There exists a signficant need for more experimental data and for deterministic mathematical models that reasonably reproduce installation effects. Reliability models that incorporate the effects of the various phenomena described herein would be useful in the interim.

REFERENCES

1. Aggarwal, D. H., "Analysis of Load Tests of Single Pile and Pile Group in Sand Under Axial Loading," M.S. thesis, Department of Civil Engineering, The University of Texas at Austin, Dec. 1980.
2. Akinmusuru, J. O., Jr., "Interaction of Piles and Cap in Piled Footings," Journal of the Geotechnical Engineering Division, A.S.C.E., Vol. 106, No. GT11, Nov. 1980, pp. 1263-1268.
3. A.R.E.A., "Steel and Timber Pile Tests; West Atchafalaya Floodway; New Orleans, Texas and Mexico Railway," Bulletin 489, American Railway Engineering Association, Sep.-Oct. 1950, pp. 140-202.
4. Arya, S., O'Neill, M. W., and Pincus, G., Design of Structures and Foundations for Vibrating Machines, Gulf Publishing Company, Houston Tex., 1979.

5. Aschenbrenner, R., "Three-Dimensional Analysis of Pile Foundations," Journal of the Structural Division, A.S.C.E., Vol. 93, No. ST1, Feb. 1967, pp. 201-219.
6. Asplund, S. O., "Pile Groups: Elastic and Plastic Design," Proceedings, Symposium on the Design of Pile Foundations, International Assoc. for Bridge and Structural Eng. 1960, pp. 77-87.
7. Audibert, J. M. E., Dover, A. B., Thompson, G. R., and Hubbard, J. L., "Geotechnical Engineering for Guyed Tower Offshore Structure," Proceedings, Civil Engineering in the Oceans IV, A.S.C.E., Sep. 1979, pp. 820-847.
8. Awoshika, K., and Reese, L. C., "Analysis of Foundation with Widely Spaced Batter Piles," Research Report 117-3F, Center for Highway Research, The University of Texas at Austin, Feb. 1971.
9. Banerjee, P. K., and Davies, T. G., "Analysis of Pile Groups Embedded in Gibson Soil," Proceedings, Ninth I.C.S.M.F.E., Vol. 1, 1977, pp. 381-386.
10. Banerjee, P. K., and Davies, T. G., Discussion; Numerical Methods in Offshore Piling, I.C.E., London, 1980, pp. 201-202.
11. Banerjee, P. K., and Driscoll, R. M., "Three-Dimensional Analysis of Raked Pile Group," Proceedings, Institution of Engineers, U.K., Vol. 61, Part 2, Dec. 1976, pp. 653-671.
12. Barden, L., and Monckton, M. F., "Tests on Model Pile Groups in Soft and Stiff Clay," Geotechnique, Vol. XX, March 1970, pp. 94-96.
13. Bartolomey, A. A., Dalmatov, B. I., Doroshkvitch, N. M., Sorochan, E. A., and Fedorovsky, V. G., "The Work of Pile Foundations and the Surrounding Soil," Proceedings, Ninth I.C.S.M.F.E., Vol. 1, 1977, pp. 387-390.
14. Beatty, C. I., "Lateral Test on Pile Groups," Foundation Facts, Raymond International, Inc., Vol. VI, No. 1, 1970, pp. 18-21.
15. Beredugo, Y. O., "An Experimental Study of the Load Distribution in Pile Groups in Sand," Canadian Geotechnical Journal, Vol. 3, Aug. 1966, pp. 145-166.
16. Berezantzev, V. G., Kristoforov, V. S., and Golbukov, V. N., "Load Bearing Capacity and Deformation of Piled Foundations," Proceedings, Fifth I.C.S.M.F.E., Vol. II, 1961, pp. 11-15.
17. Blanchet, R., Tavenas, F., and Garneau, R., "Behavior of Friction Piles in Soft Sensitive Clays," Canadian Geotechnical Journal, Vol. 17, May 1980, pp. 203-224.
18. Blessey, W. E., "Pile Foundation Loads and Deformations—Mississippi River Deltaic Plain," Proceedings, Conference on Design and Installation of Pile Foundations and Cellular Structures, Envo Pub. Co., Lehigh Valley, Pa., 1969, pp. 1-25.
19. Bogard, D., and Matlock, H., "Procedures for Analysis of Laterally Loaded Pile Groups in Soft Clay," Proceedings, Specialty Conference on Geotechnical Engineering in Offshore Practice, A.S.C.E., April 1983.
20. Bozozuk, M., Fellenius, B. H., and Samson, L., "Soil Disturbance From Pile Driving in Sensitive Clay," Canadian Geotechnical Journal, Vol. 15, Aug. 1978, pp. 346-361.
21. Brand, E. W., Muktabhant, C., and Taechathummarak, A., "Load Tests on Small Foundations in Soft Clay," Proceedings, Specialty Conference on the Performance of Earth and Earth-Supported Structures, A.S.C.E., Vol. I, Part 2, 1972, pp. 903-928.

22. Bridges, J. W., and King, D. M., "The Analysis of Offshore Platforms as Large Space Frames with Coupled Interaction to Non-Linear Pile Foundations," Preprints, Fourth Offshore Technology Conference, Vol. II, 1972, pp. 721-736.
23. Brinch Hansen, J., "A Theory for Skin Friction on Piles," Bull. No. 25, Danish Geotechnical Institute, 1968, pp. 5-12.
24. Brod, E. J., Fowler, J. N., and Boston, L. A., "A Unified Technique for the Analysis of Structure-Soil-Pile Systems," Proceedings, Seventh Offshore Technology Conference, Vol. II, pp. 199-208.
25. Broms, B. B., "Settlement of Pile Groups," Proceedings, Specialty Conference on the Performance of Earth and Earth-Supported Structures, A.S.C.E., Vol. III, 1972, pp. 181-199.
26. Bryant, L. M., and Matlock, H., "Three-Dimensional Analysis of Framed Structures with Nonlinear Pile Foundations," Proceedings, Ninth Offshore Technology Conference, Vol. III, pp. 599-606.
27. Butterfield, R., and Banerjee, P. K., "The Elastic Analysis of Compressible Piles and Pile Groups," Geotechnique, Vol. XXI, March 1971, pp. 43-60.
28. Butterfield, R., and Banerjee, P. K., "The Problem of Pile Group-Cap Interaction," Geotechnique, Vol. XXI, June, 1971, pp. 135-142.
29. Butterfield, R., and Ghosh, N., "A Linear Elastic Interpretation of Model Tests on Single Piles and Groups of Piles in Clay," Numerical Methods in Offshore Piling, I.C.E., London, 1980, pp. 109-118.
30. Chaudihuri, K. P. R., Ghataora, G. S., and Symons, M. V., "Uplift Resistance of Model Piles and Pile Groups," Proceedings, Second International Conference on Numerical Methods in Offshore Piling, I.C.E., London, 1982.
31. Chen, H. L., "Predictions of Harvey Pile Test Using PILPG1," Unpublished Report, University of Houston, Tex., Oct. 1982.
32. Chow, W. Y., and Herbich, J. B., "Scour Around a Group of Piles," Proceedings, Tenth Offshore Technology Conference, Vol. IV, 1978, pp. 2243-2252.
33. Chung, K. Y. C., and Stoll, R. D., "Dynamic Pile Interaction," Paper Presented Before A.S.C.E. Convention and Exposition, Preprint No. 80-038, April, 1980.
34. Clausen, C. J. F., Aas, P. M., and Almeland, I. B., "Analysis of the Pile Foundation System for a North Sea Drilling Platform," Behaviour of Offshore Structures, Vol. 1, Hemisphere Publishing Corp., 1982, pp. 95-104.
35. Cooke, R. W., Bryden-Smith, D. W., Gooch, M. N., and Sillett, D. F., "Some Observations of the Foundation Loading and Settlement of a Multi-Storey Building on a Piled Raft Foundation in London Clay," Proceedings, Institution of Civil Engineers, Vol. 70, Part 1, Aug. 1981, pp. 433-440.
36. Cooke, R. W., Price, G., and Tarr, K., "Jacked Piles in London Clay: A Study of Load Transfer and Settlement Under Working Load Conditions," Geotechnique, Vol. XXIX, June, 1979, pp. 113-147.
37. Cooke, R. W., Price, G., and Tarr, K., "Jacked Piles in London Clay: Interaction and Group Behavior Under Working Conditions," Geotechnique, Vol. XXX, June, 1980, pp. 97-136.
38. Coyle, H. M., and Sulaiman, I. H., "Bearing Capacity of Foundation

Piles: State of the Art," Highway Research Record 333, Highway Research Board, Washington, D.C., 1970, pp. 87-102.

39. Crooks, J. H. A., Matyás, E. L., and McKay, H. M., "Excavation Slope Stability Related to Pore-Water Pressure Variations During Piling," Canadian Geotechnical Journal, Vol. 17, May 1980, pp.225-235.

40. D'Appolonia, D. J., and Lambe, T. W.,"Performance of Four Foundations on End Bearing Piles," Journal of the Soil Mechanics and Foundations Division, A.S.C.E., Vol. 97, No. SM1, Jan. 1971, pp. 77-93.

41. Das, B. M., Seeley, G. R., and Smith, J. E., "Uplift Capacity of Group Piles in Sand," Journal of the Geotechnical Engineering Division, A.S.C.E., Vol. 102, No. GT3, March 1976, pp. 282-286.

42. Davisson, M. T., "Lateral Load Capacity of Piles," Highway Research Record 333, Highway Research Board, Washington, D.C., 1970, pp. 104-112.

43. Davisson, M. T., and Salley, J. R., "Model Study of Laterally Loaded Piles," Journal of the Soil Mechanics and Foundations Division, A.S.C.E., Vol. 96, No. SM5, Sep. 1970, pp. 1605-1627.

44. Denman, K. J., Nicholls, R. A., and Symons, M. V., "Model Studies of Negative Friction on Pile Groups," Proceedings, Conference on Large Ground Movements and Structures, Ed. by J. G. Geddes, Pentech Press, London and Plymouth, 1977, pp. 252-271.

45. Desai, C. S., Johnson, L. D., and Hargett, C. M., "Analysis of Pile-Supported Gravity Dock," Journal of the Geotechnical Engineering Division, A.S.C.E., Vol. 100, No. GT9, Sep. 1974, pp. 1009-1029.

46. Diaz, B. E., "Determination of Forces, Desplacements, and Soil Reactions of a Group of Piles," Proceedings, Eighth I.C.S.M.F.E., 1973, 2.1, pp. 83-88.

47. Diaz, G. M., "Measured Response of Pile Supported Machines," Proceedings, Symposium on Deep Foundations, A.S.C.E., 1979, pp. 75-100.

48. Dutt, R. N., and O'Neill, M. W., "Torsional Behavior of Model Piles in Sand," Proceedings, Specialty Conference on Geotechnical Practice in Offshore Engineering, A.S.C.E., April 1983.

49. Ealy, C. D., Personal Communication, 1982.

50. Eustis, J. B., "How to Design Pile Groups," Pile Foundations Know-How, American Wood Preservers Institute, 1969, pp. 36-40.

51. Evangelista, A., Pellegrino, A., and Viggiani, C., "Variability Among Piles of the Same Foundation," Proceedings, Ninth I.C.S.M.F.E., Vol. 1, 1977, pp. 493-500.

52. Feagin, L. B., "Lateral Pile-Loading Tests," Transactions, A.S.C.E., Vol. 63, No.8, 1937, pp. 236-254.

53. Feagin, L. B., "Performance of Pile Foundations of Navigation Locks and Dams on the Upper Mississippi River," Proceedings, Second I.C.S.M.F.E., Vol. IV, 1948, pp. 98-106.

54. Feagin, L. B., "Lateral Load Tests on Groups of Battered and Vertical Piles," Symposium on Lateral Load Tests on Piles, A.S.T.M. STP 154, 1953, pp. 12-20.

55. Feld, J., Discussion of "Timeber Friction Pile Foundations," Transactions, A.S.C.E., Vol. 108, 1943, pp. 143-144.

56. Fellenius, B., "Results of Tests on Piles at Gotheburg Railway Station," Bulletin No. 5, Geotechnical Department, Swedish State

Railways, Nov. 1955.
57. Fellenius, B. H., "Downdrag on Piles in Clay Due to Negative Skin Friction," Canadian Geotechnical Journal, Vol. 9, Nov. 1972, pp. 323-337.
58. Fellenius, B. H., and Samson, L., "Testing of Drivability of Concrete Piles and Disturbance to Sensitive Clay," Canadian Geotechnical Journal, Vol. 13, May 1976, pp. 139-160.
59. Flaate, K., "Effects of Pile Driving in Clays," Canadian Geotechnical Journal, Vol. 9, Feb. 1972, pp. 81-88.
60. Focht, J. A., Jr., and Koch, K. J., "Rational Analysis of the Lateral Performance of Offshore Pile Groups," Preprints, Fifth Offshore Technology Conference, Houston, Tex., Vol. 2, pp. 701-708, 1973.
61. Focht, J. A., III, "A Comparison Between Measured and Calculated Pile Group Behavior Under Lateral Loading," Master's Report, Department of Civil Engineering, The University of Texas at Austin, March, 1982.
62. Francis, A. J., "Analysis of Pile Groups with Flexible Resistance," Journal of the Soil Mechanics and Foundations Division, A.S.C.E., Vol. 90, No. SM3, May 1964, pp. 1-32.
63. Garg, K. G., "Bored Pile Groups Under Vertical Load in Sand," Journal of the Geotechnical Engineering Division, A.S.C.E., Vol. 105, No. GT8, Aug. 1979, pp. 939-956.
64. Geddes, J. D., "Stresses in Foundation Soils Due to Vertical Subsurface Loading," Geotechnique, Vol. XVI, Sep. 1966, pp. 231-235.
65. Geddes, J. D., "Boussinesq-Based Approximations to the Vertical Stresses Caused by Pile-Type Subsurface Loadings," Geotechnique, Vol. XIX, Dec. 1969, pp. 509-514.
66. Girault, P., "Settlement of Some Piled Structures in Mexico," Proceedings, Specialty Conference on Performance of Earth and Earth-Supported Structures, A.S.C.E., Vol. I, Part 2, 1972, pp. 1185-1205.
67. Greenfield, F. C., "Early Settlement of Tall Buildings Founded on Piles in London Clay," Proceedings, Conference on Behaviour of Piles, I.C.E., London, 1971, pp. 71-78.
68. Ha, H. B., and O'Neill, M.W., "Field Study of Pile Group Action; Appendix A; PILGP1 Users' Guide," Report No. FHWA/RD-81/003, Federal Highway Administration, March 1981.
69. Ha, H. B., Thompson, H., and Datta, B., "Automated Coupled-Interaction Analysis of Piled Offshore Structures," Preprint, Texas Section, A.S.C.E., San Antonio, Tex., Spring 1980.
70. Hagarty, D. J,, and Garlanger, J. E., "Consolidation Effects Around Driven Piles," Proceedings, Specialty Conference on Performance of Earth and Earth-Supported Structures, A.S.C.E., Vol. I, Part 2, 1972, pp. 1207-1222.
71. Hanna, T. H., "Model Studies of Foundation Groups in Sand," Geotechnique, Vol. XIII, Dec. 1963, pp. 334-351.
72. Hanna, T. H., "The Measurement of Pore Water Pressures Adjacent to a Driven Pile," Canadian Geotechnical Journal, Vol. IV, Sep. 1967, pp. 313-325.
73. Hays, C. O., Jr., and Matlock, H., "A Nonlinear Analysis of a Soil Supported Frame," Preprints, Fourth Offshore Technology Conference, Vol. II, 1972, pp. 737-752.

74. Hill, J. L., "User's Guide: Computer Program for Optimal Design and Analysis of Pile Foundations (PILEOPT)," Instruction Report K-81-5, U.S.A.E. Waterways Experiment Station, 1981.
75. Holloway, D. M., Moriwaki, Y., Finno, R. J., and Green, R. K., Lateral Load Response of a Pile Group in Sand," Proceedings, Second International Conference on Numerical Methods in Offshore Piling, I.C.E., 1982.
76. Hrennikoff, A., "Analysis of Pile Foundations with Battered Piles," Transactions, A.S.C.E., Vol. 115, 1950, pp. 351-374.
77. Ito, T., and Matsui, T., "Methods to Estimate Lateral Force Acting on Stabilizing Piles," Soils and Foundations, Vol. 15, Dec. 1975, pp. 43-59.
78. Kaynia, A. M., and Kausel, E., "Dynamic Behavior of Pile Groups," Proceedings, Second International Conference on Numerical Methods in Offshore Piling, I.C.E., 1982.
79. Kezdi, A., "Bearing Capacity of Piles and Pile Groups," Proceedings, Fourth I.C.S.M.F.E., Vol. II, 1957, pp. 46-51.
80. Kim, J. B., and Brungraber, R. J., "Full-Scale Lateral Load Tests of Pile Groups," Journal of the Geotechnical Engineering Division, A.S.C.E., Vol. 102, No. GT1, Jan., 1976, pp. 87-105.
81. Kim, J. B., Singh, L. P., and Brungraber, R. J., "Pile Cap-Soil Interaction from Full-Scale Lateral Load Tests," Journal of the Geotechnical Engineering Division, A.S.C.E., Vol. 105, No. GT5, May 1977, pp. 643-653.
82. Kishida, H., "Ultimate Bearing Capacity of Piles Driven Into Loose Sand," Soil and Foundation, Vol. VII, No. 3, 1967, pp. 20-27.
83. Kishida, H., and Meyerhof, G. G., "Bearing Capacity of Pile Groups Under Eccentric Loads in Sand," Proceedings, Sixth I.C.S.M.F.E., Vol. II, 1965, pp. 270-274.
84. Ko, H.Y., Personal Communication, Oct. 1982.
85. Kocsis, P., Lateral Loads on Piles, Peter Kocsis Pub., Chicago, Illinois, 1968.
86. Koerner, R.M., and Partos, A., "Settlement of Building on Pile Foundation in Sand," Journal of the Geotechnical Engineering Divison, A.S.C.E., Vol. 100, No. GT3, March, 1974, pp. 265-278.
87. Koizumi, Y., and Ito, K., "Field Tests with Regard to Pile Driving and Bearing Capacity of Piled Foundations," Soil and Foundation, Vol. VII, No. 3, 1967, pp. 30-53.
88. Kondner, R. L., "Friction Pile Groups in Cohesive Soil," Journal of the Soil Mechanics and Foundations Division, A.S.C.E., Vol. 88, No. SM3, June 1962, pp. 117-149.
89. Kriger, G. A., "Modeling of Piled Foundations," Proceedings, Twelfth Offshore Technology Conference, Vol. II, 1980, pp. 117-128.
90. Leonards, G. A., "Settlement of Pile Foundations in Granular Soil," Proceedings, Specialty Conference on Performance of Earth and Earth-Supported Structures, A.S.C.E., Vol. I, Part 2, 1972, pp. 1169-1184.
91. Lo, K. Y., and Stermac, A. G., "Induced Pore Pressures During Pile Driving Operations," Proceedings, Sixth I.C.S.M.F.E., Vol. II, 1965, pp. 285-289.
92. Mahar, L. J., and O'Neill, M. W., "Geotechnical Characterization of Desiccated Clay," Journal of the Geotechnical Engineering Division, A.S.C.E., Vol. 109. No. GT1, Jan. 1983.

93. Maniliou, I., Botea, E., and Constantinescu, A., "Behaviour of Pile Foundations Submitted to Lateral Loads," Proceedings, Ninth I.C.S.M.F.E., Vol. 1, 1977, pp. 637-640.
94. Masters, F. M., "Timber Friction Pile Foundations," Transactions, A.S.C.E., Vol. 108, 1943, pp. 115-140.
95. Matlock, H., "Design of Single Piles," Proceedings, Specialty Conference on Geotechnical Practice in Offshore Engineering, A.S.C.E., April 1983.
96. Matlock, H., Bogard, D., and Cheang, L., "A Laboratory Study of Axially Loaded Piles and Pile Groups Including Pore Pressure Measurements," Behavior of Offshore Structures, Vol. 1, Hemisphere Publishing Corp., 1982, pp. 105-121.
97. Matlock, H., and Foo, S. H. C., Discussion of "Full-Scale Lateral Load Tests of Pile Groups," by Kim and Brungraber, Journal of the Geotechnical Engineering Division, A.S.C.E., Vol. 102, No. GT12, Dec. 1976, pp. 1291-1292.
98. Matlock, H., Ingram, W. B., Kelley, A. E., and Bogard, D., "Field Tests of the Lateral-Load Behavior of Pile Groups in Soft Clay," Proceedings, Twelfth Offshore Technology Conference, Vol. IV, 1980, pp. 163-174.
99. Matlock, H., Lam, I. P. C., and Cheang, L. C. C., "Evaluation of Concepts for Guyed Tower Foundations," Proceedings, Thirteenth Offshore Technology Conference, Vol. IV, 1981, pp. 319-330.
100. Matsui, T., Hong, W. P., and Ito, T., "Earth Pressures on Piles in a Row Due to Lateral Soil Movements," Soils and Foundations, Vol. 22, June 1982, pp. 71-81.
101. Mattes, N. S., and Poulos, H. G., "Settlement of Single Compressible Pile," Journal of the Soil Mechanics and Foundations Division, A.S.C.E., Vol. 95, No. SM1, Jan. 1969, pp. 189-207.
102. McClelland, B., "Design and Performance of Deep Foundations," Proceedings, Specialty Conference on Performance of Earth and Earth-Supported Structures, A.S.C.E., Vol. II, 1972, pp. 111-144.
103. McClelland, B., "Design of Deep Penetration Piles for Ocean Structures," Journal of the Geotechnical Engineering Division, A.S.C.E., Vol. 100, No. GT7, July 1974, pp. 709-747.
104. Meyerhof, G. G., "Compaction of Sands and the Bearing Capacity of Piles," Journal of the Soil Mechanics and Foundations Division, A.S.C.E., Vol. 85, No. SM6, Nov. 1959, pp. 1-29.
105. Meyerhof, G. G., "Bearing Capacity and Settlement of Pile Foundations," Journal of the Geotechnical Engineering Division, A.S.C.E., Vol. 102, No. GT3, March 1976, pp. 197-228.
106. Meyerhof, G. G., Mathur, S. K., and Valsangkar, A. J., "The Bearing Capacity of Rigid Piles and Pile Groups Under Inclined Loads in Layered Sand," Canadian Geotechnical Journal, Vol. 18, Nov. 1981, pp. 514-519.
107. Meyerhof, G. G., and Rajan, G., "The Bearing Capacity of Rigid Piles Under Inclined Loads in Sand III: Pile Groups," Canadian Geotechnical Journal, Vol. 10, Aug. 1973, pp. 428-438.
108. Meyerhof, G. G., and Sastry, V. V. R. N., "Bearing Capacity of Piles in Layered Soils. Part 1. Clay Overlying Sand," Canadian Geotechnical Journal, Vol. 15, May 1978, pp. 171-182.
109. Meyerhof, G. G., and Sastry V. V. R. N., "Bearing Capacity of Piles in Layered Soils. Part 2. Sand Overlying Clay," Canadian Geotechnical Journal, Vol. 15, May 1978, pp. 183-189.

110. Mindlin, R. D., "Force at a Point in the Interior of a Semi-Infinite Solid," Physics, Vol. 7, May 1936, pp. 195-202.
111. Misteth, E., "Forces in Piles Supporting Piers," Proceedings, Third Budapest Conference on Soil Mechanics and Foundation Engineering, Budapest, Hungary, 1968, pp. 588-606.
112. Moorhouse, D. C., and Sheehan, J. V., "Predicting Safe Capacity of Pile Groups," Civil Engineering, A.S.C.E., Vol. 38, Oct. 1968.
113. Mora, F., "Static Analysis of Pile Groups: A Soil-Pile Model," M. S. thesis, Rensselaer Polytechnic Institute, June 1981.
114. Murthy, V. N. S., and Shrivastava, S. P., "Analysis of Pile Group Subjected to Vertical and Lateral Loads," Preprints, Fourth Offshore Technology Conference, Vol. II, 1972, pp. 709-720.
115. Myslivec, A., Simek, J., and Kysela, Z., "Interaction of Foundations and Foundation Soil," Proceedings, Ninth I.C.S.M.F.E., Vol. 1, 1977, pp. 671-674.
116. Nair, K., "Load Settlement and Load Transfer Characteristics of a Friction Pile Subject to a Vertical Load," Proceedings, Third Pan American Conference on S.M.F.E., Vol. I, 1967, pp. 565-589.
117. Nair, K., Gray, H., and Donovan, N.C., "Analysis of Pile Group Behavior," Performance of Deep Foundations, A.S.T.M. STP 444, A.S.T.M., 1969, pp. 118-159.
118. Nishida, Y., "Estimation of Group Pile Effect in Sands," Proceedings, Seventh I.C.S.M.F.E., Paper 14, Spec. Sess. 8, 1969.
119. Nogami, T., and Chen, H. L., "Application of Winkler Model in Pile Group Response Analysis," Civil Engineering Report, No. GT-TN10-82, University of Houston, Sep. 1982.
120. Nogami, T., "Dynamic Stiffness and Damping of Pile Groups in Inhomogeneous Soil," Dynamic Response of Pile Foundations: Analytical Aspects, A.S.C.E., Oct. 1980, pp. 31-52.
121. Novak, M., and Grigg, R. F., "Dynamic Experiments with Small Foundations," Canadian Geotechnical Journal, Vol. 13, 1976, pp. 372-385.
122. O'Neill, M. W., Ghazzaly, O. I., and Ha, H. B., "Analysis of Three-Dimensional Pile Groups with Nonlinear Soil Response and Pile-Soil-Pile Interaction," Proceedings, Ninth Offshore Technology Conference, Vol. II, 1977, pp. 245-256.
123. O'Neill, M. W., Ghazzaly, O. I., and Ha, H. B., "Assessment of Hybrid Model for Pile Groups," Transportation Research Record 733, Transportation Research Board, Washington D.C., 1979, pp. 36-43.
124. O'Neill, M. W., and Ha, H. B., "Comparative Modeling of Vertical Pile Groups," Proceedings, Second International Conference on Numerical Methods in Offshore Piling, I.C.E., 1982.
125. O'Neill, M. W., and Hawkins, R. A., "Pile-Head Behavior of a Rigidly Capped Pile Group," Transportation Research Record, Transporation Research Board, Washington, D.C., 1983.
126. O'Neill, M. W., Hawkins, R. A., and Audibert, J. M. E., "Installation of Pile Group in Overconsolidated Clay," Journal of the Geotechnical Engineering Division, A.S.C.E., Vol. 108, No. GT11, Nov. 1982, pp. 1369-1386.
127. O'Neill, M. W., Hawkins, R. A., and Mahar, L. J., "Load Transfer Mechanisms in Piles and Pile Groups," Journal of the Geotechnical Engineering Division, A.S.C.E., Vol. 108, No. GT12, Dec. 1982.
128. O'Neill, M. W., and Heydinger, A. G., "Observations of Full-Scale Pile Group Performance," Public Roads, U.S.D.O.T., Washington,

D.C., Dec. 1982.
129. Orrje, O., and Broms, B., "Effects of Pile Driving on Soil Properties," Journal of the Soil Mechanics and Foundations Division, A.S.C.E., Vol. 93, No. SM5, Sep. 1967, pp. 59-73.
130. Ottaviani, M., "Three-Dimensional Finite Element Analysis of Vertically Loaded Pile Groups," Geotechnique, Vol. XXV, June 1975, pp. 159-174.
131. Parker, F., Jr., and Cox, W. R., "A Method for the Analysis of Pile Supported Foundations Considering Nonlinear Soil Behavior," Research Report 117-1, Center for Highway Research, The University of Texas at Austin, 1969.
132. Peck, R. B., "Pile and Pier Foundations," Journal of the Soil Mechanics and Foundations Division, A.S.C.E., Vol. 91, No. SM2, March, 1965, pp. 33-38.
133. Pichumani, R., and D'Appolonia, E., "Theoretical Distribution of Loads Among Piles in a Group," Proceedings, Third Pan American Conference on S.M.F.E., Vol. I, 1967, pp. 547-564.
134. Poulos, H. G., "Analysis of the Settlement of Pile Groups," Geotechnique, Vol. XVIII, Dec. 1968, pp. 449-471.
135. Poulos, H. G., "Behavior of Laterally Loaded Piles: I - Single Piles," Journal of the Soil Mechanics and Foundations Division, A.S.C.E., Vol. 97, No. SM5, May 1971, pp. 711-731.
136. Poulos, H. G., "Behavior of Laterally Loaded Piles: II - Pile Groups," Journal of the Soil Mechanics and Foundations Division, A.S.C.E., Vol. 97, No. SM5, May 1971, pp. 733-751.
137. Poulos, H. G., "Load-Settlement Prediction for Piles and Piers," Journal of the Soil Mechanics and Foundations Division, A.S.C.E., Vol. 98, No. SM9, Sep. 1972, pp. 879-897.
138. Poulos, H.G., "Analysis of Piles in Soils Undergoing Lateral Movement," Journal of the Soil Mechanics and Foundations Division, A.S.C.E., Vol. 99, No. SM5, May 1973, pp. 391-406.
139. Poulos, H. G., "Lateral Load-Deflection Prediction for Pile Groups," Journal of the Geotechnical Engineering Division, A.S.C.E., Vol. 101, No. GT1, Jan. 1975, pp. 19-34.
140. Poulos, H. G., "Group Factors for Pile-Deflection Estimation," Journal of the Geotechnical Engineering Division, A.S.C.E., Vol. 105, No. GT12, Dec. 1979, pp. 1489-1509.
141. Poulos, H. G., "An Approach to the Analysis of Offshore Pile Groups," Numerical Methods in Offshore Piling, I.C.E., London, 1980, pp. 119-126.
142. Poulos, H. G., and Davis, E. H., Pile Foundation Analysis and Design, John Wiley and Sons, New York, 1980.
143. Poulos, H. G., and Mattes, N. S., "Settlement and Load Distribution Analysis of Pile Groups," Australian Geomechanics Journal, Vol. G1, No. 1, 1971, pp. 18-28.
144. Poulos, H. G., and Mattes, N. S., "Settlement of Pile Groups Bearing on Stiffer Strata," Journal of the Geotechnical Engineering Division, A.S.C.E., Vol. 100, No. GT2, pp. 185-190, Feb. 1974.
145. Poulos, H. G., and Randolph, M. F., "A Study of Two Methods for Pile Group Analysis," Research Report No. R387, School of Civil Engineering, The University of Sydney, Australia, May 1981.
146. Prakash, S., "Behavior of Pile Groups Subjected to Lateral Load," Ph.D. thesis, Dept. of Civil Eng., University of Illinois, 1962.

147. Prakash, S., and Saran, D., "Behaviour of Laterally-Loaded Piles in Cohesive Soil," Proceedings, Third Asian Regional Conference on S.M.F.E., Vol. I, Haifa, 1967, pp. 235-238.
148. Prevost, J. H., and Abdel-Ghaffar, A. H., "Centrifugal Modeling of the Dynamic Response of Piles," Proceedings, Second International Conference on Numerical Methods in Offshore Piling, I.C.E., 1982.
149. Prevost, J. H., Romano, J. D., Abdel-Ghaffar, A. H., and Rowland, R., "Dynamic Response of Laterally Loaded Piles in Centrifuge," Proceedings, Second Specialty Conference on Dynamic Response of Structures: Experimentation, Observation, Prediction, and Control, A.S.C.E., Jan. 1981, pp. 386-400.
150. Price, G., "Field Tests on Vertical Piles Under Static and Cyclic Horizontal Loading in Overconsolidated Clay," Behavior of Deep Foundations, A.S.T.M. STP 670, A.S.T.M., 1979, pp. 464-483.
151. Randolph, M. F., "PIGLET: A Computer Program for the Analysis and Design of Pile Groups Under General Loading Conditions," Soil Report TR91, CUED/D, University of Cambridge, 1980.
152. Randolph, M.F., Discussion, Numerical Methods in Offshore Piling, I.C.E., London, 1980, p. 197.
153. Randolph, M. F., "The Response of Flexible Piles to Lateral Loading," Geotechnique, Vol. XXXI, June 1981, pp. 247-259.
154. Randolph, M. F., and Poulos, H. G., "Estimating the Flexibility of Offshore Pile Groups," Proceedings, Second International Conference on Numerical Methods in Offshore Piling, I.C.E., 1982.
155. Randolph, M. F., and Wroth, C. P., "An Analytical Solution for the Consolidation Around a Driven Pile," International Journal for Numerical and Analytical Methods in Geomechanics, Vol. 3, 1979, pp. 217-229.
156. Randolph, M. F., and Wroth, C. P., "An Analysis of the Vertical Deformation of Pile Groups," Geotechnique, Vol. XXIV, Dec. 1979, pp. 423-439.
157. Reese, L. C., and Matlock, H., "Behavior of a Two-Dimensional Pile Group Under Inclined and Eccentric Loading," Proceedings, OECON, Long Beach, Cal., 1966, pp. 123-139.
158. Reese, L. C., O'Neill, M. W., and Smith, R. E., "Generalized Analysis of Pile Foundations," Journal of the Soil Mechanics and Foundations Division, A.S.C.E., Vol. 96, No. SM1, Jan. 1970, pp. 235-250.
159. Reese, L. C., and Seed, H. B., "Pressure Distribution Along Friction Piles," Proceedings, A.S.T.M., Vol. 55, 1955, pp. 1156-1182.
160. Roy, M., Blanchet, R., Tavenas, F., and La Rochelle, P. "Behaviour of a Sensitive Clay During Pile Driving," Canadian Geotechnical Journal, Vol. 18, Nov. 1981, pp. 67-85.
161. Saffery, M. R., and Tate, A. P. K., "Model Tests on Pile Groups in a Clay Soil with Particular Reference to the Behaviour of the Group When it is Loaded Eccentrically," Proceedings, Fifth I.C.S.M.F.E., Vol. II, 1961, pp. 129-134.
162. Saul, W. E., "Static and Dynamic Analysis of Pile Foundations," Journal of the Structural Division, A.S.C.E., Vol. 94, No. ST5, May 1968, pp. 1077-1100.
163. Saul, W. E., "Pile Foundation Analysis," Preprint 80-102, A.S.C.E. Convention and Exposition, Portland, Oregon, April 1980.
164. Schlitt, H. G., "Steel Pile Tests, Q Street Viaduct, Omaha,

Nebraska," Report, Nebraska Department of Roads and Irrigation, Lincoln, Nebraska, 1950.

165. Scott, R. F., Foundation Analysis, Prentice-Hall, Inc., Englewood Cliffs, N.J., 1981.

166. Sheta, M., and Novak, M., "Vertical Vibration of Pile Groups," Journal of the Geotechnical Engineering Division, A.S.C.E., Vol. 108, No. GT4, April 1982, pp. 570-590.

167. Shibata, T., Sekiguchi, H., Yukitomo, H., "Model Test and Analysis of Negative Friction Acting on Piles," Soils and Foundations, Vol. 22, June 1982, pp. 29-39.

168. Simons, H., "Bearing Capacity of Piles," International Civil Engineering Monthly, Jan. 1969, pp. 3-9.

169. Singh, A., and Prakash, S., "Model Pile Group Subjected to Cyclic Lateral Load," Soils and Foundations, Vol. 11, June 1971, pp. 51-60.

170. Singh, J. P., Donovan, N. C., and Jobsis, A. C., "Design of Machine Foundations on Piles," Journal of the Geotechnical Engineering Division, A.S.C.E., Vol. 103, No. GT8, Aug. 1977, pp. 863-877.

171. Skempton, A. W., Discussion: Piles and Pile Foundations, Proceedings, Third I.C.S.M.F.E., Vol. 3, 1953, p. 172.

172. Soderberg, L. O., "Consolidation Theory Applied to Foundation Pile Time Effects," Geotechnique, Vol. XII, 3, Sep. 1962, pp. 217-225.

173. Sowers, G. F., Martin, C. B., Wilson, L. L., and Fausold, M., "The Bearing Capacity of Friction Pile Groups in Homogeneous Clay from Model Studies," Proceedings, Fifth I.C.S.M.F.E., Vol. II, 1961, pp. 155-159.

174. Stevens,J., "Prediction of Pile Response to Vibratory Loads," Proceedings, Tenth Offshore Technology Conference, Vol. IV, 1978, pp. 2213-2223.

175. Stevens, J. B., Holloway, D. M., Moriwaki, Y., and Demsky, E. C., "Pile Group Response to Axial and Lateral Loading," Symposium on Deep Foundations, A.S.C.E., 1979, pp. 396-419,

176. Stuart, J. G., Hanna, T. H., and Naylor, A. H., "Notes on the Behaviour of Model Pile Groups in Sand," Proceedings, Symposium on the Design of Pile Foundations, International Assoc. for Bridge and Structural Eng., Stockholm, 1960, pp. 97-103,

177. Swain, C. W., "The Behaviour of Piled Foundations Supporting a Model Offshore Structure," Ph.D. dissertation, Cambridge University, Feb, 1979.

178. Tamaki, O., Mitsuhashi, K,, and Imai, T., "Horizontal Resistance of a Pile Group Subjected to Lateral Load," Proceedings, Fourth Asian Regional Conference on S.M.F.E., Vol. 1, 1971, pp. 311-315.

179. Tejchman, A. F., "Model Investigations of Pile Groups in Sand," Journal of the Soil Mechanics and Foundations Division, A.S.C.E., Vol. 99, No. SM2, Feb. 1973, pp. 199-217.

180. Toolan, F. E., and Horsnell, M. R., "Analysis of the Load-Deflexion Behaviour of Offshore Piles and Pile Groups," Numerical Methods in Offshore Piling, I.C.E., London, 1980, pp. 147-155.

181. Torstensson, B. A., "The Behavior of a Cohesion Pile Group in Soft Clay," Proceedings, Eight I.C.S.M.F.E., Vol. 1, 1973, pp. 237-242.

182. Trofimenkov, J., Discussion in "Behavior of Foundations and Structures," Proceedings, Ninth I.C.S.M.F.E., Vol. 3, pp. 370-371.

183. Valeria, J., and Oteo, C., "Dynamic Behavior of a Group of Free-

Standing Piles," Soils Under Cyclic and Transient Loading, II, Balkema, Rotterdam, 1980, pp. 627-638.

184. Vesić, A. S., "Ultimate Loads and Settlements of Deep Foundations in Sand," Bearing Capacity and Settlement of Foundations, Duke University, 1965, pp. 53-68.

185. Vesić, A. S., "Experiments with Instrumented Pile Groups in Sand," Performance of Deep Foundations, A.S.T.M. STP 444, A.S,T.M,, 1969, pp. 177-222.

186. Vesić, A. S., "Design of Pile Foundations," N.C.H.R.P. Synthesis of Highway Practice No. 42, Transportation Research Board, Washington, D.C., 1977.

187. Vesić, A. S., "Problems of Development of a Mathematical Model to Predict Performance of a Pile Group," Working Paper, Federal Highway Administration F.C.P. Review Con., Atlanta, Ga., 1977.

188. Vesić, A. S., "Pile Group Prediction Symposium, Summary of Prediction Results," Proceedings, Pile Group Predication Symposium, Federal Highway Administration, U.S.D.O.T., Washington, D.C., Nov, 1980.

189. Vetter, C. P., "Design of Pile Foundations," Transactions, A.S.C.E., Vol. 104, 1939, pp. 758-778.

190. Whitaker, T., "Experiments with Model Piles in Groups," Geotechnique, Vol. VII, Dec. 1957, pp. 147-167.

191. Whitaker, T., "Some Experiments with Model Pile Foundations in Clay," Proceedings, Symposium on the Design of Pile Foundations, International Assoc. for Bridge and Structural Engineering, Stockholm, 1960, pp. 124-139.

192. Wiesner, T. J., and Brown, P. T., "Laboratory Tests on Model Piled Raft Foundations," Journal of the Geotechnical Engineering Division, A.S.C.E., Vol. 106, No. GT7, July 1980, pp. 767-783.

193. Williams, D. J., "The Behaviour of Model Piles in Dense Sand Under Vertical and Horizontal Loading," Ph.D. thesis, Cambridge University, Oct. 1979.

194. Wolf, J. P., and von Arx, G.A., "Impedance Function of a Group of Vertical Piles," Proceedings, A.S.C.E. Specialty Conference on Earthquake Engineering and Soil Dynamics, Vol. II, 1978, pp. 1024-1041.

195. Wolf, J. P., "Dynamic Stiffness of Group of Battered Piles," Journal of the Geotechnical Engineering Division, A.S.C.E., Vol. 106, No. GT2, Feb. 1980, pp. 198-202.

196. Woodward-Clyde Consultants, Phase IV Report, Vol. IV, "Results and Interpretation of Pile Driving Effects Test Program; Existing Locks and Dam No. 26, Mississippi River, Alton, Ill.," May 1979, pp. 8.1-8.7.

197. Wright, S. G., "Laterally Loaded Pile Groups," Preprints for Short Course on Deep Foundatibns, The University of Texas at Austin, March 1982, 20 pp.

198. Yeagian, M., and Wright, S. G., "Lateral Soil Resistance—Displacement Relationships for Pile Foundations in Soft Clays," Preprints, Fifth Offshore Technology Conference, Vol. 2, pp. 663-676, 1973.

199. Zeevaert, L., Foundation Engineering for Difficult Subsoil Conditions, Van Nostrand Reinhold Company, New York, 1973, pp. 335-423.

STRESS HISTORY AND STRENGTH OF STIFF OFFSHORE CLAYS

By

Charles C. Ladd[1], F. ASCE
Amr S. Azzouz[2], A.M. ASCE

INTRODUCTION

Procedures commonly employed for the design of friction piles to support offshore structures founded on cohesive sediments generally have a high degree of empirical content, both regarding experimental methods for evaluating soil properties and analysis techniques to predict pile performance. Certain "standard" procedures have evolved over the past 30 years, mainly based on extensive experience in designing pile supported platforms for the deep "soft" plastic clay deposits encountered within the Gulf of Mexico. This approach typically places heavy emphasis on results of "strength index" tests (e.g., miniature vane, penetrometer, fall cone, UU triaxial, etc.) performed on hammered wire-line samples to develop undrained strength profiles. Although usually satisfactory for the Gulf of Mexico, this practice has produced spurious results when applied to different soil conditions such as those encountered within the North Sea or offshore California. Moreover, any exploration and testing program which employs basically empirical procedures for evaluating soil parameters involves the following inherent limitations:

1) Empirical correlations used to relate "strength index" test data pile capacity vary with the precise sampling and testing procedures, the depth of sampling, and the clay type and its stress history;

2) The data obtained frequently exhibit excessive scatter, hence giving a misleading indication of the actual in situ variability in soil properties, and may even give erroneous trends with depth; and

3) One develops little insight regarding the true nature of the foundation conditions, e.g. basic strength-deformation characteristics and how these should vary laterally and with depth, and hence each new facility requires additional (and costly) site specific exploration and testing programs.

[1]Prof. of Civil Engineering, Massachusetts Institute of Technology, Cambridge, Mass.

[2]Assistant Prof. of Civil Engineering, Massachusetts Institute of Technology, Cambridge, Mass.

The Instituto Tecnologico Venezolano del Petroleo (INTEVEP) had to develop preliminary designs for platforms at six areas offshore Venezuela involving soil deposits ranging from interlayered stiff clays and dense sands to thick strata of soft plastic clay. Recognizing the limitations associated with primary reliance on the "conventional" exploration and testing procedures described above, INTEVEP sponsored a multi year effort aimed at a more comprehensive evaluation of foundation conditions using a combination of geophysical techniques, in situ testing and "sophisticated" laboratory testing. In particular, detailed compositional analyses and SHANSEP (6) testing programs were conducted at the Massachusetts Institute of Technology (MIT) on clay samples obtained from the boring locations shown in Fig. 1.

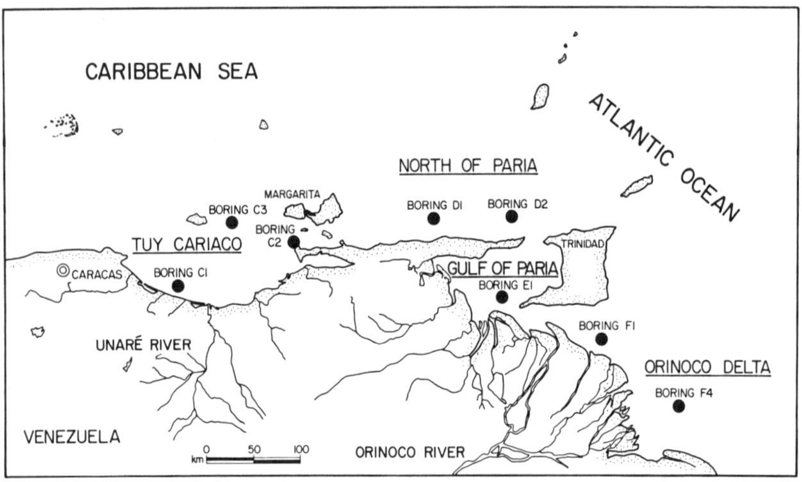

Figure 1 Boring Locations Offshore Venezuela

This paper presents stress history and normalized soil properties data used to develop undrained shear strength (c_u) profiles for five "stiff clay" deposits: three located at Tuy Cariaco; and two located at North of Paria. Simplified probabilistic techniques are used to consider the uncertainties involved in applying the SHANSEP method and the paper compares the resulting c_u profiles with those obtained from conventional strength index testing. Azzouz et.al. (1) presented data from a similar laboratory program performed on a 40 m thick deposit of soft, plastic Orinoco Clay encountered within the Gulf of Paria and Orinoco Delta. They also compared these results to those predicted from in situ electrical Dutch cone penetration and piezometer probe (13) tests run at borings E1 and F1.

SAMPLING AND LABORATORY TESTING PROGRAMS AND PROBABILISTIC ANALYSES

The exploration and sampling program was conducted during the summer of 1979 from the M/V SURVEYOR by Fugro Gulf, Inc., acting as the geotechnical consultant. The borings used a power tong rotary 5 inch OD IF drill pipe with an open-ended drill bit. A heave compensator and appropriate drilling subs and drill collars enabled use of Fugro's hydraulically operated downhole "WIP" sampler (4). This equipment pushed 2 or 3 feet long thin walled 3 inch OD tube samples at a rate of about 2 cm/sec. Most of the samples were extruded onboard for visual classification, conventional strength index tests (e.g., miniature vane, Torvane, pocket penetrometer, fall cone and some UU triaxial compression), unit weight and water content determination, etc. Those samples designated for testing at MIT (see Figs. 2-6 for their locations at the five borings being considered) were sealed inside the tubes. After being shipped to MIT, all tubes were radiographed to assess their quality regarding: variations in soil type and layering; zones of excessive disturbance; and presence of stones, coral, voids and cracks due to gas pockets and other "defects" such as shear planes. Such information proved essential in estimating the amount of suitable soil in each tube and in selecting the best portions for engineering tests.

MIT's experimental program had two main components: (1) compositional analyses to identify any unusual clay minerals and for correlation with geology and engineering properties; and (2) SHANSEP type testing to establish the stress history of the various clay strata, their consolidation characteristics and K_0 consolidated-undrained (CK_0U) shear tests to obtain "normalized" stress-strain-strength data for estimating in situ undrained strength-deformation properties. Most of the SHANSEP strength testing used the Geonor direct simple shear (DSS) device (3) since such tests require less soil, are easier to run, and yield results more appropriate for undrained stability and deformation analyses than CK_0U triaxial tests (7). Also, recent research by Randolph and Wroth (11) suggests that the failure state in the DSS test may be similar to that occurring along the shaft of friction piles in cohesive soils.

Estimates of the in situ undrained shear strength (c_u) computed from the SHANSEP test program for a DSS mode of failure used the following equation:

$$c_u = \sigma'_{vo} \cdot S \cdot (OCR)^m \qquad (1)$$

where σ'_{vo} = in situ effective overburden stress, $S = c_u\ (DSS)/\sigma'_{vc}$ for normally consolidated clay, $OCR = \sigma'_p/\sigma'_{vo}$ where σ'_p is the preconsolition pressure, and m = exponent controlling the rate of increase in c_u with overconsolidation ratio. Simplified probabilistic analyses consider the uncertainties associated with the parameters in Eq. 1 and hence in the computed strength profile. These analyses used the following relationship (2) to obtain the coefficient of variation in c_u = Cov $[c_u]$ = standard deviation divided by the estimated mean = SD $[c_u]$/E $[c_u]$:

$$\mathrm{Cov}^2\,[c_u] = \mathrm{Cov}^2\,[S] + m^2 \cdot \mathrm{Cov}^2\,[\sigma'_p] + \ln^2\,(\mathrm{OCR}) \cdot \mathrm{SD}^2\,[m] \qquad (2)$$

where the three terms reflect scatter and uncertainty in the values of S, σ'_p and m, respectively (σ'_{vo} is assumed to be deterministic). Unless otherwise stated, the magnitudes of Cov [] reflect the combined effect of the total measured scatter about the mean and a "statistical" error in the location of the mean value due to the limited number of scattered results. However, potential "bias" in the results is ignored except to exclude obviously low values of σ'_p.

Mishu et.al. (10) and Ladd et.al. (8) present detailed information regarding all test procedures and experimental data at Tuy Cariaco and North of Paria, respectively.

SOIL PROFILES, GEOLOGY AND STRESS HISTORY

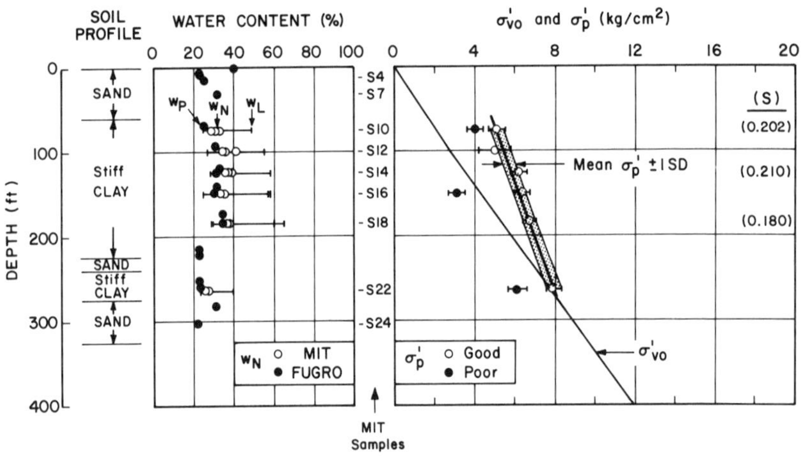

Figure 2 Soil Profile, Index Properties, and Stress History: Boring C1 (Water Depth = 201 ft)

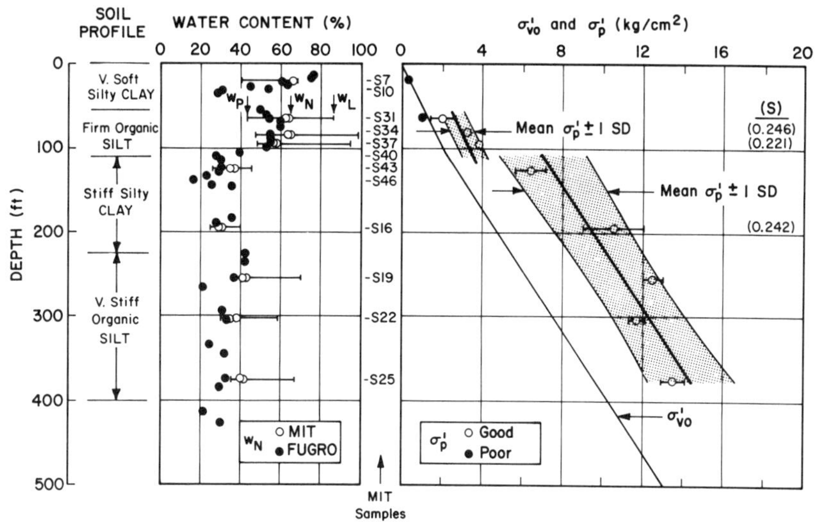

Figure 3 Soil Profile, Index Properties, and Stress History: Boring C2 (Water Depth = 425 ft)

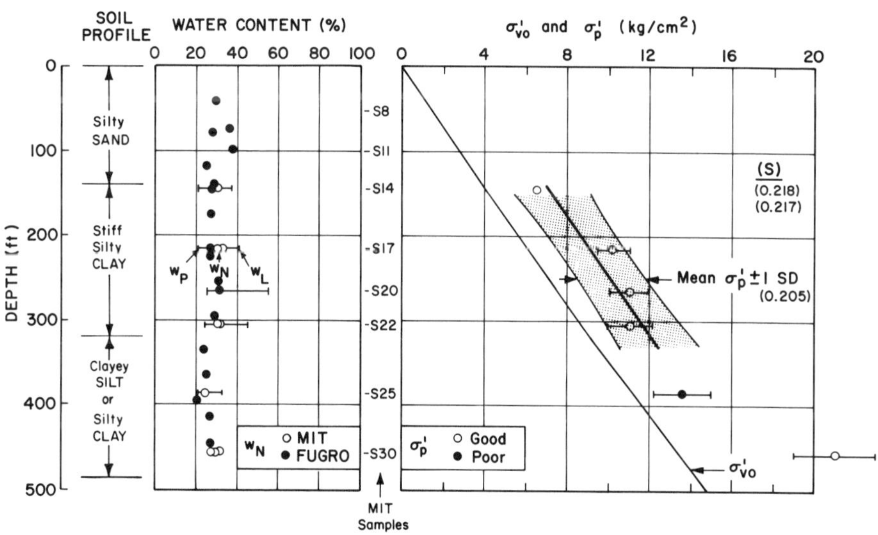

Figure 4 Soil Profile, Index Properties, and Stress History: Boring C3 (Water Depth = 341 ft)

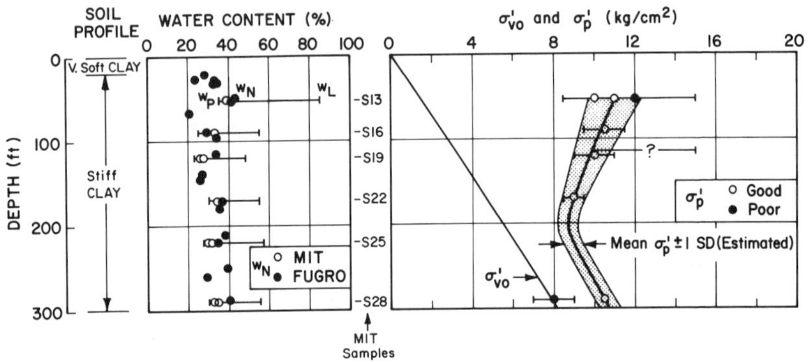

Figure 5 Soil Profile, Index Properties, and Stress History: Boring D1 (Water Depth = 180 ft)

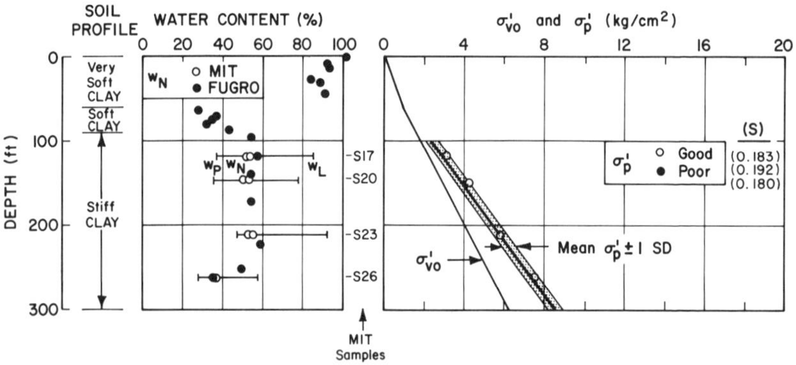

Figure 6 Soil Profile, Index Properties, and Stress History: Boring D2 (Water Depth = 351 ft)

Figures 2 through 6 present the following information for the three borings at Tuy Cariaco (C1-C3) and the two borings at North of Paria (D1 and D2): water depth; simplified soil descriptions; natural water content (w_N) and Atterberg limits (w_L, w_P); location of MIT samples; the σ'_{vo} profile computed assuming hydrostatic pore presures; and σ'_p data obtained from standard incremental oedometer tests (but using "end of primary" compression curves) or from constant rate of strain consolidation tests (12). The best estimate and range in σ'_p from individual tests used the Casagrande procedure (the tests were generally carried to a maximum stress of 30 to 50 kg/cm^2 whenever σ'_p exceeded about 6 kg/cm^2). Data from conventional strength index tests performed by Fugro and MIT will be presented and discussed after developing the SHANSEP c_u profiles.

Geology and Index Properties

The water depths at the five boring locations range from a minimum of 180 ft at D1 to a maximum of 425 ft at C2. The soil strata considered in this paper are all designated as stiff (silty) CLAY, except at C2 which also contains an upper firm organic SILT (z = 55 to 110 ft below the mudline) and a lower very stiff organic SILT. Micropaleontological dating by the Woods Hole Oceanographic Institute indicates that the firm to very stiff sediments at Tuy Cariaco were probably deposited during the middle to late Pleistocene when repeated glaciation lowered the ocean to about 450 ± 100 ft below its present level. These deposits are part of the highly faulted continental shelf and slope surrounding the Cariaco Trench (a depression about 4600 ft deep), formed chiefly from sediments from the Venezuelan mainland and surrounding islands. The North of Paria stiff clays are also Pleistocene deposits (date unknown), but were mainly formed from sediments coming from the Orinoco River.

The stiff deposits have typical plasticity indices (I_p) of 25 ± 10%, with less than 2% organic material, the exceptions being the D2 stiff clay (I_p = 45 ± 5%) and the C2 upper organic silt (I_p = 50 ± 5% with 5-6% organic matter). Illite, smectite and either chlorite (at Tuy Cariaco) or kaolinite are the main clay minerals.

Stress Histories

The stress histories developed at the five borings considered only data from "good" tests (judged from shape of compression curve and magnitude of strain at $\sigma'_{vc} = \sigma'_{vo}$). The estimated mean σ'_p profiles were determined as follows: least squares linear regression through the data points at C1 and D2; a similar analysis at C2 and C3, but requiring that this line be parallel to σ'_{vo} based on the supposition that the precompression must have resulted from prior erosion of overburden during past sea level lowerings; and "judgement" at D1. Likewise, standard statistical techniques were used to compute the resulting ± standard deviation bands in σ'_p (reflecting the combined effect of the total scatter about the mean and a statistical error in the location of the mean), except that judgement was again used at D1.

The resulting stress history profiles indicate the following:

1) Precompression due to desiccation and/or lowering of the water table down to z = 275 ft at C1 and to z ≃ 220 ft at D1:

2) Precompression due to prior overburden (explicitly assumed, but considered reasonable) at C2 and C3. But note the much lower degree of precompression within the upper firm organic silt at C2 compared to the underlying materials, which suggests two different periods of erosion from sea level lowering.

3) An exceptionally well defined linear variation in σ'_p from four excellent quality tests (two incremental and two CRSC) for the 200 ft thick stiff clay deposit at D2. The authors cannot readily explain why this deposit has a small, but essentially constant OCR. The measured σ'_p/σ'_{vo} = 1.35 corresponds to about two to three cycles of secondary compression, which is inconsistent with the recent increase in overburden stress due to deposition of the overlying soft materials within the past 10,000 years. Madsen (9) suggests that wave action, which produces cyclic shear stresses within clay sediments, may cause an otherwise normally consolidated deposit to behave as an overconsolidated clay. But it is difficult to imagine wave action producing a constant OCR throughout a 200 ft thick deposit. Finally, prior overburden removal is not indicated since this would cause a constant precompression (= $\sigma'_p - \sigma'_{vo}$).

In summary, each of the stiff deposits at the five sites appears to have a unique stress history: a uniform amount of precompression at C3 and a stepped precompression at C2, resulting from overburden erosion at both locations; a decreasing precompression with depth at C1, presumably caused by desiccation; an almost constant σ'_p versus depth profile at D1 resulting from a lowered "water table" and desiccation; and a small, constant OCR profile at D2. These stress histories are all consistent with geological events which occurred from sea level lowerings during Pleistocene glaciation, except for the D2 deposit which may have undergone some sort of chemical bonding.

DIRECT SIMPLE SHEAR TEST DATA AND SHANSEP STRENGTH PROFILES

The CK_0UDSS program mainly involved tests run on specimens reconsolidated into the virgin compression range (laboratory σ'_{vc}/in situ σ'_p = 1.7 ± 0.3) to measure $S = c_u/\sigma'_{vc}$ for normally consolidated clay, where c_u is defined as the maximum applied horizontal shear stress (τ_h). The specimens (35 cm^2 by about 2 cm thick) were sheared at a strain rate of 5% per hr while varying the vertical effective stress (σ'_v) to maintain constant volume. The peak strength typically occurred at γ_f = 10 ± 2%, followed by slight to moderate strain softening. The sample depths and measured values of S are shown on the right hand side of Figs. 2-6, i.e., three tests at each boring, except none at D1 (high capacity DSS membranes were not then available at MIT). Two "multi stage" tests were run to measure the m exponent in Eq. 1 (specimen sheared to peak at OCR = 1, reconsolidated to a higher σ'_{vc} and then rebounded to OCR = 4 for second shearing).

Figure 7 plots S versus plasticity index from the above test program plus data from 13 tests on the soft Orinoco Clay, mainly from borings E1 and F1 in Fig. 1. No trend exists (nor with mineralogical composition) and half of the data falls well below the empirical S vs. I_p correlation developed for other sedimentary clays. However, the results do show a consistent relationship with the obliquity angle $\theta = \arctan \tau_h/\sigma'_v$. Figure 7 plots S versus θ data at the undrained strength,

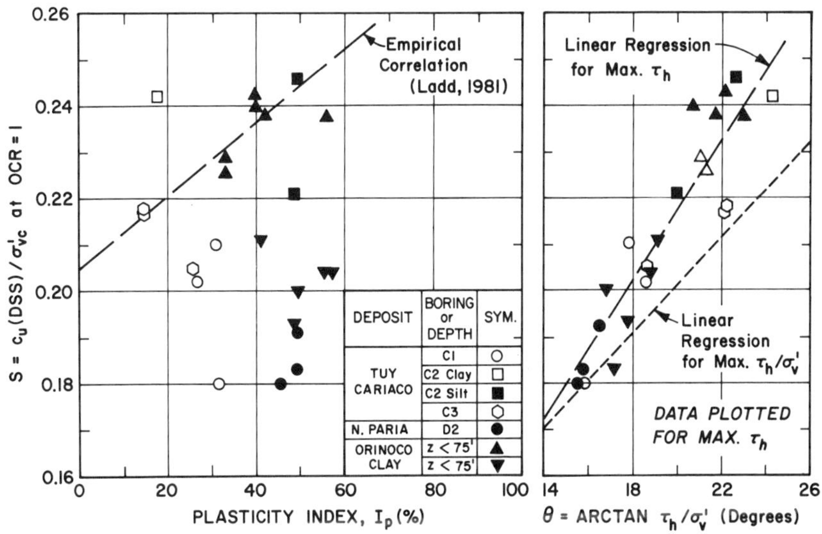

Figure 7 Normally Consolidated c_u (DSS)/σ'_{vc} vs. Plasticity Index and Obliquity Angle for Venezuelan Offshore Clays

i.e. when τ_h reached its maximum value. This figure served as a check on the quality of any given test. An equally consistent correlation also exists between S and θ at maximum obliquity, i.e. at τ_h/σ'_v maximum.

Figures 8 through 12 plot the SHANSEP c_u (DSS) profile at each boring location obtained using Eq. 1 to compute the mean strength and Eq. 2 to compute the ± one standard deviation bands. The various input parameters were obtained as follows:

1) Mean S = E[S] computed from test data for each deposit rounded off to nearest 0.005 (values shown in Figs. 8-12), except for boring D1 where E[S] was <u>estimated</u> from the S vs. θ at maximum obliquity correlation using $\overline{\phi'}$ = 20° measured from CU triaxial compression tests.

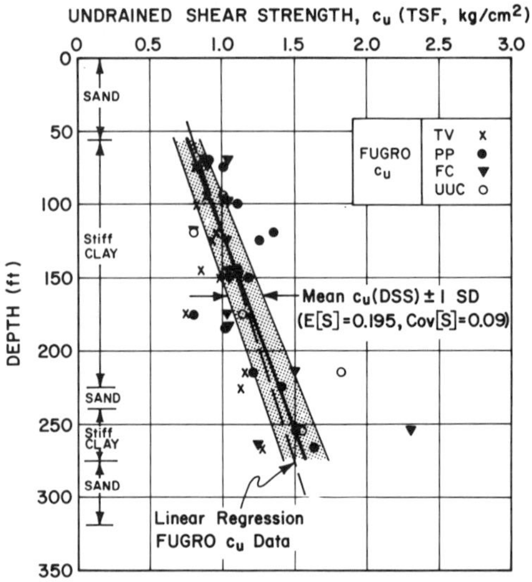

Figure 8 Comparison of Undrained Strength Data: Boring C1

2) Coefficient of variation of S computed using the equation:

$$\mathrm{Cov}[S] = \frac{\left[\, SD^2\,[S]\left(1 + \frac{1}{n}\right)\right]^{\frac{1}{2}}}{E[S]} \tag{3}$$

where SD[S] is the measured standard deviation from n tests. This technique considers the total scatter about the mean as well as the statistical uncertainty in the mean value of S. It obviously requires at least two tests, and hence Cov [S] was (conservatively?) assumed equal to 0.10 for the C2 and D1 stiff clays. Figures 8-12 show the other values of Cov [S], which ranged from 0.04 to 0.09

3) Mean OCR and coefficient of variation in σ'_p obtained from the data presented and discussed in the previous section.

4) Mean and standard deviation of the exponent m selected as E[m]=0.80 and SD[m] = 0.05 for all deposits. These values are mainly based on c_u (DSS)/σ'_{vc} versus OCR data for five sedimentary clays (Fig. 25 of Ladd et.al., 1977) which show m = 0.81 ± 0.052 SD for OCR = 2 to 8.

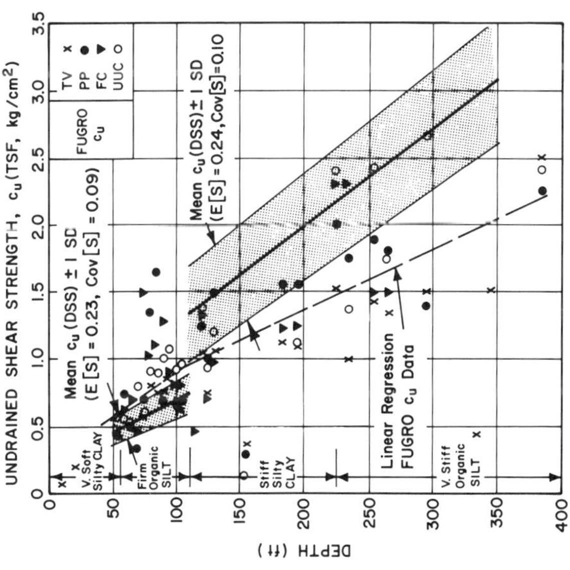

Figure 9 Comparison of Undrained Strength Data: Boring C2

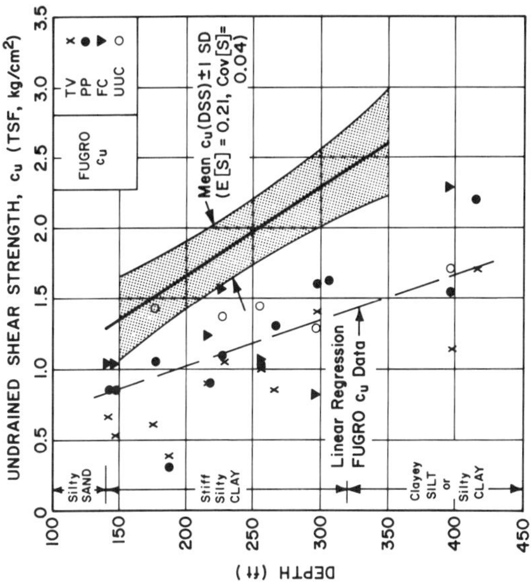

Figure 10 Comparison of Undrained Strength Data: Boring C3

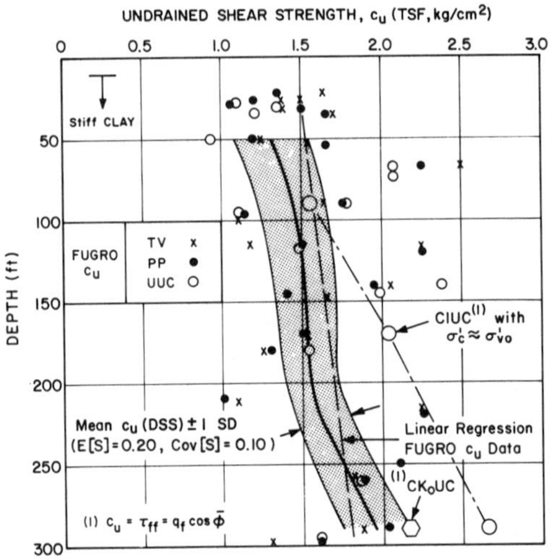

Figure 11 Comparison of Undrained Strength Data: Boring D1

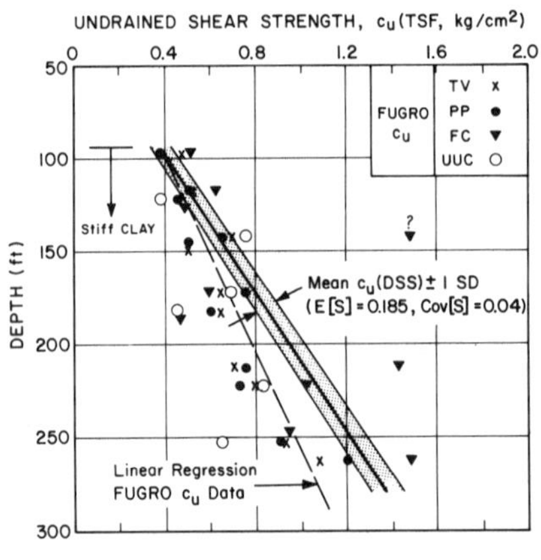

Figure 12 Comparison of Undrained Strength Data: Boring D2

The ± SD bands providing the total scatter in c_u(DSS) are thought to have the following physical significance: if one obtained new samples and ran additional consolidation and DSS tests using the same sampling and testing procedures, then about two-thirds of the newly computed values of c_u(DSS) should fall within these bands. The uncertainty associated with the location of the mean, which would be more significant for predicting axial pile capacity, has a standard deviation about half as large as those plotted on the figures.

COMPARISON OF UNDRAINED STRENGTH DATA

Figures 8 through 12 also plot individual "strength index" test data obtained by Fugro onboard the drilling vessel and linear regression lines through these data (some extreme values were not plotted). Compared to the SHANSEP c_u(DSS) profiles, two facts are clearly evident.

First, the scatter in the conventional c_u data always exceeds the total uncertainty in the SHANSEP profiles, especially at borings C2 and D1. Second, the difference between the mean strengths from the two approaches varies considerably from deposit to deposit and/or with depth. This difference is minor at C1 and D1, but certainly very significant at C2 and C3. Results for these latter deposits are especially interesting. The strength index data at C2 indicate no change in the mean c_u at the interface between the firm organic silt and the underlying stiff clay, whereas c_u(DSS) increases from 0.75 to 1.35 kg/cm^2 due to the much larger preconsolidation pressure (see Fig. 3). Moreover, conventional practice is considered to overestimate c_u by about 30% in the upper 55 ft thick layer which is very plastic and contains shell fragments, while the reverse situation applies to the very thick lower deposit. At boring C3 (Fig. 10), essentially all of the onboard strength data plot below the lower limit of the c_u(DSS) bands and the mean line yields $c_u/\sigma'_{vo} = 0.18 \pm 0.02$. This ratio is certainly too low for a low plasticity, calcareous silty clay deposit which must be significantly precompressed (see Fig. 4). In areas with less sand overburden, such strengths would presumably have a costly impact on the design of pile supported platforms.

Figure 11 plots c_u values from three triaxial compression tests run on specimens isotropically consolidated to the effective overburden stress, a common practice with some organizations. These CIUC tests yield c_u increasing almost linearly with depth. At z = 90 ft, which is highly overconsolidated (Fig. 5), the strength falls within the c_u(DSS) band. But at the two greater depths having lower OCR values (about 1.9 and 1.3), the CIUC tests give strengths judged much too high (this trend is considered typical for CIUC tests consolidated to $\sigma'_c = \sigma'_{vo}$). In contrast, the CK_0UC test at z = 290 ft reconsolidated using the SHANSEP technique yielded results as would be expected for a slightly anisotropic clay.

SUMMARY AND CONCLUSIONS

This paper presents results from 1-D consolidation and SHANSEP K_o consolidated-undrained direct simple shear (CK_0UDSS) tests performed on WIP push samples of stiff cohesive sediments from five borings located off the northern coast of Venezuela (C1-C3, D1 and D2 in Fig. 1) having water depths of 180 ft to 425 ft. The thickness of these Pleistocene deposits, composed of low to high plasticity silty clays and organic silts, ranged from 150 ft to 300 ft. Radiography proved very helpful in estimating the amount of suitable soil in each tube and in selecting the best portions for engineering tests.

The stress histories developed at the five sites show very different preconsolidation pressure (σ'_p) profiles within the various deposits. Some are precompressed due to prior erosion of overburden (Figs. 3 and 4), others indicate significant desiccation (Figs. 2 and 5) and one low OCR deposit (Fig. 6) may have undergone some sort of chemical bonding. However, except for the last case, these stress histories are all consistent with geological events which occurred from sea level lowerings during Pleistocene glaciation.

Undrained shear strength (c_u) profiles were computed for each deposit for the DSS mode of failure using Eq. 1:

$$c_u = \sigma'_{vo} \cdot S \cdot (OCR)^m \tag{1}$$

where $S = c_u(DSS)/\sigma'_{vc}$ for normally consolidated clay and $(OCR)^m$ controls the rate of increase in c_u with overconsolidation ratio. Simplified probabilistic analyses yielded ± standard deviation bands in the computed SHANSEP strength profiles considering the total scatter in the values of S, σ'_p and m. Figures 8 through 12 compare these profiles with results from conventional strength index tests (e.g. Torvane, fall cone, UU triaxial, etc.) performed onboard the drilling vessel. This comparison shows:

1) Scatter in the conventional c_u data generally much larger than the total uncertainty in the SHANSEP profiles;

2) Differences between the mean strengths from these two approaches which were small at two borings (Figs. 8 and 11), moderate at one (Fig. 12), and very significant (by +30% to -40%) at the other two locations (Figs. 9 and 11).

Hence conventional practice generally gave a misleading picture of soil variability, either over-or underestimated the mean strength at several locations and even yielded an erroneous trend with depth at one boring (Fig. 9). For these reasons, the authors would use the SHANSEP c_u(DSS) strength profiles for preliminary design and for correlations with any pile load tests. Moreover, the stress histories and normalized soil properties developed for these sites, plus the areal geology, provide a rational basis for predicting conditions at nearby locations, especially if combined with in situ piezocone penetration testing.

ACKNOWLEDGEMENTS

This research was sponsored by INTEVEP. The authors are indebted to numerous present and former INTEVEP personnel (especially J. Butenko, E. Gujardo, J. Hedberg, H. Perez La-Salvia, J. Rodriguez and E. Schwartz) for administrative and technical support and to Fugro Gulf for supervising the offshore vessel.

The MIT laboratory studies were conducted mainly by R.T. Martin, Senior Research Associate, and Graduate Assistants R. Mishu, A.M. Malek, B.E. Novich and D.R. Spikula. Other MIT students deserving credit for their assistance include D. O'Neill, J. Nasir, S. Rojas, P. Sereda and E. Urdaneta. Finally, Dr. J.T. Germaine, Geotechnical Laboratory Director, provided technical advice regarding testing equipment and procedures.

REFERENCES

1. Azzouz, A.S., Baligh, M.M., and Ladd, C.C. "Cone Penetration and Engineering Properties of the Soft Orinoco Clay," Proceedings, Third International Conference on Behavior of Offshore Structures, M.I.T., Cambridge, Mass., Vol. 1, 1982, pp. 161-180.

2. Baecher, G.B. and Ladd, C.C., "Simple Geotechnical Data Analysis," submitted for publication, Journal of the Geotechnical Engineering Division, ASCE, 1983.

3. Bjerrum, L. and Landva, A., "Direct Simple Shear Tests on Norwegian Quick Clay," Geotechnique, Vol. 16, No. 1, 1966, pp. 1-20.

4. Fugro Gulf, Inc., "Description of Field Techniques M/V Surveyor, Offshore Venezuela," Report to INTEVEP in Venezuela, April 1979, 23 p.

5. Ladd, C.C., "Discussion on Laboratory Shear Devices, Laboratory Shear Strength of Soil," American Society for Testing and Materials, Special Technical Publication 740, 1981, pp. 643-652.

6. Ladd, C.C. and Foott, R., "New Design Procedure for Stability of Soft Clays," Journal of the Geotechnical Engineering Division, ASCE, Vol. 100, No. GT7, 1974, pp. 763-786.

7. Ladd, C.C., Foott, R., Ishihara, K., Schlosser, F., and Poulos, H.G., "Stress-Deformation and Strength Characteristics," Proceedings, 9th International Conference on Soil Mechanics and Foundation Engineering, Tokyo, Japan, Vol. 2, 1977, pp. 421-494.

8. Ladd, C.C., Malek, A.M., Martin, R.T., and Mishu, R., "Evaluation of Compositional and Engineering Properties of Offshore Venezuelan Soils -Volume 2: North of Paria Stiff Clays," Massachusetts Institute of Technology, Department of Civil Engineering, Report No. R81-21, 1981, 217 p.

9. Madsen, O.S., "Wave-Induced Pore Pressures and Effective Stresses in a Porous Bed," Geotechnique, Vol. 28, No. 4, 1978, pp. 377-393.

10. Mishu, R., Ladd, C.C., Martin, R.T., and Spikula, D.R., "Evaluation of Compositional and Engineering Properties of Offshore Venezuelan Soils - Volume 3: Tuy Cariaco Clays," Massachusetts Institute of Technology, Department of Civil Engineering, Report No. R82-31, 1982, 332 p.

11. Randolph, M.F., and Wroth, C.P., "Application of the Failure State in Undrained Shear to the Shaft Capacity of Driven Piles," Geotechnique, Vol. 31, No. 1, 1981, pp. 143-157.

12. Wissa, A.E.Z., Christian, J.T., Davis, E.H., and Heiberg, S., "Consolidation at Constant Rate of Strain," Journal of the Soil Mechanics and Foundation Engineering Division, ASCE, Vol. 97, No. SM10, Paper 8447, 1971, pp. 1393-1413.

13. Wissa, A.E.Z., Martin, R.T., and Garlanger, J.E., "The Piezometer Probe," Proceedings, ASCE Specialty Conference on In Situ Measurement of Soil Properties, Raleigh, N.C., Vol. 1, 1975, pp. 536 545.

VALDEZ SILTS: A CHALLENGE IN DESIGN OF OFFSHORE FACILITIES

J. P. Singh[1] MASCE and D. W. Quigley[2] MASCE

INTRODUCTION

Valdez was subjected to strong shaking during the March 28, 1964, Prince William Sound earthquake. The old town of Valdez was severely damaged and much of the waterfront was destroyed by a submarine landslide and its associated seiche (2). Post-earthquake investigations showed that the landslide was caused by liquefaction of soils. These soils consisted primarily of loose to medium-dense sands to a depth of about 600 feet.

Port Valdez lies in a deep glaciated U-shaped valley that is connected by the Valdez Narrows and Valdez Arm to Prince William Sound and the Gulf of Alaska (see Figure 1). The site described in this paper is near Solomon Gulch on the south shore of Port Valdez (see Figure 2). A two-berth marine terminal was to have been constructed on the site and connected by pipelines to a proposed refinery complex five miles away.

The foundation conditions at the marine terminal site differ from those of the old town of Valdez. The tidal flat and submarine slope at the terminal site are blanketed with as much as 60 feet of fine-grained glaciofluvial sediments known locally as Valdez silts, rather than with the sandy soils that were present at the old town of Valdez. This sediment, which would be classified as a clayey silt on the basis of the Unified Soil Classification System, has been shown by electron microscopy to be a rock flour consisting of very fine unweathered mineral fragments of the predominant rocks in the area. The plasticity of this material varies from zero to an apparent plasticity of up to 8. However, with no clay minerals present in these soils, the plasticity index is a function of grain size rather than a true classification of the soil as clayey silt.

The behavior of these silts under cyclic loading conditions is different from that of sands in that no catastrophic loss of strength occurs during cyclic triaxial tests. The silts can undergo

---

1 Principal Engineer/Principal Seismologist, Harding Lawson Associates, Novato, CA

2 Principal Engineer, Harding Lawson Associates, Houston, TX

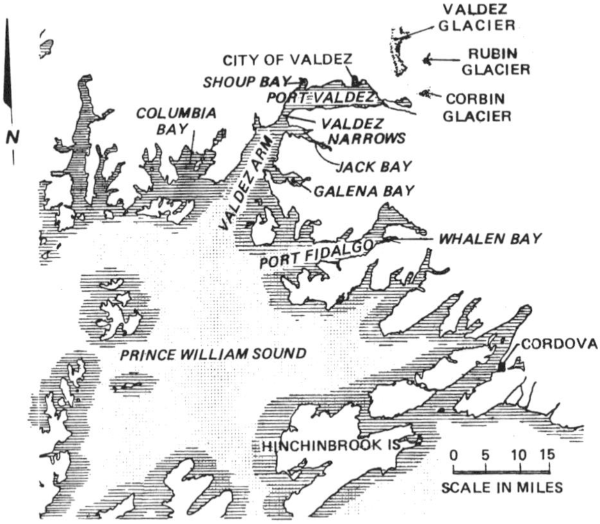

Fig. 1 Regional map

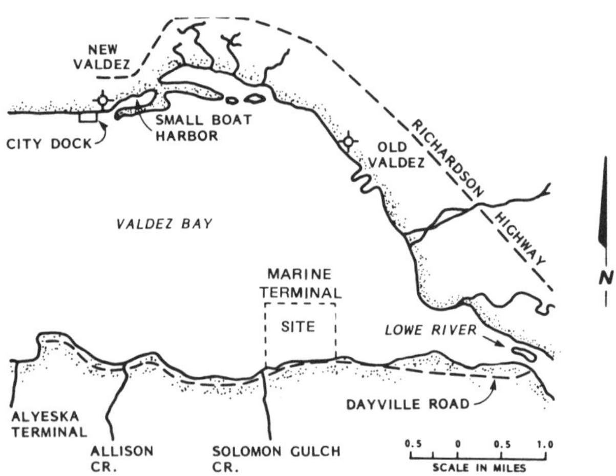

Fig. 2 Marine Terminal Site

significant cyclic mobility* without a measurable loss of strength. This unique behavior of the silts suggests that the conventional liquefaction analysis may not be appropriate for this material. Besides the liquefaction behavior, the post-cyclic stress-strain behavior of Valdez silt differs considerably from rigid plastic behavior. Therefore, Newmark's (6) method of evaluating permanent earthquake-induced slope displacement is not valid for this material as Donovan and Singh (3) suggested.

DESIGN EARTHQUAKES

The major historical event that most significantly influences the estimates for a maximum expected earthquake for the Valdez area is the magnitude 8.5+ March 28, 1964, earthquake that was centered beneath Prince William Sound about 80 km west of Valdez. The earthquake was caused by a major displacement on the Aleutian megathrust which dips northward from the Gulf of Alaska at a shallow angle. This thrust forms the juncture between the underthrust northward-migrating Pacific Plate and the North American plate, which also includes the southeastern Alaska plate. Numerous large earthquakes have occurred along this underthrust which extends westward along the Aleutian chain.

Between 1896 and 1934, several large earthquakes ranging in magnitude from 6 to 7.5 occurred in this area. They were more or less uniformly distributed in the Prince William Sound-Gulf of Alaska area with a few centered northwest of Valdez. Most are probably related to the activity in the megathrust, although some may have been caused by displacement in the St. Elias fault east of Valdez or by the Castle Mountain fault northwest of Valdez.

The results of seismic risk analysis based on the seismotectonics and earthquake activity in the area are shown on Figure 3. The figure shows acceleration, expressed as percent of gravity, plotted versus the probability of that level of mean peak surface bedrock acceleration being equalled or exceeded during the estimated 30-year life of the terminal.

Members of the design team selected two levels of earthquakes as a basis for establishing the criteria for design of the marine terminal. The two levels, namely, operating and survival level earthquakes, have 20 percent and 3 percent probability, respectively, of not being exceeded during the estimated 30-year life of the terminal. The values of mean peak bedrock surface acceleration corresponding to the operating and survival level earthquakes are 0.17 and 0.34 g. For the liquefaction studies described herein, these bedrock accelerations were modified for the effect of local soil conditions (10).

---

*The term "cyclic mobility" is often used instead of "liquefaction" to describe soils that deform a limited amount during cyclic loading without a sudden and/or complete loss of strength.

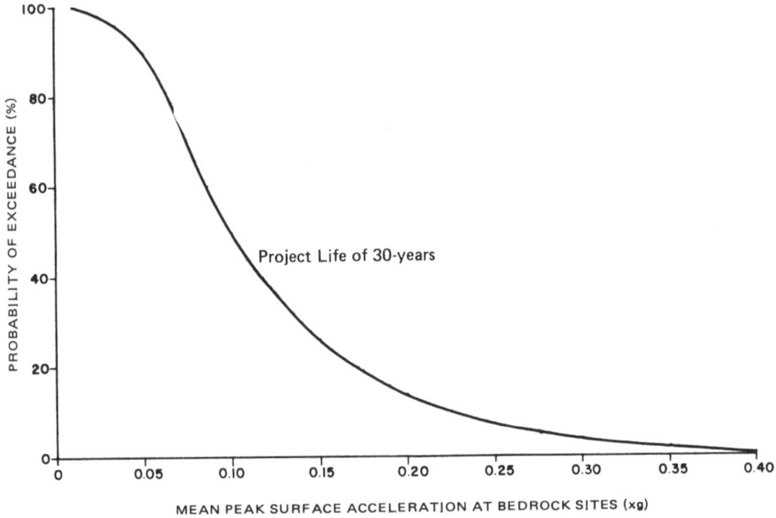

Fig. 3 Seismic risk evaluation

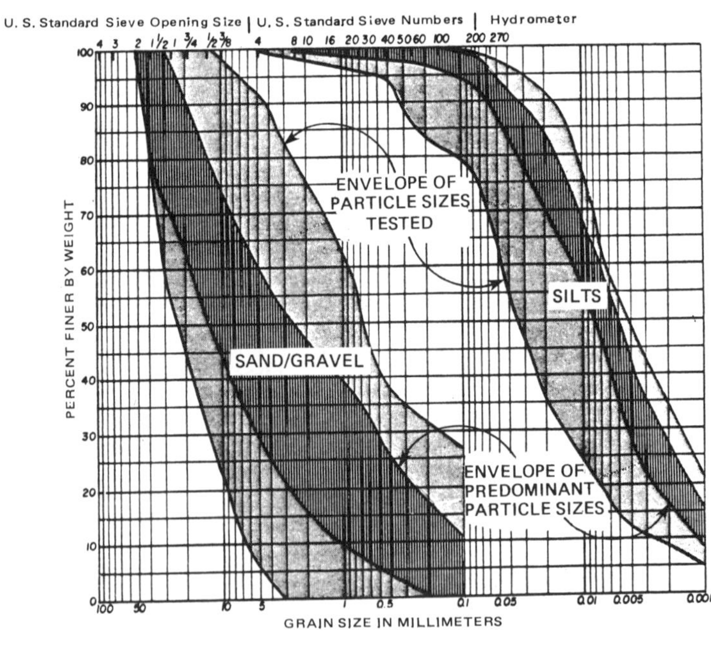

Fig. 4 Soil gradation curve

DESIGN STORM

The design storm criteria selected for this project were:

Maximum wave height = 10 feet
Significant wave height = 5.35 feet
Wave period = 4.3 seconds
Wave length = 95 feet

GEOLOGY

General

Port Valdez is a 17-mile-long glacial re-entrant bay, or fjord, in the northeast part of Prince William Sound (see Figure 1). The fjord is an east-west trending, steep-walled, flat-bottomed trough with a maximum depth of about 780 feet below Mean Sea Level (1). Valdez Arm and Valdez Narrows connect Port Valdez to Prince William Sound via a right angle bend through the fjord.

These features are indented into the south flank of the east-west trending Chugach Mountains and terminate in the valleys of the Columbia, Shoup, Valdez, and Corbin glaciers. It is generally believed that during the Pleistocene period, these remnant ice masses formed a single glacier that was responsible for the formation of Prince William Sound and adjacent embayments.

Bedrock flanking and in the floor of the Port Valdez trough is composed of a series of graywackes, slates, phyllites, and schists collectively referred to as the Valdez Group (4). Valdez Group sediments were deposited in an offshore wedge during the Mesozoic Era or early Cenozoic Era. Port Valdez occupies a steep east-west trending flexure, or monocline, a structural feature that influenced the flow paths of glacial ice and consequent orientation of the fjord.

Melting glaciers produce large volumes of sediment-charged water that discharge into Port Valdez. Large gravel and sand outwash deltas have formed at the glacial stream mouth, while large quantities of suspended silt are deposited through flocculation in the saline waters of the fjord. A high bedrock sill between Port Valdez and Valdez Narrows effectively prevents significant sediment interchange with Prince William Sound (12).

The old town of Valdez, which was destroyed by the 1964 Alaska earthquake, was located on the east shore, poorly consolidated, alluvial delta complex formed by the Valdez Glacier stream and the Lowe River (see Figure 2). Liquefaction of these poorly consolidated materials produced the submarine slide that caused slumping of almost 100 million cubic yards of sediment to the base of the delta, resulting in the collapse of the Valdez dock area. The new town of Valdez is situated on the Mineral Creek alluvial fan and delta, along the northeast side of Port Valdez (see Figure 2). This new site is considered to be less prone to future earthquake-induced damage.

Marine Terminal Site

The marine terminal site is located on the southeast side of Port Valdez, about 1.5 miles west of the Lowe River outlet, and extends 2000 feet offshore. The tidal flats are 600 to 1200 feet wide. Beyond is a steep drop-off into water 80 feet deep in the vicinity of the tanker berths and several hundred feet deep further offshore.

Surface sediments on the tidal flats consist of coarse to fine silt with trace amounts of fine sand. This material represents post-glacial deposition of the fine-grained components of glacial outwash into an encroaching marine embayment. As the Valdez glacier receded, the eroded valley was filled with seawater to the depth dictated by ambient sea level. Silt-size particles are discharged into the waters of Port Valdez from glacial meltwater streams, where they flocculate and settle out up to a few miles from their source. Sediment input is seasonal, with the largest amount being contributed during the high meltwater discharges of the warmer summer months. The average annual sedimentation rate in the area of the marine terminal has been estimated at approximately 2 inches per year (11). In the general vicinity of the site, the silt can be as much as 200 feet thick, providing good agreement between the sedimentation rate and the estimated time that the Valdez glacier receded from Port Valdez (slightly over 1,000 years ago). At the marine terminal site itself, the silt is 60 feet thick.

The silts rest unconformably over relatively well-consolidated outwash sands and gravels deposited directly from the Valdez glacier when it occupied Port Valdez. In the general vicinity of the marine terminal site, the sand and gravel unit is more than 400 feet thick and is thought to represent a sequence of several periods of glacial advance and recession (1). At the marine terminal site, this unit has a maximum thickness of about 140 feet and rests directly on the metamorphosed bedrock of the Valdez Group.

SITE CONDITIONS

The marine terminal site was investigated using exploration borings and bathymetric, geophysical, and diving surveys.

The bathymetric survey shows that the fjord is characterized by (1) a nearly flat shelf that extends about 600 to 1200 feet from the shoreline, with elevations ranging from +6 to -4 feet MLLW (Mean Lower Low Water), (2) a fjord slope that has a gentle gradient of about 10 horizontal to 1 vertical (10:1) between Elevations -4 and about -35 feet, and a steep gradient of about 2 horizontal to 1 vertical beyond Elevation -35 feet to the sea floor, and (3) an irregular bottom floor that underlies most of the fjord. The shelf break around Elevations -2 and -4 correlates with the normal tide range of about 12 feet in that it lies about 8 to 10 feet below the mean tide.

Seismic reflection subbottom profiles and their correlations with the logs from exploration borings show that the bedrock of the Valdez Group at the site is overlain by an upper silt layer and a sand and gravel layer.

1. Upper silt layer: The upper silt layer forms the bottom sediments of the fjord; of glacial outwash origin, it is locally known as the Valdez silt or "rock flour". It varies from coarse silt to some clay size particles (see Figure 4). The material varies from non- to slightly-plastic. Cyclic triaxial compression tests on "undisturbed" samples of the silt indicate that it is susceptible to large deformation caused by increased pore pressures generated by cyclic loading.

The silt thickness contour map developed from geophysical and test boring results (see Figure 5) shows that the silt is from 0 to 60 feet thick. The seismic profiles show that the irregularities in silt thickness are indicative of scarps (active slumping), slump debris, and potentially unstable material. The location of these slump features is shown on Figure 6. A sample seismic profile depicting these features is shown on Figure 7.

Potentially unstable material is generally located along the upper portion of the steep slopes between Elevation -7 and -60 below MLLW. The slopes within this zone range from 10:1 (horizontal to vertical) to about 2:1 and the slope surface varies from smooth to undulating with occasional discontinuous scarps 1 to 3 feet high. SCUBA diving observations show that small scale slumping is manifested as slump channels with head walls at elevations around -50 to -55 feet. The linear presence of unstable material along the gentle gradient (10:1) of the fjord slope between Elevations -6 and -30 feet and local flows onto the steep gradient (2:1) up to Elevation -60 suggest that the silt layer in this zone is probably undergoing creep and localized slumping. For example, an area of active slump was identified between Elevation -7 and -60 feet (see Figure 6).

The existence of slump debris along Elevations -100 to greater than -200 feet indicates slumping and/or slope failures from the upper slopes in the past. Immediately upslope of both of the slump debris areas, the silt layer is noticeably thin or missing, suggesting that the zone of silt may have slid to form the debris (see Figure 8). By contrast, on some of the steeper slopes, there appears to be a large accumulation of silts that may be a result of active shallow slumping (see Figure 9). However, due to the slope steepness, the distinct features may have been obliterated by localized creep.

2. Sand and gravel layer. This layer is a very dense mixture of sand and gravel containing cobbles and some boulders. A comparison of the gradation of these materials with respect to the overlying silt is shown on Figure 4. The thickness of the layer varies from practically nonexistent near the shoreline to several feet thick about 50 feet out from the shore, where it begins increasing to a maximum thickness of 140 feet. As shown on Figure 8, the dense sand and gravel are exposed on the slope where silt has slid off.

STABILITY OF SUBMARINE SLOPES

The submarine slopes are relatively steep in the area of the proposed dock facilities (see Figures 5 and 6) with a maximum inclination of about 2:1 (horizontal to vertical). Analyses of these slopes were made to evaluate static and dynamic stability under

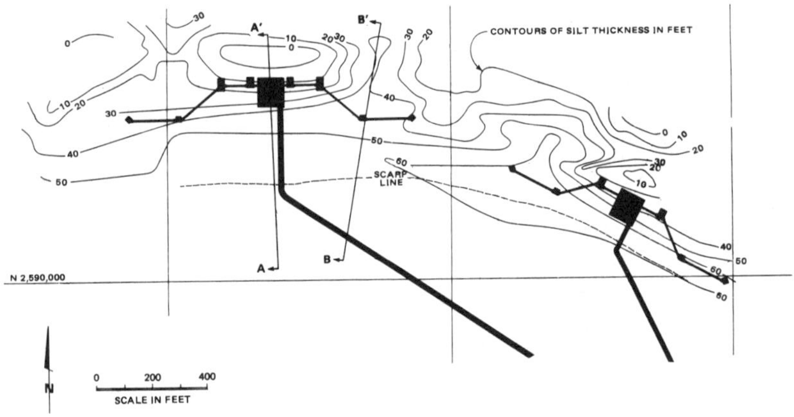

Fig. 5 Silt thickness contour map

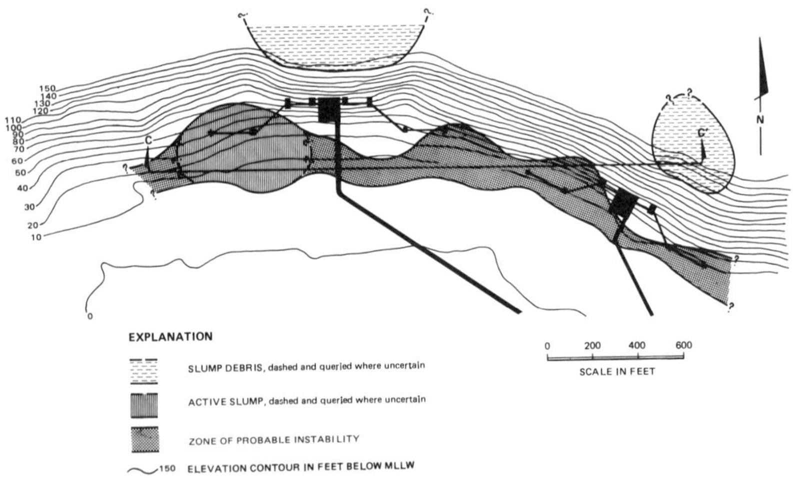

Fig. 6 Slump features

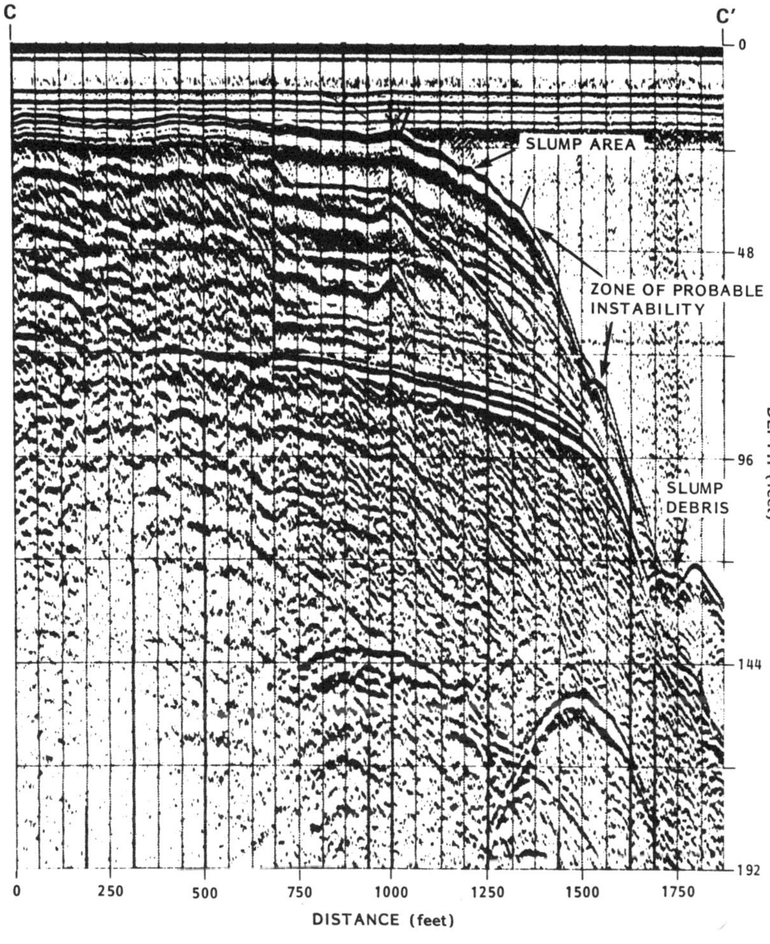

Fig. 7 Seismic profile C-C'

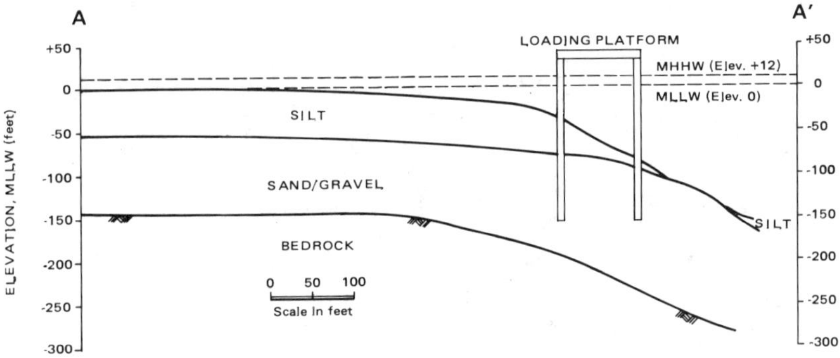

Fig. 8 Cross section A-A'

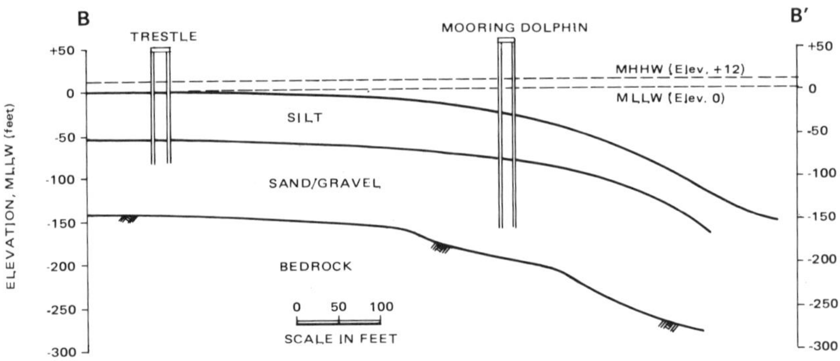

Fig. 9 Cross section B-B'

seismic loading and wave-loading conditions. The stability analyses were primarily concerned with the uppermost silt layer.

Static Stability

To evaluate the static stability of the submarine slopes, the static strength of the silts was determined using laboratory test data obtained from consolidated-drained direct shear tests and unconsolidated-undrained triaxial strength tests with pore pressure measurements. The results of these tests yielded an effective friction angle of 35° to 40° for the silt layer. Using these results, it was found that the submarine slopes are stable under static conditions at slope angles less than 30° (1.7 horizontal to 1 vertical). This is consistent with the observed slope inclinations which are generally no steeper than 2:1 in upper silt layers.

Dynamic Stability

There is evidence that the silts have slumped or slid in numerous locations at the site. For example, (1) a relatively continuous scarp showing offsets in the silt beds near the top of the submarine slope suggests active slumping; and (2) accumulations of slump deposits at the base of the steeper slopes are indicative of sliding.

Because the slumping appears to be dominant along flatter slopes in shallow water and the sliding appears to be dominant along steeper slopes at greater depth, the liquefaction under both wave loading and seismic loading conditions was carefully evaluated.

Wave-Induced Liquefaction

Wave-induced liquefaction of the upper silts was evaluated using the method suggested by Rahman (7). This method is based on the results of cyclic strength tests and wave loading analyses for the levels of assumed design storms.

Wave-induced liquefaction analyses were performed using the design storm criteria described earlier and the cyclic strengths* representative of the silts (shown on Figure 10). The results in terms of depth of liquefaction versus still water depth for the significant wave and the maximum wave are plotted on Figure 11. Because of the uncertainty involved in evaluating cyclic shear stresses due to wave loadings, and cyclic strengths of silts, an extrapolated possible upper bound to the depth of liquefaction is shown on Figure 11. Where water depths are less than 30 feet, it can be seen that storm loadings (10-foot maximum wave height) could induce liquefaction to a maximum depth of about 15 feet in the silt.

---

*For the purposes of these analyses, liquefaction has been defined as 10 percent peak-to-peak axial strain, as measured in our cyclic triaxial compression tests but modified for simple shear and field conditions.

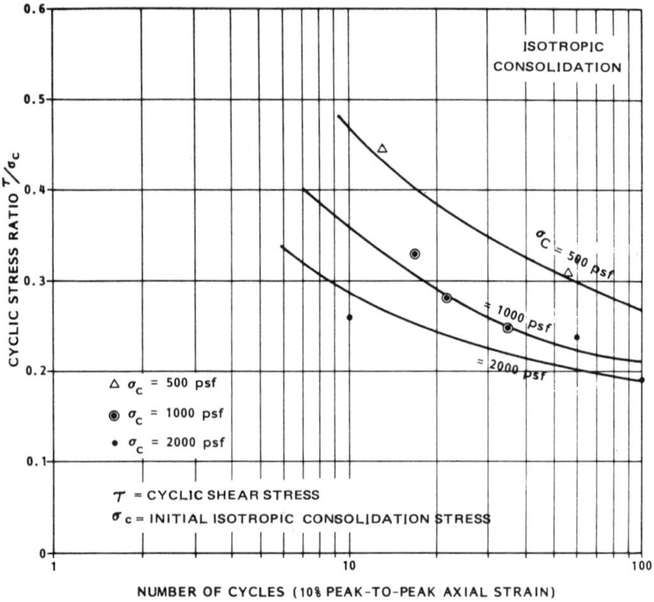

Fig. 10 Summary of stress controlled cyclic triaxial tests

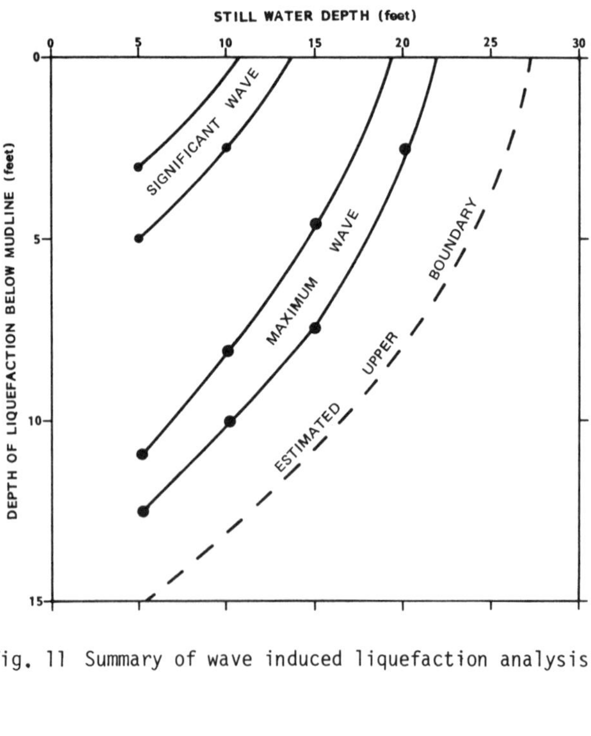

Fig. 11 Summary of wave induced liquefaction analysis

To find possible field evidence of wave-induced liquefaction, available boring logs and the geophysical information obtained from the region of still water depths to 30 feet were carefully examined. Several of the borings showed that the silts in the upper 15 to 20 feet were soft to medium stiff as compared to relatively stiff silts encountered at the mud line in borings drilled outside the slump areas in still water depths greater than 30 feet. On the other hand, the evaluation of geophysical subbottom profiling shows a zone of probable unstable silts located in this general region between Elevation -6 and -30 feet below MLLW, with areas of localized slumping to Elevation -60 (see Figure 6).

The field evidence of weaker soils and zones of probable unstable material along the gentle gradients of the fjord slopes suggests that the silts in this region are undergoing wave-induced liquefaction as predicted by the analyses. The localized slumping onto the steeper slopes appears to result from liquefied materials creeping from gentle slopes onto these steep slopes.

Earthquake-Induced Liquefaction

The earthquake-induced liquefaction potential for the silt layer was evaluated using both empirical and analytical methods (9).

Empirical Method. The empirical method is based primarily on the observed performance of sand deposits during previous earthquakes. It correlates the estimated cyclic stress ratio (τ/σ_0') to normalized SPT value, N_1, for an effective vertical pressure of 1 ton per square foot. In the cyclic stress ratio, τ is the average cyclic shear stress and σ_0' is the effective overburden pressure of the soil involved (9). Figure 12 shows a correlation (developed by Seed, 1976) between N_1 and τ/σ_0' for sand deposits during past earthquakes.

The field penetration resistance, N, of the upper silt soils corrected to N_1 versus depth is shown on Figure 13. N_1 ranges from 4 to 35 with the majority of the data points falling within the range of 6 to 20. In contrast, the N_1 required to prevent liquefaction is 11 to 15 during the operating earthquake and 20 to 22 during the survival earthquake. These values of N_1 were computed by making the proper adjustments for the soil types and effects of overburden pressure. To account for soil type, the silts were considered to be about 50 percent stronger than sands (8).

A comparison of the N_1 values required to prevent liquefaction with the measured values (see Figure 13) shows that for level ground conditions:

1. A major portion of the silts will liquefy during the operating level earthquake.
2. Almost all silty soils will liquefy during the survival earthquake.

Analytical Approach. The analytical method is based on the results of cyclic strength tests and estimated cyclic shear stresses for the levels of assumed ground motion. The results of the analytical approach show that:

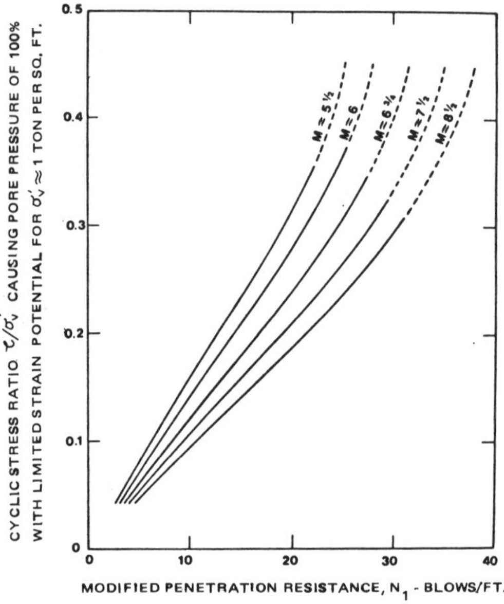

Fig. 12 Evaluation of liquefaction potential for different magnitude earthquakes (after Seed, 1976)

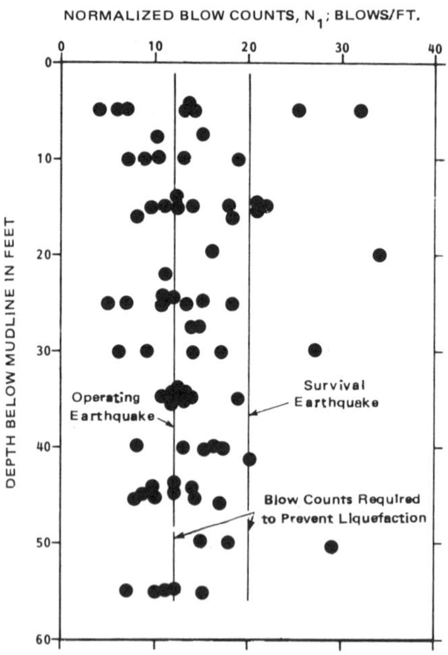

Fig. 13 Summary of earthquake induced liquefaction analysis

1. Where the mud line is horizontal, a peak bedrock acceleration below the site of greater than 0.17 g could cause liquefaction for the entire depth of silt layer.

2. Where the ground is sloping, liquefaction could occur for the full depth of silt when peak bedrock accelerations exceed about 0.22 g. The higher accelerations required to cause liquefaction under sloping ground conditions are a result of the presence of initial shear stresses on a horizontal plane of the soil element which exhibits higher resistance to liquefaction.

COMPARISON OF CYCLIC BEHAVIOR OF VALDEZ SILTS WITH SANDS AND CLAYS

It is important to note the differences in the cyclic behavior of the Valdez silts as compared to the more well-documented cyclic behavior of sands. Unlike sands, the Valdez silts do not show a catastrophic loss of strength in the cyclic strength tests. Even when pore water pressure approaches the confining pressure, the samples continue to deform gradually to large strains, remaining intact but behaving like an accordian. This behavior shows that the Valdez silts can undergo significant cyclic mobility without a measureable loss in strength. This behavior is also evident from the post-cyclic strength tests where the post-cyclic peak strength is comparable to the pre-cyclic peak strength. The stress-strain characteristics, however, are altered significantly. In post-cyclic strength tests, the peak strengths are mobilized at very large strains (on the order of 20 percent or more) as compared to lower strains (on the order of 10 percent) for the pre-cyclic case. Furthermore, for the post-cyclic case, the soils do not exhibit much resistance up to a strain of about 7 to 10 percent.

This unique post-cyclic behavior of the Valdez silts differs considerably from rigid plastic behavior, suggesting that the methods of evaluating permanent earthquake-induced slope displacement as proposed by Newmark (6) are not valid for this material.

STRUCTURAL DESIGN PROBLEMS

Tidal Flat

The results of the liquefaction studies for the tidal flat area show that:

1) Storm wave loadings (10-foot maximum height) could induce liquefaction to a maximum depth of about 15 feet in the silt.

2) Earthquake-caused peak bedrock acceleration of greater than about 0.17 g below the site could cause liquefaction for the full depth of silt.

Because of the unique behavior, the silts under liquefied conditions will continually creep downhill on relatively flat slopes (i.e., slopes steeper than about 1 percent). This creeping soil mass produced by the wave loading condition would gradually increase lateral loads on any piles installed at the site. Therefore, piles located in the tidal flats were designed for reduction in lateral resistance due to softening as well as for increase in lateral loads due to downhill creep.

Submarine Slopes

The results of the liquefaction studies for the submarine slopes indicate that, in a survival-level earthquake, when bedrock accelerations exceed about 0.22 g, major slides could occur that might extend to the full depth of the silt (50 feet). Foundations for structures sited on the submarine slopes would probably be subjected to lateral forces caused by liquefaction-induced slides or downslope deformations in the upper silt layer, and would therefore have to be designed to resist these lateral forces. For a major slide extending to the full depth of the silt, the following soil pressures, to be applied over the projected area of the piles, were estimated for use in sizing the piles.

Depth Below Mud Line (feet)	Lateral Soil Pressure (ksf)
0	1
30	8
50	11

Using these design loads, the pipe pile sizes were estimated as (1) 30- to 42-inch-diameter, 1-1/2-inch wall thickness to resist the loads to a depth of 15 feet in the tidal flats; and (2) 84- to 108-inch, 4-inch wall thickness to resist loads to a depth of 50 feet on the submarine slopes.

Although the depth of sliding in the submarine slope during the survival earthquake could involve the full thickness of the silt layer, only the key structures were designed to resist this "worst" condition. Other structures, which, if lost, could economically be repaired and would not significantly endanger life, were designed for a lesser depth of sliding. Because of the creeping soil mass produced by the storm wave loading, all structures were designed for at least 15 feet of sliding.

ALTERNATIVE DESIGN SCHEMES

The following alternative design schemes were considered because the large-diameter piles proposed for the terminal significantly increased its cost.

a. Dredging. Dredging was considered to remove slide-susceptible silts so that the submarine slopes would be more stable. Dredging possibilities considered varied from removal of most of the silt on the slopes beneath and immediately behind the structures to limited dredging at the major structures such as the loading platforms. Limited dredging was considered to remove enough silt to significantly reduce the depth of silt slides and the corresponding soil loads that would act against foundation piles.

b. Floating docks. Floating docks, similar to the one built at the Alyeska Tanker Terminal, just west of the project site, were considered. This scheme was considered to eliminate the problems of designing the foundation piles of such important structures as loading platforms and berthing dolphins to withstand large slide forces. These structures could be anchored back on the tidal flats beyond areas of possible slope instability caused by seismic loading. The anchors for these structures would be designed to resist creep resulting from wave-induced loading. Other nonimportant structures, such as trestles, mooring dolphins and anchor piles, which would not endanger life in an earthquake and could economically be repaired, could be constructed in the slide zone produced by the seismic loading.

c. Cellular sheet pile docks. Cellular sheet pile docks, essentially gravity structures, were considered for the loading platforms. They would have to be capable of resisting the large lateral loads imposed by the liquefied silts.

CONCLUSIONS

Valdez silts are a rock flour consisting of very fine mineral fragments of the rocks predominant in the area. The plasticity of the material varies from zero to an apparent plasticity of up to 8. Since no clay minerals are present, the plasticity index is a function of grain size.

The cyclic behavior of Valdez silts under cyclic loading conditions is different from that of sands in that no catastrophic loss of strength occurs during cyclic triaxial strength tests. The silts can undergo significant mobility without a measurable loss in strength. Under dynamic loading conditions, this unique behavior will cause large movements along sloping ground. Therefore, conventional liquefaction analyses or the methods of computing permanent earthquake-induced slope displacements are not valid for this material. Because the soils can develop post-cyclic peak strengths (at large strains) almost equal to pre-cyclic peak strengths (at small strains), small movements resulting from slumping or large movements resulting from sliding would impose very large loads on foundation piles. These loads would require the use of very large diameter foundation piles.

REFERENCES

1. Carlson, P. R., and B. F. Molnia, "Minisparker profiles and sedimentologic data from R/V Acona cruise in the Gulf of Alaska and Prince William Sound," U.S. Geological Survey Open File Report 78-381, 1978.

2. Coulter, H. W., and R. R. Migliaccio, "Effects of the earthquake of March 27, 1964 at Valdez, Alaska," U.S. Geological Survey Professional Paper 542-C, 1966.

3. Donovan, N. C., and S. Singh, "Liquefaction criteria for the Trans-Alaska Pipeline," ASCE Annual Convention, Philadelphia, Pennsylvania, Preprint 2752, 1976.

4. Moffit, F. H., "Geology of the Prince William Sound Region, Alaska," U.S. Geological Survey Bulletin 989-E, 1954.

5. Naidu, A. S., "Science in Alaska," Alaskan Geological Society Proceeding No. 27, 1976.

6. Newmark, N. M., "Effects of Earthquakes on Dams and Embankments," Geotechnique 15, No. 2, 1965.

7. Rahman, M. S., "Analyses of Wave-Induced Liquefaction in Relation to Offshore Construction," Thesis submitted in partial fulfillment of requirements for Ph.D., University of California, Berkeley, 1977.

8. Seed, H. B., personal communications, 1981.

9. Seed, H. B., "Evaluation of Soil Liquefaction Effects on Level Ground During Earthquakes," ASCE Annual Convention, Philadelphia, Pennsylvania, Preprint 2752, 1976.

10. Seed, H. B., R. Murarka, J. Lysmer, I. M. Idriss, "Relationships Between Maximum Acceleration, Maximum Velocity, Distance from the Source and Local Site Conditions for Moderately Strong Earthquakes," Earthquake Engineering Research Center, Report No. EERC 75-17, College of Engineering, University of California, Berkeley, 1975.

11. Sharma, G. D., and D. C. Burbank, "Geological Oceanography, In Environmental Studies of Port Valdez," University of Alaska Institute of Marine Science, 1973.

12. von Heune, R., G. G. Shore, and E. Reimnitz, "Geological interpretation of seismic profiles in Prince William Sound, Alaska," Geological Society of America, 1967.

YIELD BOUNDARY FOR CYCLIC LOADING OF A SILTY CLAY

By A.F.L. Hyde[1] and S.J. Ward[2]

INTRODUCTION

The great water depths and hostile environment of the North Sea oil fields has led to the development of the large scale gravity structure. This type of structure, manufactured in a sheltered location on the coastline and then floated out and placed in position by ballasting, relies for its stability on the bearing capacity of the underlying soil. Under wave loading the foundation soil is subjected to stress conditions which vary according to position beneath the structure, (Fig. 1). Beneath the centre of the structure the stresses can be modelled by some form of cyclic simple shear apparatus while at the edges of the foundation the stress regime is more accurately depicted by a cyclic triaxial test (1).

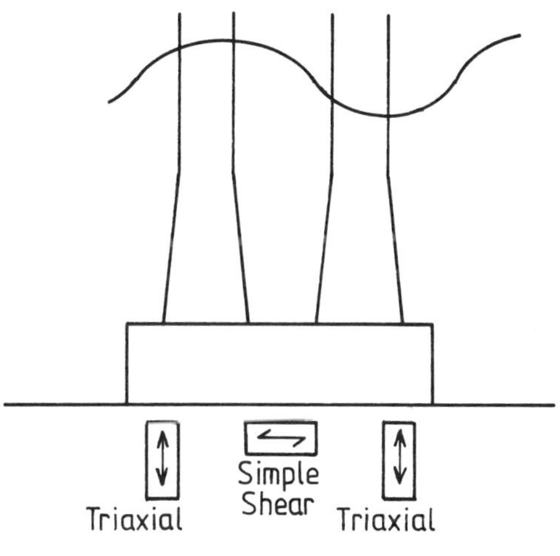

Fig. 1 Stress conditions beneath a gravity structure

[1]Lecturer, Dept. of Civil Engineering, Loughborough University of Technology, United Kingdom.

[2]Research Student, Dept. of Civil Engineering, Loughborough University of Technology, United Kingdom.

Design methods involving data from cyclic simple shear tests have been developed (1,2,3,4) and are commonly used. These methods allow an estimation of the build-up of both cyclic pore pressures and cyclic shear strains beneath the structure. However it is well established that failure also occurs in clays under undrained cyclic triaxial loading at peak deviator stresses that are far less than the conventionally determined compressive strength (5,8). The authors therefore believe that the designer should take some account of this phenomenon. A stability criterion for initially heavily overconsolidated clays subjected to cyclic loading has already been proposed (6). This stability criterion is based on the state boundary surface on the 'dry' side of the critical state, sometimes known as the Hvorslev surface (10). This paper describes repeated load triaxial tests carried out on a silty clay with pore pressure measurements which demonstrate that normally consolidated and lightly overconsolidated clays can undergo a process of overconsolidation through the development of pore pressures, a phenomenon already observed for marine clays (7). Samples subsequently reach equilibrium or fail on the Hvorslev surface in the same way as heavily overconsolidated soil. The dry side of the state boundary surface therefore forms an empirically determined yield and consequent failure surface for cyclic effective stress paths of soils of all stress histories. The Hvorslev surface is easily determined using conventional monotonic strain controlled triaxial tests with pore water pressure measurements.

EXPERIMENTAL PROCEDURE

Sample preparation

The soil was a Keuper Marl obtained in powdered form. Samples were prepared by processing the powder through a rotary classifier. Particles larger than silt size 0.063 mm were discarded. The resulting soil had a liquid limit of 36% and plastic limit of 19%. The powdered material was mixed with de-aired distilled water to form a slurry with a moisture content of 62%. The resulting slurry was poured into 100 mm diameter moulds for one dimensional consolidation using a vertical pressure of 180 kPa. Drainage was allowed from each end of the mould. After 72 hours the soil was extruded into 37 mm diameter sample tubes. The extruded samples were cut to a length of 75 mm placed in a triaxial cell and isotropically consolidated to the various stress histories shown in Fig. 2. Samples were prepared to overconsolidation ratios of 1, 4, 10 and 20. Overconsolidated samples were prepared by preconsolidation at a cell pressure of 700 kPa and then gradually applying a back pressure to the drainage line to produce the required final effective consolidation stress. The normally consolidated samples were prepared at an effective consolidation pressure of 700 kPa with no back pressure.

Frictionless end platens were used which took the form of polished stainless steel lubricated with silicon grease and covered with a latex membrane.

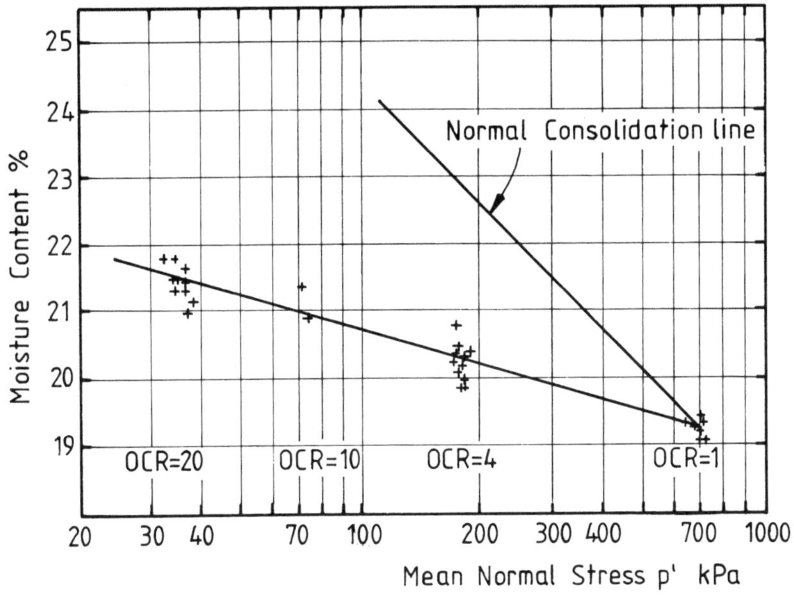

Fig. 2 Consolidation stress histories

Drainage of the samples was through filter paper side drains to porous polytetrafluoroethylene (PTFE) rings mounted in the base platen. In addition a porous PTFE disc was inserted in the centre of the frictionless base platen in order to improve the quality of pore water pressure measurements. Pore water pressures were measured by a transducer connected to the drainage line on the triaxial cell base. The pore pressure response was measured by determining the 'B' parameter (11). This was found to be >0.99 for the overconsolidated samples and >0.9 for the normally consolidated samples, the difference being because of the back pressure applied to overconsolidated samples.

Loading Equipment

The loading of samples was carried out using an electro-pneumatic apparatus which gave a quasi-sinusoidal variation in deviator stress. The deviator stress was monitored using a load cell located on the loading ram inside the triaxial cell. The cell pressure was monitored by a pressure transducer attached to the cell pressure inlet.

The vertical deformation of the samples was measured by means of a linear variable displacement transducer (LVDT) attached to the loading ram. Corrections were made to allow for deformations in the apparatus and lubricated membranes at each end of the sample.

Samples were tested at a frequency of 0.1 Hz as this was judged to be the order of magnitude of frequencies to be expected under offshore wave loading. The deviator stress was varied from a small positive value to prevent bounce of the loading ram.

Test Programme

Three types of triaxial test were carried out. Firstly consolidated undrained strength tests were performed using monotonic strain controlled loading. The purpose of these tests was to establish the nature of the stable state boundary surface for the soil, and they were therefore performed at each of the four stress histories. These tests were followed by two series of undrained cyclic tests consisting of single stage constant stress amplitude tests and multi-stage increasing stress amplitude tests. These tests were carried out at three representative overconsolidation ratios of 1, 4 and 20. The purpose of these tests was to compare states of equilibrium and failure under cyclic loading with the stable state boundary surface.

ANALYSIS OF THE EXPERIMENTAL RESULTS

The applied and effective principal stresses have been expressed in terms of the basic parameters,

deviator stress or principal stress difference $q' = \sigma_1' - \sigma_3'$

mean normal effective stress $p' = \frac{1}{3}(\sigma_1' + 2\sigma_3')$

mean normal applied stress $p = \frac{1}{3}(\sigma_1 + 2\sigma_3)$

where $\sigma' = \sigma - u$
and u = pore water pressure.

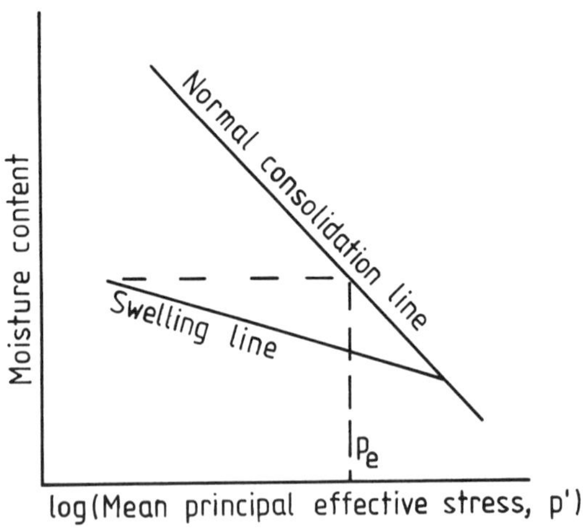

Fig. 3 Equivalent pressure p_e

A further parameter p_e was introduced and is defined as the pressure of the point on the normal or virgin consolidation line which is at the same moisture content as the sample itself. This parameter is related to sample volume or moisture content and therefore allows the results from samples with different stress histories to be normalised onto a single plane by combining it with the effective stress parameters p' and q' (Fig. 3).

Undrained strength tests

The object of the undrained strength tests was to establish the position of the stable state boundary surface and the values of the critical state soil parameters. The normalised effective stress paths are shown in Fig. 4.

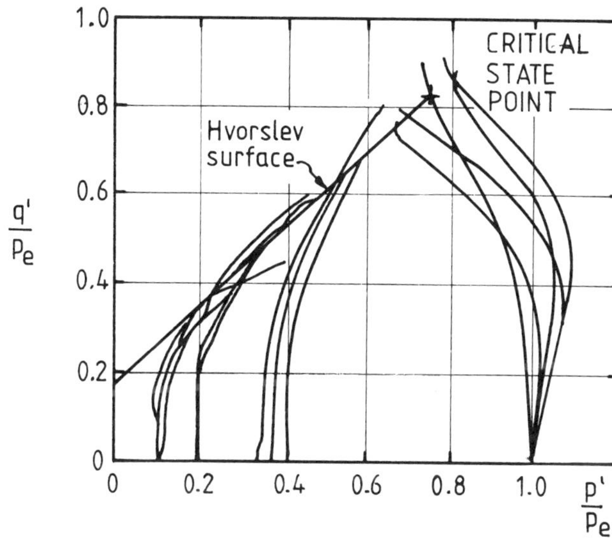

Fig. 4 Normalised effective stress paths for undrained strength tests

It can be seen that the overconsolidated samples have failed prior to reaching the critical state and that these failures have occurred on a surface often defined as the 'Hvorslev' surface (9). Failure was defined as being the point at which the peak value of q' was reached.

The normally consolidated samples have however reached the critical state. The normally consolidated samples developed large positive pore pressures during shearing and are therefore defined as being on the 'wet' side of the critical state. Samples with an OCR of 4 developed little pore pressure and could be defined as having a critical void ratio and forming the boundary between dilatant and non-dilatant stress histories. The more heavily overconsolidated samples with OCR's of 10 and 20 developed negative pore pressures and are defined as being on the 'dry' side of the critical state. The critical state point has been defined as the average end point of the stress paths for the normally consolidated samples. At the critical state,

$$\frac{q'}{p_e} = 0.81$$

$$\frac{p'}{p_e} = 0.75$$

The 'Hvorslev' surface

The effective stress paths for the samples consolidated to the 'dry' side of the critical state initially rise vertically and then curve towards the critical state. The Hvorslev surface is therefore defined as the straight line tangential to the stress paths on the dry side and passing through the critical state and is shown on Fig. 4. It has already been established (6) that this surface forms a stability criterion for heavily overconsolidated clay subjected to repeated triaxial loading. Further and more recent testing by the authors indicates that this is also true for normally and lightly overconsolidated soil.

The surface is simply defined by subjecting normally and heavily overconsolidated samples of the soil to routine undrained triaxial tests with pore water pressure measurements. The data from these tests is then normalised onto a single plane by determining the virgin isotropic consolidation curve and hence values of p_e.

Cyclic load tests

Repeated load tests were carried out at OCR's of 1, 4 and 20 and were carried out undrained with pore water pressure measurements. As it is not possible to show the complete effective stress path for a large number of cycles the cyclic effective stress paths are shown by drawing the total stress path and showing the movement of p'/p_e at the peak value of q'/p_e. This is illustrated in Fig. 5.

Failure for this type of test was defined as the point at which vertical strains in excess of 15% occurred. Once the process of failure commenced large strains occurred over a relatively small number of cycles and a shear plane was clearly visible in the sample. However for stable samples the maximum plastic strain measured was 3%.

Single stage tests were carried out at a constant peak deviator stress for up to 10,000 cycles or failure whichever occurred earlier.

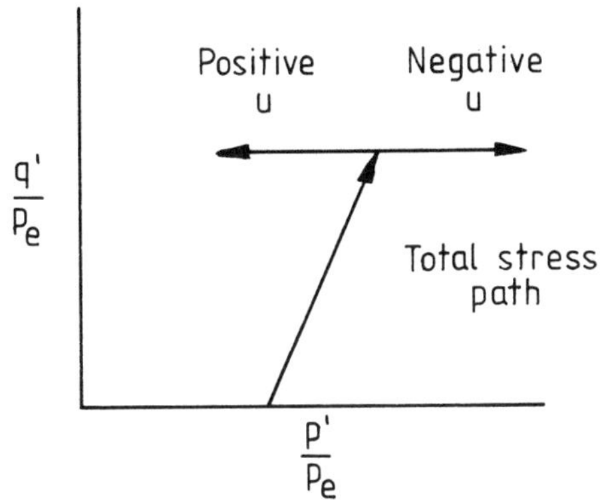

Fig. 5 Representation of cyclic loading effective stress paths

The results of these tests are shown in Fig. 6. Considering first the normally consolidated samples, the total stress path initiates from $p/p_e = 1$. At a very early stage in the tests large positive pore pressures were generated which had the effect of decreasing p'/p_e and moving the effective cyclic stress path towards the 'dry' side of the critical state. At values of q'/p_e up to 0.44 the pore pressures reached apparent equilibrium and the effective stress path stabilised at some point within the yield envelope. If q'/p_e was increased above this value the effective stress path continued to move towards the Hvorslev surface and samples then failed in this region. Cyclic loading therefore was leading to failure at 54% of the critical state value of q'/p_e of 0.81 achieved under monotonic strain controlled loading. A similar phenomenon occurred for the samples consolidated to an OCR of 4, with most samples developing large positive pore pressures under cyclic loading. Sub-failure equilibrium was achieved for samples cycled up to $q'/p_e = 0.4$. At deviator stresses above this value the effective stress path moved on to the Hvorslev surface and failure occurred. Generally therefore normally consolidated and lightly overconsolidated samples cycled at or above about 50% of their peak strength failed on the 'dry' side of the stable state boundary surface.

Pore pressure movements at an OCR of 20 were much smaller and failure therefore for these samples only occurred when the total stress path approached the Hvorslev surface. This finding agrees with previous work carried out on a different Keuper Marl (6).

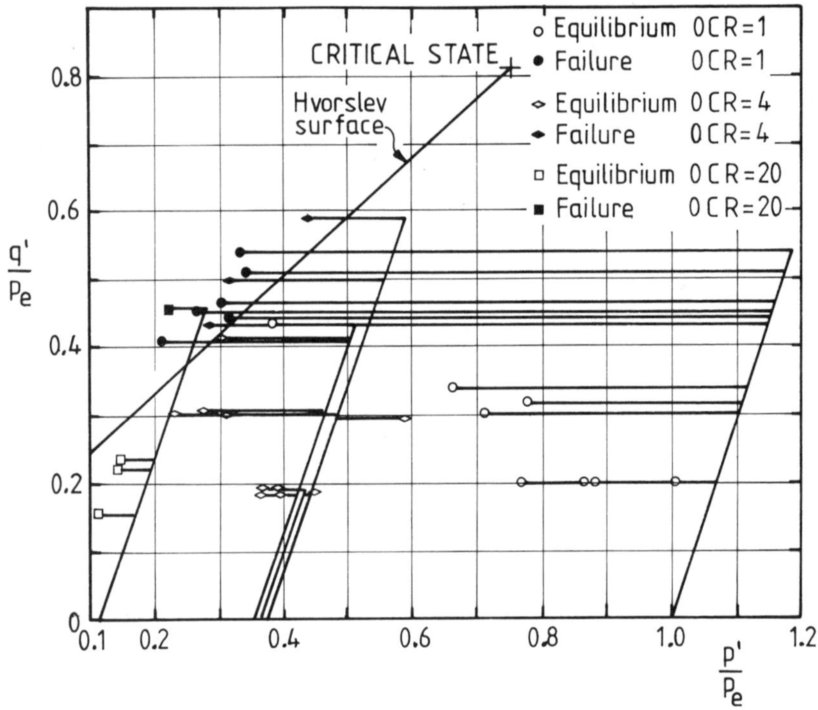

Fig. 6 Cyclic effective stress paths for single stage tests

A further series of samples were tested using multi-stage loading. In these tests samples were subjected to ever-increasing peak deviator stresses until failure occurred. 300 cycles were performed at each level before increasing the peak deviator stress. Once again it can be seen (Fig. 7) that the effective stress path moved towards the dry side of the critical state and that samples failed in the region of the Hvorslev surface. It should be noted however that samples subjected to multi-stage loading generally failed at lower peak deviator stresses than those subjected to single stage loading. For normally consolidated soil this was as low as $q'/p_e = 0.35$ and for an OCR of 4 failure occurred at $q'/p_e = 0.28$.

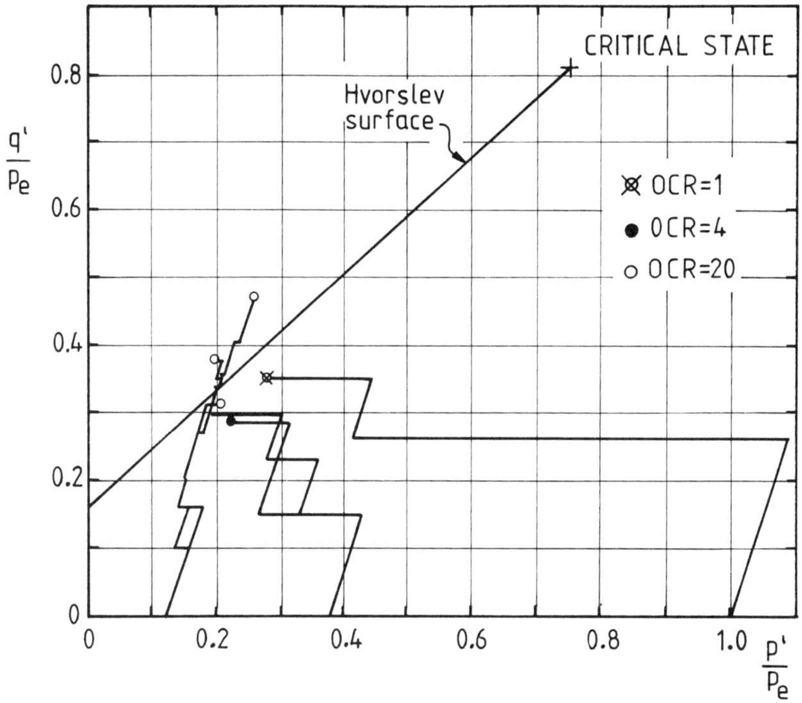

Fig. 7 Cyclic effective stress paths for multi-stage tests

The end points of the effective stress paths of all the samples that failed under cyclic loading are shown in Fig. 8. It can clearly be seen that the Hvorslev surface forms a stability criterion for samples of all stress histories and that failure occurs as the effective stress path approaches this boundary. This graph also illustrates the point already made that samples subjected to multi-stage loading failed at lower cyclic stresses than those subjected to single stage loading.

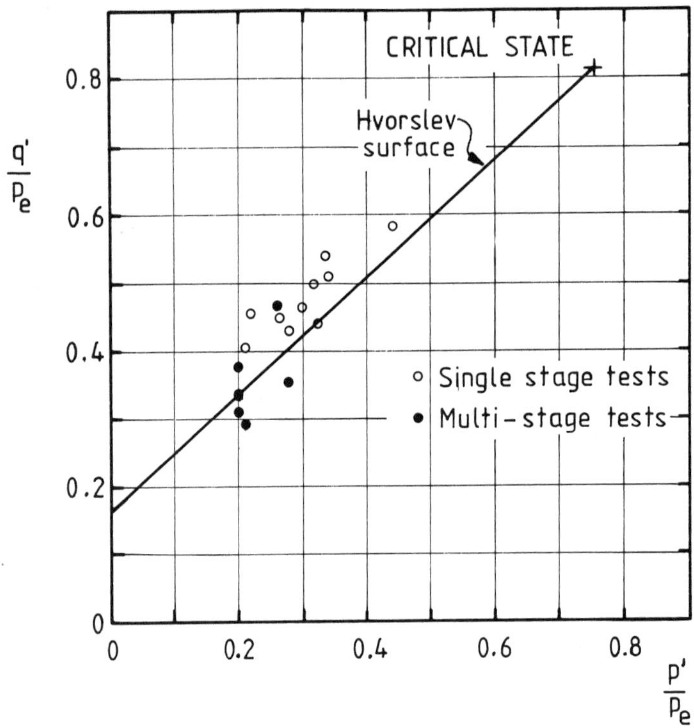

Fig. 8 Failure of cyclically loaded samples relative to Hvorslev surface

Pore water pressures

We shall define the pore pressure parameter, A, (11)

as $\frac{u_{max}}{q'_r}$

where u_{max} is the maximum pore pressure after 10,000 cycles or failure

and q'_r is the cycled value of q',

and a cyclic stress ratio

as $\frac{q'_r}{q'_{cs}}$

where q'_{cs} is the value of q' at the critical state.

Considering the single stage tests it can be seen that the 'A' value for normally consolidated samples with a cyclic stress ratio of <0.43 lies between 0.9 and 1.5 after 10,000 cycles. As the cyclic stress ratio increases then the pore pressures developed under cyclic loading increase until 'A' values in excess of 2 are achieved for cyclic stress ratios above 0.5. These values of pore pressure push the effective stress path onto the Hvorslev surface and result in sample failure. A similar rapid increase of the 'A' value occurs for samples with an OCR of 4 as the cyclic stress ratio is increased. As the effective stress paths for these samples start nearer to the Hvorslev surface then the corresponding pore pressures required to reach failure are smaller. In this case failure occurs when the 'A' value reaches 0.78 at a cyclic stress ratio of 0.5. At greater cyclic stress ratios the 'A' value required for failure decreases further. For example a sample tested with a cyclic stress ratio of 0.73 failed when the 'A' value reached 0.26.

Heavily overconsolidated samples (OCR = 20) can develop initial positive pore pressures equal to 0.2 q' peak (6), thereafter the pore pressure becomes negative. Therefore providing the effective stress path does not approach the Hvorslev surface in the initial stages of the test failure is unlikely.

CONCLUSIONS

The 'Hvorslev' surface on the 'dry' side of the critical state determined from monotonic undrained strength tests acts as a boundary between stable and unstable states for soil samples of any stress history subjected to cyclic triaxial loading.

Cyclic effective stress paths of soil consolidated to the wet side of the critical state migrate towards the 'Hvorslev' surface and either fail on the surface or remain below the surface in a state of equilibrium. The pore pressures generated in these samples increase with the cyclic stress to strength ratio.

Heavily overconsolidated soil does not generate large pore water pressures under cyclic loading and therefore failure does not occur if the total or applied stress path does not approach the Hvorslev surface.

REFERENCES

1. Andersen, K.H., "Behaviour of Clay Subjected to Undrained Cyclic Loading", Proceedings of the International Conference on the Behaviour of Offshore Structures, Norway, August 1976, Vol. 1, pp. 392-403.
2. Andersen, K.H., Brown, S.F., Foss, I., Pool, J.H., and Rosenbrand, W.F., "Effect of Cyclic Loading on Clay Behaviour", Design and Construction of Offshore Structures, Institution of Civil Engineers, London, 1977, pp. 75-79.
3. Bjerrum, L., "Geotechnical Problems involved in Foundations of Structures in the North Sea", Geotechnique, Vol. 23, No.3, 1973, pp. 319-358.

4. Foss, I., Dahlberg, R., and Kvalstad, T., "Foundation Design for Gravity Structures with Respect to Failure in Cyclic Loading", Proc. 10th Annual Offshore Technology Conference, Houston, Texas, 1978, OTC 3114, pp. 535-545.
5. Hyde, A.F.L., "Repeated Load Triaxial Testing of Soils", Ph.D. Thesis, University of Nottingham, 1974.
6. Hyde, A.F.L., "Stability of an Overconsolidated Silty Clay Under Repeated Loading", Proceedings of the Tenth International Conference on Soil Mechanics and Foundation Engineering, Stockholm, 1981, Vol. 1, pp. 635-640.
7. Koutsoftas, D.C., "Effect of Cyclic Loads on Undrained Strength of Two Marine Clays", Journal of the Geotechnical Engineering Division, ASCE, Vol. 104, GT5, May 1978, pp. 609-620.
8. Lashine, A.K.F., "Deformation Characteristics of a Silty Clay Under Repeated Loading", Proceedings of the Eighth International Conference on Soil Mechanics and Foundation Engineering, Moscow, 1973, Vol. 1, pp. 237-244.
9. Parry, R.H.G., "Triaxial Compression and extension tests on remoulded saturated clay", Geotechnique, Vol. 10, 1960, pp. 166-180.
10. Roscoe, K.H., Schofield, A.N., and Wroth, C.P., "On Yielding of Soils", Geotechnique, Vol. 8, No. 1, 1958, pp. 22-53.
11. Skempton, A.W., "The pore-pressure coefficients A and B", Geotechnique, Vol. 4, 1954, pp. 143-147.

DYNAMIC SEDIMENT PROPERTIES, MISSISSIPPI DELTA

By Sibel Pamukcu,[1] SM.ASCE, J. K. Poplin,[2] M.ASCE, J. N. Suhayda,[3] M. T. Tumay,[4] M.ASCE

INTRODUCTION

Marine sediment instabilities are known to be a major design concern for offshore structures. The integrity of oil platforms, subbottom storage tanks, offshore pipelines, and various other structures of such nature depend highly on the stability of the ocean floor on which they are built (32). The severity of the problem is determined by the local soil properties and loading conditions. Dynamic loading of the ocean floor sediments by waves is a primary interest in design due to concerns about progressive reduction and degradation of soil stiffness, liquefaction and cyclic mobility effects under cyclic loading. Several theoretical investigations have addressed the problem of wave-sea bottom interaction. Many of these investigations require knowledge of the shear modulus values and profiles beneath the mudline (7,8,10,16,25,26). A number of laboratory and theoretical investigations conducted on the dynamic behavior of soft saturated clays have been reported during the past years (15,30,34,36,37). These studies show the constitutive relations for clays are generally nonlinear and hysteretic in nature. Very little dynamic data, however, is available with regard to marine sediments and most of the experimental data that is available concentrate on San Francisco Bay Mud.

The main objective of this report is to present data of the dynamic properties of sediments of the Mississippi Delta region under cyclic loading conditions. The experimental information gathered and reported here provides new understanding of the dynamic response of soft marine sediments under cyclic wave loading and gives insight into the dependency of this response on burial depth and changing soil properties of the sediment layers.

There are theories that provide means of analyzing marine sediment instability:

1. Infinite slope stability analysis (20,22)

---

[1] Graduate Research Assistant, Department of Civil Engineering, Louisiana State University, Baton Rouge, Louisiana.

[2] Associate Professor, Department of Civil Engineering and Construction, Louisiana State University, Baton Rouge, Louisiana.

[3] Associate Professor, Department of Civil Engineering, Louisiana State University, Baton Rouge, Louisiana.

[4] Professor, Department of Civil Engineering, Louisiana State University, Baton Rouge, Louisiana.

2. Spontaneous liquefaction due to progressive pore pressure build up and softening under cyclic wave loading (3,18,19,23,29)
3. Simple rheological model analysis (35)
4. Numerical analysis (21,29,39)
5. Wave-propagation and viscoelastic subbottom interaction analysis (17,25,26,33)

Due to the extremely complex nature of the sliding process, the first theory does not prove to be valuable for the geographical area of concern in this study. The second theory does not correlate with the problem due to the cohesive nature of the material. However, in-situ and experimental findings of existing high pore water pressure values within some cohesive underwater sediments (2,14,22) allows one to attribute the initiation of these slides to an "analogous liquefaction". The third theory attempts to bring a numerical solution method that would be applicable to all soils, provided rheological principles model the material behavior. Solutions by this theory lack readily available data for workability. The last theory is a numerical solution and is currently being used in platform design procedures for offshore Mississippi Delta. The solution technique is sensitive to the input data of the initial tangent shear modulus and ultimate undrained shear strength ratio, (G_{max}/s_u). Consequently, realistic estimation of the initial shear modulus is required to obtain reliable response profiles of soft soil layers using this model.

A key point in the behavior of marine sediments is that cyclic loading experiments on soft saturated clays exhibit progressive moduli reduction and degradation depending on strain amplitude, frequency and duration of loading. These observations correlate with progressive softening or cyclic mobility (5,6). Cyclic mobility can be defined as the initial stage of progressive effective stress reduction prior to liquefied state. Castro and Poulos (5) explained the differences in the mechanics of liquefaction and cyclic mobility and showed that the initial state defined by void ratio, confining pressure and shear stress is the critical factor that govens the development of both phenomena. While liquefaction is considered to develop in cohesionless soils either under static or cyclic loading, cyclic mobility can develop in any type of soil under sufficient magnitude and number of cyclic loading. Dynamic behavior of soft marine clays is closely associated with this phenomenon as observed through various dynamic tests of the material. As a result, the submarine mudflows of the Mississippi Delta may be attributed to possible cyclic mobility of the high pore pressure bearing sediments of the area.

Current critical needs which have been identified are:

1. Knowledge of existing pore pressure values which have been observed to be very high through various investigations.
2. Knowledge of the progressive softening that can be related to liquefaction or cyclic mobility phenomenon.
3. More information concerning the dynamic properties of marine sediments and of particular interest the variation of dynamic moduli with depth.

In this paper, detailed dynamic response information and quantitative data which are useful in various soil response models for Mississippi Delta sediments are presented and discussed.

INVESTIGATION

Geographic and Geological Information of Site

Samples for this investigation were taken from a single borehole located near Pass A Loutre of the Mississippi River Delta at a water depth of 10 m. The boring penetrated to 110 m below mudline. Samples for laboratory analysis were taken at selected depths throughout the boring.

Due to fast deposition of the Mississippi River, great thicknesses of very soft, underconsolidated clays (31) have accumulated in the delta. This accumulation is promoted by large scale delta front movements in forms of mudflows and turbidity flows. Seaward from the river mouth, the mudflow strata overlie an earlier sedimentary sequence of continental shelf marine clays which are stiff and likely to be overconsolidated. Mudflow strata are gassy from biogenic methane production and highly underconsolidated due to deposition being faster than consolidation. When water depths in front of the distributary bar of the river decrease after mudflows, silty and sandy sediments are deposited directly onto the soft clay seafloor which in turn form a "crustal" zone, typically found at many locations in the Mississippi Delta.

Samples of the top soil layer from 0 to 7 m were very soft, olive-colored clay with "spongy" cross-sectional appearances indicating gaseous nature. Traces of organic matter (fiberous plant parts) were encountered in these samples. From 7 to 14 m, the material showed silty to sandy formations in thin layers. Below this depth samples were again very soft and dark colored gaseous clay down to 60 m. Between 60 and 70 m samples indicated sand formation and they were not tested for dynamic properties. Below 70 m they were much firmer with traces of shell fragments and lighter in color compared to the gaseous clay above 60 m. Further down at 100 m, fragments of shells and other hard substances were scattered within the clay matrix.

Index Properties of the Samples

The results of the density and plasticity analysis are given graphically in Figures 1 and 2, respectively. The sandy formation from 7 to 14 m of depth shows an increase in dry and bulk unit weights, typical of crust formation. Figure 2 shows the natural water content and plasticity range variation with depth. The natural water contents of the samples above 60 m are higher or very close to their liquid limits which is an indication of underconsolidation. The firmer samples from below 60 m had natural water contents close to their plastic limits which is an indication of overconsolidation (4).

Using the plasticity information, other properties that are related to the stress-strain nature of the soil samples were estimated. These parameters were lateral earth pressure at rest coefficient K_o and Poisson's ratio μ. The empirical relation between plasticity index I_p, μ and K_o (Brooker and Ireland formula (4)) and then the general Hooke's law were utilized to determine these values, respectively. The values of Poisson's ratio averaged 0.38, and ranged from 0.40 to 0.34 for the entire depth of core.

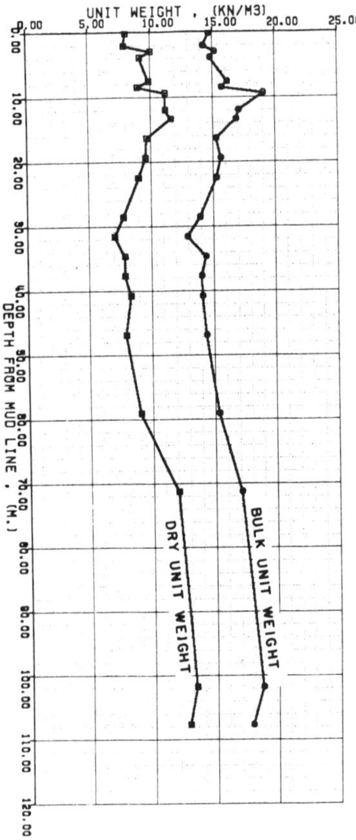

Fig. 1 Bulk and dry unit weights profile.

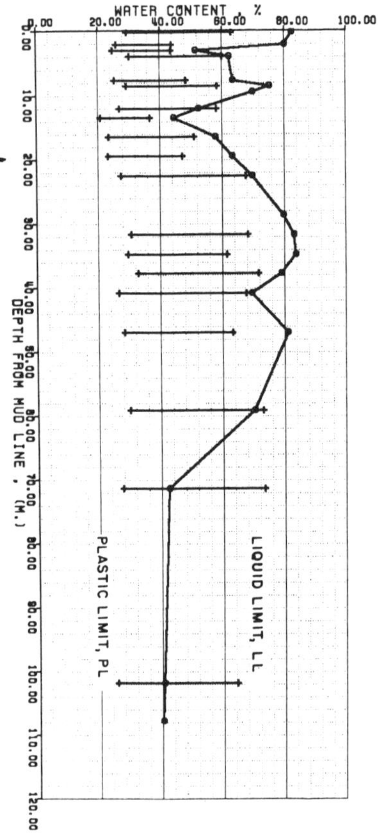

Fig. 2 Water content and plasticity profile.

Dynamic Experiments

The dynamic properties of the samples were determined using a Drnevich resonant column apparatus operating in the torsional vibration mode. Details of test procedures and data analysis are available from published sources (9,12,21). Samples, recovered in hard plastic tubes of outside diameter 5.74 cm and inside diameter of 5.37 cm, were trimmed to diameter and length of 3.80 cm and 7.50 cm, respectively. All the samples, after being properly mounted in the device, were confined at their calculated effective overburden stresses for a period of 24 hours under undrained conditions. The following list of reasons accounted for this practice.

1. Failure or yielding of the material in-situ have been theoretically assumed to occur under undrained conditions since rapid cyclic loading of clay does not allow pore pressure dissipation readily.
2. The recovered samples were presumed to be only slightly disturbed from the original site conditions. This conclusion was derived due to the insensitive nature of the material (sensitivity ratio runs from 1 to 4).
3. In order to determine the initial stiffness, presumed to exist in-situ, the samples should not be consolidated. Otherwise, consolidation would naturally stiffen thus strengthen the soft soil and tests would exhibit higher initial stiffness moduli values unrepresentative of in-situ conditions.

Resonant column test data resulted in:

1. A series of shear modulus and damping ratio versus strain amplitude correlations for samples of various depths along the soil column.
2. Shear modulus versus number of cycles of loading correlations at various strain amplitudes for a few number of samples to observe the relative significance of degradation of soil stiffness.
3. A series of cyclic stress-strain diagrams and depth profiles of shear modulus at different strain amplitudes.
4. Families of G/s_u versus depth curves.
5. One normalized modulus reduction curve for all of the samples tested.

Evaluation of Resonant Column Test Data

Curve Fitting. Hyperbolic curves were fitted to the shear modulus versus shear strain data (13). Figures 3, 4 and 5 present typical fitted curves and the data points for three different depth samples on semi-logarithmic scale. Curve fitting was done using a hyperbolic formula of the form

$$G = \frac{\gamma - \gamma_1}{A' + B'\gamma} + G_1 \qquad (1)$$

where γ is the shear strain amplitude, G is the shear modulus, (γ_1, G_1) is a reference point in the actual data and A' and B' are the constants obtained through linear regression of the data (24). The G versus γ data were generated and plotted as $(\gamma - \gamma_1)/(G - G_1)$ versus γ and subjected to linear regression analysis which yielded the A' and B' constants. Table 1 presents the A' and B' values calculated and used in curve fitting procedure of G versus γ along with γ_1, G_1 and standard deviation values. A very useful outcome of curve fitting was being able to estimate G_{max} or initial tangent shear modulus corresponding to zero shear strain. As observed from equation (1) at zero shear strain G_{max} can be calculated with the proper value of A' and the corresponding coordinates of the reference point. In Figure 3, 4 and 5, G is normalized by G_{max}.

Hyperbolic curve fitting agreed well with most of the data. However, the firmer samples below 60 m of depth showed some

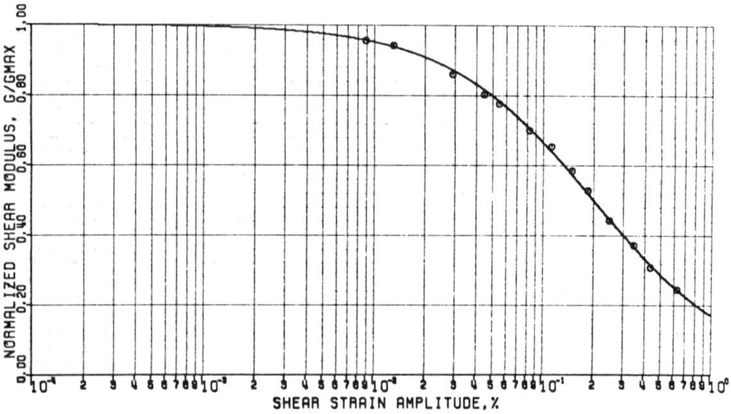

Fig. 3 The variation of shear modulus with strain amplitude, sample depth = 2.10 m, G_{max} = 962 kPa.

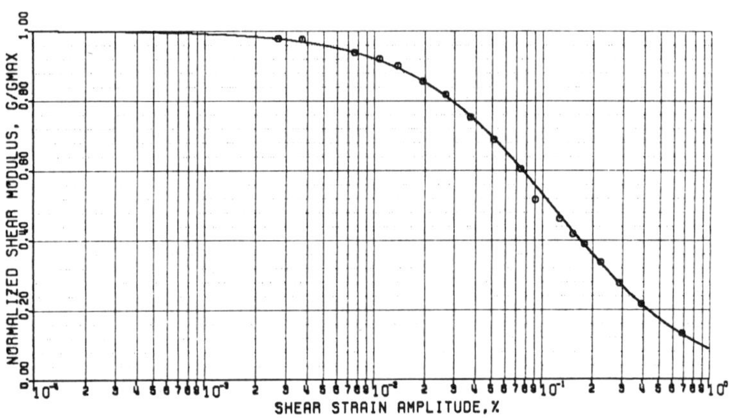

Fig. 4 The variation of shear modulus with strain amplitude, sample depth = 16.50 m, G_{max} = 3325 kPa.

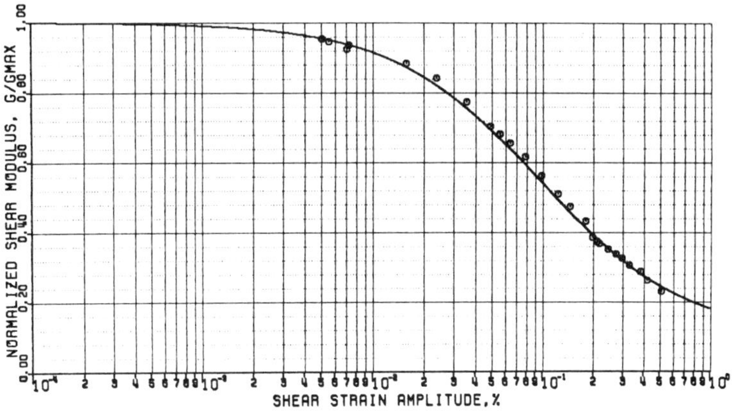

Fig. 5 The variation of shear modulus with strain amplitude, sample depth = 40.78 m, G_{max} = 2173 kPa.

Table 1

Regression Constants for Curve Fitting of G versus γ

Depth From Mudline (m)	x_1	y_1	A' (10^{-5})	B' (10^{-4})	Standard Deviation (10^{-5})
0.244	0.01433	912.00	-13.915	-11.490	1.442
2.100	0.00905	920.00	-21.830	-10.934	1.941
2.980	0.00618	1,279.00	- 1.788	- 9.654	1.224
3.900	0.00977	1,007.00	- 8.440	-13.370	3.458
7.560	0.00315	2,256.00	- 2.636	- 5.974	0.959
8.470	0.00359	2,229.00	- 3.908	- 4.520	0.582
9.390	0.00279	3,481.00	- 1.140	- 4.190	0.173
12.040	0.00499	1,726.00	- 5.796	- 6.168	1.040
13.350	0.00242	2,940.00	- 1.076	- 3.486	0.391
16.400	0.00272	3,249.00	- 3.607	- 3.010	0.212
19.440	0.00193	4,303.00	- 1.409	- 2.440	0.581
22.500	0.00317	2,413.00	-52.741	- 4.280	0.743
28.600	0.00419	1,777.00	- 7.408	- 6.215	1.755
31.640	0.00377	1,942.00	- 8.543	- 4.899	0.870
34.680	0.02292	1,361.00	- 8.807	- 7.364	0.633
37.730	0.00286	2,577.00	- 4.806	- 4.182	0.757
40.780	0.00511	2,075.00	- 5.195	- 5.397	0.941
46.880	0.01452	3,069.00	- 2.035	- 3.793	0.149
59.070	0.00312	4,853.00	- 2.164	- 2.331	0.267
71.320	0.000377	26,881.00	- 0.744	- 0.284	0.032
101.800	0.000515	36,531.00	- 0.037	- 0.536	0.019
107.650	0.000234	69,437.00	- 0.225	- 0.112	0.007

disagreement which may be attributed to the overconsolidation effects observed for these samples. The modulus reduction curves indicated some general trends. The shear modulus value at 0.0001 percent strain amplitude is approximately equal to G_{max}, and at 1.0 percent strain amplitude the shear modulus is reduced to about 10 to 20 percent of its maximum value. When plotted on normal scale and extended to high strain amplitude levels, the relationship between G and γ can be approximated as linear.

Various Representations of Shear Modulus Data. The modulus reduction curves were further used to develop a family of curves representing different strain amplitudes in a depth profile analysis of shear moduli, as illustrated in Figures 6a and 6b. In Figure 6a only the top 60 m of soil column is considered and shear moduli values corresponding to shear strain levels ranging from zero to 1.0 percent are presented. The increase in shear modulus between depths 10 to 20 m is consistent with the silty sandy formation at those depths. A sharp decrease in shear modulus below 20 m and a gradual increase between 40 and 60 m are also observed. For shear strains between 0.1 and 0.5 percent, shear moduli decrease more than other intervals for most depths. Figure 6b contains the same data for upper 60 m at a different scale for shear modulus with data of deeper samples superimposed. The sharp increase below 60 m indicates a transition into a layer of stiffer material. Since the data from the sand layer between 60 and 70 m was not available, the straight line approximation as shown may not be warranted. Compared to the strong variation in the firmer layer, the shear moduli for the top layer can safely be estimated to be constant for some ranges of depth.

Degradation of Shear Moduli with Number of Cycles of Loading. The degradation affects of number of cycles of loading is investigated by observing the change in G with number of cycles of loading (N) for a few randomly chosen samples. Figures 7, 8, 9, and 10 represent these relations. Some of the samples were loaded at single strain amplitude while others at several different amplitudes. Figure 9 shows that the degradation is more pronounced at higher shear strain loadings. However, in Figure 10, the curves seem to have identical slopes, showing that the degradation is approximately the same for both of those strain amplitudes of loading. In addition, curves for shallower samples (Figures 7 and 8) displayed a leveling effect at high number of cycles, which may correspond to a steady state or no further degradation after a critical value of N. Since the values between the first and the hundredth cycle were not measured, mainly due to experimental limitations, the range is represented by dotted lines in all curves and may not represent the actual variation.

Degradation in shear moduli between 100 to 10,000 cycles of loading ranged from 5 to 25 percent. However, 10,000 cycles corresponded to approximately 1.0 hour duration of loading while the total time involved during the resonant column measurements was only a few minutes. Thus, assuming degradation effects during testing to be negligible is warranted. Therefore, no corrections with regard to degradation were introduced into the results.

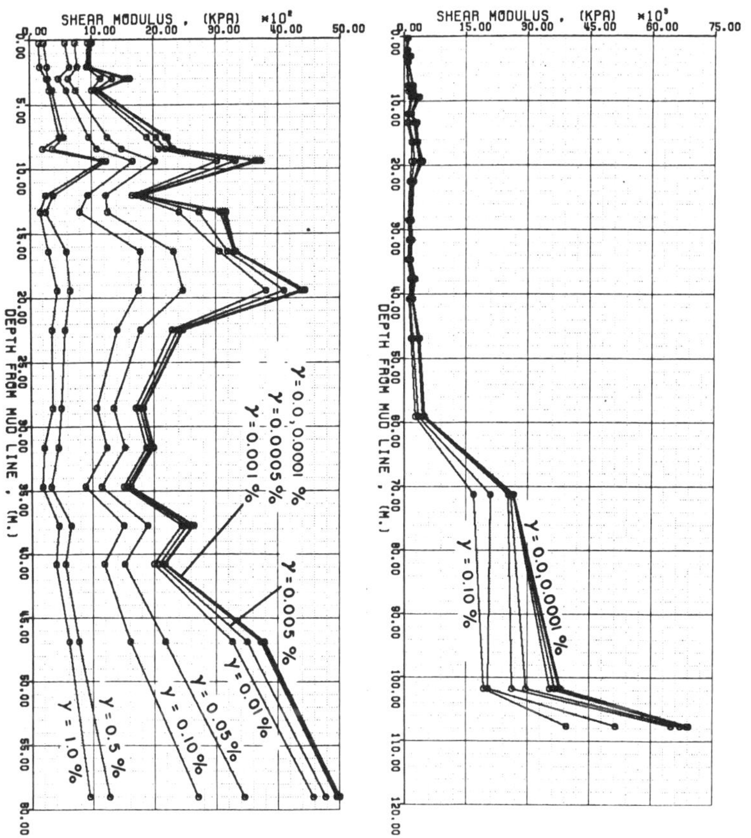

(a) (b)

Fig. 6 The variation of shear modulus with depth.

<u>Damping Ratio Versus Shear Strain Relations</u>. The initial resonant column data also yielded information about the variation of damping ratio, D with shear strain amplitude. Typical relations are given graphically in Figures 11 and 12. Curves of D versus γ displayed some interesting features that are not observed in curves of the same type in soil dynamics literature (11,13,28,34). These differences may be attributed to either,

1. Experimental scatter,
2. Method used to estimate the damping ratio, namely the magnification factor method (12), or
3. Soft and weak nature of the samples.

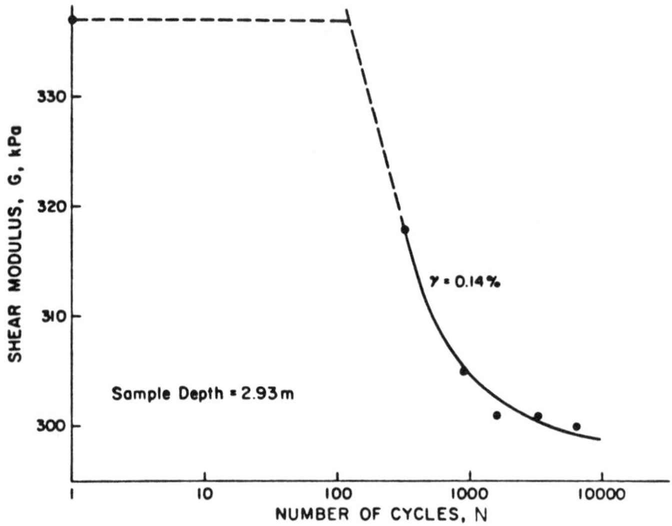

Fig. 7 Variation of shear modulus with number of cycles.

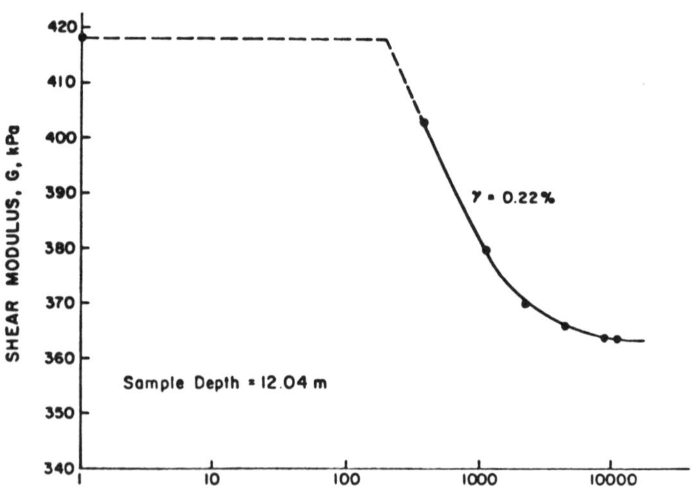

Fig. 8 Variation of shear modulus with number of cycles.

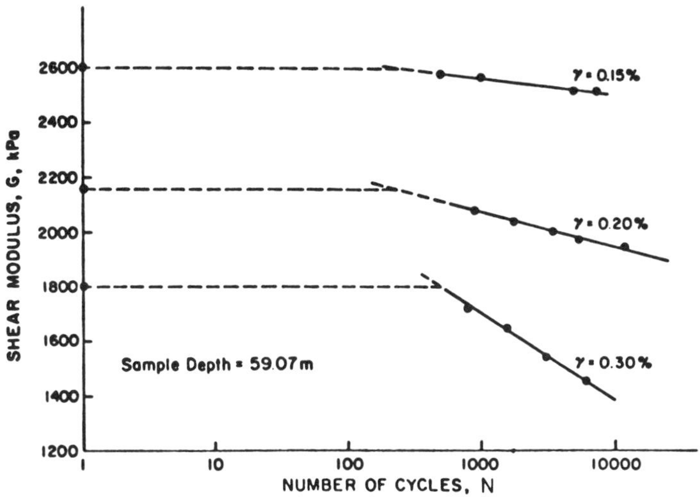

Fig. 9 Variation of shear modulus with number of cycles.

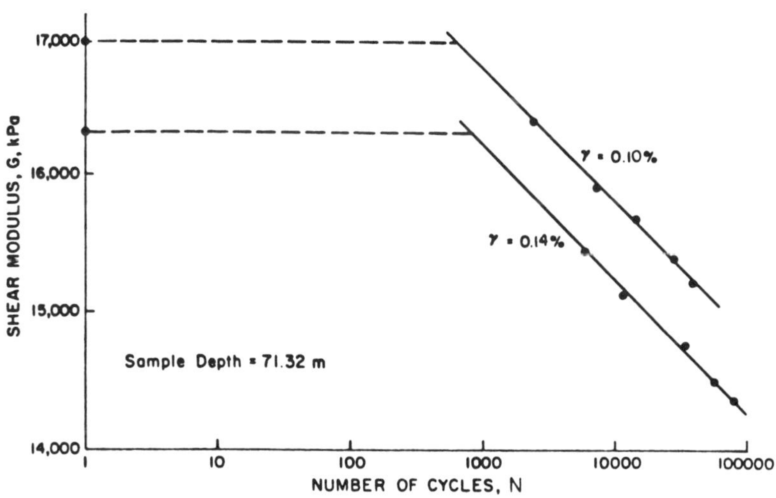

Fig. 10. Variation of shear modulus with number of cycles.

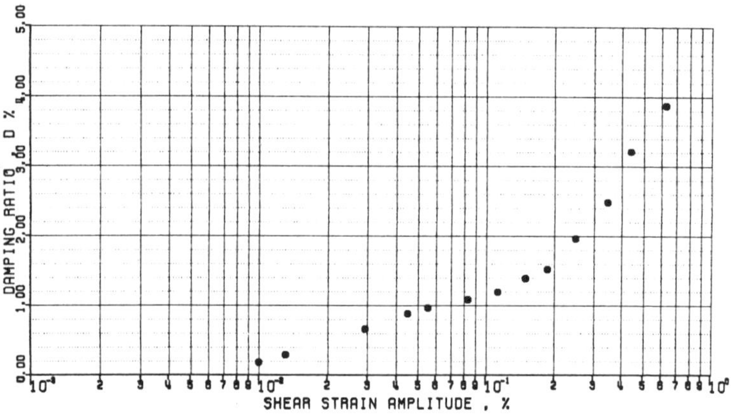

Fig. 11 The variation of damping ratio with strain amplitude, sample depth = 2.10 m.

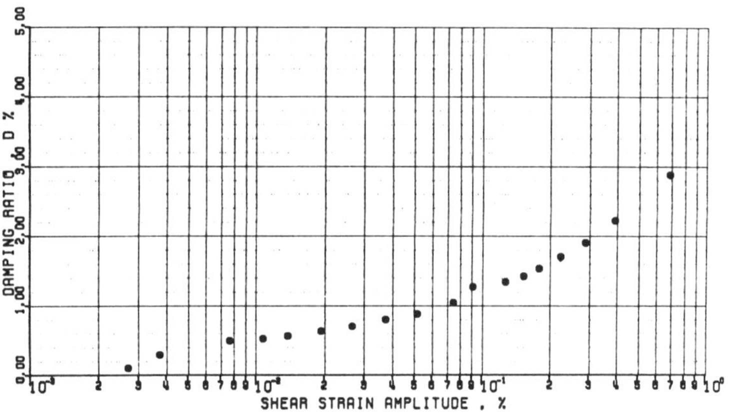

Fig. 12 The variation of damping ratio with strain amplitude, sample depth = 16.50 m.

The curves tended to have several inflection points and they did not have a noticeable independence of strain amplitude below 0.001 percent, as observed from Figures 11 and 12. However, they all increased with increasing shear strain amplitude. The maximum damping ratios at shear strains within the vicinity of 0.5 percent ranged from 0.6 to 20 percent with a mean value of 3.5 percent. The firmer samples displayed relations which were more consistent with D versus γ curves encountered in literature (34). The average damping ratio of 3.5 percent at 0.5 percent shear strain is regarded to be a rather low value with respect to the soft material tested. Ordinarily, they would be

Table 2

Regression Constants for Curve Fitting of τ Versus γ

Depth From Mudline (m)	A (10^{-4})	B (10^{-1})	Standard Deviation (10^{-5})
0.244	9.761	7.538	2.595
2.100	10.418	5.124	3.302
2.990	10.900	5.615	22.666
3.900	12.047	3.400	14.114
7.560	4.864	4.730	2.588
8.470	4.689	4.496	4.919
9.390	3.417	2.154	4.827
12.040	5.626	4.559	5.011
13.350	3.108	7.866	25.027
16.400	2.915	2.819	1.638
19.440	2.630	2.877	4.806
22.500	3.841	3.119	12.120
28.600	5.692	3.058	1.685
31.640	4.972	3.397	1.761
34.680	6.175	5.112	5.165
37.730	3.872	2.606	1.431
40.780	5.047	3.030	4.426
46.880	2.750	3.674	0.458
59.070	2.007	1.511	0.937
71.320	0.369	0.220	0.108
101.800	0.284	0.770	0.102
107.650	0.135	0.121	0.074

expected to have higher damping ratios. Since no comparisons are available from the literature for resonant column testing of these soft sediments at this time, the damping ratios obtained in this study should only be considered as tentative values.

Cyclic Stress-Strain Relationship. Resonant column test results yielded a third set of data. Using the general relation between shear modulus and shear strain, and the assumption made with regard to degradation of the backbone curve that the degradation of shear modulus with increasing number of cycles of loading during the resonant column test is negligible, cyclic shear stress versus shear strain amplitude relations were generated for each sample. Hyperbolic curve fitting techniques were used to describe a hyperbolic form relating τ and γ cyclic, where τ = shear stress amplitude and γ = shear strain amplitude. When γ/τ is plotted versus γ, a straight line with the following equation results:

$$\frac{\gamma}{\tau} = A + B\gamma \tag{2}$$

Through linear regression, the constants A and B were determined and fitted curves could be produced. Table 2 lists the A and B constants used in this curve fitting procedure along with the standard deviation values. Figures 13 and 14 show typical fitted cyclic stress-strain curves and the actual data points associated with them. Most of the data were represented well by the curves.

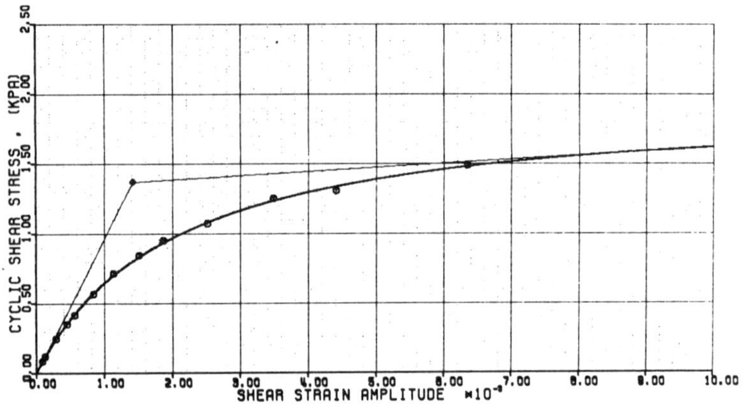

Fig. 13 The variation of shear stress with strain amplitude, sample depth = 2.10 m.

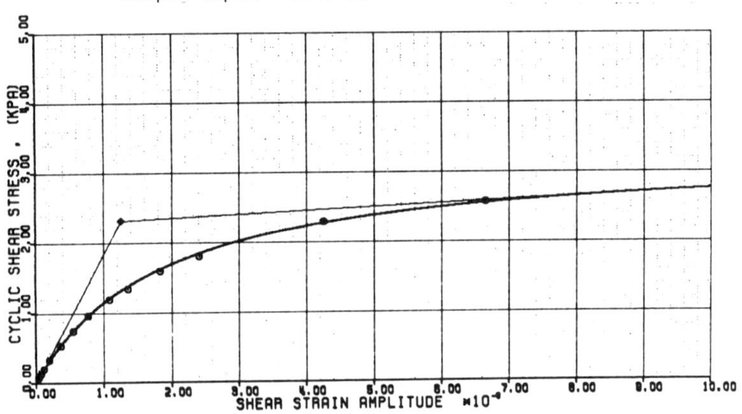

Fig. 14 The variation of shear stress with strain amplitude, sample depth = 28.59 m.

The stress-strain data was re-analyzed using a material model with strain-hardening characteristics, namely Von Mises material model. The observed "bulging" failure mode of most samples in the undrained triaxial tests strongly indicated yielding. One of the well known theories by which the failure of soils by yielding are generally expressed is Von Mises criterion. Most of the solutions to problems involving large-displacements and large-strains utilize yielding characteristics of the materials (1). Hence, in addition to the unique failure model of the soil samples, utilizing the data in a large-strain numerical solution for mudflows in a later study justified the choice of data analysis with a yield criterion. The bilinear representations are also illustrated in Figures 13 and 14. The yield point was designated (γ_y, τ_y), namely cyclic shear strain and

Table 3

Shear Strength Data of Core Samples

Depth From Mudline (m)	Max. Shear* Modulus G_{max} (kPa)	Max. Cyclic* Shear Strength τ_{max} (kPa)	Cyclic Yield* Shear Strength τ_y (kPa)	Undrained+ Shear Strength s_u (kPa)
0.244	1,015.00	1.73	1.04	5.40
2.100	961.50	1.95	1.37	8.06
2.980	1,624.80	1.78	1.27	6.45
3.900	1,122.80	2.94	1.70	8.12
7.560	2,232.50	2.11	1.75	7.50
8.470	2,320.70	2.22	1.81	---
9.390	3,725.40	4.64	3.53	10.16
12.040	1,812.00	2.19	1.71	9.26
13.350	3,165.00	1.27	1.17	13.41
16.400	3,324.50	3.55	2.89	11.80
19.440	4,439.70	3.48	2.94	14.60
22.500	2,473.00	3.21	2.54	---
28.600	1,833.50	3.27	2.32	---
31.640	1,986.00	2.94	2.23	9.70
34.680	1,621.00	1.95	1.54	7.14
37.730	2,636.50	3.84	2.93	8.91
40.780	2,173.30	3.30	2.43	---
46.880	3,782.70	2.72	2.35	11.77
59.070	4,997.00	6.62	2.12	17.22
71.320	26,475.50	45.62	33.25	102.31
101.800	37,918.00	---	---	---
107.650	69,163.30	82.55	66.88	143.20

*Values from Resonant Column Test
+Values from Undrained Shear Strength Test

cyclic shear stress at yield. In addition to Von Mises material model representation, the maximum cyclic shear stress τ_{max} was also evaluated using the hyperbolic relationship where the reciprocal of constant B was equal to τ_{max}. It was found that τ_{max} values did not deviate significantly from the τ_y values estimated graphically.

Shear strength analysis through conventional unconsolidated undrained triaxial tests yielded the undrained shear strength, s_u, for a number of samples. The complete shear strength data, along with the G_{max} values are listed in Table 3. The undrained shear strength values are considerably higher than both of the cyclic shear strength values as observed from the results in Table 3. These results are in disagreement with the findings of Whitman (38) who showed that for clays cyclic shear strength values are higher than static shear strength values. One of the possible causes of the contradiction may be due to the different experimental procedures used. Resonant column test data are limited to small strain amplitudes and therefore is a unique procedure compared to static tests or even other types of cyclic tests that can be extended to much higher strain amplitudes of loading.

Using the shear strength data, the family of curves for shear moduli depth profile were normalized and recompiled. The results are given in Figures 15a and 15b. Once again, Figure 15a shows the family of G/s_u versus depth curves for the top 60 m of the soil column and Figure 15b shows the same curves on a different scale for the total

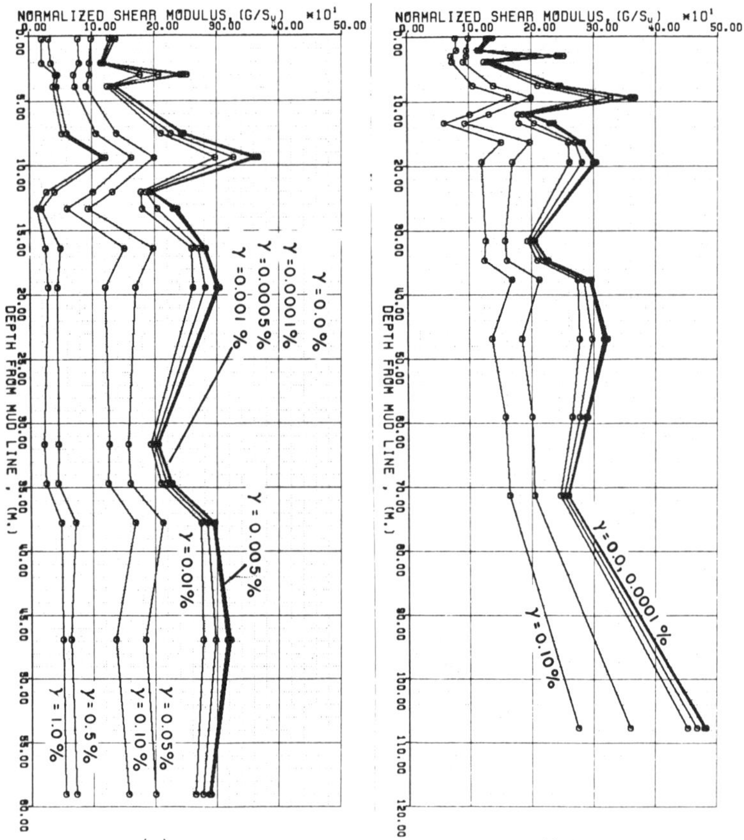

(a) (b)

Fig. 15 The variation of normalized shear modulus with depth.

depth. It is observed that G/s_u ratios for zero percent shear strain amplitude ranges from 140 to 470 for the entire depth of material. The sharp change in material property is not observed below 60 m of depth when G values are normalized by s_u. G/s_u ratios almost as high as the deeper values are also encountered around 10 m of depth. The lowest values of G/s_u correspond generally to near the surface of the soil column, while for low shear strain amplitudes the rest of the data tends to range between 200 and 300 with a mean value of 250 for the top 70 m.

The same normalization was done with τ_y values also. As expected, the G/τ_y values were much higher that G/s_u values. The new

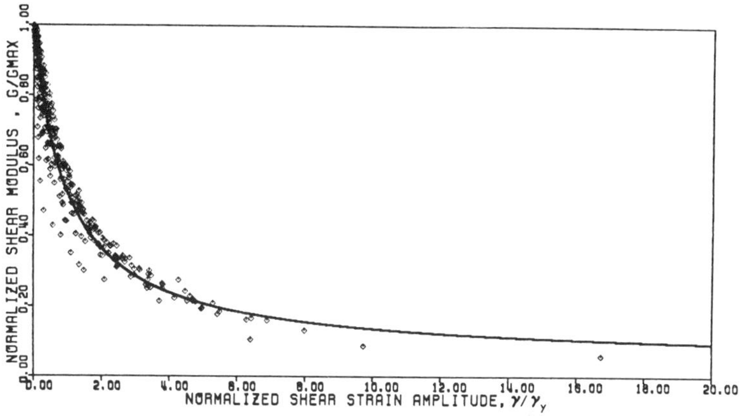

Fig. 16 The variation of shear modulus with strain amplitude.

set of data ranged between 600 and 2,700 for the low shear strain amplitudes, with a mean value of 1,000.

One yield state parameter, γ_y , shear strain at yield, obtained from the bilinear representation of the nonlinear shear stress-strain curves was used for normalizing shear strain in order to represent all the shear modulus data with a single curve. The procedure was similar to that used by Hardin and Drnevich (13). However, in that work shear strain was normalized by reference strain, γ_r defined as τ_{max}/G_{max}. Figure 16 illustrates the results of this new procedure. By normalizing the shear modulus data in the form of G/G_{max} and γ/γ_y, the complete data could be combined on a single graph. The curve through the data points is a hyperbolic fit by the method discussed earlier.

CONCLUSIONS

A number of conclusions were drawn from the study presented herein. Mudflows in the Gulf of Mexico are potentially hazardous to the offshore structures located in the area. Foundation analysis for these structures are conducted with regard to the dynamic properties of sea-floor soils and continued research provides different solutions to obtain the design parameters. The existing design methods are limited either by inability to define the actual conditions closely or by lack of reliable data. In this study, the effort mainly concentrated on presenting more realistic data which can further be used in numerical modeling to understand and predict mudflows.

The resonant column apparatus can be effectively used to obtain dynamic properties of marine sediments in the laboratory. The G_{max}/s_u ratios obtained through this method tended to range between 140 to 470 for Boring B-6 Main Pass Block 70 in the Gulf of Mexico, from the mud surface to a depth of 110 m of soil column. The average value of the ratio was estimated to be 250 for the top 70 m of the material. The maximum cyclic shear strength was estimated to be approximately 3.0 kPa, while the yield cyclic shear strength was around 2.0 kPa for the upper 60 m of soil column. On the other hand, the undrained shear strength, measured by triaxial tests, ranged between 5.0 to 14.0 kPa for the same column of material.

Dynamic shear modulus values were between 900 and 5,000 kPa for the top 60 m and increased sharply to much higher values as much as 70,000 kPa at around 110 m of depth from mudline. The missing data from between 60 to 70 m created some difficulties in assessing the transition between the two distinct zones of stiffness.

Note that the G/s_u variation for the total depth of the material tested is not as large as the variation of dynamic shear modulus of the same depth of material. The former range is between 140 to 470 while the latter is between 900 and 70,000 kPa. The relatively small variation of dynamic shear modulus for the top 60 m of soil layer is significant. It shows that a great thickness of material overlaying a much stiffer base can be assumed to be relatively homogeneous in terms of mechanical behavior. The transition from the softer material to the stiffer base is not evident with G/s_u versus depth representations (Figure 15b). Another result to be noted is the large difference between G_{max}/s_u and G_{max}/τ_y values, the mean of which are 250 and 1,000, respectively. Note that G/τ_y is the ratio of two dynamic measurements, while G/s_u is the ratio of a dynamic and a static measurement.

The damping ratio of the upper soft sediments had a mean value of 3.5 percent at 0.5 percent shear strain amplitude. The average minimum damping ratio for the entire material of Boring B-6 was estimated to be approximately 0.15 percent. These values were found to be low with respect to the soft nature of the soil samples tested.

The number of cycles (N) versus G curves were used in order to predict if any significant amount of degradation occurred during the cycling of the soft samples for shear moduli measurements. It was found that the maximum degradation in shear moduli for high strain amplitude loadings was not above 25 percent for 10,000 cycles, which in turn corresponded to about 1.0 hour of duration of loading. The strain amplitudes were large and the cycling frequencies were on the order of 2 to 7 Hz for these sets of tests.

Sample disturbance is a handicap and must be acknowledged. Confining the samples for 24 hours prior to testing proved to be a sensible way to partially overcome disturbance effects because all of the samples displayed 5 to 15 percent higher shear modulus values then the initial values after confinement. This behavior cannot readily be explained since the samples were undrained during confinement, but may be attributed to the gaseous nature of the soil samples. Another complicating effect on the test results is the lack of data on excess pore water and gas pressures that are known to exist at high levels in the sediments of the Mississippi Delta. Therefore, the actual effective pressure is predicted to be lower than the calculated effective overburden pressure (i.e., total stress minus hydrostatic stress) used to confine the samples. Hence, the softening effect of sample disturbance is once more compensated by the use of higher confining pressures. However, these procedures are approximate and actual measurements of in-situ pore water and gas pressures are needed in order to make better assumptions and obtain more realistic results.

The overall effort of this study has been to provide a set of dynamic property data of the soft, underconsolidated sediments of the

Gulf of Mexico and to highlight the experimental and analytical methods used to obtain them. These data can be used in existing numerical analysis methods to simulate the mechanics of mudflows by their static and kinematic variables dependent on the depth from the mudline (1,22,23).

ACKNOWLEDGEMENT

The financial assistance provided by the Office of Naval Research and the Environmental Protection Agency is gratefully acknowledged. We would also like to thank the oil company which provided the core samples.

REFERENCES

1. Bathe, K. J., Wilson, E. L., and Iding, R. H., "NONSAP, A Structural Analysis Program for Static and Dynamic Response of Nonlinear Systems," Report No. UCSESM 74-3, Structures and Materials Research Department of Civil Engineering Division of Structural Engineering and Structural Mechanics, University of California, Berkeley, Feb., 1974.
2. Bennett, R. H., and Faris, J. R., "Ambient and Dynamic Pore Pressures in Fine Grained Submarine Sediments: Mississippi Delta," Applied Ocean Research, Vol. 1, No. 3, 1979, pp. 115-123.
3. Booker, R. J., Rahman, M. S., and Seed, H. B., "GADFLEA - A Computer Program for the Analysis of Pore Pressure Generation and Dissipation During Cyclic or Earthquake Loadings," Report No. EERC 76-24, Earthquake Engineering Research Center, University of California, Berkeley, Oct., 1976.
4. Bowles, J. E., Foundation Analysis and Design, McGraw-Hill Book Company, New York, NY, 1977.
5. Castro, G., "Liquefaction and Cyclic Mobility of Saturated Sands," Journal of the Geotechnical Engineering Division, ASCE, No. GT6, Proc. Paper 11388, June, 1975, pp. 511-569.
6. Castro, G., and Poulos, S. J., "Factors Affecting Liquefaction and Cyclic Mobility," Journal of the Geotechnical Engineering Division, ASCE, No. GT6, Proc. Paper 12994, June, 1977, pp. 501-516.
7. Coleman, J. M., Suhayda, J. N., and Dawson, T. H., "Determination of Elastic Shear Modulus of Marine Sediments from Wave Theory and Field Measurements," 12th Annual Offshore Technical Conference, OTC 3811, May 5-8, 1980.
8. Dawson, T. H., Suhayda, J. N., and Coleman, J. M., "Correlation of Field Measurements with Elastic Theory of Sea-Floor Response to Surface Waves," 13th Annual Offshore Technology Conference, OTC 3973, May 4-7, 1981.
9. Drnevich, V. P., "Resonant Column Testing - Problems and Solutions," Dynamic Geotechnical Testing, ASTM Special Technical Publication 654, pp. 384-398.
10. Dumas, F., and Lee, K. L., "Cyclic Movements of Offshore Structures on Clay," Journal of the Geotechnical Engineering Division, ASCE, No. GT8, Proc. Paper 15655 , August, 1980, pp. 877-897.

11. Hardin, B. O., "The Nature of Damping in Sands," Journal of the Soil Mechanics and Foundation Division, ASCE, No. SM1, January, 1965, pp. 63-97.
12. Hardin, B. O., "Suggested Methods of Test for Shear Modulus and Damping of Soils by the Resonant Column," ASTM STP 479, American Society for Testing and Materials, 1970, pp. 516-529.
13. Hardin, B. O., and Drnevich, V. P., "Shear Modulus and Damping in Soils: Design Equations and Curves," Journal of the Soil Mechanics and Foundations Division, ASCE, No. SM7, Proc. Paper 9006, July, 1972, pp. 667-692.
14. Herman, H. G., and Houston, W. N., "Response of Sea-Floor Soils to Combined Static and Cyclic Loading," Eighth Annual Offshore Technology Conference, OTC 2428, May 3-6, 1976.
15. Idriss, I. M., Dobry, R., and Singh, R. D., "Nonlinear Behavior of Soft Clays During Cyclic Loading," Journal of the Geotechnical Engineering Division, ASCE, No. GT12, Proc. Paper 14265, December, 1978, pp. 1427-1447.
16. Kagawa, T., and Kraft, L. M., Jr., "Seismic P-Y Responses of Flexible Piles," Journal of the Geotechnical Engineering Division, ASCE, No. GT8, Proc. Paper 15656, August, 1980, pp. 889-918.
17. King, J. B., Dunlap, W. A., and Schapery, R. A., "Characterization of Viscoelastic Properties of Submarine Sediments," A Report from Mechanics and Materials Research Center, Texas Engineering Experiment Station, Texas A & M University, April, 1976.
18. Lee, K. L., and Focht, J. A., "Liquefaction Potential at Ekofisk Tank in North Sea," Journal of the Geotechnical Engineering Division, ASCE, No. GT1, Proc. Paper 11054, Jan., 1975, pp. 1-18.
19. Martin, G. R., Lam, I., and Tsai, Chan-Feng, "Pore Pressure Dissipation During Offshore Cyclic Loading," Journal of the Geotechnical Engineering Division, ASCE, No. GT9, Proc. Paper 15706, Sept. 1980, pp. 981-995.
20. Mitchell, J. K., Tsui, K. K., and Sangrey, D. A., "Failure of Submarine Slopes Under Wave Action," Proceedings 13th Coastal Engineering Conference, Vancouver, Vol. II, 1972, pp. 1515-1542.
21. Pamukcu, Sibel, "Dynamic Analysis of Sediments of Gulf of Mexico and Numerical Simulation of Mud Flows," M.S. Thesis, Civil Engineering Department, Louisiana State University, Baton Rouge, Dec., 1981.
22. Prior, D. B., and Suhayda, J. N., "Application of Infinite Slope Analysis to Subaqueous Sediment Instability, Mississippi Delta," Engineering Geology, Vol. 14, 1979, pp. 1-10.
23. Rahman, M. S., Seed, H. B., and Booker, J. R., "Pore Pressure Development Under Offshore Gravity Structures," Journal of the Geotechnical Engineering Division, ASCE, No. GT12, Proc. Paper 13411, Dec., 1977, pp. 1419-1435.
24. SAS User's Guide, SAS Institute, Inc., 1979.
25. Schapery, R. A., "Wave-Sea Bottom Interaction Study (Phase One), Part I: Theory and Results," A Report from Mechanics and Materials Research Center, Texas Engineering Experiment Station, Texas A & M University, March, 1974.
26. Schapery, R. A., "Wave-Sea Bottom Interaction Study (Final Report), Part I: Theory and Results," A Report from Mechanics and

Materials Research Center, Texas Engineering Experiment Station, Texas A & M University, March, 1974.
27. Schapery, R. A., Riggins, M., and Dunlap, W. A., "Movement of Marine Clay Sediments During Cyclic Loading," A Report from Mechanics and Materials Research Center, Texas Engineering Experiment Station, Texas A & M University, March, 1974.
28. Seed, H. B., and Idriss, I. M., "Soil Moduli and Damping Factors for Dynamic Response Analysis," Report No. EERC 70-10, University of California, Berkeley, Sept., 1970.
29. Seed, H. B., and Rahman, M. S., "Analysis of Wave-Induced Liquefaction in Relation to Ocean Floor Stability," Report No. UCB/TE-77/02, Department of Civil Engineering, Institute of Transportation and Traffic Engineering, University of California, Berkeley, May, 1977.
30. Singh, R. D., Dobry, R., Doyle, E. H., and Idriss, I. M., "Nonlinear Seismic Response of Soft Clay Sites," Journal of the Geotechnical Engineering Division, ASCE, No. GT9, Proc. Paper 16493, Sept., 1981, pp. 1201-1218.
31. Sterling, J. H., Jr., and Bryant, W. R., "Engineering Properties of Shallow Sediments in West Delta and South Pass Outer Continental Shelf Lease Areas, Offshore Louisiana," Technical Report 77-4-T, Department of Oceanography, Texas A & M University, May, 1977.
32. Sterling, G. H., and Stohbeck, E. E., "The Failure of South Pass 70 "B" Platform in Hurricane Camille," Fifty Annual Offshore Technology Conference, OTC 1898, April 29 - May 2, 1973.
33. Stevenson, H. S., "Vane Shear Determination of Viscoelastic Shear Modulus of Submarine Sediments," M.S. Thesis, Civil Engineering Department, Texas A & M University, College Station, Dec., 1973.
34. Stokoe, K. H., II, "Dynamic Properties of Offshore Silty Samples," 12th Annual Offshore Technology Conference, OTC 3771, May 5-8, 1980.
35. Suhayda, J. N., and Prior, D. B., "Explanation of Submarine Landslide Morphology by Stability Analysis and Rheological Models," Tenth Annual Offshore Technology Conference, OTC 3171, May 8-11, 1978.
36. Thiers, G. R., and Seed, H. B., "Strength and Stress-Strain Characteristics of Clays Subjected to Seismic Loading Conditions," Vibration Effects of Earthquakes on Soils and Foundations, ASTM STP 450, American Society of Testing and Materials, 1968.
37. Tsai, Chen-Feng, Lam, I., and Martin, R. M., "Seismic Response of Cohesive Marine Soils," Journal of the Geotechnical Engineering Division, ASCE, No. GT9, Proc. Paper 15708, Sept., 1980, pp. 997-1012.
38. Whitman, R. V., "The Response of Soils to Dynamic Loadings," Defense Atomic Support Agency, U.S. Army Waterways Experiment Station, Report No. 26, 1970.
39. Wright, S. G., "Analysis for Wave Induced Sea-Floor Movements," Eighth Annual Offshore Technology Conference, OTC 2427, May 3-6, 1976.

LIST OF SYMBOLS

A', B' - Hyperbolic equation constants
A, B - Hyperbolic equation constants
D - Damping ratio
G - Shear modulus
G_1 - Reference shear modulus
G_{max} - Maximum shear modulus
I_p - Plasticity index
K_o - Coefficient of lateral earth pressure
N - Number of cycle of loading
s_u - Undrained shear strength
γ - Shear strain
γ_1 - Reference shear strain
γ_r - Reference shear strain
γ_y - Shear strain at yield
μ - Poisson's ratio
τ - Shear stress
τ_{max} - Maximum shear stress
τ_y - Shear stress at yield

DYNAMIC FLUID-STRUCTURE-SOIL INTERACTION

Jean H. Prevost[(1)], M. ASCE, Thomas J.R. Hughes[(2)], M.ASCE

ABSTRACT

A general finite element model is proposed for dynamic fluid-structure-soil foundation interaction problems. The procedure allows for a realistic and optimal modeling of (1) the fluid subdomain and its free surface; and (2) the two-phase saturated soil foundation.

1. INTRODUCTION

The need for new oil and gas supplies has led to the recent development of numerous new offshore fields. At many of these sites, the waves are larger and the water depths are greater than in areas previously explored. Further, many of the sites are earthquake prone. Fixed offshore structures have thus had to be designed to withstand these difficult site conditions.

Seismic events and major storms may develop high dynamic forces in the fluid-structure-foundation interacting system and it is of importance that they be predicted accurately for design purposes. The complexity of the problem dictates that the method of analysis be numerical in nature. To make accurate predictions, a realistic analytical model of the complete fluid-structure-foundation system is essential. The behavior of the foundation soil is highly nonlinear and path-dependent, and it is of importance that such behavior be taken into account. Further, soil is a two-phase system consisting of a solid (soil skeleton) and a fluid (pore-fluid) phase. Fluid saturation of an otherwise inviscid porous soil skeleton introduces a time dependence into the response to applied loads. A realistic modeling of the soil foundation therefore also requires that the effect of the transient flow of the pore-fluid through the soil skeleton voids be incorporated.

(1) Associate Professor, Department of Civil Engineering, Princeton University, Princeton, NJ 08544.

(2) Professor, Division of Applied Mechanics, Department of Mechanical Engineering, Stanford University, Stanford, CA 94305.

The capability to solve the dynamic fluid-structure-foundation interaction problem therefore depends upon the ability to realistically model (1) the surrounding fluid domain; (2) the two-phase soil foundation system; and (3) upon the ability to interface the fluid, structural and soil subdomains. It is the purpose of this paper to define an appropriate finite element model for the fluid-structure-foundation system.

2. FLUID DOMAIN

In modeling the fluid, either Lagrangian or Eulerian descriptions are possible, in general. A Lagrangian description is often preferred for constrained fluid problems in which only small motions take place. On the other hand, an Eulerian description is best suited for problems in which the mesh would become too distorted if required to follow the flow. However, in fluid-structure-foundation interaction problems, it is necessary to have a more versatile description of the fluid domain. This stems from the requirements of (1) realistically modeling the free-surface; and (2) appropriately interfacing with the structural and foundation subdomains, which are naturally treated in a Lagrangian mode. This had been achieved in finite-difference codes by arbitrary Lagrange-Euler formulations (see e.g. [1]), and a great deal of research in finite elements has recently been devoted to the development of appropriate schemes for fluid-structure interaction (see e.g. (2,3,5]).

The surrounding fluid domain is hereafter described by using a mixed Euler-Lagrange formulation developed by Hughes and co-workers [5] in which each degree of freedom at each node may be selectively treated in an Eulerian or Lagrangian fashion. The procedure allows for an optimal representation of the free-surface by allowing horizontal degrees of freedom to be treated in an Eulerian mode and vertical degrees of freedom in a Lagrangian mode. It also allows for appropriate interfacing with structural and foundation subdomains. Details about the numerical implementation of the mixed Euler-Lagrange formulation are reported in [5,7].

3. FOUNDATION SOIL DOMAIN

Soils consist of an assemblage of particles with different sizes and shapes which form a skeleton whose voids are filled with water and air or gas. The word "soil", therefore, implies a mixture of assorted mineral grains with various fluids. Hence, soil in general must be looked at as a two (saturated soils) or multiphase material whose state is to be described by the stresses and displacements (velocities) within each phase. Fluid saturation of an otherwise inviscid porous soil skeleton introduces a time dependence into the response to applied loads. When the rate of loading is much smaller than the pore-fluid diffusion rate, the pore fluid pressure remains constant in each material element and the response is said to be "fully drained". In that case, the pore fluid pressures only depend on the hydraulic conditions and are independent of the soil skeleton response to external loads, and a single phase description of soil media is therefore adequate. Similarly, a single phase description of soil media is also adequate when the loading rate is much higher than the diffusion rate. In that case, no flow of the pore-fluid can take place, the fluid mass content in each material element remains constant, and the fluid follows the motion of the soil skeleton. The response is then said to be "fully undrained". However, in intermediate cases in which some flow can take place, there is an interaction between the soil skeleton strains and the pore-fluid flow. The solution of these problems requires that the saturated soil medium be analyzed by incorporating the effects of the transient flow of the pore-fluid through the soil skeleton voids, and therefore requires that a two-phase continuum formulation be used for the soil medium.

The foundation soil domain is hereafter described as a two-phase system consisting of a solid and a fluid phase [10]. Each phase is regarded as a continuum, and each follows its own motion. Details about the model and its numerical implementation are reported in [11,12].

4. TIME INTEGRATION OF THE SEMI-DISCRETE FINITE ELEMENT EQUATIONS

Direct time integration of the resulting (nonlinear) semi-discrete finite element equations is achieved by using the implicit-explicit algorithm proposed by Hughes, et al [4], based on the Newmark family of methods [9].

5. NUMERICAL EXAMPLES

The procedures described previously have been incorporated into the computer code DYNA-FLOW [11]. In the following, a few simple examples are presented which illustrate the versatility of the proposed procedure in solving dynamic fluid-structure-soil interaction problems. In these calculations, four nodes bilinear isoparametric plane elements (see e.g. [13] for details) are used for the fluid and the foundation soil domains. The structure is modelled by using beam elements. Fluid and solid subdomains are interfaced by using sliding elements [6].

Figure 1 shows the two-dimensional finite element idealization of an offshore structure and its soil foundation. The offshore structure is modelled by using 8 beam elements and is embedded in its soil foundation over a depth $d = 1.35\ \mathrm{m}$. The material properties are as follows:

BEAM:		
	$E = 2.0 \times 10^{10}$ Pa	(Young's modulus)
	$\gamma = 0.30$	(Poisson's ratio)
	$\rho = 8.0 \times 10^{3}\ \mathrm{kg/m^3}$	(Mass Density)
	$I = 1.0\ \mathrm{m}^4$	(Bending Moment of Inertia)
	$A = 1.0\ \mathrm{m}^2$	(Cross Sectional Area)

TABLE 1

Soil-Structure Interacting System:
Eigenfrequencies

	Drained Soil		Undrained Soil	
Mode Number	Frequency (Hz)	Period (sec)	Frequency (Hz)	Period (sec)
1	7.813×10^{-2}	12.80	8.201×10^{-2}	12.19
2	3.270	2.967×10^{-1}	3.945	2.535×10^{-1}

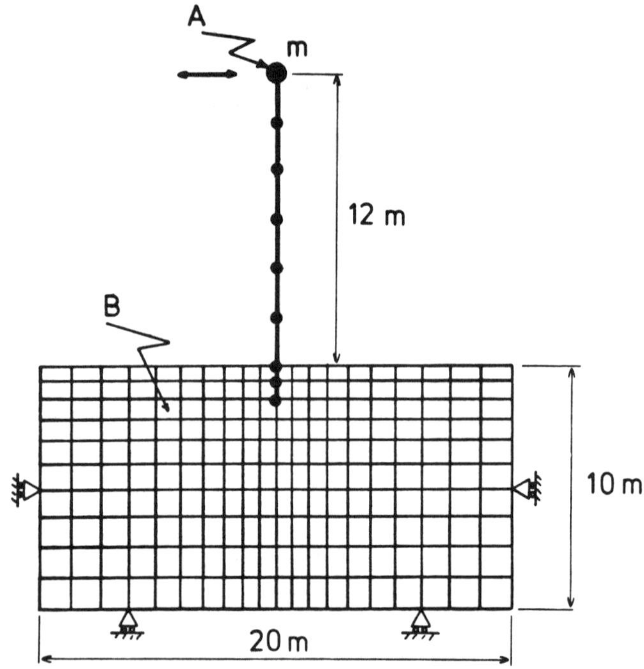

Figure 1

Soil-Structure Interacting System - Finite Element Mesh

The superstructure is modelled by adding a nodal mass $m = 1.0 \times 10^3$ kg at the top of the structure.

The soil foundation is modelled by using 200 four-node bilinear isoparametric elements, and the soil skeleton is idealized as linear elastic. The material properties are as follows:

SOIL:		
	$E = 1.0 \times 10^8$ Pa	(Young's Modulus, Solid Phase)
	$\gamma = 0.0$	(Poisson's Ratio; Solid Phase)
	$\rho_s = 4.0 \times 10^3$ kg/m^3	(Mass Density; Solid Phase)
	$\rho_w = 1.0 \times 10^3$ kg/m^3	(Mass Density; Fluid Phase)
	$n = 0.50$	(Porosity)
	$k = 10^{-4} - 10^{-10}$ m/s	(Permeability)

The resulting eigenfrequencies of the soil-structure interacting system, for both fully drained and undrained conditions in the soil foundation, are summarized in Table 1.

Figures 2 and 3 illustrate the influence of drainage conditions on the resulting response of the soil-structure interacting system. Loading is achieved by subjecting the superstructure to an impact and by letting the system free-vibrate thereafter. Fig. 2 shows the computed horizontal displacement time history of the top node (node A in Fig. 1) for various drainage conditions in the foundation. In Fig. 2a, $k = 10^{-10}$ m/s and no substantial drainage has time to take place within the time window of the plot. In Fig. 2b, $k = 10^{-4}$m/s and the resulting drainage highly damps the oscillations. Fig. 3 shows the computed excess pore-fluid pressure time histories at location B (Fig. 1) for $k = 10^{-10}$ m/s (Fig. 3a) and $k = 10^{-4}$ m/s (Fig. 3b), respectively.

Fig. 4 shows the two-dimensional finite element representation of a fluid-structure-soil interacting system. The offshore structure and its foundation are the same as the one previously described. A wave is generated in the fluid reservoir by prescribing the displacement time history of the fluid along the left-hand boundary of the fluid domain. The fluid is assumed inviscid. The fluid properties are as follows:

FLUID:		
	$\lambda = 1.0 \times 10^8$	(Penalty Parameter; see e.g. [8])
	$\mu = 0.0$	(Shear Viscosity)
	$\rho = 1.0 \times 10^3$ kg/m^3	(Mass Density)

The fluid domain is modelled by two rows of elements through the depth, and the split Euler-Lagrange used for the various fluid degrees

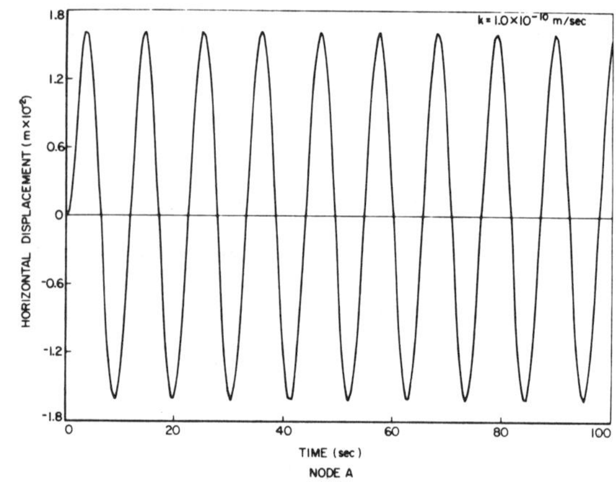

Figure 2a

Computed Horizontal Displacement Time History
(Node A)

$k = 1.0 \times 10^{-10}$ m/sec

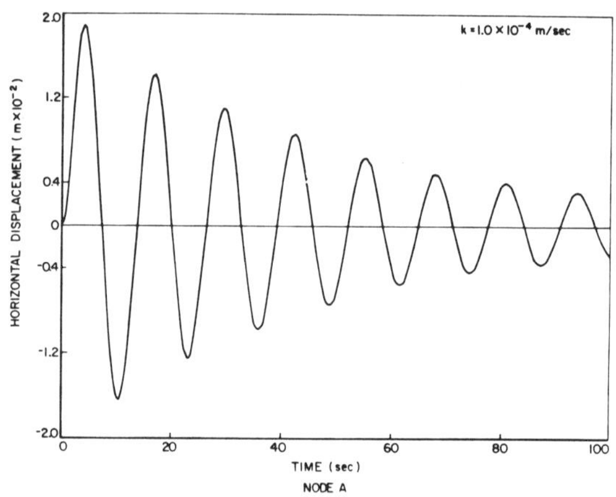

Figure 2b

Computed Horizontal Displacement Time History
(Node A)

$k = 1.0 \times 10^{-4}$ m/sec.

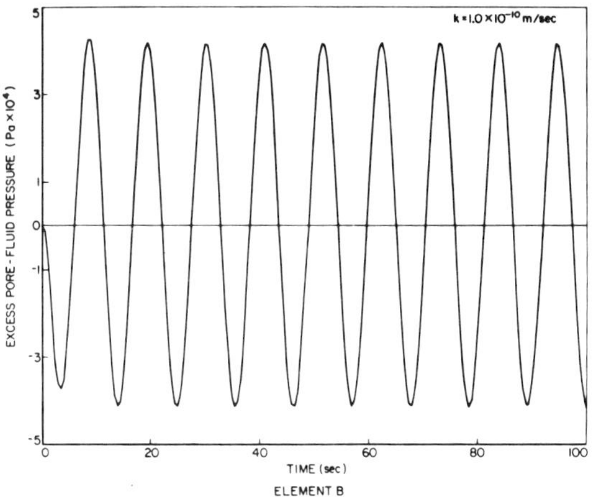

Figure 3a

Computed Excess Pore-Fluid Pressure Time History in Soil Foundation (Element B)

$k = 1.0 \times 10^{-10}$ m/sec

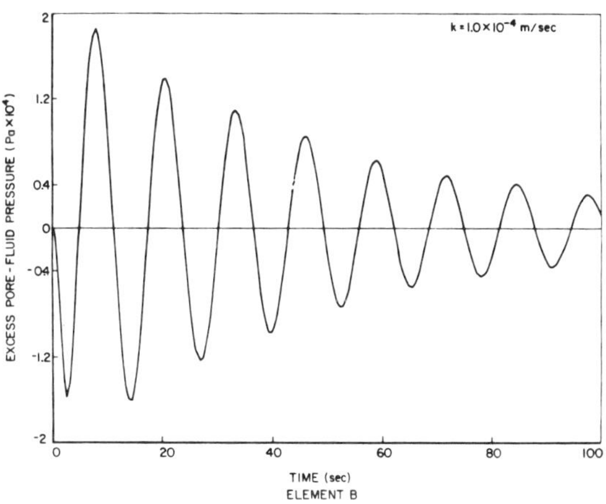

Figure 3b

Computed Excess Pore-Fluid Pressure Time History in Soil Foundation (Element B)

$k = 1.0 \times 10^{-4}$ m/sec

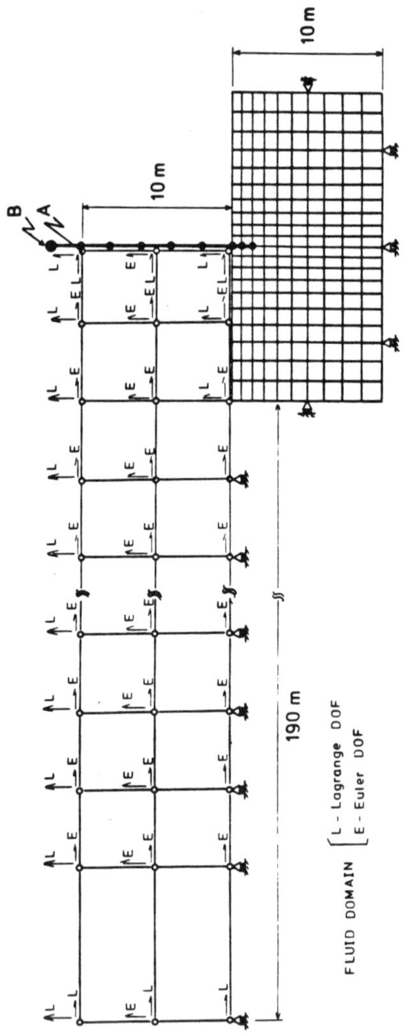

Figure 4

Fluid-Structure-Soil Interacting System
Finite Element Mesh

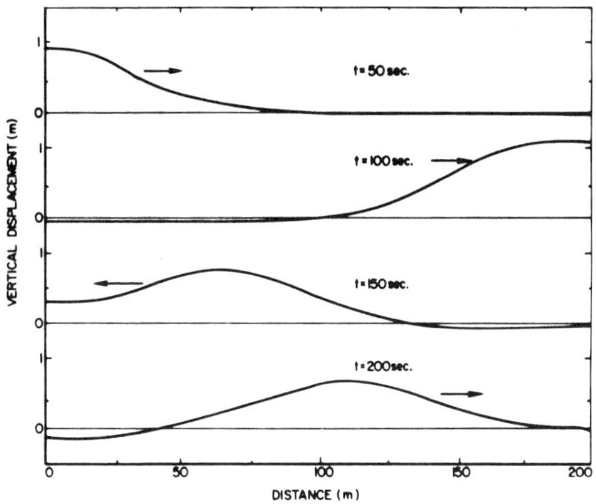

Figure 5

Free-Surface Profile at Various Instant of Time

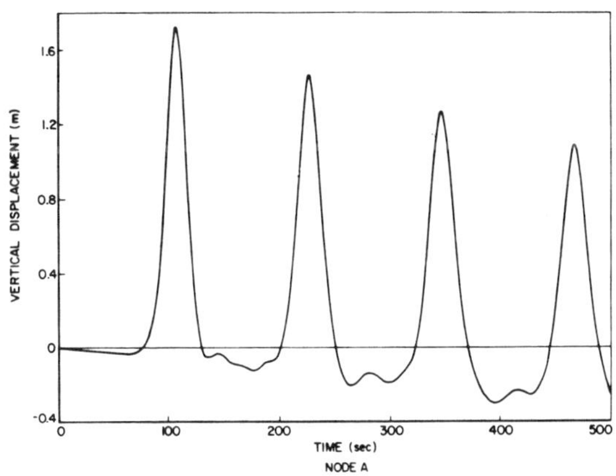

Figure 6

Computed Vertical Displacement Time History of Fluid Node A on Free-Surface at the Fluid-Structure Interface of freedom is shown in Fig. 4. The wave-generation function was designed to produce a solitary wave within the reservoir. Fig. 5 shows the free-surface profile at various instants of time. The wave first hits the structure at $t \approx 105$ sec. Fig. 6 shows the vertical displacement time history of the fluid node A (Fig. 4) on the free surface at the interface between the fluid and the structure. Fig. 7 shows the horizontal displacement time history of node B on top of the structure (Fig. 4). A fully drained situation was assumed to prevail in the soil foundation for the calculations reported in Figs. 5-7.

6. CONCLUSIONS

In this paper, a general finite element model is proposed to analyze dynamic fluid-structure-soil foundation interaction problems. It is shown that the procedure allows for a realistic and optimal modeling of (1) the fluid subdomain and its free-surface; and (2) the two-phase saturated soil foundation.

7. ACKNOWLEDGEMENTS

Computer time was provided by Princeton University Computer Center.

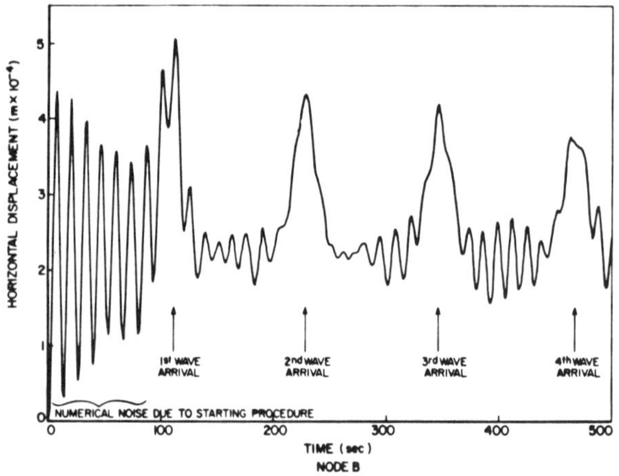

Figure 7

Computed Horizontal Displacement Time History of Node B on Top of Structure

REFERENCES

1. Amsden, A., and Hirt, C., "YAQUI: An Arbitrary Lagrangian Eulerian Computer Program for Fluid Flow at all Speeds," Rept. LA-5100, Los Alamos Scientific Laboratory, Los Alamos, NM, 1973.

2. Belytschko, T., et al., "Quasi-Eulerian Finite Element for Fluid-Structure Interactions," Jnl. Pressure Vessel Tech., ASME, Vol. 102, 1980, pp. 62-69.

3. Donea, J., et al., "Lagrangian and Eulerian Finite Element Techniques for Transient Fluid-Structure Interaction Problems," Transactions, Fourth SMIRT Conf., San Francisco, CA. 1977.

4. Hughes, T.J.R., et al., "Implicit-Explicit Elements in Nonlinear Transient Analysis," Comp. Meths. Appl. Mech. Engrg., Vol. 17/18, 1979, pp. 159-182.

5. Hughes, T.J.R., et al., "Lagrangian-Eulerian Finite Element Formulation for Incompressible Viscous Flows," Comp. Meths. Appl. Mech. Engrg., Vol. 29, 1981, pp. 329-349.

6. Liu, W.K., "Development of Finite Element Procedures for Fluid-Structure Interaction" Rept. EERL 80-6, California Institute of Technology, Pasadena, CA. 1980.

7. Liu, W.K., and Ma, D.C., "Computer Implementation Aspects for Fluid-Structure Interaction Problems," Comp. Meths. Appl. Mech. Engrg., Vol. 31, 1982, pp. 129-148.

8. Malkus, D. and Hughes, T.J.R., "Mixed Finite Element Methods-Reduced and Selective Integration Techniques: A Unification of Concepts," Comp. Meth. Appl. Mech. Engrg., Vol. 15, 1978, pp. 63-81.

9. Newmark, N.M., "A Method of Computation for Structural Dynamics," Jnl. Eng. Mech. Div., ASCE, Vol. 85, No. EM3, 1959, pp. 67-94.

10. Prevost, J.H., "Mechanics of Continuous Porous Media," Int. Jnl. Engrg. Sci., Vol. 18, 1980, pp. 787-800.

11. Prevost, J.H., "DYNA-FLOW: A Nonlinear Transient Finite Element Analysis Program," Dept. Civil Eng., Princeton Univ. 1981.

12. Prevost, J.H., "Nonlinear Transient Phenomena in Saturated Porous Media," Comp. Meths. Appl. Engrg., Vol. 30, 1982, pp. 3-18.

13. Zienkiewicz, O.C., The Finite Element Method, 3rd Edition, McGraw-Hill, London, 1977.

SHEAR STRENGTH INTERPRETATION FOR GULF OF MEXICO CLAYS

By G. W. Quiros[1], A.M. ASCE; A. G Young[2], A.M. ASCE;
J. H. Pelletier[3], A.M. ASCE; J. H-C Chan[4]

INTRODUCTION

The primary purpose of an offshore geotechnical investigation is to obtain soils data, such as undrained shear strength (S_u), that can be used for foundation design. With few exceptions, the $\phi = 0$ concept (17) has been used for design in cohesive soils. Correct results from the empirical procedures based on the $\phi = 0$ method depend on using the most appropriate S_u data. Unfortunately, design practice includes a large number of different sampling techniques and laboratory tests such as the unconfined compression test (UC), unconsolidated undrained triaxial compression test (UU), and miniature vane (MV), that may complicate interpretation of S_u. In recent years, other techniques such as in situ measurements with the field vane (FV) and cone penetrometer (CPT) are becoming more widely available to obtain S_u data. Application of the Normalized Soil Parameter (NSP) concept (11, 13) has also gained much acceptance as another method to estimate S_u. In many offshore areas, the variation in S_u caused by the wide range of different measurement methods creates uncertainty in determining appropriate S_u values to use in foundation design. In light of these considerations, comparison studies were performed at three sites in the Gulf of Mexico to gain an understanding of how to acquire more reliable and consistent S_u data.

A number of other factors influencing the interpretation of S_u may occur during the data gathering operation (drilling, sampling, and testing). Young et al (18) described how soil samples could be disturbed to varying degrees by: (1) boat motion and movement of the drill pipe; (2) drilling operations; (3) sampling technique; (4) stress relief during sample recovery; (5) sample extrusion; (6) sample handling, packaging, and transportation procedures; (7) sample storage;

[1]Senior Geotechnical Engineer, McClelland Engineers, Inc., Houston, TX;

[2]Engineer Manager, McClelland Engineers, Inc., Houston, TX;

[3]Civil Engineer, Shell Oil Company, Houston, TX;

[4]Senior Staff Engineer, Conoco Inc., Houston, TX

(8) laboratory testing procedures; and (9) geologic and physico-chemical properties of the soil. The influence of these factors if not carefully monitored may also complicate the interpretation of S_u, since the factors often result in more scatter in the measured S_u values. For these reasons, the appropriate selection of S_u will be the most difficult step in the foundation design analysis and is subject to the largest error.

At three sites, S_u data were acquired on wire-line percussion (driven) samples as well as pushed samples. The field vane was used in the studies to gain an understanding of in situ strength characteristics. Finally, the NSP approach was applied to both driven and pushed samples to compare with all the other types of laboratory data. The primary purposes of this paper are to: (1) present comparisons of the S_u data obtained by these different sampling and testing techniques, and (2) make recommendations for obtaining the most consistent and reliable S_u data for offshore foundation design in the Gulf of Mexico.

GENERAL SITE CONDITIONS

The three sites are situated offshore Louisiana on the upper continental slope. Although Sites 1 and 2 and Sites 2 and 3, respectively, are about 40 and 180 miles (64 to 290 km) apart in water depths ranging from 800 to 1500 ft (244 to 457 m), the depositional histories are very similar. The predominant soil consists of normally consolidated, highly plastic clays to final boring penetrations of about 500 ft (152 m). Figures 1 through 3 show the boring logs for the three sites. The boring logs show the measured undrained shear strengths, moisture contents, Atterberg limits, and interpreted shear strength profiles. The interpreted undrained strength profiles shown on the logs were chosen based on the following factors: (1) the nearly equal values of S_u measured by the FV and MV on pushed samples, (2) good agreement between UU tests on pushed samples and the data of Item 1, (3) good agreement between maximum values of the UC data on pushed samples and the data of Items (1) and (2), and (4) the close agreement between the NSP strength profiles obtained on pushed samples and the data from Items (1), (2), and (3).

Plasticity characteristics in the three soil borings are very similar with plastic limits ranging between 25 and 35 and liquid limits generally ranging between 65 and 100. The natural moisture contents decrease with penetration from about 125 percent at the seafloor to about 35 at terminal boring depth of 500 ft (152 m).

FIELD INVESTIGATION PROCEDURES

The borings were drilled and sampled with a Failing 2000 rotary drilling rig operating through a centerwell in the deck of a semisubmersible drilling rig. Soil samples were obtained through the 4.5-in. (11.4-cm) IF drill pipe at 3-ft (1-m) intervals to about 50 ft (15 m) and at 10-ft (3-m) intervals below 50 ft (15 m). The crews obtained pushed samples by latching a 3.0-in.-OD (76-mm), 2.85-in.-ID (72-mm) thin-wall sampling tube into the drill bit and using the weight of the drill pipe to advance the sampler into the soil. Percussion sampling was performed by driving a 2.25-in.-OD (57-mm), 2.125-in.-ID (54-mm) thin-wall tube into the soil using a 175-lb (79-kg) weight on a wire

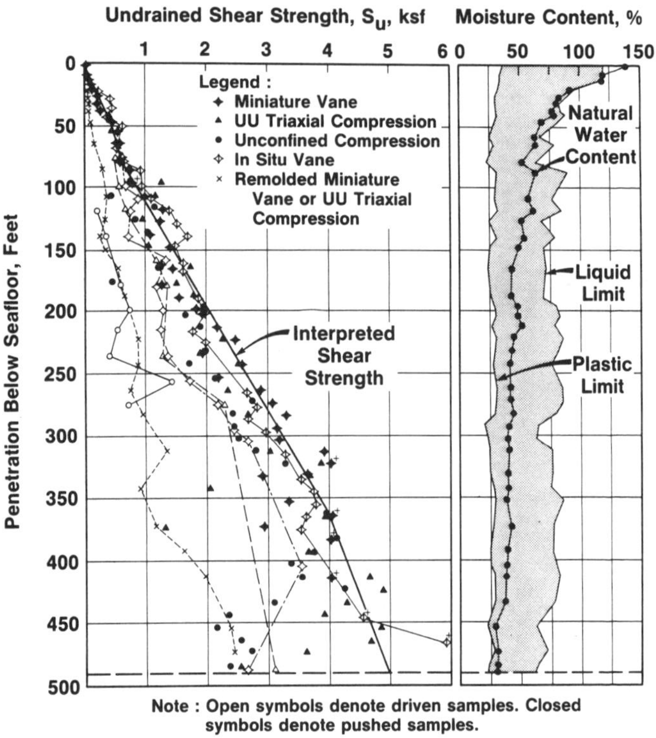

Fig. 1 - Soil conditions at Site 1

line. Most samples were recovered using the push sampling technique. Driven samples were obtained at selected intervals in the borings about 1 to 2 ft below the pushed samples.

In situ strength measurements were also made with a wire-line operated Remote Vane (7). The four-bladed vane is pushed 3 to 5 ft (1 to 1.5 m) into the soil below the bottom of the borehole, and is rotated by an electric motor. S_u data are computed from the maximum torque-rotation recorded during the test.

MV and UC tests were run in the field on both the pushed and driven samples. The remainder of the samples were packaged and transported to the laboratory for additional strength and classification testing.

RESULTS OF CONVENTIONAL TESTING PROGRAM

General Testing Procedures

Since 1962, the conventional practice for conducting field investigations in the Gulf of Mexico and in many other areas around the world has been to obtain 2.25-in.-OD (57-mm), thin-wall tube samples

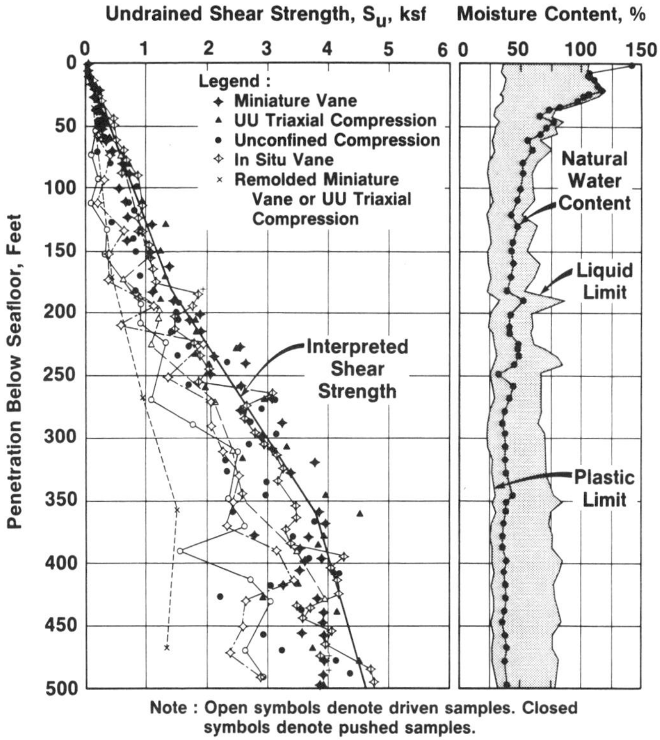

Fig. 2 - Soil conditions at Site 2

with a wire-line activated hammer. This method provides a simple and economical means of compensating for vessel motion of an oil field supply vessel held on location by a four-point mooring spread. The deleterious effects of this sampling technique upon sample quality were first described by Emrich (6); however, the increased cost associated with acquiring higher quality samples has prolonged the use of percussion sampling.

Laboratory testing in the field generally has been limited to MV and UC tests. MV tests are usually performed 2 to 3 in. (50-76 mm) from the bottom of the sampling tube while the sample is still in the tube. Upon extrusion, a UC test is conducted on a soil specimen taken from between 9 to 15 in. (23 to 38 cm) from the sample bottom. UU triaxial and other classification tests are usually performed in the laboratory on the remaining specimens. These procedures were also used for the field investigations at Sites 1, 2, and 3.

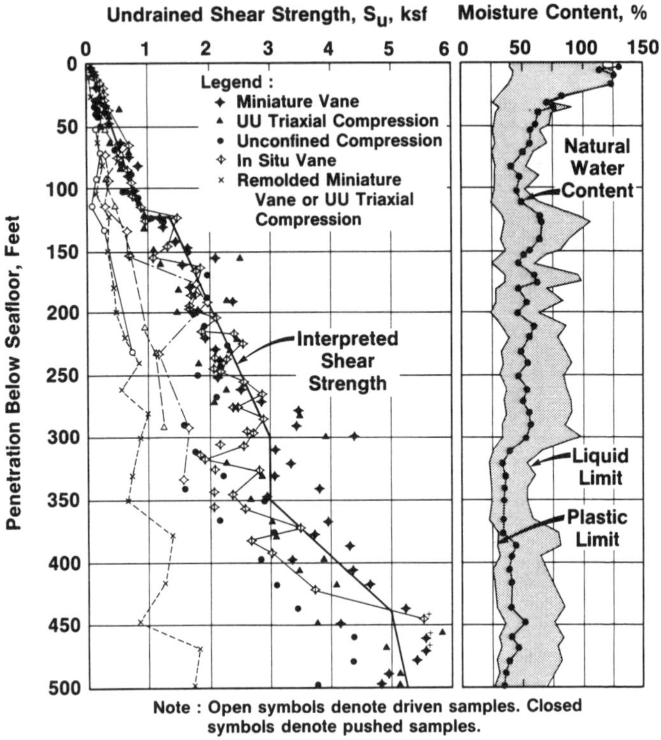

Fig. 3 - Soil conditions at Site 3

Unconfined Compression Tests

For the three sites, results of UC tests performed on both pushed and driven samples are presented in Figs. 1 through 3. In these figures, the driven sample data are connected to enhance their distinction from other strength data. Figures 1 through 3 illustrate that UC strengths on driven samples are markedly lower than those obtained on pushed samples. In fact, most of the UC strength data for driven samples from Sites 1 and 3 fall on the remolded strength line. The UC pushed and driven data are replotted in Fig. 4 to illustrate the variations in the strength data. This comparison indicates a unique correlation for the three sites does not exist between UC strengths from pushed samples, $(S_u)_{UCP}$, and UC strengths from driven samples, $(S_u)_{UCD}$. The ratio of $(S_u)_{UCP}$ to $(S_u)_{UCD}$ averages about 3.3 for Site 1, 1.7 for Site 2, and 2.1 for Site 3. These varying ratios, therefore, suggest that adjustment factors applied to driven samples at one site may not be valid at another site. The lower strengths for driven samples probably result mostly from sample disturbance. Structural disturbance or

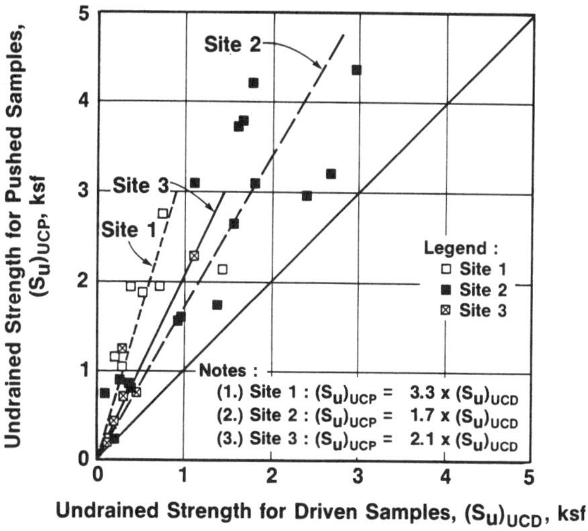

Fig. 4 - Comparison of unconfined compression test results for pushed and driven samples

breakage of particle bonds induced by the driving process also causes a large reduction in effective stresses as indicated in an earlier study by Ladd and Lambe (12).

The effects that sample disturbance has on undrained strength tests are most noticeable in the UC test data. Of the three conventional laboratory tests performed in this study, the UC data in Figs. 1 through 3 are the lowest. Compared to the interpreted shear strength profile, average UC strengths for driven samples are 30 to 80 percent lower. Pushed samples yielded average UC strengths about 5 to 15 percent lower than the interpreted strength profile to a depth of 300 to 400 ft (91 to 122 m). Below that depth range, the difference between UC data and data from other strength tests increases because the UC data fall off suddenly. This behavior in UC data has been observed by the authors in many other offshore borings. The authors believe it results from (1) increased structural disturbance associated with the higher sampler driving resistance in the stronger clays and (2) release of total stresses during sample recovery which induces disturbance because of the resultant release of in situ shear stresses.

Miniature Vane Tests

The MV strength data presented in Figs. 1 through 3 have been replotted in Fig. 5 to better illustrate that the data from driven samples are lower than the pushed samples. The differences in MV strengths on the two sample types are not as dramatic as the UC strength data. Figure 5 shows that the ratio of MV strength on pushed samples, $(S_u)_{MVP}$, to MV strength on driven samples, $(S_u)_{MVD}$, are 1.4, 1.5, and 2.0 for Sites 1, 2, and 3, respectively.

The smaller differences between the MV strengths than the UC strengths on pushed and driven samples probably results from two major factors. First, MV tests are performed by inserting the vane blade into the center of soil specimen, which is the least disturbed portion of the specimen. On the other hand, the UC test is performed on the whole soil specimen without trimming. Consequently, the UC test measures an average strength of the least disturbed central portion of the specimen and the most disturbed outer edges. Second, the MV test is usually conducted while the sample is still in the sampling tube, whereas UC tests are performed after the sample has been extruded from the tube. Thus, the MV test is performed in the soil specimen while it is partially confined. In contrast, the UC specimen is tested without confinement and after the sample has been further disturbed by the extrusion process.

The extrusion process may be more severe for the smaller driven samples because a larger percentage of the area is affected as compared to a pushed sample. The fact that driven samples yield lower MV strengths than pushed samples suggests that even the least disturbed portion of the smaller sample is disturbed as compared to the pushed sample; therefore, trimming the smaller driven specimens will reduce, but not eliminate, sample disturbance.

UU Triaxial Tests

For both pushed and driven samples, Figs. 1 through 3 show that UU triaxial tests typically yield strengths similar to MV strengths. A comparison of UU strengths for pushed and driven samples in Fig. 6 suggests that UU strengths for pushed samples, $(S_u)_{UUP}$, are about 1.3

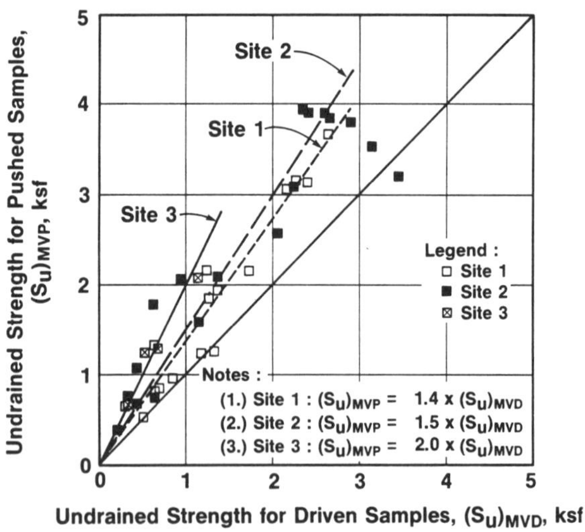

Fig. 5 - Comparison of miniature vane test results for pushed and driven samples.

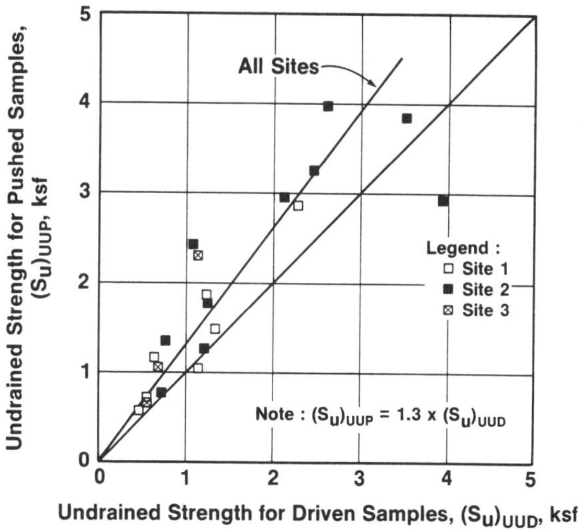

Fig. 6 - Comparison of UU triaxial test results for pushed and driven samples.

times the UU strengths for driven samples, $(S_u)_{UUD}$, for all the sites. The lack of distinctly different UU strength correlations between pushed and driven samples for each site, as was observed for the UC and MV tests, may result from the small number of UU triaxial tests performed on driven samples at Sites 1 and 3.

UU triaxial tests performed on pushed samples typically yield S_u values that are consistent with the strengths interpreted from the MV and FV data. This consistency in S_u data from the three different tests can be observed in Figs. 1 through 3 and allows one to select a higher shear strength profile. If only percussion samples had been recovered in Sites 1 through 3, the shear strength profiles interpreted for the three sites would have been very conservative.

Disturbance in Driven Soil Samples

Thus far, conventional laboratory strength tests have shown that samples recovered by the wire-line percussion sampling technique will yield lower shear strengths than samples recovered by push sampling. The field engineers and technicians noted during the field work at the three sites that they could actually see and feel the better quality of the pushed samples in contrast to the driven samples. The better quality of pushed samples was also observed in X-ray radiographs.

A number of factors may contribute to the disturbance of driven samples. One major factor is that the dynamic process of percussion sampling breaks particle bonds, thereby increasing the disturbance associated with total stress relief and gas expansion. Another factor is the smaller diameter (2.125-in.-ID or 54 mm) of the driven samples as compared to the diameter (2.85-in.-ID or 72 mm) of the pushed samples as previously described (8). A third important source contributing to sample disturbance during percussion sampling involves the human factor. As water depths increase beyond 800 ft (244 m), control of the wire-line hammer becomes increasingly difficult for a driller because of the increasing weight of the wire line. The lack of hammer control will cause: (1) variations in hammer stroke; (2) an increase in number of blows required to achieve a particular tube penetration as compared to shallow water sites; and (3) in some cases, partial extraction and reinsertion of the sampling tube during the hammering process. The human factor probably is most responsible for the different ratios of pushed sample strength to driven sample strength at each site, depicted in Figs. 4 and 5.

In Situ Vane Tests

Field Vane Measurements. Uncertainty in S_u data from laboratory strength tests has given impetus to development of in situ tests such as the FV. Ehlers et al (7) and Koutsoftas and Fischer (9) showed that FV data have less scatter, thereby providing a better definition of the interpreted shear strength profile. The FV used in this study is the fifth generation of the device described by Doyle et al (5). This study provided one of the few opportunities to compare in situ S_u data from the FV with laboratory strength data on pushed samples. The FV data in Figs. 1 through 3 show good agreement with the MV and UU strength data on pushed samples. This observation is confirmed in Figs. 7 and 8, where the FV data are compared to MV and UU strength measurements on

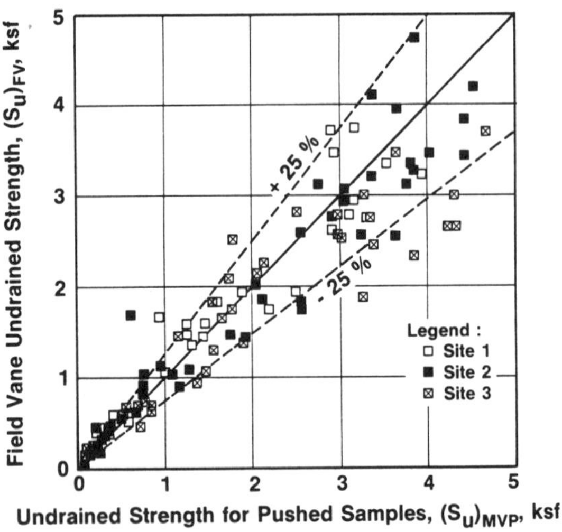

Fig. 7 - Comparison of in situ vane tests with miniature vane tests for pushed samples.

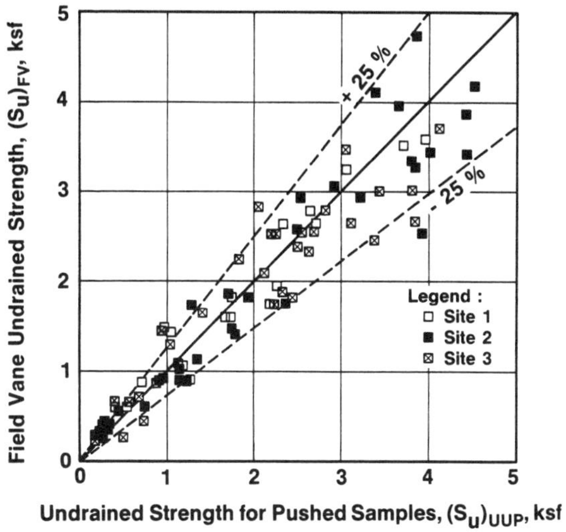

Fig. 8 - Comparison of in situ vane tests with UU triaxial tests for pushed samples.

pushed samples, respectively. For these sites, the ratio of FV strength, $(S_u)_{FV}$, to MV and UU strengths is equal to about 1. Koutsoftas and Fischer (9) observed a similar correlation between FV and UU strengths on high quality pushed samples.

The close correlation between the FV data and MV and UU data on pushed samples holds for these sites where the sediments are not highly expansive due to methane gas concentrations. The authors have obtained similar types of data in gassy sediments of the Mississippi River Delta and found that the FV strength data are higher than the MV and UU data on pushed samples. We believe this difference is caused by the large amount of expansion of the soil specimens out of the sampling tube caused by the gas coming out of solution.

Adjustment Factors. Adjustment factors such as those recommended by Bjerrum (3) and Ehlers et al (7) were not applied to the FV strength data. If Bjerrum's adjustment factors had been used, the FV measurements would have been reduced by about 10 to 30 percent to correspond to the 30 to 70 percent range in plasticity index typical of the clays at the three study sites. Ehlers et al (7) recommended adjusting the FV strength measurements by a factor of 0.75 to produce an S_u comparable to that expected from unconfined compression tests on high-quality samples. However, the net result of applying an adjustment factor of 0.75 to the FV data would be an FV strength profile that falls considerably below the UC strength profiles for the three sites. Applying these adjustment factors to the FV data, therefore, will yield S_u values that are significantly lower than the strengths measured on pushed samples. Based on these results, the authors believe that adjustment factors such as those described above are not unique values and may vary for different marine sediments. It should also be noted that Bjerrum's adjustment factors were correlated to embankment failures which would impose different loading conditions than may occur with other foundation problems.

Factors Affecting FV, MV, and UU Strength Correlations. The S_u data in Figs. 7 and 8 contain a reasonable amount of data scatter ranging within ±25 percent. This intensity of data scatter should be expected due to: (1) random measurement error resulting in differences between in situ and conventional laboratory strength tests, (2) changes in soil fabric and structure, and (3) differences in the amount of sample expansion between specimens. Random variations in FV tests are typically less than S_u generally observed in the conventional laboratory strength tests. Consequently, the MV and UU tests influenced the data scatter in Figs. 7 and 8 more than the FV test.

Changes in soil fabric and structure with depth within soil deposits also lead to variations in S_u measurements. Rowe (16) has indicated that the structure of a soil reflects, to some degree, all facets of the sediment's composition and depositional history. The sediments at the three study sites were deposited during periods of changing sea levels. Since changes in rates of sedimentation, turbulence, temperature, currents, suspended soil particle size, and concentration of suspended sediments generally accompany sea level changes, one should expect variations in soil composition and structure with depth. These variations were confirmed by X-ray diffraction and radiography analyses performed on a limited number of soil specimens from the three sites.

The third factor contributing to data scatter is sample expansion. Soil samples recovered at the three sites below about 110-ft (34-m) penetration expanded by varying amounts while still in the sampling tube. During sampling operations, field engineers and technicians noted that sample expansion beyond the tip of the sampling tube ranged from 0.125 in. (3 mm) to 1.5 in. (38 mm). The greater the amount of sample expansion, the lower laboratory S_u values tend to be. Thus, variations in the amount of sample expansion will induce different degrees of sample disturbance and, in turn, cause variations in the S_u data. The effects of disturbance associated with sample expansion may be overcome when pressurized samplers (4) become readily available for offshore use. Presently, however, the disturbance effects due to sample expansion may only be reduced by performing strength tests on the least disturbed portions of the sample. Young et al (18) and Bea and Aurora (2) have addressed the subject of avoiding disturbed portions of the soil sample when selecting specimens for laboratory strength tests.

Normalized Laboratory Strength Measurements

Description of Method. In recent years, the Normalized Soil Parameter (NSP) concept is gaining much wider acceptance as a method of estimating S_u for foundation design. For this study, the SHANSEP (Stress History and Normalized Soil Engineering Properties) technique described by Ladd and Foott (11) was employed at the three sites. SHANSEP is a sophisticated laboratory strength testing technique in which measured S_u data is not used directly in design.

In applying the SHANSEP method, high quality undisturbed soil samples are consolidated in oedometer tests to determine the stress history of the soil profile. The overconsolidation ratio (OCR) values and strength tests which best model in situ conditions are selected to study the soil's normalized behavior. Soil specimens in appropriate strength tests are consolidated up to the virgin compression line where the vertical stress, $\bar{\sigma}_{vc}$, is greater than the maximum previous consolidation pressure, $\bar{\sigma}_{vm}$. (Consolidation to the virgin compression line decreases sample disturbance effects and results in a test specimen of known OCR = 1.) The measured undrained shear strength is normalized with respect to the consolidation stress, $\bar{\sigma}_{vc}$. Soils which exhibit normalized strength behavior result in a relatively constant $S_u/\bar{\sigma}_{vc}$ value at different consolidation pressures, usually greater than 1.5 to 2 times $\bar{\sigma}_{vm}$. In order to obtain an $S_u/\bar{\sigma}_{vc}$ versus OCR relationship, similar to those shown in Figure 9, further strength tests are performed on specimens consolidated to the virgin compression line and unloaded to obtain specific OCR's.

These laboratory test data are used to determine a shear strength profile. A plot of maximum past pressure ($\bar{\sigma}_{vm}$) compared with the in situ effective vertical stress ($\bar{\sigma}_{vo}$), as shown in Fig. 10, establishes OCR values with depth. The OCR values throughout the profile are then used with the plot, as shown in Fig. 9, to establish $S_u/\bar{\sigma}_{vo}$ with depth.

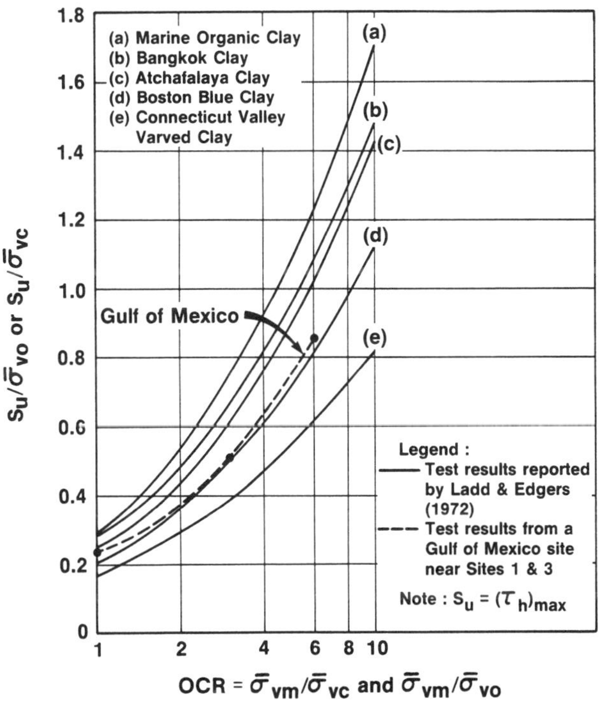

Fig. 9 - Undrained strength ratio vs. overconsolidation ratio from CK_0U direct simple shear tests.

At this point, the plot of $S_u/\bar{\sigma}_{vo}$ is multiplied by the plot of $\bar{\sigma}_{vo}$ versus depth to establish an S_u profile with depth. Thus, the method requires a knowledge of the in situ vertical effective stresses and $\bar{\sigma}_{vm}$ values that are essential from high quality oedometer tests.

Results from Consolidation Tests. The stress histories of the soil deposits at the three sites were investigated by performing constant-rate-of-strain (CRS) consolidation tests on selected soil samples. The CRS test specimens were taken from samples acquired by both the pushed and wire-line percussion sampling techniques. The preconsolidation pressures interpreted from CRS tests are presented in Fig. 10 along with the computed vertical effective stress profile.

Based on the geology of the region encompassing Sites 1 through 3, the soil deposits should be normally consolidated. Figure 10 illustrates that pushed samples generally gave a good estimate of the in situ vertical effective stress. In contrast, driven samples generally yielded lower preconsolidation pressures due to the disturbed shape of the consolidation stress-strain curves. In areas of the world where little is known about the geology of the region, driven samples could therefore lead to misinterpretation of the soil stress history. This misinterpretation, in turn, would result in an inaccurate estimate of in situ S_u by the SHANSEP concept. For this reason, Ladd and Foott (11) have indicated that, "Good undisturbed samples are . . .a major requirement" to obtain reliable results from the SHANSEP method.

Results of Laboratory Strength Tests. In applying the SHANSEP method at the three sites, a suite of different strength tests was performed to check the potential variation in the computed S_u data. For Site 1, CIU triaxial compression and CK_oU direct simple shear (CK_oUDSS) were performed; for Site 2, only CIU triaxial tests; and for Site 3, CK_oUDSS and CK_oU triaxial compression tests. Most of these tests were performed on both pushed and driven soil specimens. Figures 11 through 13 present the results of these tests in graphical form. Table I summarizes the averages from these laboratory test results.

TABLE I - Summary of Laboratory Strength Tests for OCR = 1

Average Strength Ratio, $S_u/\bar{\sigma}_{vc}$ or $S_u/\bar{\sigma}_c$

Site	Type of Test	Pushed Samples	Driven Samples
1	CK_oUDSS	0.23	0.24
	CIU Triaxial	0.26	0.28
2	CIU Triaxial	above 350 ft: 0.27	0.31
		below 350 ft: 0.24	0.31
3	CK_oUDSS	0.23	0.28
	CK_oU Triaxial	0.27	-

Note: Consolidation pressures for these tests ranged from 1.5 to 2.5 times the computed in situ vertical effective stress.

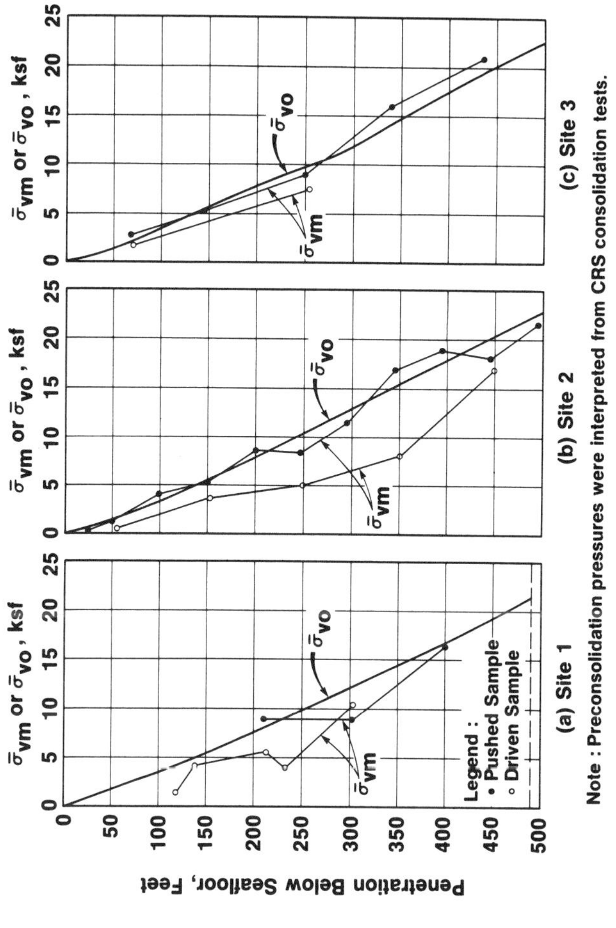

Fig. 10 - Vertical effective stress

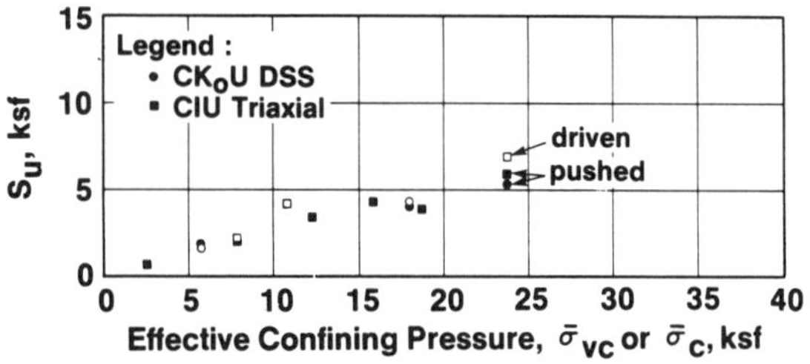

Fig. 11 - Site 1: results of CIU triaxial and CK_0U DSS tests.

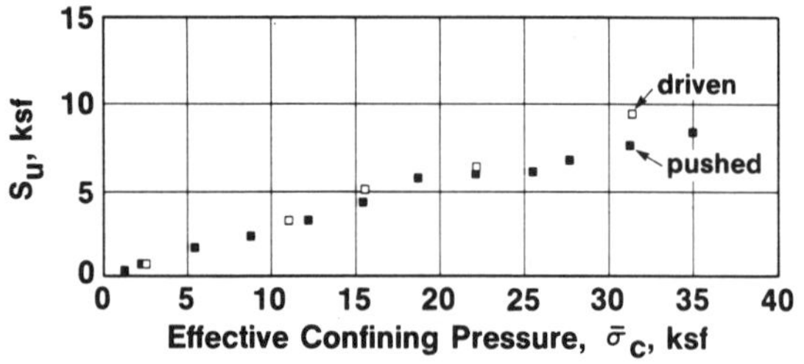

Fig. 12 - Site 2: results of CIU triaxial tests.

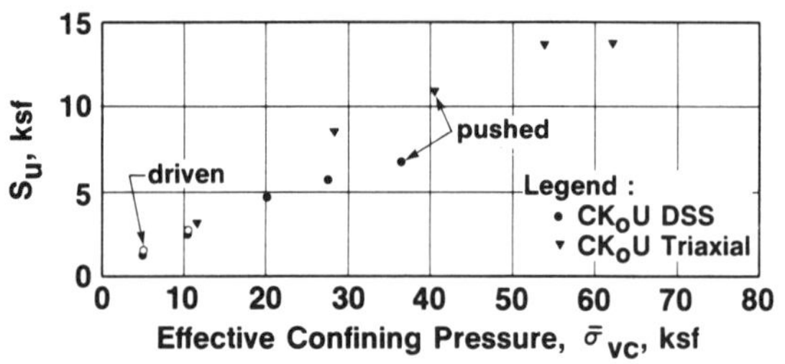

Fig. 13 - Site 3: results of CK_0U triaxial and CK_0U DSS tests.

Discussion of Laboratory Results. The results in Table I indicate that driven samples tend to yield strength ratios 4 to 29 percent higher than pushed samples when both tests are consolidated to the same pressures. It is apparent from these results that the driven specimens may not have been consolidated to high enough pressures so as to adequately reduce or overcome the effects of sample disturbance.

The need to consolidate soil specimens to pressures high enough to overcome disturbance raises another question regarding the application of the NSP concept to driven samples. If the percussion sampling technique can induce different degrees of disturbance in samples, then the selection of one laboratory consolidation pressure to overcome disturbance may not be adequate for all the driven samples throughout the boring. Consequently, using a consolidation pressure of 1.5 to 2.5 times $\bar{\sigma}_{vm}$ in situ may not be high enough to overcome sample disturbance in driven samples. With pushed samples, however, one would not have to contend with disturbance caused by dynamic sampling process.

Application of the NSP Concept. To gain an appreciation for how S_u based on SHANSEP may differ between pushed and driven samples, we used the strength ratios from Table I in conjunction with the measured values of $\bar{\sigma}_{vm}$ presented in Fig. 10. Comparing the interpreted S_u profile for each site with the results in Fig. 14 illustrates that (1) tests performed on driven samples generally underestimate the shear strength, (2) tests on pushed samples generally yield strengths nearly equal to or greater than the interpreted S_u line, (3) CIU and CK_oU triaxial tests give strengths 13 to 17 percent greater than CK_oUDSS tests, (4) CIU and CK_oU triaxial tests on pushed samples yield strengths nearly equal to or greater than the interpreted S_u and (5) CK_oUDSS tests usually indicate strengths equal to or slightly less than the interpreted strength line. A comparison of Figs. 1, 2, and 3 with Figs. 14(a), 14(b), and 14(c), respectively, show that the in situ S_u values interpreted from pushed sample data using the SHANSEP concept fall within the upper and lower bounds of the conventional laboratory strengths for pushed samples. This consistency in the results gives the engineer increased confidence for selecting a higher design S_u profile as more representative of in situ conditions. In contrast, the S_u profile based on SHANSEP for driven samples is generally greater than the S_u profile from MV, UC, and UU tests on driven samples and less for pushed samples. Consequently, interpretation of the design S_u becomes more subjective.

In a discussion of the SHANSEP method, Mesri (14) indicated that oedometer tests involve ". . . as much uncertainty as the measurement of undrained strength itself." Since the preconsolidation pressures interpreted from the CRS consolidation tests were used to compute the SHANSEP strengths shown in Fig. 14, these S_u values give an indication of variations in oedometer test results relative to strength measurements. The results presented in Fig. 14 reveal that scatter in the oedometer test data is of the same intensity as variations observed in conventional laboratory S_u tests.

Effect of Shear Strength on Pile Capacity

The influence of the driven and pushed S_u profiles upon the computed axial pile capacity of a 72-in.(183-cm)-diameter pile at Site 2 is illustrated in Fig. 15. Pile capacity was computed using the method described in API RP 2A, Para. 1, Sec. 2.6.4b (1). The driven S_u profile was based on the MV and UU strengths on driven samples presented in Fig. 2. The pushed S_u profile was the interpreted shear strength profile marked on this same figure.

For an ultimate load of 12,500 kips, the pushed and driven S_u profiles indicate the required capacity is obtained at a depth of 355 and 410 ft, respectively. The 55-ft difference in pile penetration translates into savings in both steel and pile installation costs. The authors believe the cost savings in the foundations for offshore structures, resulting from high quality pushed samples and in situ vane tests, will be large in comparison to the slightly higher costs for performing a high quality geotechnical investigation.

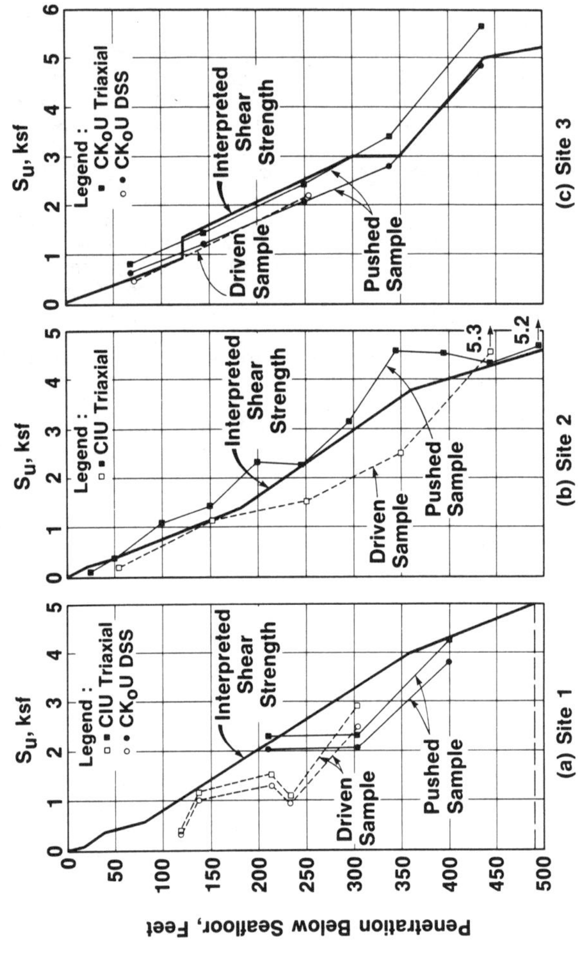

Note : S_u was computed by multiplying average $S_u/\bar{\sigma}_c$ or $S_u/\bar{\sigma}_{vc}$ from strength tests by preconsolidation pressure interpreted from CRS consolidation test.

Fig. 14 - Shear strength based on NSP concept

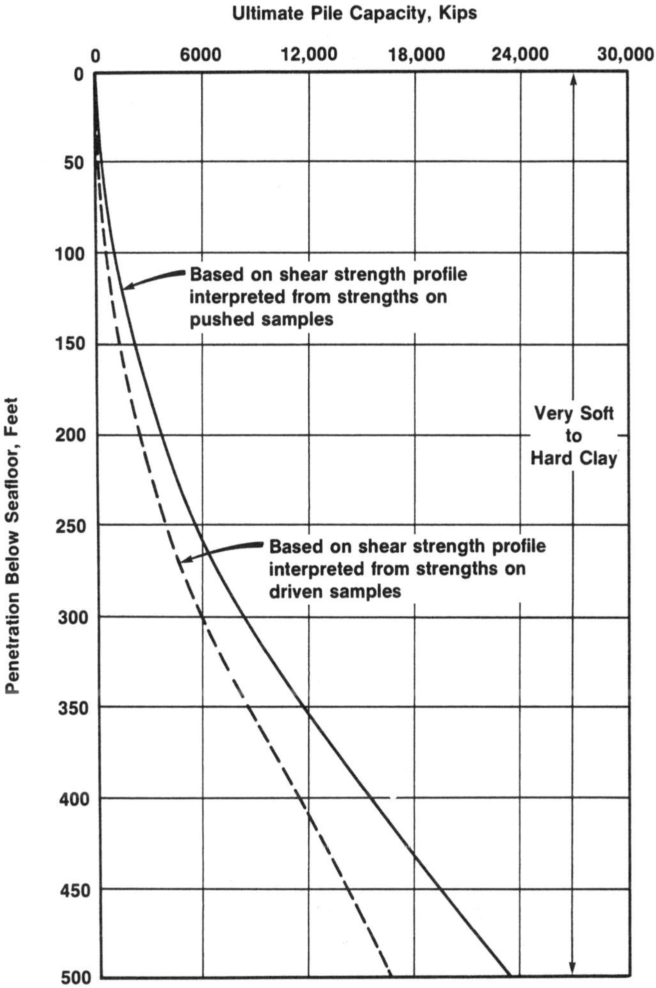

Fig. 15 - Comparison of pile capacity curves, 72-in.-diameter pipe piles, API RP 2A (January 1982) method

The authors recognize the importance of using consistent strength data with the empirical pile capacity procedures as were used to calibrate with the available pile load test data. However, a wide variety of sampling and testing procedures (15) were used at the test pile sites, and the authors question whether the empirical design procedures can be judged at this time to be valid for one particular sampling and testing technique over some other. We believe the significant difference in the predicted and the observed capacities from the pile load tests can be reduced only by using consistent and better quality sampling and testing procedures, as described in this paper. The authors recognize that some slight adjustment in the empirical procedures might be required if high quality data is obtained at the test pile locations. However, the authors believe a change is needed from the conventional offshore sampling and testing procedures that have been providing overly conservative foundation designs for more than 20 years because of the disturbance effects upon S_u. The improved sampling and testing procedures described here will provide a base of high quality and consistent data that can be continually used to improve offshore foundation design practice.

CONCLUSIONS

This paper presents a wide variety of S_u data from three sites in the Gulf of Mexico located on the upper continental slope in water depths ranging from about 800 to 1500 ft (244 to 457 m). The laboratory testing programs included conventional strength tests such as UC, MV, and UU tests on both pushed and driven samples. FV data were also obtained at the three sites. CIU triaxial, CK_oU triaxial, CK_oUDSS tests, and CRS consolidation tests were performed so S_u data could be estimated by the NSP concept. The results from the three sites led the authors to the following conclusions:

(1) Selection of S_u data for foundation design may be complicated by many factors during an offshore geotechnical investigation. The sampling technique can have a significant influence upon providing reliable S_u data. The authors believe high quality pushed samples and in situ vane tests will yield higher, more consistent S_u data that is more reliable for foundation design.

(2) Close agreement exists between the S_u data obtained with the FV data and UU and MV data obtained on pushed samples. The FV test results in less scatter in the S_u data than the MV and UU tests on pushed samples, because the laboratory test data are affected by differences in degree of sample disturbance.

(3) Depending on the laboratory test, pushed samples yield strengths 1.3 to 3.3 times greater than driven samples. UC tests on pushed and driven samples typically measure the lowest values of S_u as compared to MV and UU tests. UC tests are affected the most by the reduction of effective stresses in the soil specimens due to sampling disturbance.

(4) Strength adjustment factors to correct for differences between the various types of S_u data are not unique values. Use of a strength adjustment factor for the FV data as proposed by

Bjerrum (3) and Ehlers et al (7) will result in an underestimate of the S_u profile as indicated by conventional lab tests and NSP approach performed on pushed samples.

(5) The NSP method did not overcome driving induced sample disturbance. $S_u/\bar{\sigma}_{vc}$ data from CK_oUDSS, CIU, and CK_oU tests indicate higher strength ratios for driven samples than pushed samples. Preconsolidation pressures from CRS tests are lower for the driven samples than the pushed samples; therefore, driven samples could lead to misinterpretation of the soil stress history. Low preconsolidation pressures from driven samples will also result in low estimates of S_u using the SHANSEP approach; thus, good high quality pushed samples are recommended for this approach.

(6) The authors believe the data presented here illustrate the consistency in S_u data obtained from: (a) the FV test, (b) MV, UU, UC tests on pushed samples, and (c) SHANSEP data on pushed samples. This consistency makes interpretation of the S_u profile less subjective and also allows the engineer to interpret a higher, more reliable S_u profile. These factors translate directly into cost savings in the pile foundations from reductions in steel and pile installation time. For this reason, the authors recommend adopting pushed sampling and in situ vane testing procedures for offshore geotechnical investigations instead of the commonly used percussion sampling procedures. These new procedures will provide an improved data base that can be continually used to improve overly conservative foundation design based on the lower quality data.

REFERENCES

1. American Petroleum Institute (1982), Recommended Practice for Planning, Designing and Constructing Offshore Platforms, API RP 2A, 13th Edition.

2. Bea, R.G. and Aurora, R.P. (1981), "A Simplified Evaluation of Seafloor Stability," Proceedings, 13th Offshore Technology Conference, Houston, Vol. 1, pp. 223-239.

3. Bjerrum, L. (1972), "Embankments on Soft Ground," Proceedings, Specialty Conference on Performance of Earth and Earth-Supported Structures, ASCE, Vol. II, June, pp. 1-54.

4. Denk, E.W., Dunlap, W.A., Bryant, W.R., Milberger, L.J., and Wheelan, T.J., III (1981), "A Pressurized Core Barrel for Sampling Gas-Charged Marine Sediments," Proceedings, 13th Offshore Technology Conference, Houston, Vol. 4, pp. 43-52.

5. Doyle, E.H., McClelland, B., and Ferguson, G.H. (1971), "Wire-Line Vane Probe for Deep Penetration Measurements of Ocean Sediment Strength," Proceedings, 3rd Offshore Technology Conference, Houston, Vol. 1, pp. 21-32.

6. Emrich, W.J. (1971), "Performance Study of Soil Sampler for Deep-Penetration Marine Borings," Proceedings, Sampling of Soil and Rock, ASTM STP 483, pp. 30-50.

7. Ehlers, C.J., Young, A.G, and Focht, J.A., Jr. (1980), "Advantages of Using In Situ Vane Tests for Marine Soil Investigations," Proceedings, International Symposium on Marine Soil Mechanics, Mexico City.

8. Hvorslev, M.J. (1972), Subsurface Exploration and Sampling of Soils for Civil Engineering Purposes, Report - Research Project, Committee on Sampling and Testing, Soil Mechanics and Foundation Division, ASCE, Waterways Experiment Station, Vicksburg, p. 109.

9. Koutsoftas, D. and Fischer, J.A. (1976), "In-situ Undrained Shear Strength of Two Marine Clays," Journal, Geotechnical Engineering Division, ASCE, Vol. 102, No. GT9, pp. 989-1005.

10. Ladd, C.C. and Edgers, L. (1972), Consolidated-Undrained Direct Simple Shear Tests on Saturated Clays, Research Report 72-82, Department of Civil Engineering, Massachusetts Institute of Technology, Cambridge, Massachusetts.

11. Ladd, C.C. and Foott, R. (1974), "New Design Procedure for Stability of Soft Clays," Journal, Geotechnical Engineering Division, ASCE, Vol. 100, No. GT7, pp. 763-786.

12. Ladd, C.C., and Lambe, T.W. (1963), "The Strength of Undisturbed Clay Determined from Undrained Tests," Special Technical Publication No. 361, National Research Council of Canada-American Society of Testing and Materials Symposium on Laboratory Shear Testing of Soils, pp. 342-371.

13. Lambe, T.W. and Whitman, R.V. (1969), Soil Mechanics, Wiley, New York, p. 451.

14. Mesri, G. (1975), Discussion: "New Design Procedures for Stability of Soft Clays," Charles C. Ladd and Roger Foott, Journal, Geotechnical Engineering Division, ASCE, Vol. 101, No. GT4, pp. 409-411.

15. Olson, R.E. and Winter, D.G. (1981), Review and Compilation of Pile Test Results: Axial Pile Capacity, Geotechnical Engineering Report GR81-14, Department of Civil Engineering, The University of Texas, Austin, Texas.

16. Rowe, P.W. (1972), "The Relevance of Soil Fabric to Site Investigation Practice," Geotechnique, Vol. 22, No. 2, pp. 195-300.

17. Skempton, A.W. (1948), "The ϕ = 0 Analysis and Its Theoretical Basis," Proceedings, 2nd International Conference on Soil Mechanics and Foundation Engineering, Rotterdam, Vol. 1, pp. 72-78.

18. Young, A.G, Quiros, G.W., and Ehlers, C.J. (1983), "Effects of Offshore Sampling and Testing on Undrained Soil Shear Strength," Accepted for Presentation to Offshore Technology Conference, May 1983, Houston (OTC 4465).

DEEP WATER SOIL INVESTIGATION:PLANNING TO RESULTS

By

Tor-Inge Tjelta, [1] MNIF
Peter H. Watt, [2] MICE
Derek W. F. Senner, [2] MICE

INTRODUCTION

A deep water soils investigation was completed in the remote, hostile waters of the Norwegian Sea, (Block 7120/8), Fig. 1., during June and July 1982. Recent publications summarize the overall requirements for deep water soil investigations (1) and provide guidance for estimating time and costs (2) based on an analysis of soil investigation records from the North Sea during 1978 to 1981. The information presented supplements these recent publications providing guidance for planning future deep water soil investigations.

The continuing requirement for hydrocarbon energy sources and rapid offshore technology development during the past decade resulted in Norwegian Government approval during 1980 for offshore exploration drilling north of latitude 71°N in the Norwegian Sea. Presently the hostile arctic environment limits drilling operations to the relatively short summer season. However, future operations are being planned for all year access.

The initial exploration programmes identified promising gas reserves and potential commercial development is being considered. Specific soil data were required by Statoil (Den norske Stats Oljeselskap a.s.) for correlation with shallow seismic records over the general area for feasibility studies, preliminary platform engineering, and economic assessment of field development.

FIELD DEVELOPMENT PROPOSALS

Gas reservoir information indicated that feasibility soil investigations were desirable at five locations covering an area about 3 by 8 km. Water depths of the five sites vary from 240 to 280 metres, as shown on the bathymetric map, Fig. 2.

The platform types being considered in feasibility studies for economic assessment of field development plans for Block 7120/8 include:

Concrete Tripod Gravity Structure
Steel Tripod Gravity Structure
Piled Steel Jacket
Tension Leg Platform

In the initial phase, emphasis has been placed on foundation design assessment for a tripod concrete gravity structure, Fig. 3. Because of the deep water, sub sea completions for development wells are also under consideration.

---

1 Den norske Stats Oljeselskap a.s. (Statoil), Norway.
2 McClelland Engineers, London.

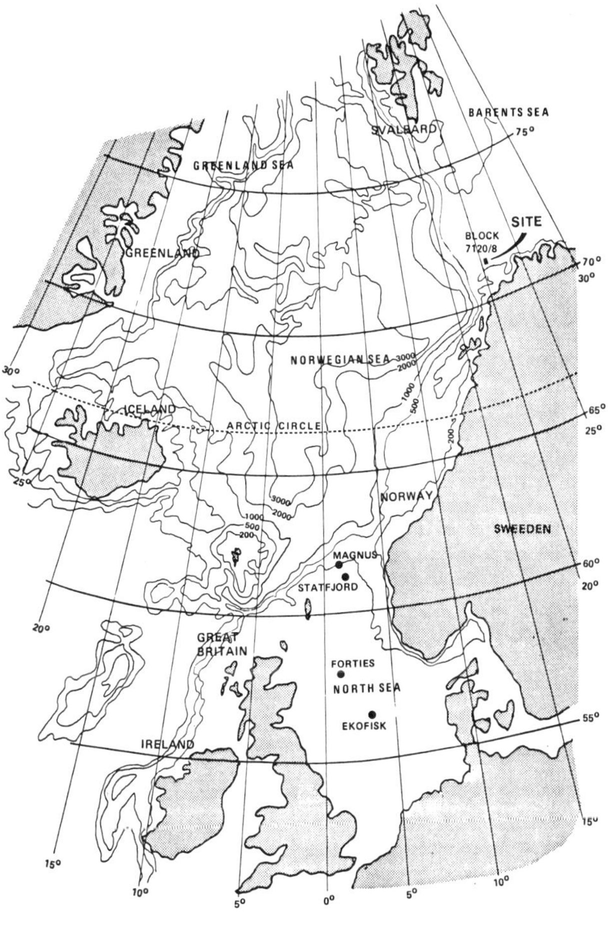

Fig. 1. - Location Plan

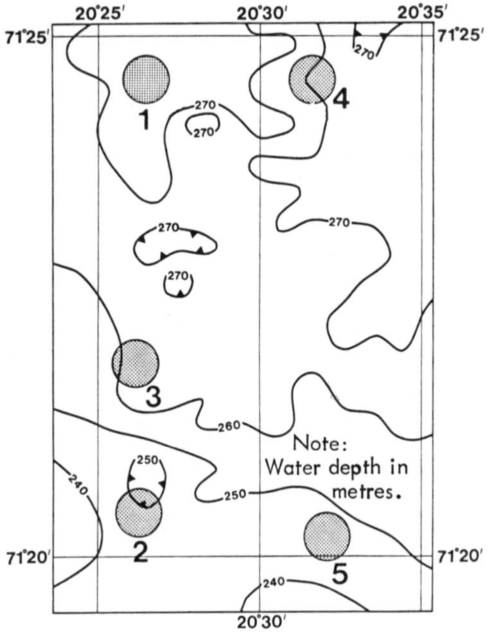

Fig. 2. - Site Plan and Bathymetry

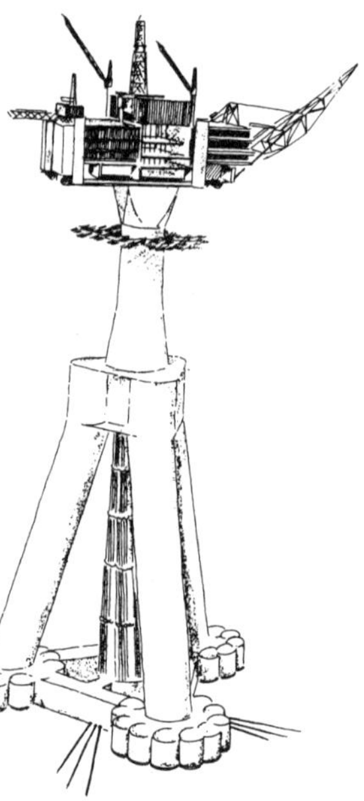

Fig. 3. - Tripod Concrete Gravity Structure

GEOTECHNICAL DATA REQUIREMENTS

For the feasibility study, sufficient soil information was required to assess foundation suitability. For gravity structures information was primarily required on bearing capacity, sliding resistance, skirt penetration resistance, soil response to cyclic loading and settlement. Deep water gravity structures attract substantial horizontal forces, and hence overturning moments, from wind, waves and subsea currents. The sea level loading will be predominantly cyclic. Large spread bases (tripod concept) will be required to resist overturning moments, the size of these bases depending on soil conditions.

Pile supported structures require axial capacity and installation assessments. For tension piles, the consequences of long term creep, progressive failure and cyclic loading were to be addressed.

PLANNING

The project was planned to provide the information needed for feasibility studies within a specified time frame and guidance for possible additional soil investigations. Fig. 4 summarises the planned and achieved time scale of activities for this deep water site investigation. Fig. 5 indicates the approximate percentage cost of each activity. The dominant cost is the offshore investigation phase, taking 79 percent of the total budget in only 8 percent of the time. Because of project dependence on successful offshore performance all available information on the soils to about 100-m penetration was thoroughly reviewed prior to field work commencement.

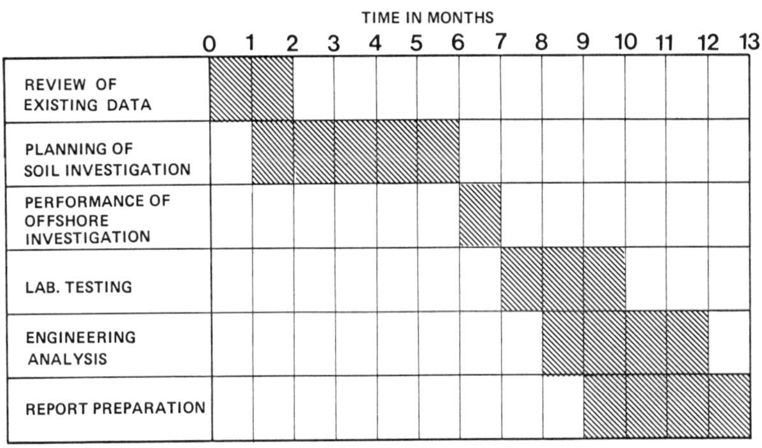

Fig. 4. - Breakdown of Project Activities

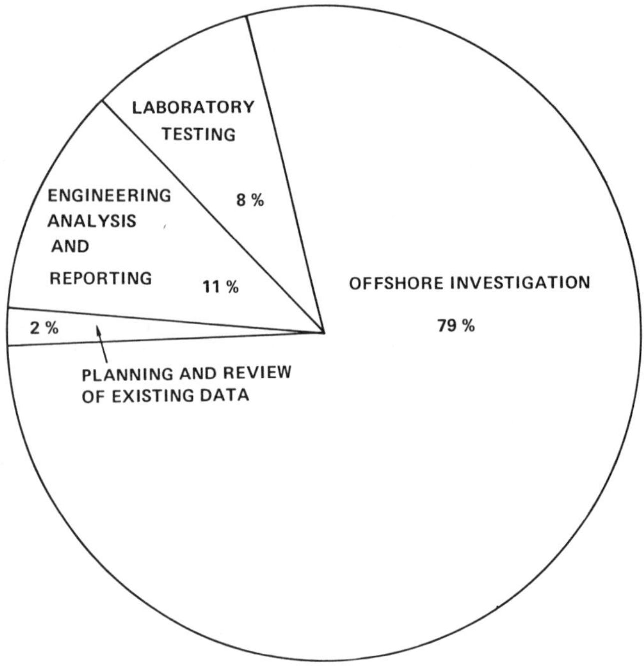

Fig. 5. - Breakdown of Project Costs

Cost. The total cost for the field investigation, all associated laboratory testing and preliminary foundation analyses was NOK 10 M, (USD 1.5 M). This investment to provide feasibility level soil data is readily justified when the foundation of a deep water structure can represent up to 40 percent of the total structure cost.

Available Site Information. The five possible platform locations were initially selected considering the most probable reservoir drainage opportunities. A review of existing geophysical data and exploration drilling records generated further constraints on these locations and also provided insight into potential problems.

Prior to the soil investigation a limited amount of soil data was available from shallow seismic surveys, gravity and vibrocore samples and from exploration drilling records. This information, although valuable for planning the soil investigation, was inadequate to provide confident assessment of foundation conditions for evaluating the relative merits of different platform concepts.

Several geophysical surveys were performed covering the areas shown on Fig. 6. These surveys, conducted in complience with government regulations, provided information on possible dangers to life and equipment from hazards such as shallow gas pockets or anchoring problems for semi submersibles. Geophysical tools used include:

- deep towed boomers
- analogue and digital sparkers
- high-frequency profilers
- side scan-sonar
- echo sounders

Geophysical data interpretation indicated soils to about 250 m below the seafloor were predominantly clays, with the principal geological units identified as:

Geological Unit	Thickness, m	Description
A	2 - 35	Soft to stiff clay
B	20 - 100	Hard sandy clay
C	200	Hard clay with sand lenses

Numerous diffraction hyperbolae observed in Unit A are evidence of seafloor iceberg scour marks; some may indicate boulders. Iceberg scour marks covering the whole study area, Fig. 7, range in width from 10 to 200 m and are 2 to 15 m deep, with relatively steep sides. Recent, softer infill soils were readily identified on the geophysical records. Within the scour marks, the geophysical records identified local zones of very strong reflectors, interpreted as probable boulder beds, or in some cases large single boulders. Fig. 8 shows an example record. These strong reflectors severely limited the amount of signal penetration and made geophysical record interpretation difficult. Areas free from these features were selected for the soil investigations.

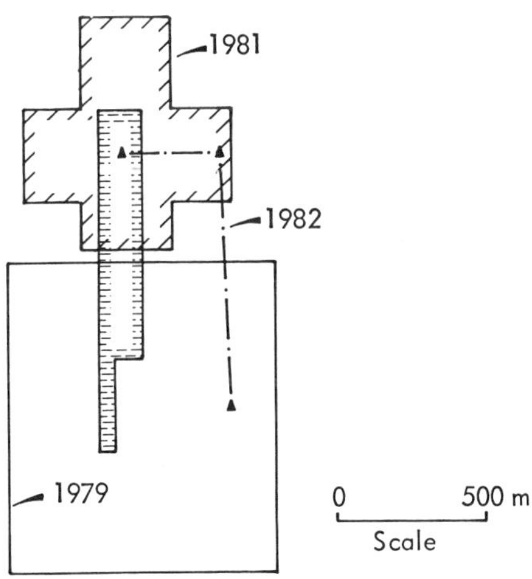

Fig. 6. - Plan of Geophysical Surveys

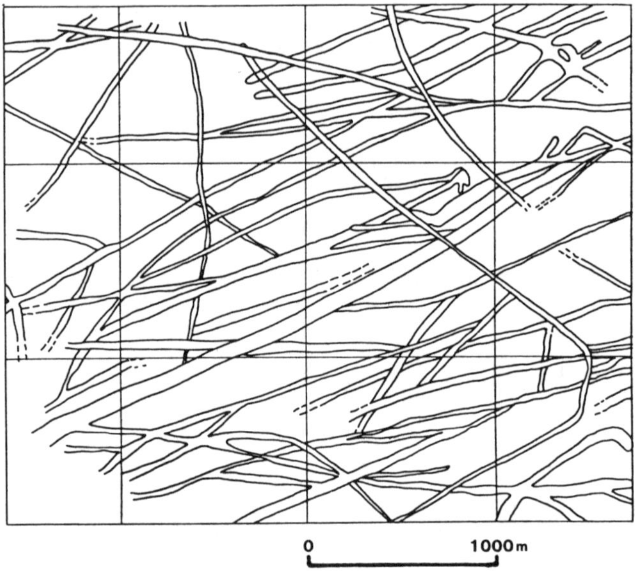

Fig. 7. - Iceberg Scour Marks in Study Area Mapped Using Sidescan Sonar

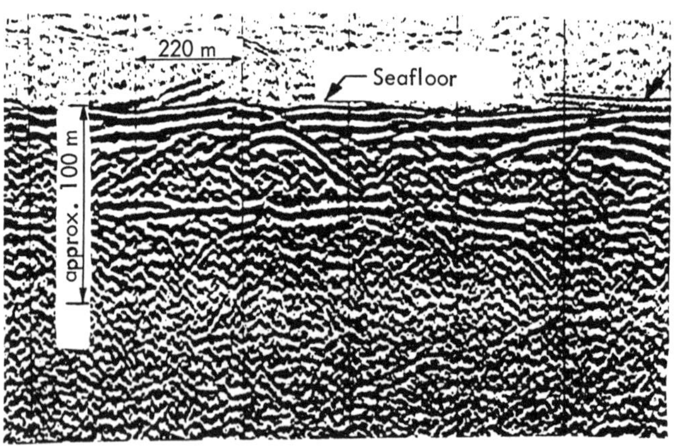

Fig. 8. - Geophysical Record of Infilled Scour Mark

Major and minor channelling was identified from the geophysical data within Unit B; in these areas lithological and geotechnical properties may vary locally. Boring locations were selected outside areas of major channelling. The upper sediments of Unit C were interpreted as having hard or compact lithologies such as sandstones or claystones; these sediments were within the soil borings target penetration.

Exploration drilling records confirmed the possible presence of boulders within the upper 80 m of the soil profile. Statoil's Exploration Department's opinion was that soil borings would be difficult, and at some locations might be impossible because of boulders. Locally exploration well drilling was proving troublesome in this upper 80 m zone, with very high bit wear, slow penetration rates, local drilling refusal resulting in hole restart, and in one case the drill string twisting off.

Drilling Equipment Soil investigation planning was dominated by the anticipated difficult drilling and sampling conditions. To ensure adequate boring penetration, the following additions to normal soil drilling equipment were provided:

- Down the pipe boulder breaking hammer chisels.
- Centre plug latch-in units with hard rock formation tricones to drill ahead of the standard open centre drag bit.
- Open centre drag bits fitted with four rock roller cutting cones which would also accomodate the centre plug units.
- Standard rock formation tricones to drill a blind pilot hole in difficult ground.
- A piggy-back drilling system using HQ and NQ rods running a wire line core barrel down the centre of an API main string.

The cost of these additions was NOK 200000, or (USD 35 000). Except for the standard rock formation tricone all of the above items were deployed to achieve penetrations to 80 m in difficult ground. Only at two locations did a boulder bed cause total refusal at 14 m and 38 m penetration respectively.

CONTRACT ORGANISATION

The soil investigation was organized by Statoil with separate contractors providing a drillship, soil testing and engineering services, and offshore navigation, Fig. 9.

Because of site remoteness and the inherent high transportation cost, the soil investigation was carefully planned to identify potential problems. The drillship M.V. Pholas, a dynamically positioned vessel of 6636 tons displacements and 99 metres in length, was used to provide support for the drilling, sampling and in situ testing operations. Drilling services were performed as a subcontract to the drillship. The geotechnical consultant was responsible for supervising drilling operations and performing in situ cone penetration testing and sampling using their subsea, remote controlled seafloor jack. Offshore and onshore laboratory testing and engineering analysis were also performed by the geotechnical consultant. High accuracy position fixing, $\pm$ 10 m some 90 km offshore, was achieved using Syledis. Contractor co-ordination was the responsibility of Statoils offshore representative. Detailed daily reports were sent ashore to carefully monitor progress and costs.

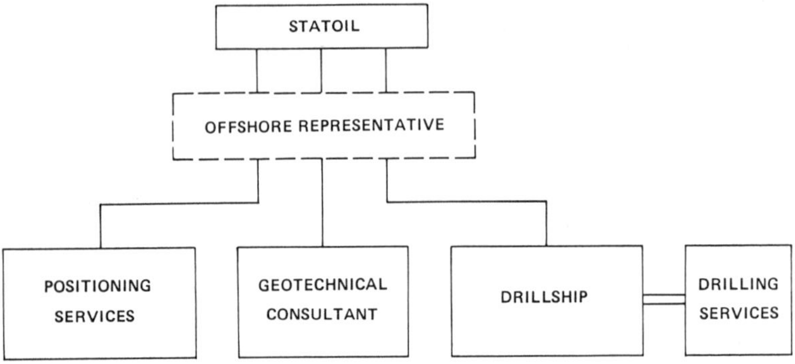

Fig. 9. - Contract Organization

OFFSHORE PERFORMANCE

The soil investigation was performed between June 13th and July 12th 1982, including all port mobilization time, travelling to site, positioning and working. Five possible platform sites were investigated, Fig. 3. In total, 539 metres were drilled with 232 samples and 279 metres of cone penetration testing. This is equivalent to 1.3 hours per metre over the total job time.

The most successfull drill bit was a open centre drag bit shaped to roll boulders away from the tip. Average bit life was 49 metres, however, at one location a bit completely wore away in one hour drilling for 1.5 metres.

Downtime for repair and maintenance of the geotechnical equipment over the investigation was 6.8 percent. The majority of this being related to equipment recovery from the seafloor which in deep water usually takes longer than effecting the repair. As is typical in large scale investigations deploying a wide range of interrelated equipment provided by different contractors, most of the downtime occurred in the first few days as the system was "field adjusted".

Weather conditions were favourable during most of the offshore period. However, wind force 8 and heavy swell was experienced on one occasion. Using the dynamic positioning system referenced to a taut wire input and acoustic beacons on seafloor, the drillship maintained station with zero weather downtime during the operations.

The use of electronic instrumentation available on the remote controlled, sub-sea hydraulic power pack that operated the seafloor jack enabled high quality push sampling and cone penetration testing to be performed in the 245 to 270 metres water depths. In addition to enhancing control of all downhole operations, the system provided an accurate tidal record and vital operational data on jack tilt and roll during deployment and drilling operations.

The difficult soil conditions exposed the geotechnical sampling and testing equipment to damage, especially as it was used to its operational limit. The total cost of field damaged equipment and consumed items was NOK 100 000, (USD 15 000).

An accurate daily log of field costs was recorded for the entire drillship operation. This enabled the final costs to be controlled to within five percent of the budget.

LABORATORY TESTING

Typically there are three phases in the laboratory testing programme for a major soil investigation. The first phase is performed offshore as samples are recovered while the second and third phases are performed onshore. The overall programme was designed to provide sufficient information for feasibility analysis for different foundation concepts.

Phase 1: Offshore Testing. All specimens were extruded from the sample tubes offshore for quality assessment and detailed soil description. Offshore testing included water content determination and shear strength estimates of clay. Quick undrained triaxial compression tests were performed with samples often confined within 15 minutes of recovery.

Phase 2: Classification Tests. For the predominantly clay soils found at the five sites classification tests were run to provide index soil properties to aide assessment of in situ stresses (4, 5, 6, 7). These tests included Atterberg Limit determinations, grain size analysis, water content determination, unconsolidated-undrained triaxial compression tests and consolidation tests. This phase cost 25 percent of the laboratory testing budget and took about four weeks to complete.

Phase 3: Detailed Testing. Laboratory testing for the gravity structure was more extensive than that required for conventional piled structures. Both triaxial compression and extension tests were performed on samples consolidated anisotropically or isotropically to either stresses representing in situ stress conditions or the stress state imposed by structure loading. The test programme provided information on the active and passive shear strength profiles (8) required in stability analyses. Static simple shear tests provided static horizontal shear strength data. Soil strength degredation due to cyclic loading was evaluated by performing cyclic simple shear tests. Typical results from the cyclic tests, are compared with results from North Sea locations and published information, Fig. 10.

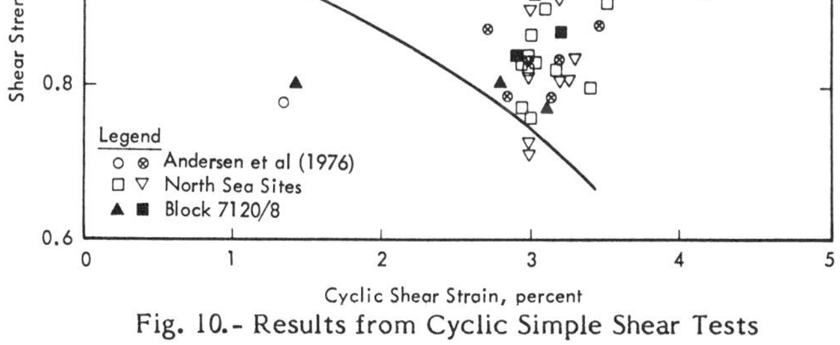

Fig. 10.- Results from Cyclic Simple Shear Tests

As part of the testing programme, micropetrographic, micropalaeontological and palynological tests were made to determine mineralogy and sediment age. This was useful not only for detailed description of the soils, but for correlation with published geological data (10, 11) and comparison with mineralogy of known soils. The results provide the soils engineer with a better understanding of the behaviour of the soils and a basis for interpolating and extrapolating geotechnical conditions between and outside the investigated sites.

Phase three laboratory testing took 2 months and was about 75 percent the laboratory testing budget.

SOILS ENCOUNTERED

The sediments encountered at the five sites were similar to those anticipated at the planning stage. Marine clay (Stratum I: Holocene) and glacial and glacio-marine clay (Stratum II: Pleistocene) overly a claystone (Stratum III: Eocene). Strata thicknesses are:

		Strata Thickness, m		Elevation, m
Site 1	Water Depth, m	Stratum I	Stratum II	Strata I-II
1	270	6	36	276
2	245	25	55	270
3	260	18	52	278
4	270	10	25	280
5	250	11	6+	261

Measured undrained shear strength and cone point resistance plotted against elevation, Fig. 11, show the foundation soils are relatively strong and suitable for the structure concepts being considered. Laboratory testing indicates no unusual behaviour.

Marine Clays. Plasticity and gradation characteristics of the marine clay (Stratum I), Figs. 12 and 13, are similar to those of marine clays encountered in the North Sea. The predominant clay mineral is illite with smaller quantities of chlorite and kaolinite and traces of smectite.

Glacial and Glacio-Marine Clay. Mineralogy is similar to the marine clay but with slightly greater percentage of chlorite and smectite. As is typical of glacial deposits, the composition of these soils is highly variable, Figs 12 and 13. There are numerous silt, sand, gravel and cobble inclusions throughout. These clays are hard, incompressible and overconsolidated and are similar to glacial clays encountered in the North Sea.

Claystone. The cristobolite found in the claystone may indicate volcanic origin, and the clay minerals present, the nature of the microfloral assemblages and the distinctive bedding suggest a marine depositional environment. These factors plus the early Eocene age are consistent with the material being a volcanic tuff deposit.

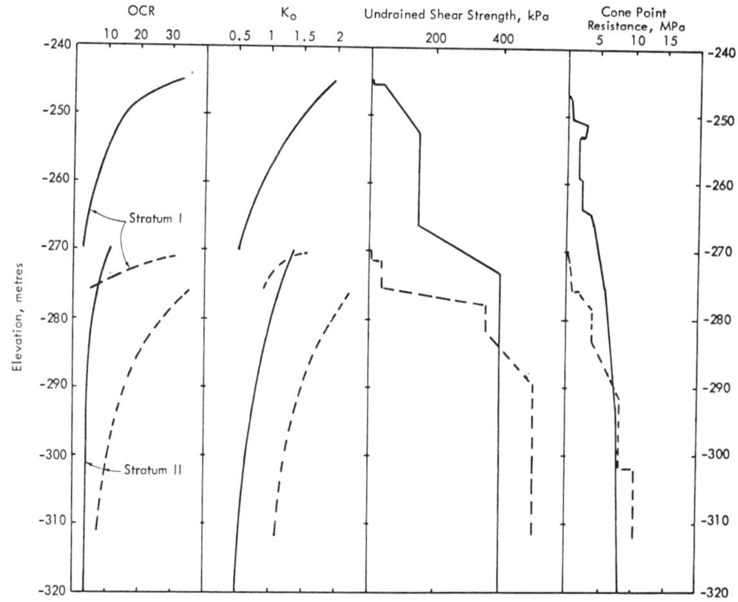

Fig. 11. - Typical Soil Properties

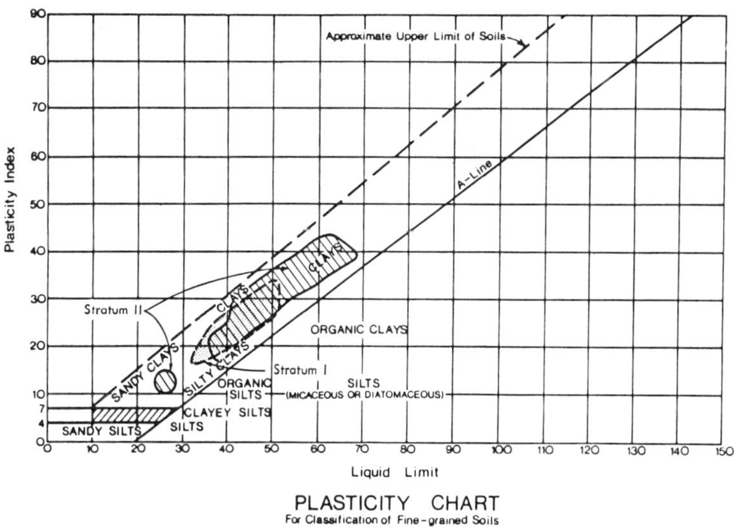

Fig. 12. - Plasticity Characteristics

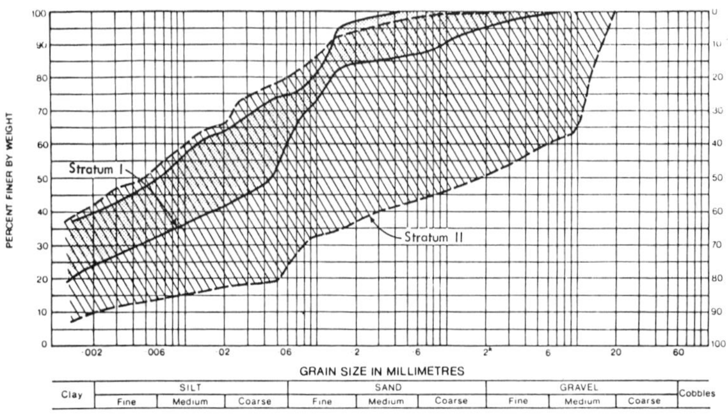

Fig. 13. - Grain Size Curves

CONCLUSIONS

(1) This investigation was successful in providing the required soil information within the time table and budget allocated. As the first large scale deep water investigation undertaken by Statoil it provides an important reference source for planning future offshore operations.

(2) Deep water increases the costs of soil investigations for every offshore operation in the following ways:

- More sophisticated and more expensive equipment is required.
- Deployment of downhole tools takes more time.
- Recovery of equipment for repair takes more time.
- Slower operations are more exposed to interruptions from weather and breakdown.

(3) Extensive planning is required for the relatively short but very expensive offshore operation. Shallow seismic records, data from geological references and exploration drilling experience provide valuable information at the planning stage.

(4) The geotechnical equipment available for deep water operations has good performance characteristics. Equipment development could enhance the quality and performance of such soil investigations. Items for consideration are:

- Provision of a TV on the seafloor jack to monitor operations.
- Provision of current direction and velocity devices on the seafloor jack to assist in deployment of the jack and to provide useful environmental data.
- Provision of a remote position monitoring device for the seafloor jack.
- Further development of in situ testing and sampling equipment. Sample disturbance due to stress relief increases with water depth. For sensitive soils, as may be found on the continental slopes, enhanced in situ testing devices are required. For very hard layers special sampling equipment is needed.

ACKNOWLEDGEMENT

The authors are very greatful to colleagues at Statoil and McClelland Engineers respectively for support in preparing this paper. Also to all those who contributed to this successful soil investigation, crews from Coe Metcalf Shipping Ltd., owner of the drillship, Seacore providing drilling services, McClelland Engineers S.A. the geotechnical consultant, A/S Geoteam providing position fixing and Det norske Veritas by N.F. Braathen acting as Statoil's offshore representative during part of the work.

REFERENCES

1. Campbell, Kerry J., Dobson, Mark B., and Ehlers, Clarance J., "Geotechnical and Engineering Geological Investigations of Deep Water Sites," Fourteenth Annual Technology Conference, Vol. 2, May, 1982, pp.
2. Amundsen, T. and Lauritsen, R., "Time and Cost Planning for Offshore Soil Investigations", Boss-82 Conference, Boston, August, 1982.
3. Bea, R.G., Dover, A.R., Audibert, J.M.E. "Pile Foundation Design Considerations for Deepwater Fixed Structures", Boss-82 Conference, Boston, August, 1982.
4. Skempton, A. W., Discussion of "Planning and Design of the New Hong Kong Airport," Proceedings, Institution of Civil Engineers, London, Vol. 7, 1957, pp. 305-307.
5. Brooker, E. W. and Ireland, H. O., "Earth Pressures at Rest Related to Stress History," Canadian Geotechnical Journal, Vol. 2, No. 2 May, 1965, pp. 1-15.
6. Ladd, C. D. and Edgers, L., "Consolidated-Undrained Direct-Simple Shear Tests on Saturated Clays," MIT Department of Civil Engineering, Research Report No. R72-82, July, 1972, pp.
7. Koutsoftas, D. and Fischer, J. A., "Earth Pressures at Rest Related to Stress History," Canadian Geotechnical Journal, Vol. 13, No. 2, 1976, pp. 1-15.
8. Andersen, A., Berre, T., Kleven, A. and T. Lunne, "Procedures Used to Obtain Soil Parameters for Foundation Engineering in the North Sea," Marine Geotechnology, Vol. 3, No. 3, 1979, pp. 201-266.

9. Andersen, K. H., Brown, S. F., Foss, I., Pool, J. H. and Rosenbrand, W. F., "Effect of Cyclic Loading on Clay Behaviour," Proceedings, Conference on the Design and Construction of Offshore Structures, The Institution of Civil Engineers, London, 1976, pp. 75-79.

10. Gunleiksrud, T. and Rokoengen, K., "Regional Geological Mapping of the Norwegian Continental Shelf with Examples of Engineering Applications," Proceedings, Conference on Offshore Site Investigation, Society of Underwater Technology, London, 1979, pp. 23-35.

11. Rokoengen, K., Bugge, T. and Lofaldi, M., "Quaternary Geology and Deglaciation of the Continental Shelf off Troms, North Norway," Boreas, Oslo, Vol. 8, 1979, pp. 217-227.

DISPLACEMENTS AND STRAINS AROUND PROBES IN SAND

By John L. Davidson,[1] M ASCE and Alireza Boghrat[2]

INTRODUCTION

In the past decade the use of insitu testing for geotechnical purposes, both onshore and offshore, has increased dramatically. Compared to the traditional drilling, sampling and laboratory testing procedures, insitu testing has several important benefits. These are that insitu tests are performed in the natural environment of moisture and stress, with reduced disturbance and on large volumes of soil. They are also generally quicker and cheaper, relative to the quantity of data acquired. New pieces of insitu test equipment with increasing sophistication and capabilities are continually coming onto the market.

This paper seeks to compare and evaluate two standard shapes of insitu test equipment. The first shape, the 'cone', is a 35.7 mm diameter cylindrical rod with a 60° apex cone, Figure 1. This is the shape of the standard electric Cone Penetration Test tip (14) and of many of the available piezocones (e.g. 7, 10, 16, 17), Figure 2. The second shape, the 'blade', is a flat plate 235 mm x 95 mm x 14 mm, with a bevelled lower end and a threaded connector, Figure 1. This is the shape of the Marchetti dilatometer (12, 13) and of the University of Florida piezoblade (3), Figure 3.

Both cone penetration tests and piezocone tests are standard offshore investigation tools, performed from a ship or barge, from a sea floor system (9) or from a deep submersible (11). Dilatometer tests have also been performed from barges. Dr. Marchetti has tested in water depths up to 30 meters to total depths of 60 meters. The device was also used recently by Schmertmann and Crapps, Inc. during site investigation for the Sunshine Skyway bridge in Tampa, Florida.

Since these types of tests are performed by penetrating a solid probe into the ground, a certain disturbance of the soil results. The magnitude, location and type of disturbance are all important in evaluating the usefulness and reliability of the test.

This paper describes a laboratory testing technique which allows a determination of the displacements and strains around a penetrated

---

1. Associate Professor, Department of Civil Engineering, University of Florida, Gainesville, Florida.

2. Graduate Assistant, Department of Civil Engineering, University of Florida, Gainesville, Florida.

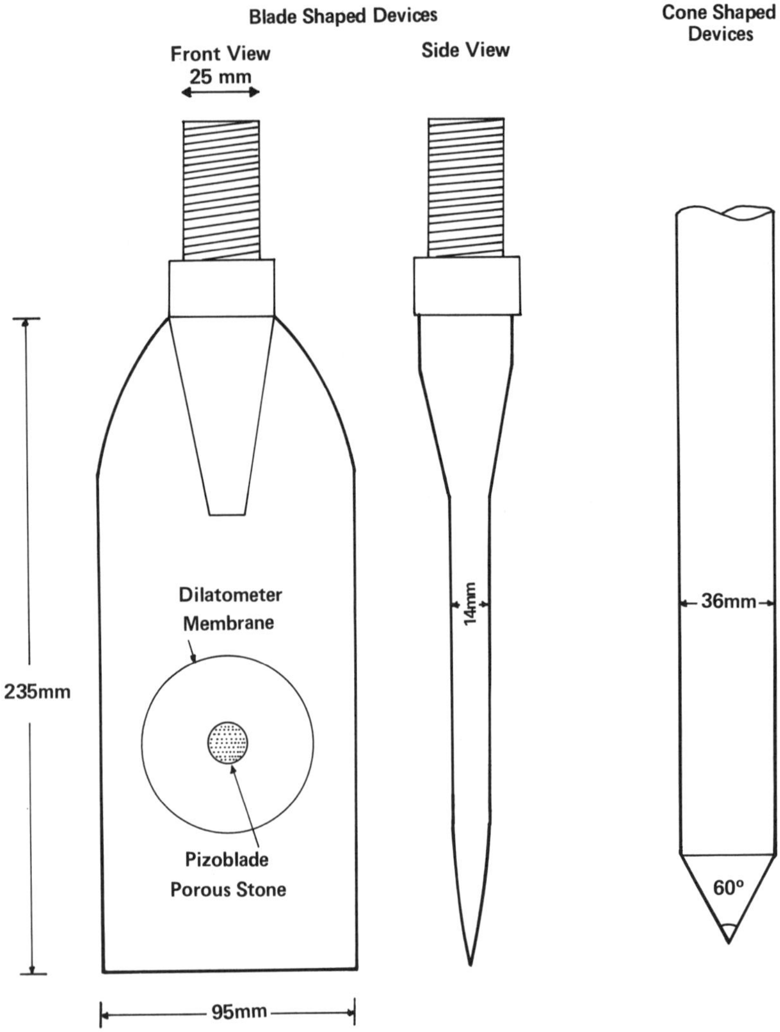

Figure 1. Sketches of the Blade and Cone

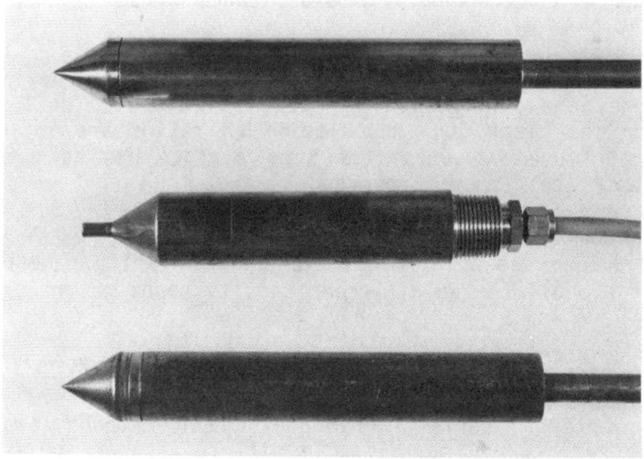

Figure 2. Electric Cone Tip, Wissa Piezocone and Fugro Piezocone

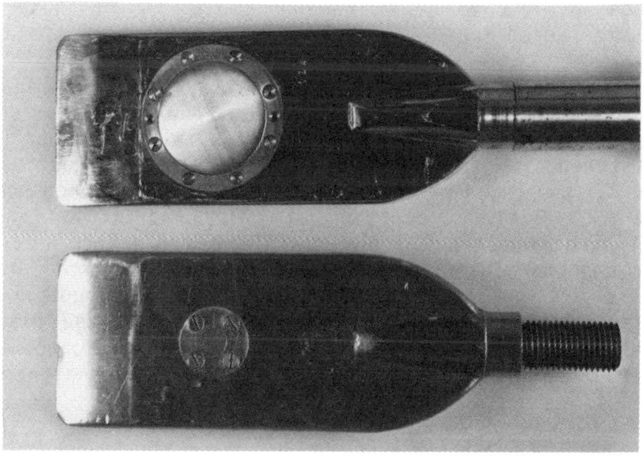

Figure 3. Marchetti Dilatometer and University of Florida Piezoblades

probe. The two standard shaped probes were tested in this manner, and are compared with regard to the disturbance caused around each during insertion.

The tests reported in this paper are all of shallow penetration - for the cone, the penetration tip depth to diameter ratios are only 6 and 8 - for the blade, the depth to width ratios are 16. A few tests have been performed in which the sand surface has been surcharged to simulate depth, but are not reported herein.

This paper is an extension of an earlier paper (6) in which the testing technique was briefly described and in which penetration of the cone shaped probe into two different density sands was reported.

PSEUDO STEREO TECHNIQUE

In conventional aerial stereo photography methods, two photographs of an object are taken from slightly different camera positions. These photographic plates are then positioned in a stereo projector and a stereoscopic image obtained. Figure 4 illustrates the taking of such photographs and the corresponding stereo image obtained by a simultaneous projection of the plates. The height of the object in the stereo image is due to the difference in locations of the corresponding points in the two photographs, called x-parallax. This is the basic principle of stereo photography.

Parallax can also be produced by keeping the camera position stationary and photographing the displacement of an object. Figure 5 illustrates a fixed camera photographing a translated object and the stereo image obtained by a simultaneous projection of these plates. The height of the stereo image depends on the magnitude of the x-movement.

If the object undergoes a general planar displacement, i.e., movement components in two directions, only the component of displacement along the line joining the projectors contributes to the elevated stereo image. In order to completely define an actual displacement, two measurements must be made. The component of the displacement is first measured in any particular direction by positioning the photographic plates in the projector so that the line connecting the two plates is parallel to that direction. The plates are then rotated 90° in order to determine the perpendicular component of movement. The actual magnitude and direction of the displacement is then obtained by a vector sum of these normal components. This technique, called pseudo stereo photgraphy, was pioneered at the University of Southampton, U. K. (1, 4).

TEST EQUIPMENT AND PROCEDURE

A wooden container with one glass wall was constructed, with inside dimensions of length 100 cm, width 50 cm and height 65 cm. The inside volume, measuring to the level to which the sand was usually placed, was 0.267 m^3 (9.44 ft^3). A square grid or static datum plane was scored on the inside face of the glass. The middle third of the glass length had

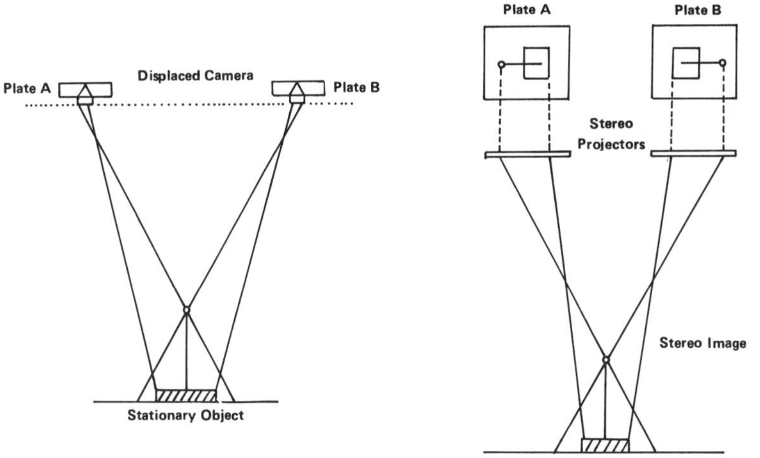

Figure 4. Stereo Image - Displaced Camera

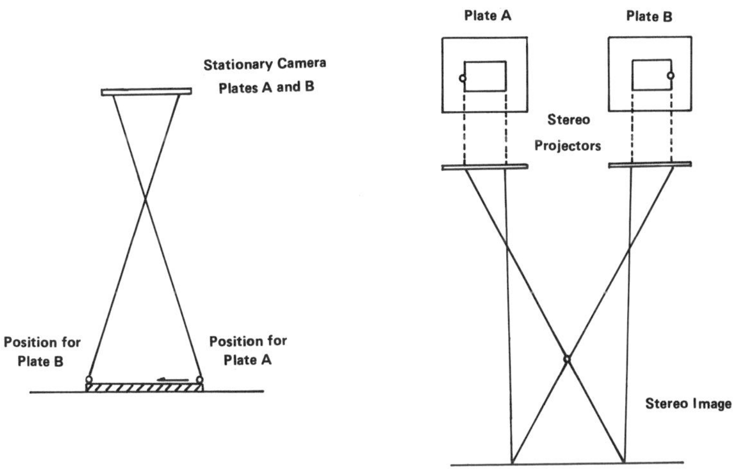

Figure 5. Stereo Image - Displaced Object

a 15 mm x 15 mm square grid, while the two outer thirds had 30 mm x 30 mm grids. Before each test, the grid was darkened with a black marker, so that it could be more easily seen on the photographs and in the stereo image.

The sand used in the research was a blend of four graded sands; 8/20, 20/30, 30/64 and EGS, from the Edgar Plastic Kaolin Company, Florida. The pairs of numbers, e.g. 20/30, signify the sieve size range in which 75% or more of the soil particles lie. The blend had an effective size D_{10} of 0.2 mm and a unformity coefficient C_u of 7. The sand was placed in the container and tests performed at five different densities, obtained by different filling and mechanical compaction procedures.

Dummy cone and blade probes were machined from stainless steel, then cut longitudinally in half, Figure 6. The test procedure consisted of penetrating a probe into the sand with the flat, cut face against the glass, Figure 7. A small hydraulic jack was employed. A series of 6 or 8 photographs were taken - one prior to penetration, and the rest during penetration - from a rigidly fixed camera. Figure 8 shows the test set-up.

The three-dimensional problems are thus studied by physically cutting the probes and viewing a two-dimensional plane - a radial plane in the case of the axisymmetrical cone and a plane strain plane in the case of the blade. Direct shear tests were performed to measure the friction between the sand and the glass of the container wall. For a dense sand ($\phi=40^0$) a friction angle of 14° was found for smooth glass and 16.5° for a plate scored with the 15 mm grid (5).

To avoid possible distortion of parallax due to quality of photographic paper or the enlargement process, photographic positives were used rather than prepared prints. The positive slides were sandwiched between thin, clear photographic glass plates and used directly in the stereo plotter. The camera was fixed in position relative to the model, with its optical axis normal to the displacement plane. A B & J orbit view camera with a 4" x 5" plate holder was used with Ektachrome 50 Professional Tungsten Film. After each test the container was carefully emptied, the sand weighed and the bulk density determined.

For analysis of each pair of test photographs, a Fotocartigrafo Nistri Model VI double-projection anaglyphic plotter was used, Figure 9. This type of plotter employs the anaglyph principle of projection of light of two complementary colors, red and blue-green, to create the stereo image. The overlapping image created by projecting through two photographic plates is viewed through glasses with one red and one blue-green lens. Since each eye sees only one of the photographs of the pair, the three-dimensional effect is created. The portion of the photographs which is of interest is projected onto the white surface of a moveable platen assembly. A small white dot on the platen which, when viewed through the glasses, appears to be floating in space, can be moved vertically and placed on the surface of the 3-D model.

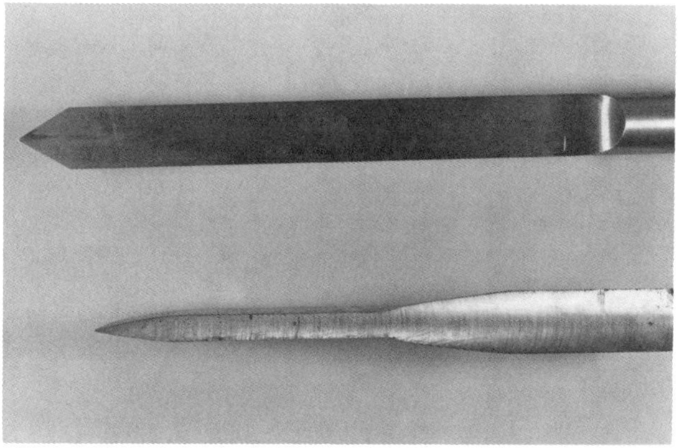

Figure 6. The Test Probe and Blade

Figure 7. Probe Penetrated into Sand

Figure 8. Test Set Up

Figure 9. Double Projection Anaglyphic Plotter

Figure 10 shows two pairs of stereo photographs. Two readings are taken at each grid point. The floating dot is first placed on the grid and a reading from the platen dial recorded. The dot is then placed on the sand, which may appear above or below the stationary datum grid, and a second reading taken. The difference in these readings is a measure of how far the sand moved at that point, during the time between photographs, in the direction parallel to the line of the projectors.

The grid scored on the glass was numbered in finite element terminology with 1234 node points and 1213 elements. At each node point where sand movement had occurred, the two floating dot readings were recorded and subsequently entered as data for a computer reduction program. Each pair of photographs was viewed twice to provide orthogonal displacements which could be summed to give the actual movement vector.

Displacements were determined by

$$U \text{ or } W = C \cdot \Delta h$$

where

U = horizontal displacement in millimeters
W = vertical displacement in millimeters
Δh = difference in floating dot grid and sand readings
C = constant which is a function of the stereo plotter set-up
= 0.012958 for the cone study
= 0.016402 for the blade study

$$D = \sqrt{(U^2 + W^2)}$$

$$\theta = \tan^{-1} (W/U)$$

where

D = actual displacement
θ = angle of displacement from horizontal

Volumetric strains were defined for each element by

$$\varepsilon_V = \frac{\Delta V}{V}$$

where

ε_V = volumetric strain
ΔV = change in volume
V = original volume

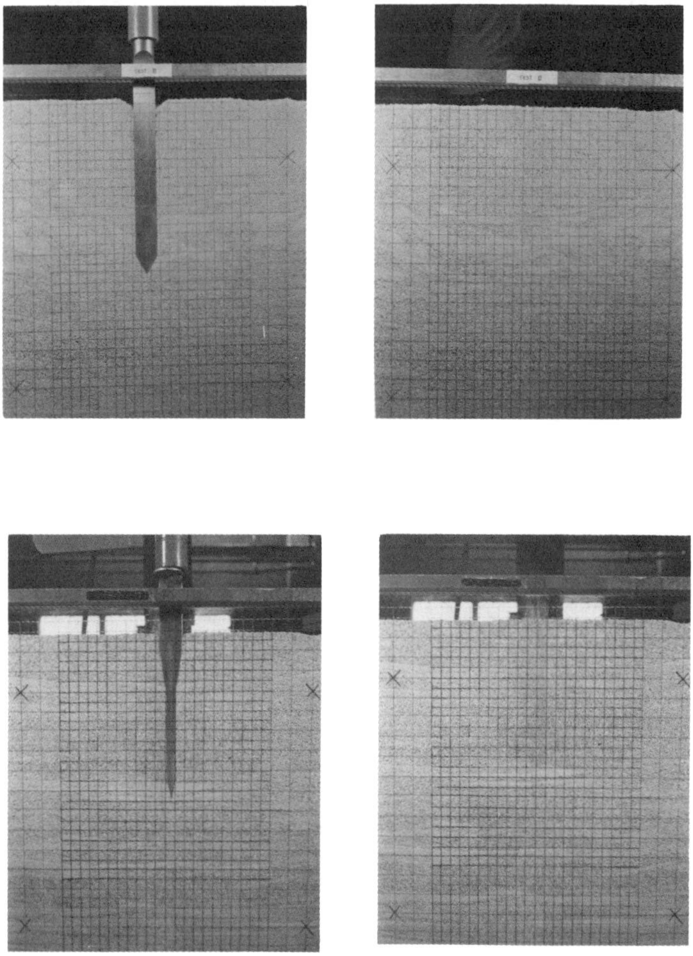

Figure 10. Two Stereo Pairs

In the case of the blade shaped device, plane strain conditions were assumed to apply, since the width to thickness ratio was 6.8. The volumetric strain was therefore defined as

$$\varepsilon_v = \frac{A - A_o}{A_o}$$

where

A_o = original area of a grid element
A = deformed area of an element defined by the ends of the four corner node point displacements, D.

For the case of the axisymmetrical cone the effect of the tangential strains must be included. The volumetric strain was therefore calculated using the equation

$$\varepsilon_v = \frac{V - V_o}{V_o}$$

where

V_o = the volume found by revolving the original underformed square grid element around the axis of symmetry
V = the volume found by revolving the deformed quadrilateral element around the same axis

A decrease in area or volume, i.e., compression, was defined as a positive volumetric strain, while an increase was defined as negative.

Shear strains were determined at each node point by

$$\gamma_{xz} = (\gamma_{xz_2} + \gamma_{xz_4})/2$$

where

γ_{xz_2} = shear strain in the 2nd quadrant grid element, node at the origin
γ_{xz_4} = shear strain in the 4th quadrant grid element, node at the origin

$$\gamma_{xz_i} = \alpha_i + \beta_i$$

where

i = 2 or 4

α = the change in direction of the horizontal side of the element at the node point, in radians

β = the change in direction of the vertical side of the element at the node point, in radians

An increase in the original right angle was defined as a negative shear strain and a decrease in angle was defined as positve.

The accuracy of the stereo measurements were checked by multiple reading of the parallaxes at several node points, especially close to the probe where grain movement is large. The variation found translates into a maximum error of ±7% in the displacement vectors (5) and corresponding errors of similar magnitude in the volumetric and shear strains. Andrawes and Butterfield (1) were able to measure individual grain displacements to an accuracy of better than 0.005 mm.

A detailed description of the test equipment and procedures, and of the computer reduction program, can be found in a report by Davidson (5).

RESULTS

In this research program, ten laboratory penetration tests were performed, using two different probes and five different density soils. In this paper, because of length restrictions, only four of these tests, the two probes in two different soils are reported. The 'loose' sand was placed, using a 3/4" funnel held 5-7 inches above the sand surface. The 'dense' sand was placed in 4 inch lifts, using the same method. Each lift was then compacted with 5 passes of a mechanical compactor.

Only one pair of photographs from each of these four tests is analyzed. One photograph is of the sand prior to insertion, the other is the last photograph taken, at final penetration, Figure 10.

Three figures are plotted for each of the four tests. These are node point displacement vectors, contours of volumetric strain and contours of shear strain. Since the probes are both symmetrical, results are plotted for only half the problem, the right hand side.

Individual tests will first be discussed and then the cone and blade will be compared.

Cone - Loose Sand (Figure 11)

All the displacements are to the right and down, away from the cone. The vectors are all directed at angles less than 45° to the horizontal, except in a bulb shaped zone beneath the tip. The largest displacement measured was approximately 4 mm. The effect of the cone was measurable outward to a distance of 125 mm from the edge of the probe and downward 135 mm below the tip.

Negative volumetric strains were detected below the tip and immediately adjacent to the shaft of the cone. Minor irregular densification was observed further from the probe.

The shear strains are primarily negative (increase in right angle) except for a number of very low positive readings directly below the cone tip. The strains are fairly uniform adjacent to the probe shaft and except for one small pocket are small, less than -2%. Adjacent to the tip is a bulb of high negative shear straining with a maximum reading of 10%.

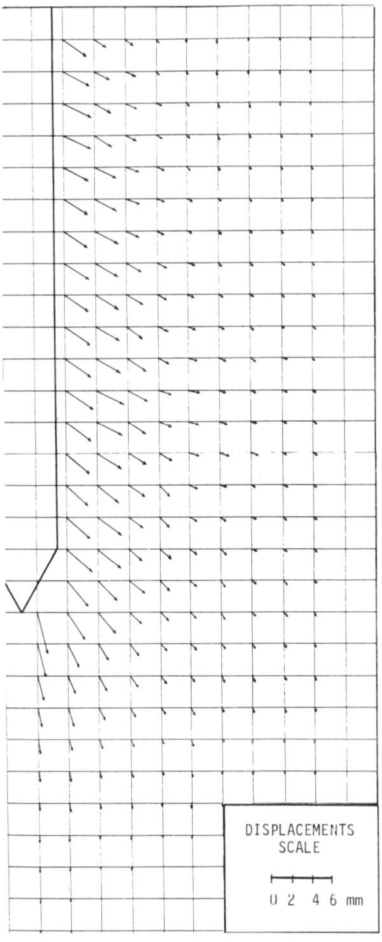

Figure 11a. Cone - Loose
Displacements

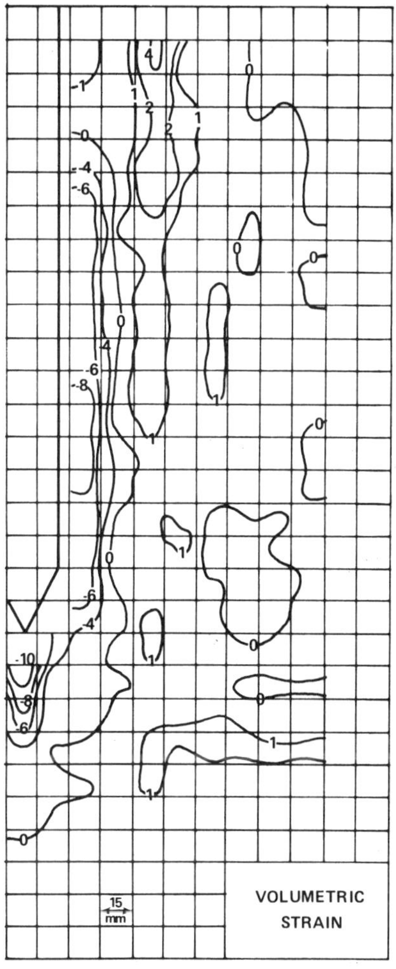

Figure 11b. Cone - Loose
Volumetric strains

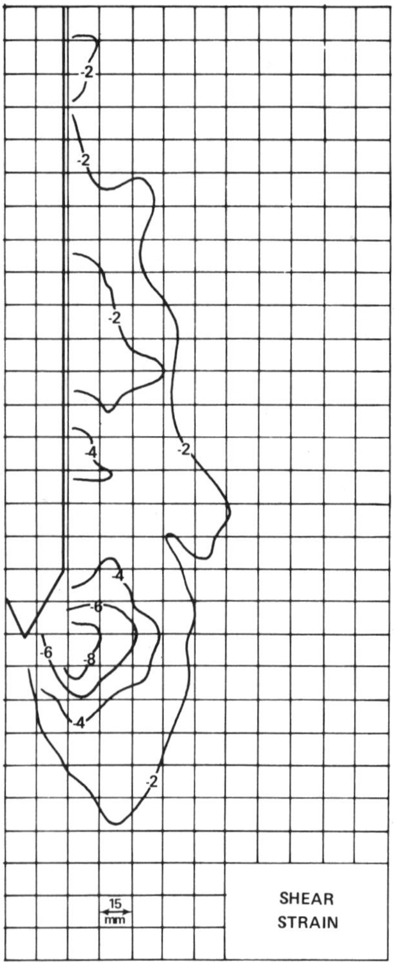

Figure 11c. Cone - Loose
Shear strains

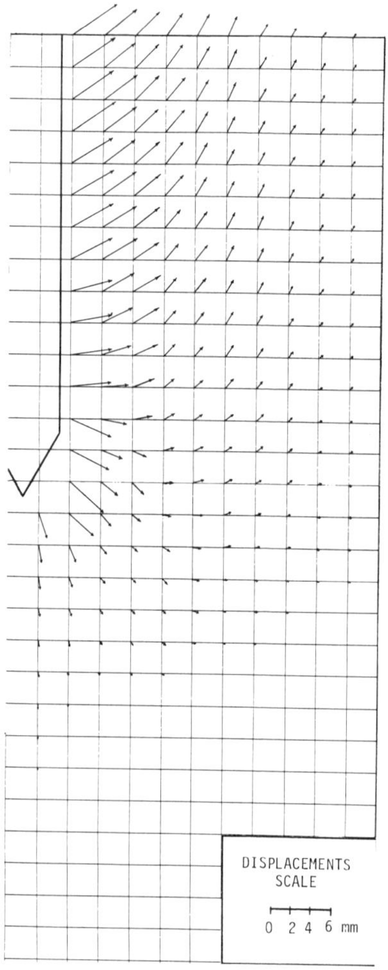

Figure 12a. Cone - Dense
Displacements

Cone - Dense Sand (Figure 12)

The displacement vectors for the dense sand show a downward and outward movement below the cone tip, and an upward and outward movement above the tip. There is a noticeable pattern of curved surfaces of particle movement from the probe to the sand surface. The maximum displacement is approximately 5 mm. Displacements were measurable horizontally to 155 mm, and vertically to 100 mm below the tip.

For this sand, negative volumetric strains are detected below the tip and along the shaft. However, unlike the loose sand, a zone of densification (ε_v = +6%) is found adjacent to the tip.

As in the loose sand, the shear strains are primarily negative. Again, a concentration of straining is found adjacent to the tip, in this case of maximum magnitude -12%. Along the shaft the strains are slightly higher (≈ -4%), becoming positive as the soil surface is approached.

Blade - Loose Sand (Figure 13)

Displacements in this loose sand are all to the right and down, away from the probe. Movements were measurable to 100 mm from the edge of the probe and downward below the tip. The maximum particle displacement adjacent to the flat surface of the probe is 2.2 mm.

The volumetric strain readings are all positive except for a few negative values directly below the tip. The contours are very uniform adjacent to the blade face, showing decreasing densification with distance from the probe.

The shear strains are all negative, except for some low positive values on the symmetry line below the tip. Adjacent to the blade's flat surface the shear strains are very uniform, decreasing from ≈ -2.5% to zero with distance from the probe.

Blade - Dense Sand (Figure 14)

The displacement vectors for this test show outward and upward movements, except below the probe tip level, where downward movement occurs. A very sharp distinction exists at the tip level. The outward movement was measurable to a distance of 160 mm from the edge of the probe and down to a depth of 90 mm below the tip. The maximum movement adjacent to the flat area of the probe is 3.6 mm.

The volumetric strain readings are all positive, except for two small zones directly below and adjacent to the tip. The contours are again very uniform in the soil adjacent to the flat surface of the blade, decreasing from 4% with distance from the probe. The contour pattern is more complex around the tip and near the width enlargement, or shoulder, at the top of the blade.

The shear strain contours show zones of both positive and negative straining. Positive shear strains are observed below the tip, and adjacent to the blade's flat surface, decreasing here from ≈1% to zero

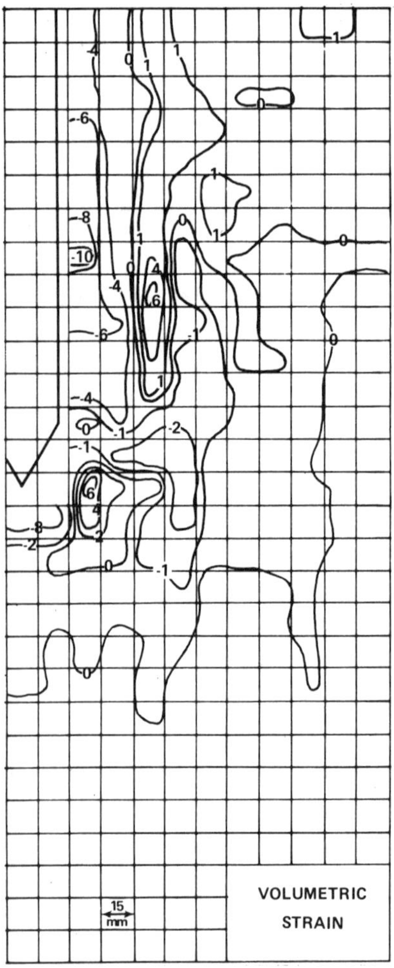

Figure 12b. Cone - Dense
Volumetric strains

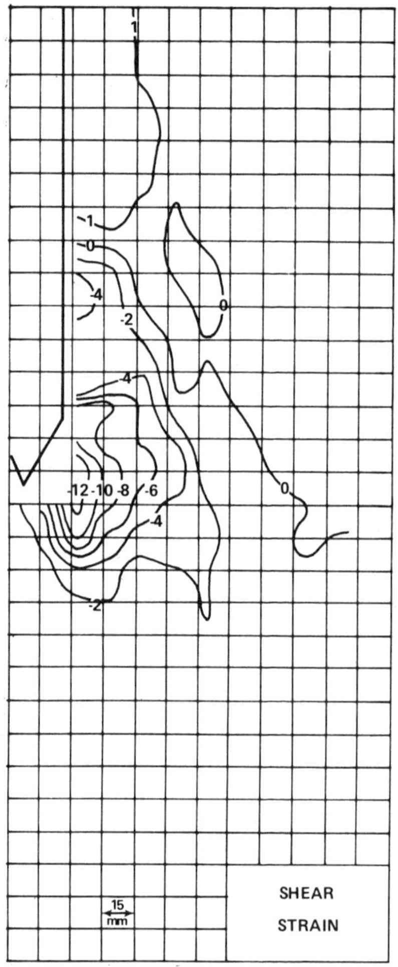

Figure 12c. Cone - Dense
Shear strains

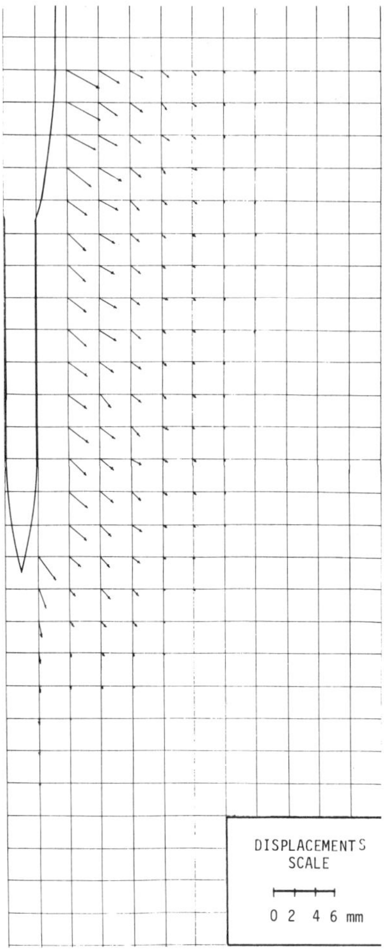

Figure 13a. Blade - Loose
Displacements

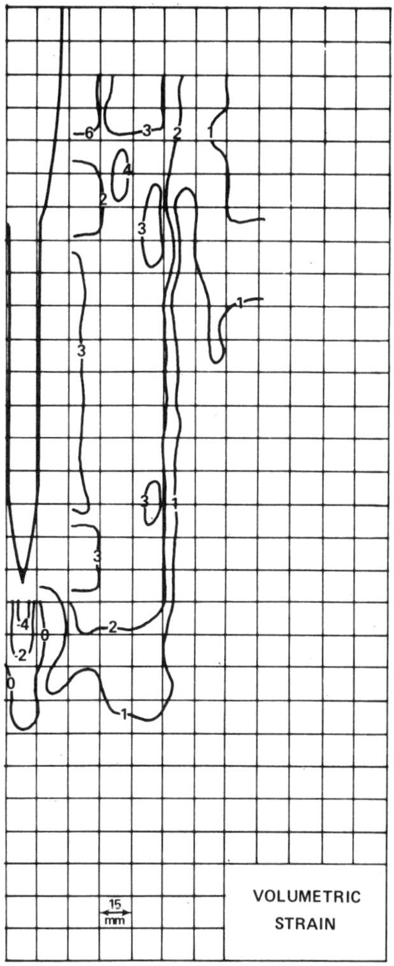

Figure 13b. Blade - Loose
Volumetric strains

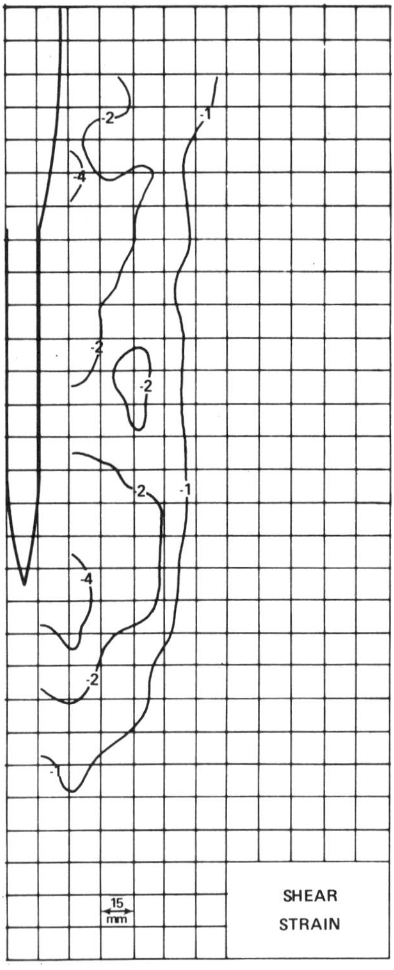

Figure 13c. Blade - Loose
Shear strains

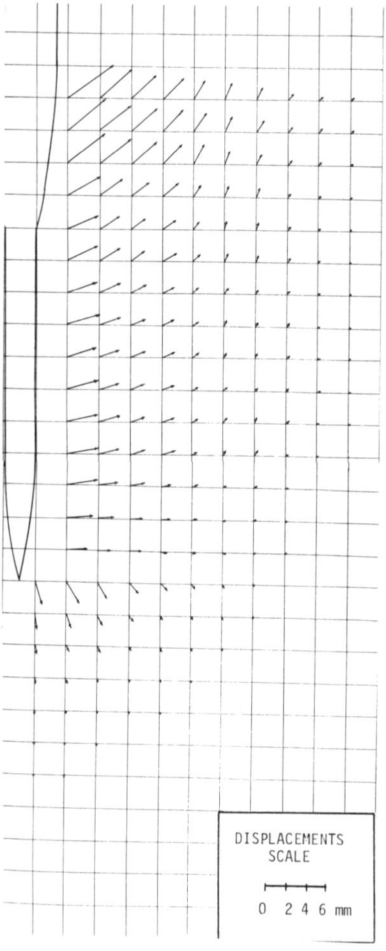

Figure 14a. Blade - Dense
Displacements

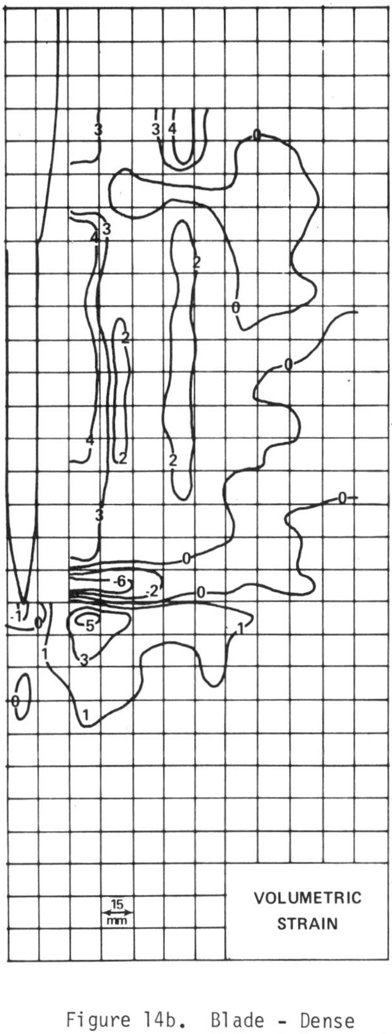

Figure 14b. Blade - Dense
Volumetric strains

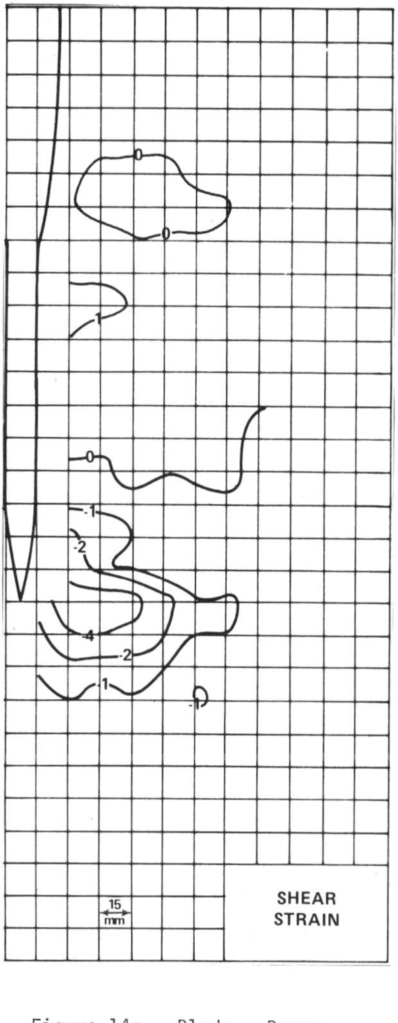

Figure 14c. Blade - Dense
Shear strains

with distance. A zone of negative shear strain, maximum value -5%, is found adjacent to the bevelled region at the bottom of the blade. A small negative zone is located near the blade shoulder.

COMPARISON OF CONE AND BLADE

Displacement

In the loose sand, both probes produced downward and outward displacement only. Near the top of the probes the movements were of the same order of magnitude. This was to be expected, since the enlarged blade width is 35 mm, while the cone diameter is 35.7 mm. Adjacent to the flat, 14 mm thick, blade surface and around the tip, the displacements were considerably smaller than for the cone shaped instrument.

Similar results were obtained in the dense sand. In general, displacements were smaller for the blade shaped device. In the dense sand, the y-direction of the movement, up or down, changes for both probes. For the blade, only the soil below the actual tip level moved downward, while in the cone the change of direction occurred just above the top of the conical tip.

Volumetric Strains

In loose sands, the cone produced an unexpected zone of loosening around the tip and shaft. Such results are somewhat corroborated by Schmertmann (14), who measured negative pore pressures with a piezocone during field testing in loose sand.

In dense sands, the cone generated high negative volumetric strains adjacent to the tip and shaft, and a zone of positive strain adjacent to the tip.

In both loose and dense sands, the blade showed very uniform volumetric strains adjacent to the flat surface and some concentrations of strain near the tip and the shoulder.

Shear Strains

For both probes, in the loose sand, shear strains were negative everywhere except directly under the tips. Adjacent to both the cone and the blade tips were high concentrations of negative shear strain. Along the cone shaft and the blade face, the shear strains were uniform and of approximately the same magnitude.

In the dense sand, both probes generated positive shear strains under the tip and along the upper shaft and blade surfaces. The cone had a very negative concentration adjacent to the tip, which extended part way up the shaft. The blade had much lower negative shear strains adjacent to the bevelled section and low uniform positive strains adjacent to the flat surface.

Conclusions

The laboratory tests were performed on solid steel dummy probes. However, to evaluate particular pieces of insitu test equipment, consideration must be given to the actual location of the measuring devices. In the blade instruments, the expanding circular membrane of the Marchetti dilatometer and the porous disc of the University of Florida piezoblade are located on the flat, 14 mm thick surface, Figure 1. In the piezocone, the porous sensing element is either a small cylindrical element at the point of the tip, (8, 10, 18) or is located on the surface of the tip or on the shaft above the tip (7, 16, 17)

For the dilatometer and piezoblade, in the region where measurements are made, displacements are small and uniform (maximum recorded values: 2.2 mm in loose, 3.6 mm in dense), volumetric strains are very uniform (maximum recorded values: 3% in loose, 4% in dense) and shear strains are very uniform (maximum recorded values: -2.5% in loose, 1.3% in dense).

Around the tip, where piezocone measurements are made, there appear to be very severe concentrations of straining. In the dense sand, volumetric strain varied around the tip from -9% to +6% and shear strains from slightly positive to less than -12%. In loose sand the shear strains varied from 0% to -10%. The volumetric strains did not vary excessively in this test, between -6% and -10%.

Since these strain contours represent a measure of soil disturbance, the blade shaped instrument, with measuring devices on the flat surface, appears to test a much less disturbed soil than do the cone shaped instruments.

The pseudo stereo photograph technique can be used to very accurately measure the displacements of sand grains at designated points in a viewed plane. In this paper both volumetric and shear strains were then calculated based on simple small strain theory, and displacement vectors and strain contours carefully drawn. Conclusions have been drawn based on the numerical values determined and on the trends visible in the plots. However, since extensive duplication of testing has not been performed, the authors do believe that the reported results should be more qualitatively than quantitatively regarded. Possible non-homogeneity and non-reproductibility of the large samples, vibrations, non-vertical penetration, random stereo reading errors etc. can not be entirely discounted.

REFERENCES

1. Andrawes, K. Z. and Butterfield, R., "The Measurement of Planar Displacements of Sand Grains," Geotechnique, Vol. 23, 1973, pp. 571-576.
2. Battaglio, M., Jamiolkowski, M., Lancellotta, R., and Maniscalco, R., "Piezometer Probe Test in Cohesive Deposits," Proceeding, ASCE National Convention, St. Louis, MO, 1981, pp. 284-301.

3. Boghrat, A., "The Design and Construction of a Piezoblade and an Evaluation of the Marchetti Dilatometer in Some Florida Soils," Ph.D. Dissertation, University of Florida, December, 1982.
4. Butterfield, R., Harkness, R. M., and Andrawes, K. Z., "A Stereo Photogrammetric Technique for Measuring Displacement Fields," Geotechnique, Vol. 20, 1970, pp. 308-314.
5. Davidson, J. L., "A study of Deformations in Sand Around a Penetrated Probe Using Stereo Photography," Report, National Science Foundation, ENG76-18849, 1980, 111 pages.
6. Davidson, J. L., Mortensen, R. A. and Barreiro, D., "Deformation in Sand Around a Cone Penetrometer Tip," Xth ICSMFE, Stockholm, Vol. 2, June, 1981, pp. 467-470.
7. deRuiter, J., "Electric Penetrometer for Site Investigation," Journal of the Soil Mechanics and Foundations Division, ASCE, Vol. 97, No. SM2, February, 1971, pp. 457-472.
8. Franklin, A. G. and Cooper, S. S., "Tests in Alluvial Sand with the PQS Probe," Xth ICSMFE, Stockholm, Vol. 2, June, 1981, pp. 475-478.
9. Ferguson, G. H., McClelland, B., and Bell, W. D., "Seafloor Cone Penetrometer for Deep Penetration Measurements of Ocean Sediment Strength," Paper No. OTC 2787, 9th Offshore Technology Conference, Houston, TX, 1977.
10. Gupta, R. C., "A Study to Establish Correlations between Piezometer Probe Test Results and Insitu Permeability," Master's Thesis, University of Florida, June, 1980.
11. Hirst, T. J., Richards, A. F., and Inderbitzen, A. L., "A Static Cone Penetrometer for Ocean Sediments," ASTM STP 501, American Society for Testing and Materials, 1972, pp. 69-80.
12. Marchetti, S., "A New In Situ Test for the Measurement of Horizontal Soil Deformability," Proceedings Conference on In Situ Measurement of Soil Properties, ASCE Specialty Conference, Raleigh, NC, Vol. 2, June, 1975, PP. 255-259.
13. Marchetti, S., "In Situ Tests by Flat Dilatometer," Journal of the Geotechnical Engineering Division, ASCE, Vol. 106, No. GT3, Proc. Paper 15290, March, 1980, pp. 299-321.
14. "Report of the Subcommittee on Standardization of Penetration Testing in Europe", Proceedings, IXth ICSMFE, Tokyo, Vol. 3, 1977, pp. 95-120.
15. Schmertmann, J. H., "Penetration Pore Pressure Effects on Quasi-Static Cone Bearing, q_c," European Conference on Penetration Testing, Stockholm, Vol. 2, 1974, pp. 345-352.
16. Tumay, M. T., Boggess, R. L., and Acar, Y., "Subsurface Investigations with Piezo-cone Penetrometer," Proceedings ASCE National Convention, St. Louis, MO, 1981, pp. 269-301.
17. Torstensson, B. -A., "Pore Pressure Sounding Equipment," ASCE Speciality Conference on Insitu Measurement of Soil Properties, Raliegh, NC, 1975, pp. 48-55.
18. Wissa, A. E. Z., Martin, R. T., and Garlanger, J. E., "The Piezo-Probe," Proceedings, ASCE Speciality Conference on In Situ Measurement of Soil Properties, Raleigh, NC, Vol. 1, 1975, pp. 536-545.

THE LPC PRESSIOPENETROMETER

By François J. Baguelin [1]
Jean-François Jézéquel [2]

The LPC pressiopenetrometer has been developed to measure soil mechanical characteristics in-situ, for offshore soil surveys, but a simplified version can be also used for onshore soil surveys.

In the present paper, the equipment and the implementation and measurement procedures are described in reference to the conditions of offshore surveys. Then, correlations are given between the results yielded by the pressiopenetrometer and those obtained from classical in-situ tests.

1. DESCRIPTION OF THE LPC PRESSIOPENETROMETER

The LPC pressiopenetrometer (Fig. 1 and 2) is operated with a vibrating unit (1), supported by a submerged, bottom sitting frame (2), such as the B.S.M.V.A. equipment shown, from Techniques Louis Ménard. Other types of vibrators or hammers can be used, once appropriate fittings are provided.

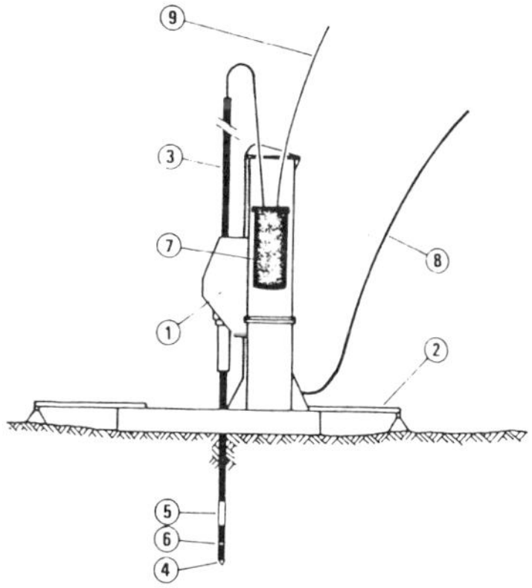

Fig. 1 - Scheme of the LPC pressiopenetrometer and of the B.S.M.V.A. vibrating unit.

[1] Technical Director, Laboratoire Central des Ponts et Chaussées, Centre de Nantes, Bouguenais, France

[2] Head Soil Mechanics Group, Laboratoire Régional des Ponts et Chaussées de Saint Brieuc, Saint Brieuc, France

The pressiopenetrometer itself has three main parts, which are connected by various electrical cables or hydraulic tubing :

- the drill rod (3), with a measurement tip at its lower end.
- the watertight chamber (7), attached to the frame, which contains various measurement apparatuses.
- the data acquisition and processing system, located on board the vessel or pontoon.

Fig. 2 - View of the pressiopenetrometer (measurement tip) and of the B.S.M.V.A. vibrating unit (CEA-TME picture)

1.1. The drill rod and the measurement tip

The drill rod, 14 m (45 ft) long, with an outer diameter of 114 mm (4 1/2 in.), consists of several elements, screwed and fixed with respect to one another. A handle is provided at the upper end, while the lower end is equipped with the measurement tip, whose outer diameter is 89 mm (3 1/2 in.). The tip includes several measurement modules (Fig. 3) :

. penetrometer cone

The force exerted on the cone by the soil is measured by electrical strain gauges (full bridge setting) and can be as high as 300 kN (≃ 65 kips) The accuracy of the cone resistance measurement is ± 50 kPa (± 0,5 tsf).

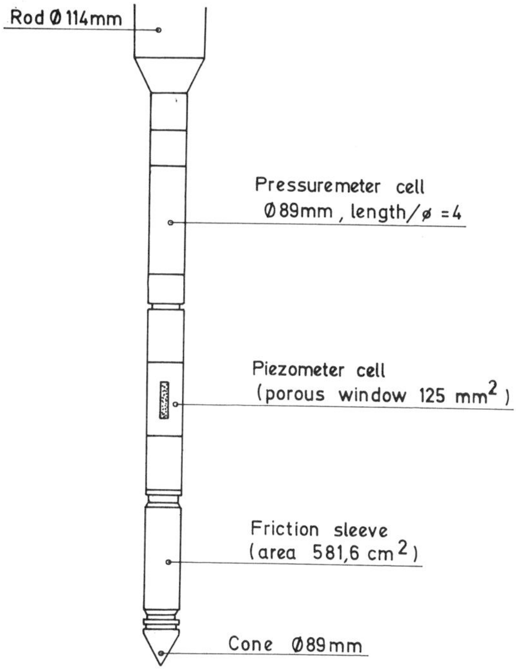

Fig. 3 - Scheme of the measurement tip

Measurements can be taken in static penetration as well as in dynamic penetration.

. friction sleeve
Located just behind the cone, it gives local side friction values during static or dynamic penetration. The sensing device also comprises electrical strain gauges.

. piezometric module
A 115 mm^2 porous window is connected to a pore-pressure transducer located in the watertight chamber (7) thanks to a semi-rigid tubing. This equipment can be used as a constant volume piezometer, to measure the pore pressure during or after penetration. But it can be also be used as a permeameter, with a constant rate flow generated by a volumetric pump located in chamber (7).

. pressiometric module
A cylindrical, inflatable cell, with a height-diameter ratio equal to 4, is connected to the volumetric pump located in chamber (7). The rubber membrane of the pressiometric cell has been especially designed to resist driving and to accomodate a twofold volume expansion.

1.2. The watertight chamber attached to the frame

This chamber is connected by various tubings or electrical cables to the measurement modules on the one hand, to various pieces of equipment located on board the vessel on the other hand.

It contains the volumetric pump used to expand the pressiometric cell or to generate a flow through the piezometric cell. This pump consists of a cylinder and a piston actuated by an electrical motor through a speed controler. With this device, monotonic or cyclic pressiometric tests can be performed within a wide range of expansion rates.

The chamber contains also a set of electro-valves, to bring the water flow in or out of the pressiometric cell, the piezometric cell or a water-tank, also located in the chamber.

The main characteristics of the volumetric pump are as follows :

- maximum injection pressure : 2 500 kPa (≃ 25 tsf),
- net injection volume : 2 500 cm³ (0,66 gallon),
- volume measurement accuracy : ± 4 cm³ ; this corresponds to ± 0,16 % for volumic strain during a pressiometric test.

The net volume of the water-tank is 20 liters (≃ 5 gallons).

1.3. The data acquisition and processing system

The system has been arranged as shown by the scheme of Fig. 4, with a programmable computer as a central unit, to allow an automatic control of measurement sequences, data acquisition and their processing in real or deferred time. In Fig. 4, the symbols A_1 to A_4 correspond to stabilized power supplies of various pressure or displacement transducers.

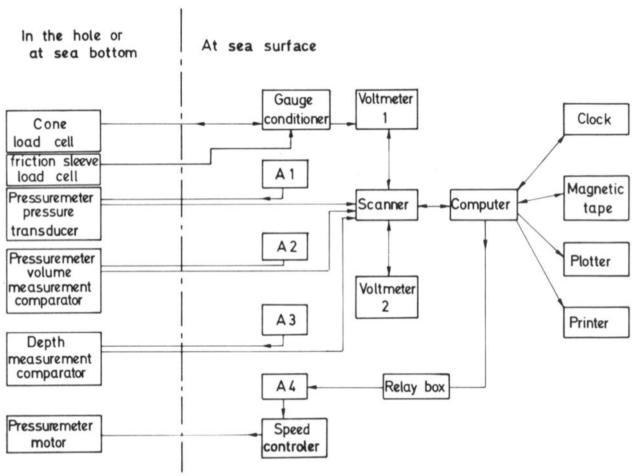

Fig. 4 - Lay-out chart of the data acquisition and processing system

It can be noticed that two voltmeters are used :

- the first one is reserved for the gauges of the penetrometer, which provide dynamic measurements and thus require high characteristics.
- the second one is used for all other measurements.

A channel "depth measurement" can be also noticed. This gives the depth of penetration of the cone. In fact, it is measured in two ways :

- the vertical displacement of the vibrating unit (1) (Fig. 1) is measured with respect to the frame (2). A rotative transducer is used.

- the hydrostatic water pressure is measured at the upper end of the rod (3), thanks to an electrical transducer.

With this double measurement, it is possible to check the penetration and extraction phases of the rod, taking into account the possible slidings of the rod within the vibrating unit. The accuracy on the absolute depth of penetration is about ± 5 cm (± 2 in.).

2. IMPLEMENTATION AND MEASUREMENT PROCEDURES

The implementation procedure described below applies to current cases. It may be modified due to particular site, vessel or hoisting conditions.

First, the rod elements are assembled and, at the lower end, the selected measurement modules (the cone and the pressiometric cell as a minimum) are attached.

The various cables and tubing run inside the rod and go to the measurement modules through a special fitting unit ; thanks to it, it will not be necessary to repeat all preparation and calibration operations for the measurement modules when they have to be separated from the rod.

Then the rod tip assembly is placed through the hydraulic clamp of the vibrating unit.

The measurement modules and the watertight chamber are then connected, as well as the chamber and the data acquisition and processing system. The various classical operations of saturation and calibration follow.

Then, the frame with the rod and measurement modules is lowered into the sea by an appropriate hoisting equipment, which must have two independant raising systems : the first one is attached to the frame and carries most of the load, the second one holds the rod close to the vertical. Once the frame is sitting correctly on the sea-bottom, the measurement modules are checked and zero-readings are taken. The driving is started.

The rod penetrates the ground in 1.25 m strokes (4 ft). At the end of each stroke, the hydraulic clamp is deflated and the vibrating unit is raised by two lateral hydraulic jacks. The hydraulic clamp is then inflated again to grab the rod, the lateral jacks are retracted and the driving is resumed.

The cone response during driving is shown in Fig. 5 as output voltage as a function of time. The peaks correspond to the maximum penetration resistance under each blow, and their values are referred to as cone dynamic resistance q_{cd} (in contrast with the classical static cone resistance, q_c). The same applies to the friction sleeve, which yields the dynamic side resistance, f_{sd}.

These two penetration parameters, q_{cd} and f_{sd}, are plotted as a function of depth, z, in real time. Fig. 6 gives an example of a penetration profile, for a site having the following stratigraphy :

- from 0 to 2 m (0-6.5 ft) : loose silt,
- from 2 to 4 m (6.5-13 ft) : gravel,
- from 4 to 12 m (13-40 ft) : coarse sand.

Special symbols indicate the depths where expansion tests (▽) or permeability tests (→) have been carried out. These tests can been performed either systematically at a predetermined vertical spacing (for instance every meter of every 1.25 m equal to the stroke), or at particular depths selected by the operator according to the penetration profile, since the cone is ahead of the permeability or pressiometric cells.

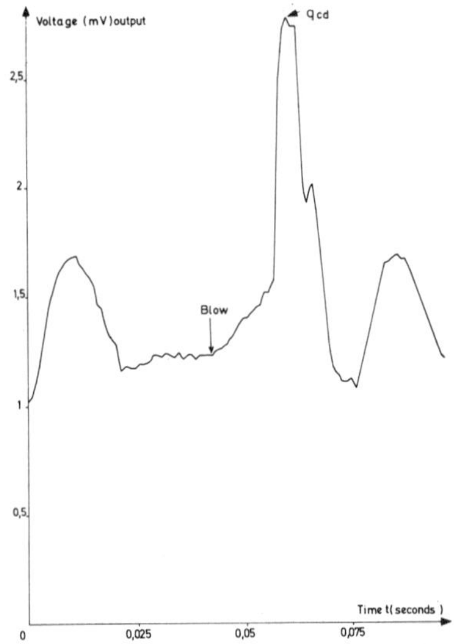

Fig. 5 - Cone response during driving : output voltage as a function of time.

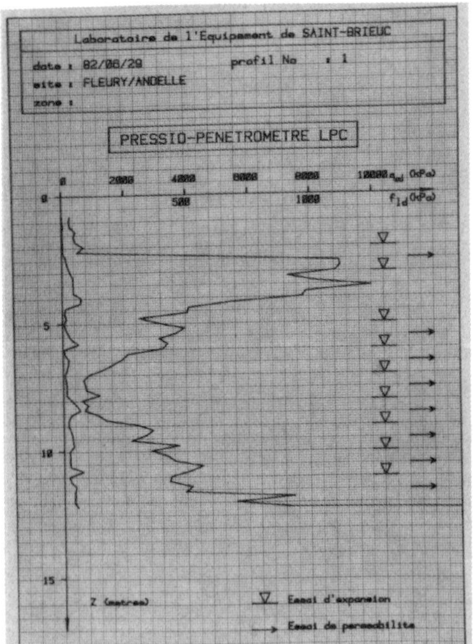

Fig. 6 - Picture of a penetration plot obtained on the Fleury-sur-Andelle site.

Note :

Horizontal axis :

upper scale : q_{cd} in kPa : 2 000 to 10 000 (i.e. 20 to 100 tsf)
lower scale : f_{ld} (i.e. f_{sd}) in kPa : 500 and 1000 (i.e. 5 and 10 tsf)

Vertical axis : z in meters

▽ indicates the location of an expansion test.

→ indicates the location of a permeability test.

To carry out an expansion test, the driving is stopped at the appropriate depth, and the operator has only to start the test which is performed automatically according to a preselected procedure. At the end of the expansion, the corrected expansion curve (p,v) is obtained as shown in Fig. 7 ; the pressiometric characteristics, the modulus E_p and the limit pressure p_l are given by the computer, once the operator has selected the quasi-linear portion of the expansion curve on the graph.

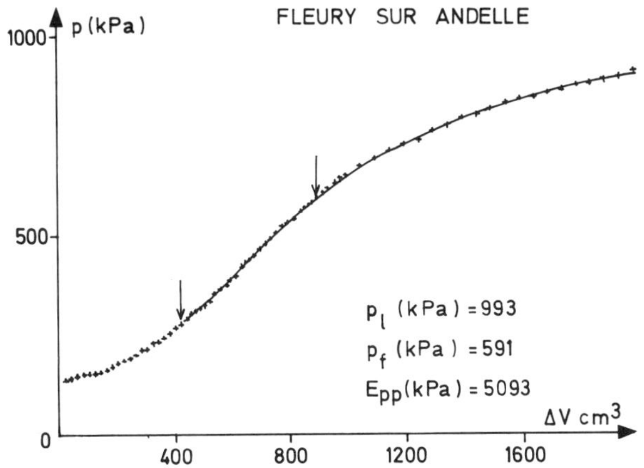

Fig. 7 - An expansion curve (type II) in Fleury-sur-Andelle chalk. (Note : 100 kPa ≃ 1 tsf)

If necessary, pore pressure readings are then taken or a permeability test is performed.

Then the pressiometric cell is deflated and the penetration is resumed.

When penetration refusal is obtained or when the rod has penetrated the soil over its full usable length (12 m ≃ 40 ft), it is dynamically retracted, following a procedure similar to the one used during insertion. Then all the equipment can be moved to another boring location by the hoisting devices ; to do so, there is no need to take it out of water.

The LPC pressiopenetrometer has been used especially to determine the characteristics of foundation soils for a jack-up rig in the following conditions : water depth : 40 to 50 m - vibrating equipment : B.S.M.V.A. of Techniques L. Ménard - length of rod and measurement tip : 15 m - average output : 3 borings per 10 hours day including 10 pressiometric tests.

3. TEST RESULTS IN TYPICAL SOILS

To apply the test results of the pressiopenetrometer, it is necessary to refer to the interpretation methods of the conventional tests, i.e. :

- static penetrometer test,
- Ménard pressuremeter test.

But the results of the pressiopenetrometer may differ from those of the conventional tests for a number of reasons, and to an extent which may vary with the type of soils. Such reasons can be :

- for the penetrometer modules (cone and friction sleeve) : the blow effect, the average penetration rate, the scale effect (diameter equal to 89 mm (3 1/2 in.), to be compared to 36 mm and 45 mm (1.4 and 1.8 in.) for conventional penetrometers),
- for the pressuremeter modules : the probe insertion, which displaces the soil ; the type of expansion (constant rate of strain); the type of cell (structure, geometry).

To study the influence of these factors and to establish correlations, a research program has been undertaken on several sites on shore; they include the classical soils : clay, silt and sand, but also some particular soils frequently encountered in offshore geotechnics : chalk and carbonated soil. The main geotechnical characteristics of the 6 sites involved are given in the Table of Fig. 8.

3.1. Penetrometer characteristics

3.1.1. Cone resistance, q_{cd}

In Fig. 9, a q_{cd} profile is compared to two profiles of q_{cG}, cone resistance obtained from the Gouda static penetrometer (dia. : 36 mm ; rate of penetration : 2 cm/s) for the Saint Jean de la Croix site. Fig. 10 shows the same profiles for the Plouasne site. It can be seen that q_{cd} is less than q_{cG} in both cases ; these cases probably represent two extremes in terms of soil response to penetration : the Saint Jean compact sand will dilate, where the Plouasne carbonated soil, which is brittle, will break and collapse.

Similar results have been obtained for the other sites. However, the static resistance has been measured either by a Gouda penetrometer -symbol q_{cG} for the cone resistance- which has a mantle cone, or by an electrical penetrometer -symbol q_c- which has a simple cone whose geometry complies with the recommended standard of the I.S.S.M.F.E. (1977). It should be noticed that q_{cG} is close to q_c for sands, but is larger for clays and silts, with ratio q_{cG}/q_c as high as 2 in some cases.

The table of Fig. 19 gives the average values of q_{cd}/q_c or q_{cd}/q_{cG} for all the sites.

3.1.2. Dynamic side friction, f_{sd}, and friction index, I_{fd}

Fig. 11 shows a graph f_{sd}, q_{cd} for 3 sites : Pont-Lévêque, Saint Jean de la Croix et Fleury-sur-Andelle. Two groups can be clearly distinguished : clays and silts which yield comparatively higher results, and sands and chalk. The corresponding values of the friction index, I_{fd} (Fig. 12) are very similar to those obtained in static penetration ($I_f < 20$ for clays and silts, 50 to 200 for sands).

3.2. Pressuremeter characteristics

3.2.1. Pressuremeter curves

The pressiopenetrometer yields two types of expansion curves :

- type I - curves without a flexure point (Fig. 13). This is the general case in clays, silts and sands (Cran - Pont Lévêque - Saint Jean

SITES	SOIL TYFE	W (%)	S_r (%)	γ (T/m³)	I_p/W_l (%)	C_u(kPa)	ϕ' (°)	σ'_c/σ'_o	C_c	Org. Mat. (%)	C_uCO_3 (%)
Cran	plastic silt	70-120	100	1,45-1,65	20/70	30	20-30	3	0,5-1	3-5	-
Plancoët	0- 5 m: lear clay 5-12 m: fine sand	30- 50 40- 50	100 100	1,7 -1,85 1,7 -1,85	20/40 NP/40	30-40 -	38 35-40	3 1,3	0,1-0,2 0,1-0,2	1-2 -	- -
Plouasne	0- 4 m: lear silt 4-12 m: carbonated soil	25- 35 30- 45	100 100	- 1,85	15/50 -	20-50 -	- 35-45	-	- < 0,1	- -	5-10 80- 9
St Jean de La Croix	clean sand and gravel	25	-	1,90	NP	-	35	-	-	-	-
Pont Lévêque	0- 2 m: gravel 2-12 m: plastic silt	20- 30	85-95	2,0	30/70	175	-	4	< 0,1	-	10
Fleury-sur-Andelle	0-5 m: fill and gravel 5-12 m: weatered chalk	25- 35	100	1,85-1,95	-	20-120	28	-	-	-	-

Fig. 8 - Main geotechnical characteristics of the 6 sites
(Note : 100 kPa ≃ 1 tsf)

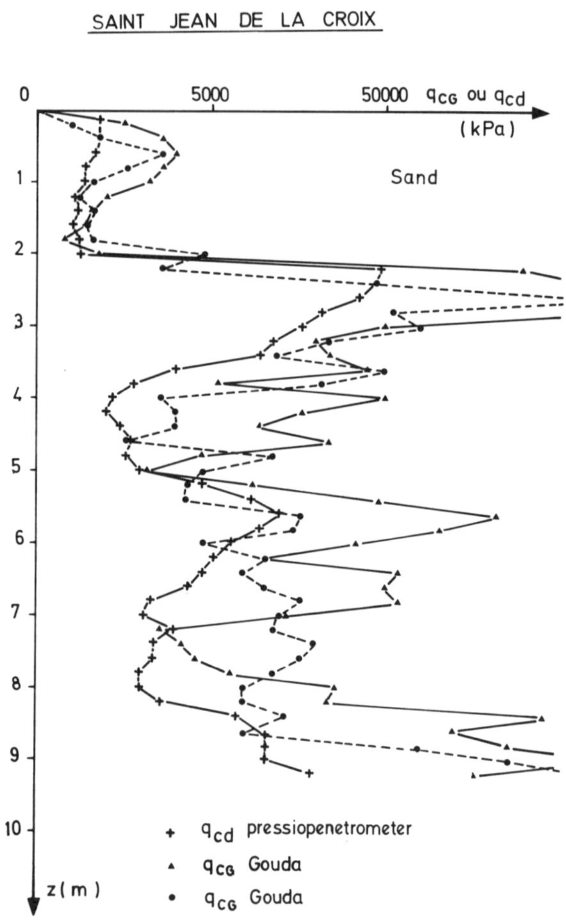

Fig. 9 - Penetration resistance profiles at Saint Jean de La Croix : q_{cG} : Gouda (static) ; q_{cd} : LPC pressiopenetrometer (dynamic) (Note : 100 kPa ≃ 1 tsf)

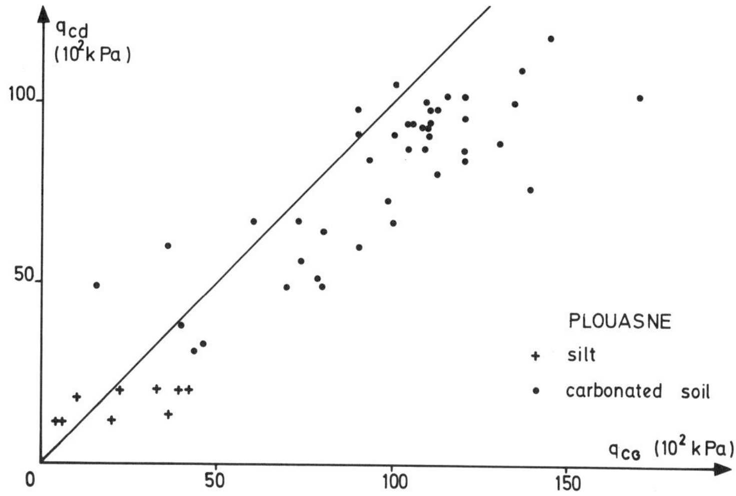

Fig. 10 - Comparison of penetration resistances obtained at Plouasne : q_{cG} : Gouda (static) ; q_{cd} : LPC pressiopenetrometer (dynamic).
(Note : 100 kPa ≃ 1 tsf)

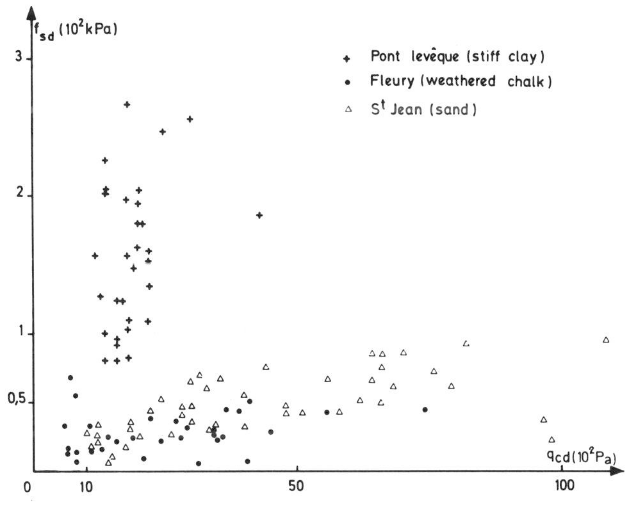

Fig. 11 - LPC pressiopenetrometer penetration results at 3 sites : friction sleeve resistance, f_{sd}, versus cone resistance, q_{cd} (Note : 100 kPa ≃ 1 tsf)

SITE	SOIL TYPE	SIDE FRICTION f_{sd} (kPa)	I_{fd}
CRAN	plastic silt	20 to 50	10 à 20
PONT L'EVEQUE	plastic silt	100 to 300	10 à 20
ST JEAN DE LA CROIX	clean sand and gravel	10 to 150	50 à 100
FLEURY-SUR-ANDELLE	weathered chalk	20 to 50	50 à 150

Fig. 12 - LPC pressiopenetrometer side friction results at 4 sites ($I_{fd} = q_{cd}/f_{sd}$)
(Note : 100 kPa ≃ 1 tsf)

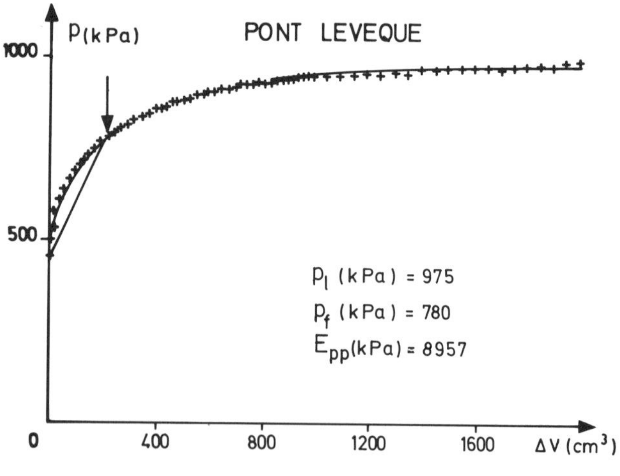

Fig. 13 - An expansion curve (type I) in Pont Lévêque plastic silt.
(Note : 100 kPa ≃ 1 tsf)

de la Croix - Plancoët). There is no initial recompression phase such as in the Ménard classical test, which is generally performed in a prebored hole. Instead, the soil, which is displaced laterally by the cone during insertion, presses against the pressiometric cell.

- type II - curves with a flexure point (Fig. 7).

These are obtained with the brittle soils (Plouasne and Fleury-sur-Andelle). Their shape is identical to the one given by the Ménard classical tests, but it should be attributed to another mechanism : the soil structure is locally broken by the cone and collapses, thus the pressiometric cell is in contact with a remolded soil of very low characteristics.

3.2.2. Limit pressure, p_l

The same definition has been used for the pressiopenetrometer test as for the Ménard test : p_l corresponds to a twofold volume increase ; the initial volume is taken as the volume of the cell at rest for type I curves, as the volume corresponding to the end of the initial recompression phase for type II curves. As a general rule, the p_l values obtained from the pressiopenetrometer are not significantly different from those of the Ménard test ; this is shown in Fig. 14 for the Cran and Pont Lévêque plastic silts.

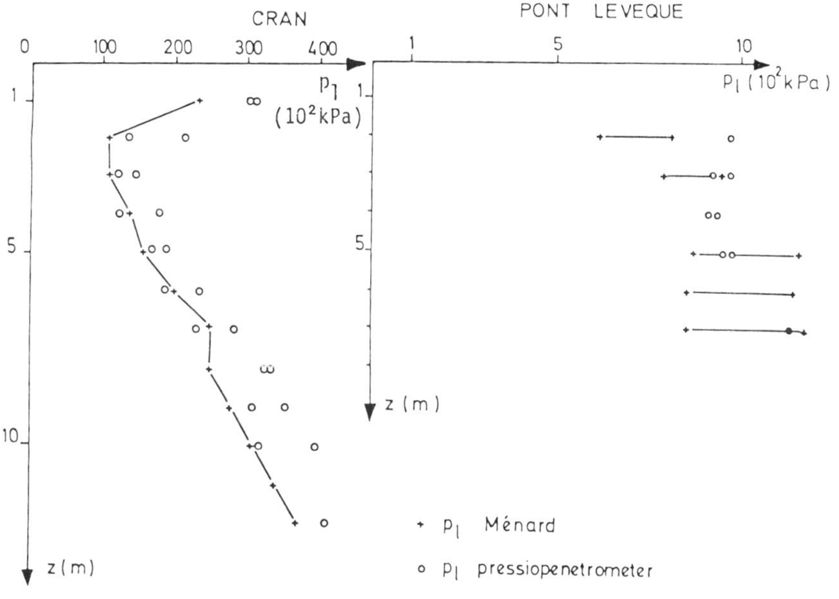

Fig. 14 - Ménard and pressiopenetrometer limit pressure at Cran and Pont Lévêque plastic silts.
(Note : 100 kPa ≃ 1 tsf)

3.2.3. Pressiometric modulus, E_{pp}

For type II curves, the modulus can be determined on the quasi-linear portion, as in the Ménard test. In chalk (Fig. 15), values are nearly identical. In contrast, the pressiopenetrometer values are less by a factor of 2 to 3 than the Ménard values in the Plouasne carbonated soil (Fig. 19).

For type I curves, the modulus is not readily determined. Trials have shown that the secant modulus at 10 % volumetric strain ($\Delta V/V_o$) yields values close to the Ménard values, as shown in Fig. 16 for the Pont Lévêque site.

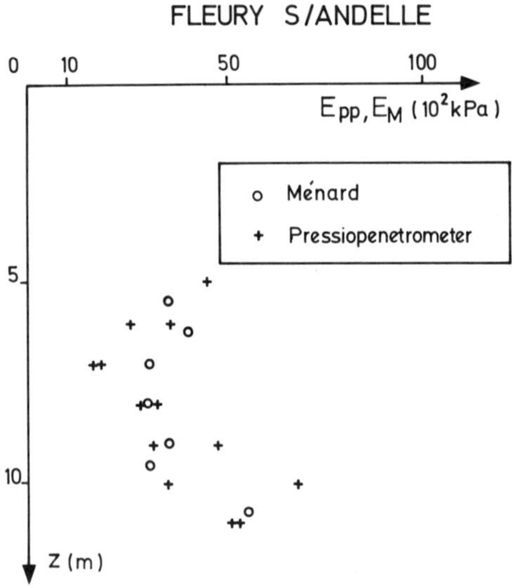

Fig. 15 - Ménard and pressiopenetrometer moduli in Fleury-sur-Andelle chalk.
(Note : 100 kPa ≃ 1 tsf)

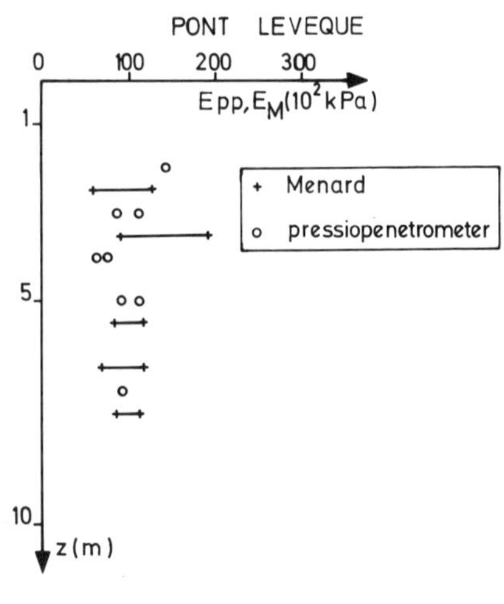

Fig. 16 - Ménard and pressiopenetrometer moduli in Pont Lévêque plastic silts.
(Note : 100 kPa ≃ 1 tsf)

3.3. Permeameter measurements

The tests carried out are constant flow rate tests (240 cm^3 per minute). Their goal is simply to indicate roughly how permeable the soil is. Three types of pressure response have been observed :

① the pressure keeps constant, equal to the initial, hydrostatic pressure : this happens for permeable soils, sands and chalk, as illustrated by Fig. 17 and Fig. 18a for Fleury-sur-Andelle.

② the pressure increases rapidly, then keeps constant : this has been observed for the Cran soft plastic silt.

③ the pressure exhibits a peak then a plateau : this has been obtained in the Pont Lévêque stiff plastic silt (Fig. 18b). This behavior should probably be attributed to an hydraulic fracturing, initiated by the preexisting fissures of the soil.

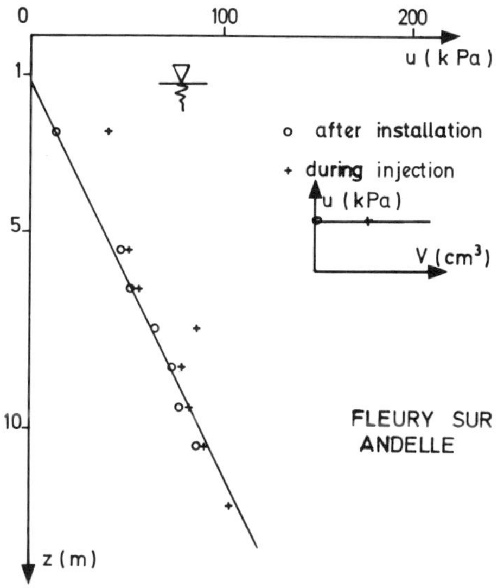

Fig. 17 - Permeability test results in Fleury-sur-Andelle chalk.
(Note : 100 kPa ≃ 1 tsf)

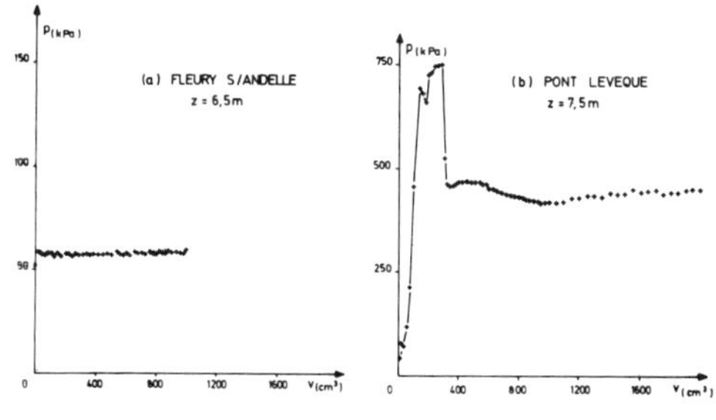

Fig. 18 - Types of pressure response in permeability tests.
(Note : 100 kPa ≃ 1 tsf)

3.4. General comparison of the results

The table of Fig. 19 summarizes the main results obtained on the 6 sites. From the comparison of the various test results, it is possible to suggest a method of soil identification related to its mechanical behavior.

The permeameter test allows the fine soils to be separated from the other types of soils. But, even when such tests are not available, the following rules can be applied :

- shape of expansion curve : type II, with an initial phase, denotes a brittle material ;
- friction index, I_{fd} : values less than 20 indicate clay or silt, more than 50 indicate sand or chalk;
- ratio q_{cd}/p_1 : values around 2 are typical of fine soils or cemented soils, while values in the range of 4 to 6 apply to granular soils, larger values indicating higher densities.

SITE	SOIL TYPE	PRESSIOMETRE		PENETROMETRE
		p_1/p_{1M}	E_{pp}/E_M	q_{cd}/q_{cG}
PLOUASNE	carbonated sand	-	0,3	0,8
PLANCOET	lean clay and fine sand	~ 1	~ 1	1,1
CRAN	plastic silt	1,1	2*	0,8
PONT L'EVEQUE	plastic silt	1,1	~ 1	0,6
ST JEAN DE LA CROIX	sand and gravel	-	-	0,7
FLEURY-SUR-ANDELLE	weathered chalk	~ 1	~ 1	0,5

Fig. 19 - Comparison of LPC pressiopenetrometer values (p_1, E_{pp}, q_{cd}) Ménard values (p_{1M}, E_M) and Gouda stable penetrometer (q_{cG}).

4. CONCLUSION

The pressiopenetrometer has a number of advantages in the implementation and for soil characterization :

- it can be implemented in deep water depths, the limitation coming from the vibro-driving equipment ;
- it does not require important reaction equipment ; for instance, the weight of the B.S.M.V.A. equipment used for 50 m of water is approximately 5 t ;
- the tests are fully automatic, and the final test results are obtained on the site as the survey proceeds ;
- because several types of measurements can be performed in the same profile, and even on the same volume of soil, it is possible to classify the soils according to their mechanical behavior from the test results, and to apply the current design rules of foundations.

The main limitations of the apparatus are the depth of penetration (about 12 m $\simeq$ 40 ft) and the penetration resistance (about 10 000 kPa $\simeq$ 100 tsf with the B.S.M.V.A. vibrating unit to the present time).

The equipment is available to the offshore engineering industry. It has been developed, and used intensively, for a particular application : the investigation of a carbonated soil which had to support jack-up rigs. Combined with borings, it can be used with much profit for other purposes, such as gravity structures or strong seabeds, pipe line supports and trenches, and so on...and more generally, for shallow soil surveys.

REFERENCES

"Recommended Standard for the Cone Penetration Test (CPT), (1977), Report of the Subcommittee on Standardization of Penetration Testing in Europe, Proceedings of the 9th International Conference on Soil Mechanics and Foundation Engineering (Tokyo, 1977), vol. 3, pp. 99-109.

CONE PENETRATION TESTING IN THE BEAUFORT SEA

By Michael G. Jefferies[1], M.ASCE, and Erik Funegard[2]

ABSTRACT

The electric cone penetrometer (CPT) has a very significant role to play in the exploration and development of hydrocarbon resources from the Canadian Beaufort Sea. Gulf Canada Resources Inc. has carried out some 1400 m (4600 ft.) of cone probings during 1982 in support of the deployment of their Mobile Arctic Caisson drill system.

The penetrometer system used could measure either end bearing and friction or end bearing and pore water pressure; both types of measurement have been used. The penetrometer system differed in two major respects from those commonly encountered. First, it was a wireless system using a digitally encoded ultrasonic system to transmit tip data to the surface recording apparatus. Second, it used mud injection behind the tip to minimize rod friction and permit single pushes of as much as 71 m (233 ft.) below the seabed with only a 5 tonne reaction.

The cone was used primarily to define the presence or absence of weak layers at potential exploration wellsites. However, in doing this, much experience has been gained with local correlation to relate cone data to more fundamental engineering parameters. Some of these correlations are reported.

INTRODUCTION

Canada is striving to become self-sufficient in oil. The continental shelf in the Beaufort Sea is an area of potential oil resources that is being actively explored by a number of companies. Gulf Canada Resources Inc. is a major operator in the effort to find and produce hydrocarbons from the Beaufort Shelf.

The exploration and development of this area cannot be based solely on conventional methods; new ideas are needed. The southern Beaufort Sea is characterized by very severe ice hazards including first year and multi-year ice features as well as ice fragments. The open water season lasts for only about three months; even during this time, there is the hazard of contact with ice fragments.

The GCRI acreage (for location, see Figure 1) is located in the shear ice zone. The shear zone separates the ice that is essentially stationary because of contact with the sea floor (called shore fast ice) and the arctic pack ice which is in continual motion. The shear zone is characterized by continuous failure of the ice. The ice will exert

---

[1]Supervising Civil Engineer, Gulf Canada Resources Inc., Calgary, Alberta, CANADA

[2]President, Western Geosystems Inc., Vancouver, British Columbia, CANADA

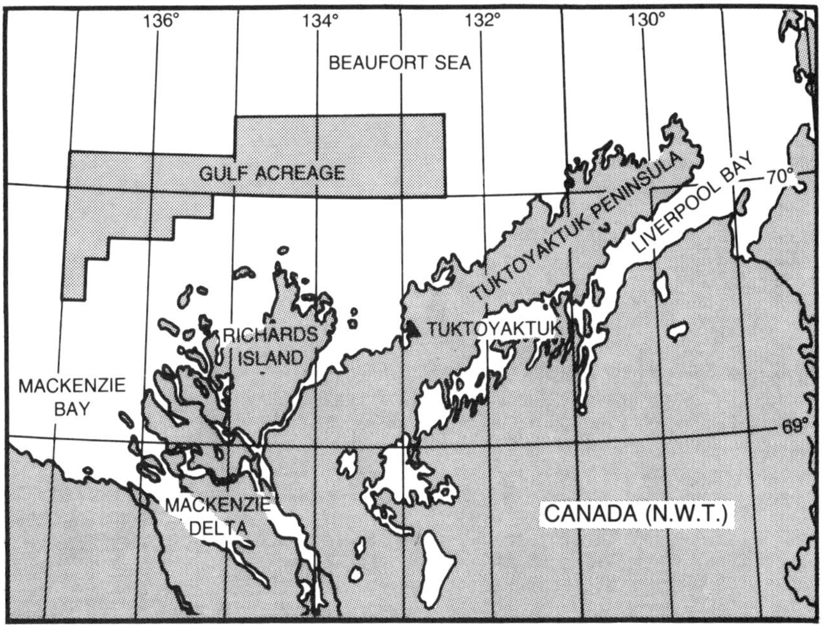

FIGURE 1 SITE LOCATION PLAN

large horizontal loads on any stationary structure such as an exploration structure

To date, artificial islands are the only year-round drilling platforms that have been employed in the Beaufort Sea. They have proved successful in resisting ice forces. The cost of constructing simple artificial islands in water depths of around 35 m (115 ft.) (which is the average for prospective wells on the GCRI acreage) is excessive. One method by which the ability of artificial islands to resist ice forces can be economically utilized in deep water is to use a caisson to retain the sand fill. The caisson substantially reduces the amount of fill required in a given water depth. But, the resulting structure is still sufficiently massive to resist the lateral loads.

THE GULF MOBILE ARCTIC CAISSON

GCRI has adopted the caisson retained island concept for one of the drill systems designed for use in the Beaufort Sea. The GCRI system is reuseable and will be used at several locations. The system has been called the Mobile Arctic Caisson, or MAC for short, in recognition of its mobility.

The MAC is a hybrid soil-steel structure. A steel annular box (caisson) with a simply-supported steel deck carries the drilling equipment, supplies and crew accommodation. The caisson rests on a hydraulically placed underwater berm, which is used to tailor the caisson to the water depth at a particular site. The caisson is towed to the drill location and installed on the previously constructed berm by water ballasting. Once the caisson is firmly resting on the berm, the inner core is filled with hydraulically placed sand. The overall system is illustrated in Figure 2.

In essence, the steel annular caisson has three functions. It serves as formwork for placing the necessary weight of sand fill to resist ice forces, it serves to armor the fill from ice attack, and it provides a permanent platform on which to locate the drill and associated equipment. The fill core carries most of the imposed horizontal load, although a small amount (5 to 10%) is transferred directly to the berm through friction on the base of the caisson. The berm in turn transfers the entire load, both horizontal and vertical, to the sea floor.

To put the MAC in perspective, it can be noted the structure is approximately 110 m (360 ft.) by 110 m in plan and stands 29 m (95 ft.) above the top of the berm. The berm surface is 21 m (69 ft.) below sea level. The entire system is designed to resist a maximum horizontal ice loading of 700 MN (70,000 tons) with a load factor of 1.5.

SHALLOW GEOLOGY OF THE CANADIAN BEAUFORT SHELF

Soil conditions in the Beaufort Sea vary greatly. A somewhat oversimplified summary is that the eastern Beaufort Shelf is largely composed of thick sand beds which are covered in part by a veneer of silty clay. The western side (in Canada) of the shelf is formed of silty clays which extend to considerable depths. A simplified E-W cross section through GCRI acreage is shown in Figure 3 to illustrate the geology.

Much about the near-surface geology of the area remains to be clarified. Nevertheless, it appears that the sediments within the zone of interest for MAC foundations (upper 100 m, 328 ft. or so) have been deposited in four eras; Recent, post-Wisconsin transgressive, Wisconsin glacial and peri-glacial and Pleistocene.

The Pleistocene deposits are deltaic and pro-delta sediments. At that time, the predecessor of the present MacKenzie river flowed into the Beaufort Sea by way of what is now called Liverpool Bay (Figure 1). The deltaic sands were, therefore, laid down on the eastern side of the GCRI acreage while pro-delta muds were deposited in the west.

The Wisconsin period was characterized by a sea level 60 to 90 m (200 to 300 ft.) below the present level, as well as by glaciation. The Wisconsin was primarily a time of erosion. But, till-like materials and apparently varved clays were laid down in at least part of the Beaufort Shelf.

The Transgressive period saw a substantial rise in sea level to that of today and the ice melting that caused this also produced a substantial erosion and reworking of the near surface soils. Most of the Wisconsin deposits appear to have been reworked in this era. Soils laid down in this period are composed of a discontinuous and variable sequence of sands, silts and clays which have been deposited in a variety of situations such as deltaic, littoral, lagoons, and channel filling.

Sediments deposited in the Recent period have been transported by the MacKenzie River which is building a delta out into the MacKenzie Bay. Pro-delta silts and clays extend well beyond the delta into the

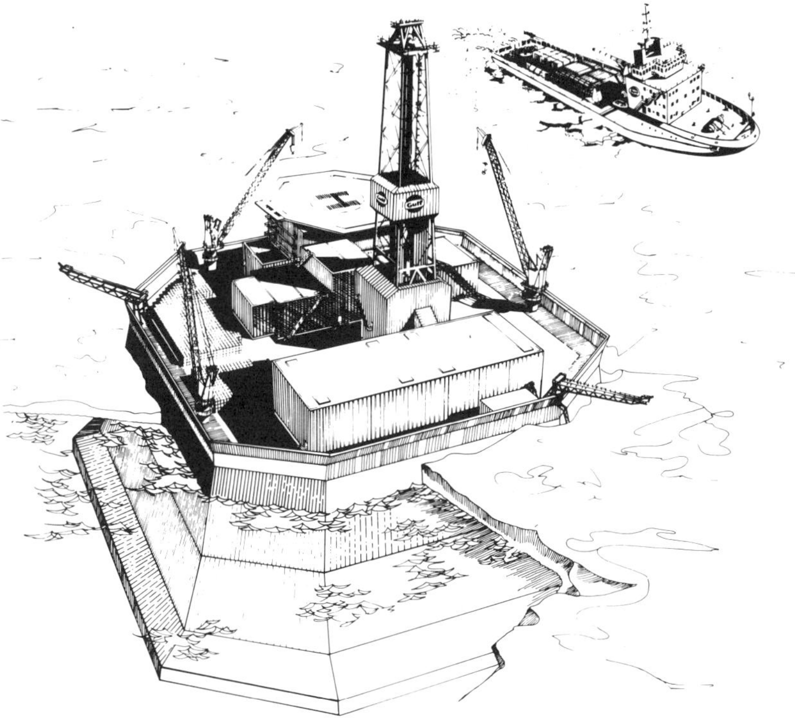

ARTIST'S IMPRESSION

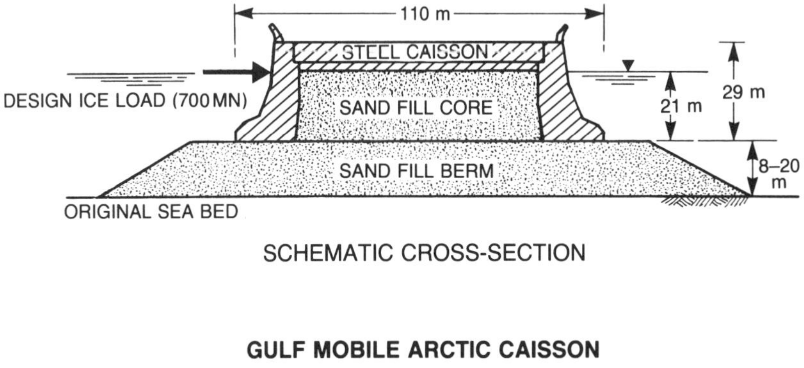

SCHEMATIC CROSS-SECTION

GULF MOBILE ARCTIC CAISSON

FIGURE 2

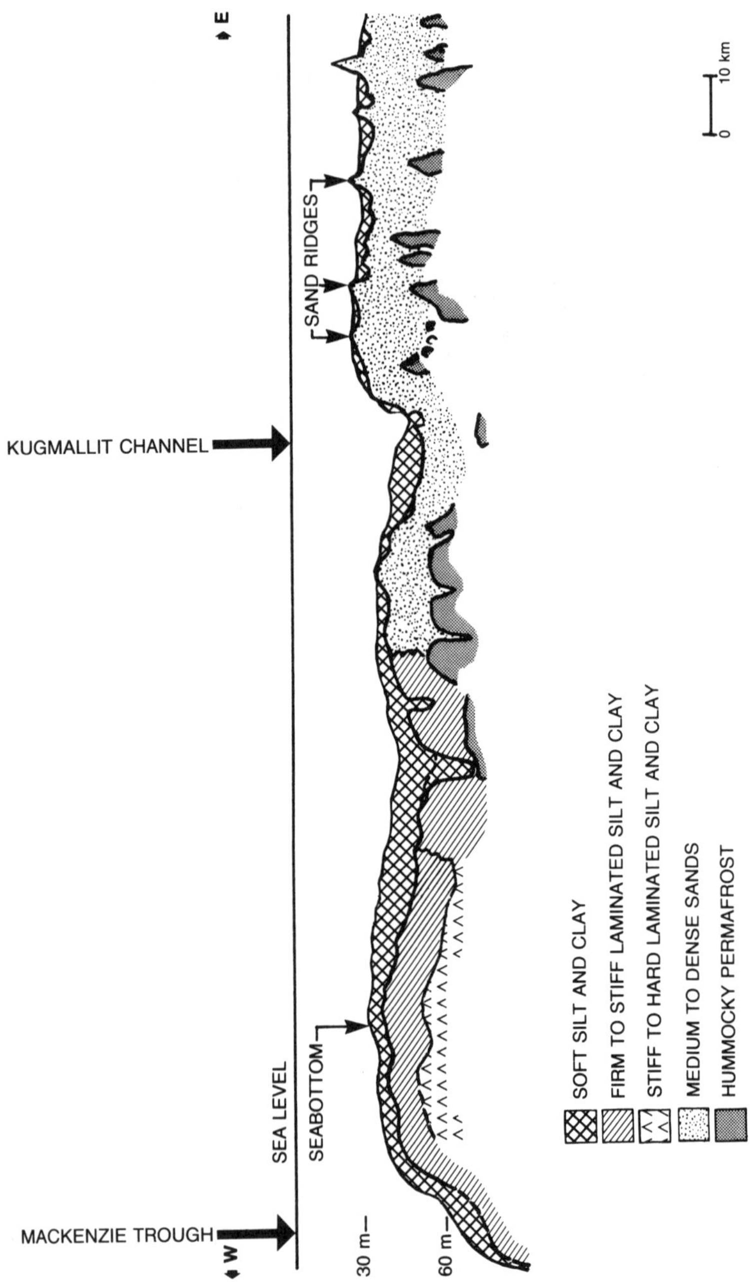

SCHEMATIC GEOLOGICAL CROSS-SECTION

FIGURE 3

Beaufort Sea. Some of these fine sediments are carried eastward by littoral drift and partially cover the shelf across the whole of the GCRI acreage. These sediments, on the GCRI acreage, are thinly laminated usually with a curved ripple-like bedding. Shear strength varies substantially both with position and with depth. Distributed pore gas (sometimes at substantial excess pressure) is commonly encountered in the sediments at depths greater than about 5 m (16 ft.) below the sea floor.

The key engineering issues to emerge from the geology of the near surface sediments are that firstly, subsurface conditions can change significantly over short distances (100 m, 328 ft.) in at least the Transgressive and Wisconsin sediments. Secondly, the depositional environment in all four eras will have usually produced horizontally bedded sediments with layers of probably quite markedly different strengths.

THE ROLE OF THE ELECTRIC CONE PENETROMETER

Two geotechnical considerations stand out with respect to the MAC. First, the stability and performance of the MAC depends directly on the behavior of uncompacted sand fill placed hydraulically underwater. The behavior of such fills is largely a function of density. It is extremely difficult to directly measure the density of underwater sand fills. The cone penetrometer has been used as a quality assurance tool for MAC hydraulic sand fills. It is a relatively simple operation as only small reactions are required to push the cone into the fill and the required penetration of the berm is 20 m (66 ft.) at the most. The interpretation of this cone data is, however, quite complicated (6) and needs to be performed by qualified personnel.

The second major geotechnical concern with the MAC is the ability of the foundation to resist the massive horizontal loading. A look at Figure 2 shows that the MAC can be thought of as a large scale shearbox. Average shear strengths may be inadequate for design of a foundation subject to so near horizontal shear loading. It is essential to evaluate the presence of weak or soft near-horizontal layers in the upper 50 m (164 ft.) or so of the sea bed. These layers could form preferred failure zones. The CPT is well suited to looking for weak layers and particularly when an end bearing/pore pressure tip is used.

A simple and often asked question is: Why not use continuously sampled boreholes in the search for weak layers? There are three reasons why the cone does the job better:

- Cone soundings are much faster than a continuous sampled borehole. This is important given the large daily cost of geotechnical drilling ships in the Arctic.
- Cone soundings produce a truly continuous record of the layering in the soils at a site. Borehole samples must be broken into sub-samples of no more than 5 cm (2 in.) with very careful logging of water content and Atterberg limits in the case of cohesive soils in order to compare with CPT logs. Detailed logging of density (for lack of a better expression) is virtually impossible in cohesionless deposits with borehole samples.
- Cone soundings provide a continuous measure of the strength of the various layers. It may not be possible to obtain this information in cohesive deposits since a triaxial test requires

a sample of statistically representative soil at least 150 mm (6 in.) long. In the case of cohesive soils in the Beaufort, definition of layering is usually not possible until the sample has been air dried for about 36 hours. Obviously, this makes testing of representative samples very difficult.

It must be noted that the CPT cannot detect shear surfaces in most instances. A continuously sampled borehole with careful structural logging is required. However, the change in bedding angles caused by soil mass movement above the shear surface shows up on shallow geophysical surveys.

The data obtained with shallow geophysical surveys is not always unambigious. The CPT is sometimes very helpful in clarifying the shallow geophysics because the CPT provides a continuous record of soil mechanical response. The contrast in mechanical properties between soil layers largely controls the record obtained with shallow reflection surveys. Thus the shallow geophysics and the CPT complement each other.

The above remarks should not be taken to suggest that the cone penetration test is superior to a sampled borehole as a site investigation technique. It is felt that both techniques are complementary and both should be routinely carried out.

How many cone soundings are appropriate at a given site? Consideration of the engineering geology and nature of the loading on the MAC indicate the following guidelines for cone probings:

- preliminary site survey at prospective MAC sites - 1 sounding
- ground truthing shallow geophysics at any location - 1 sounding
- detailed design investigation for MAC sites where the structure will be founded on sands (Eastern Beaufort) - 3 soundings. A suggested arrangement of the soundings is an equilateral triangle in plan with sides of approximately 100 m (328 ft.) (the MAC diameter).
- detailed design investigation for MAC sites where the structure will be founded on clays (western Beaufort) - 5 soundings. A suggested arrangement in plan is a sounding at each corner of the MAC and one in the centre.

A NEW ELECTRONIC CONE PENETROMETER

The CPT system used for the investigation of the GCRI acreage differs in two major respects from those used elsewhere:

- Data from the transducers in the cone is digitally encoded and transmitted as an ultrasonic signal up the cone rods to a receiver unit at the surface.
- A friction-reducing fluid is injected into the soil immediately adjacent to the cone rods thus reducing soil friction on the rods and hence the reaction required to push the cone into the soil.

The geometry of the cone tips, however, follows the standard right cylindrical (Fugro) shape and the transducer technology is conventional.

How does the new cone system work? There are two aspects to this question, the electronic and the operational.

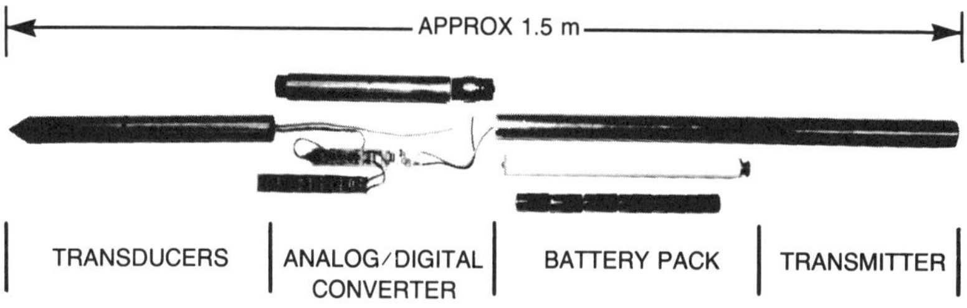

ELECTRIC CONE PENETROMETER TRANSMITTER & TIP

FIGURE 4

Digital Technology

The downhole electronic equipment has the appearance of an extended, conventional cone tip, as shown in an exploded view in Figure 4. It comprises:

- A two-channel cone tip which may be either end bearing and friction or end bearing and piezometric pressure. The transducers in the tip sense load pressure and pass a signal to ...
- an eight-channel analogue-to-digital converter of which only two channels are presently used. The scan rate of each channel is twice a second. The digitally encoded signal is then fed to ...
- a transmitter which puts out an ultrasonic pulse. The entire downhole array is powered by ...
- a battery pack with a life of about 10 hours.

The receiving equipment at the top of the hole includes the following:

- A microphone to receive the ultrasonic signal and ...
- a cone rod motion synchronizer, which provides the depth input signal to a ...
- recording unit which includes a digital-analogue converter. The analogue signal from each of the two channels is displayed (with a choice of scales) on a strip chart recorder fully synchronized for depth. In addition...
- a replaceable, solid state memory module stores the digital signal and depth synchronization for subsequent processing by computer.

The schematic layout of the cone testing system is shown in Figure 5.

Some remarks on the digital techniques are in order. Digital transmission of data eliminates the problem of 'noise' that can occur with ultrasonic signal transmission using FM or AM encoding. The digital signal is made up of two parts: one 3-bit code telling the receiver the type of measurement that will follow (e.g. tip resistance, skin friction or pore pressure), and then follows a 12-bit code containing the actual measurement. The 12-bit data code means that the resolution of the digital code is 1 in 4096 or .02%. The data is

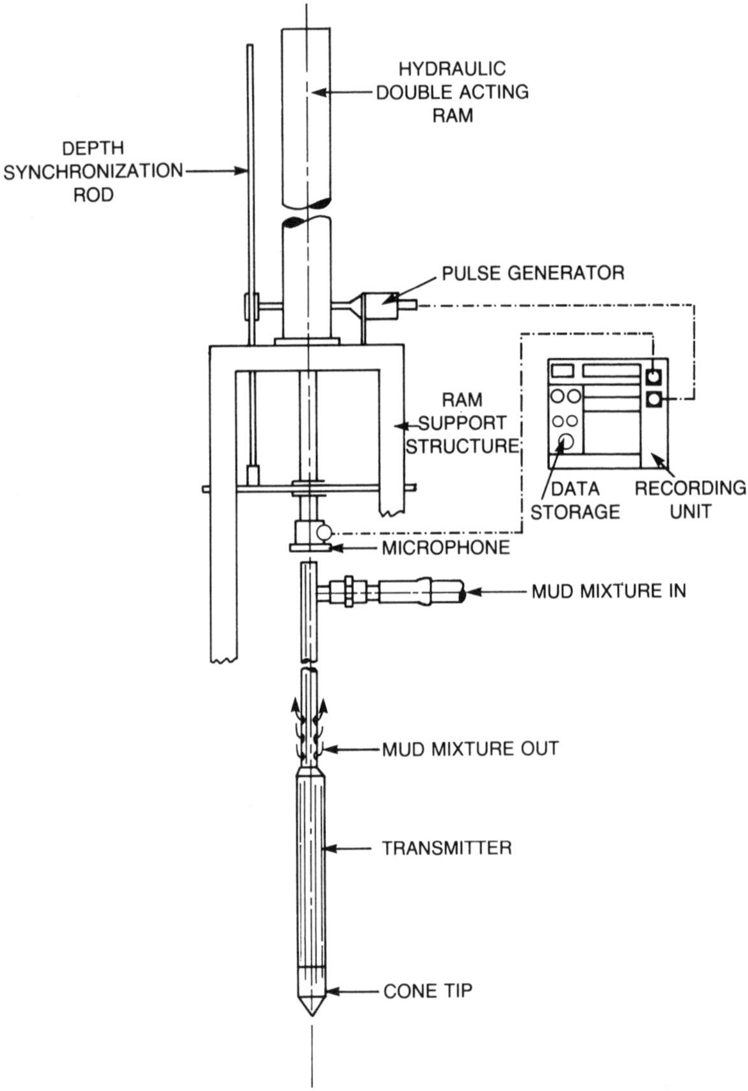

ELECTRIC CONE PENETROMETER SCHEMATIC

FIGURE 5

transmitted as a positively coded signal and then immediately as a negatively coded signal. Comparison of these signals by the surface receiving unit allows the rejection of spurious signals. The risk of accidentally introduced vibrations affecting the signal is actually less than one in a million with the present equipment. The risk translates into one bad reading in every 5000 m (16,400 ft.) of cone sounding.

The digital data is also stored, still in digital form, in solid state memory cartridges that contain no moving parts. The memory also stores: the rate of the sounding, borehole number and the calibration codes for the probe used. This permits checking at a later date that the correct calibration codes were indeed used in the field. The data cartridges currently in use can store 100 m (328 ft.) of penetration data, with one value from each of two channels stored for every 5 cm (2 in) of penetration.

Skin Friction Reduction

On the operational side, much of the performance of the system depends on the mud injection technique. Although it would be possible to pump mud through rods of a cable transmission cone system, the absence of cable simplifies the matter as no glands or slip couplings are required. Mud is injected into the soil at a steady flowrate of about 1 litre/min. from several injection ports located approximately 1.5 m (5 ft.) behind the tip. The injection ports are illustrated in Figure 5.

Trials were carried out to determine whether or not the mud injection process influenced the cone tip measurements. No effects were observed with the present configuration. Indeed, the pore pressure measured by the cone could not be changed, even when the cone was stationary and injection was proceeding as fast as the equipment would allow. The injection ports would be located even closer to the tip if the electronic package could be made shorter.

The mud used is a seawater/bentonite/polymer mixture with a Marsh cone viscosity of about 60 secs. The mud holds the soil off the cone rods thus minimizing friction. A comparison of two cone pushes adjacent to each other is shown in Figure 6. One of these pushes used mud, the other did not. It can be seen that the mud reduced the required reaction by at least 50%.

Tests and projects completed to date indicate that most soils can be penetrated to any depth (limited only by signal transmission) by using a maximum reaction weight of 5t. Achieved maximum depths to date are single pushes of 100 m (328 ft.) on land and 71 m (233 ft.) offshore in 40 m (131 ft.) of water.

OPERATIONAL EXPERIENCE

Some 1400 m (nearly 5000 ft.) of CPT soundings have been completed to date. The work was carried out from a 90 m (295 ft.) long modified freight ship, the MV Frank Broderick. The ship deployed a five-point anchor spread during all tests. Both CPT soundings and conventional soil borings were carried out through a moonpool situated approximately amidships.

A bottom founded casing system was used to isolate the CPT from vertical movements of the drillship. This casing system had three major

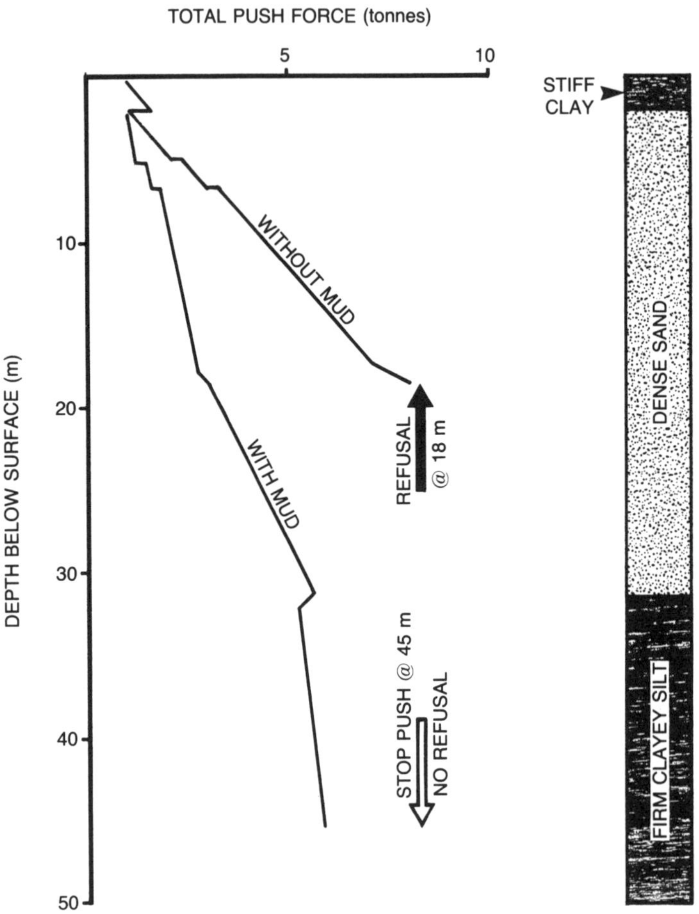

TOTAL CONE PUSH FORCES WITH AND WITHOUT MUD INJECTION

FIGURE 6

components:

- A hinged, four-leaf splayed footing (called a 'flapper'). The leafs hung near vertical when lowered through the moonpool but opened to a horizontal position on contact with the seabed. The flapper supported...
- a 223 mm (8 in.) I.D. guidepipe. The guidepipe extended from the flapper through a moonpool cover and supported...
- a hydraulic ram to push the cone. This was mounted in a frame attached to the guidepipe (Figure 5).

The casing system is illustrated schematically in Figure 7. Reaction for the cone push was provided by the dead weight of the flapper and guidepipe assembly, a total of about 3.5t. The drillship provided horizontal support to the top of the casing. Cone rod buckling in the guidepipe was prevented by nylon bushings fixed within the guidepipe at approximately 1.5 m (5 ft.) spacing.

The casing system was handled by the Simco 5000 WS all hydraulic drill rig mounted on the ship. The drill rig also provided hydraulic power for the CPT ram. The splayed footing was pivoted on perpendicular horizontal axes and allowed rotation of 5 degrees in any direction. The purpose of this was to avoid breaking the connection with the guide pipe due to the local slope of the seabed. The guide pipe was designed with a safety factor of 1.4 against buckling for a maximum water depth of 40 m (131 ft.) with a simultaneous maximum current of 3 knots. The guide pipe was formed from lengths of pipe that were standardized as 6 m (20 ft.), 3 m (10 ft.), 1.5 m (5 ft.) and 1 m (3 ft.) lengths. This allowed the casing length to be matched closely to the water depth at a given site.

The CPT was used with few problems at sites underlain by clay, except when shallow gas was encountered.

Shallow gas is found in the Beaufort at considerable excess pressures. If the CPT encounters pressurized gas, then the cone rods drop under their own weight into the soil. The use of a contact microphone necessarily means that signal loss occurs in this situation. The remedy is to preload the ram by gripping the cone rods in a frictional roller.

The average penetration achieved in clay has been about 30 m (98 ft.) below the sea bed; the maximum achieved was 71 m (233 ft.). The cut-off in all cases has been attenuation of the acoustic signal to the point where the receiving equipment ceases to operate.

Thin layers of permafrost and ice lensing within the clay have been penetrated without difficulty. However, the permafrost was "warm", typically no colder than -2.5°C (27.5°F).

Penetration of the eastern Beaufort sites, where sand is the dominant subsurface material, proved more difficult than expected. This was largely due to the upper 7 m (23 ft.) or so of the sand being extremely dense (relative density in excess of 120%). The sand was apparently uncemented as judged from undisturbed samples in adjacent boreholes. It was not always possible to push the cone through this dense layer and, in some instances the layer had to be drilled before the cone could be pushed. However, once the cone was through the dense layer, penetration proved normal. Friction values measured by the cone during attempts to penetrate this dense layer indicate the major component of the resistance was soil friction on the tip-transmitter assembly ahead of the mud injection ports.

Drilling ahead of the cone was not possible with the bushings in

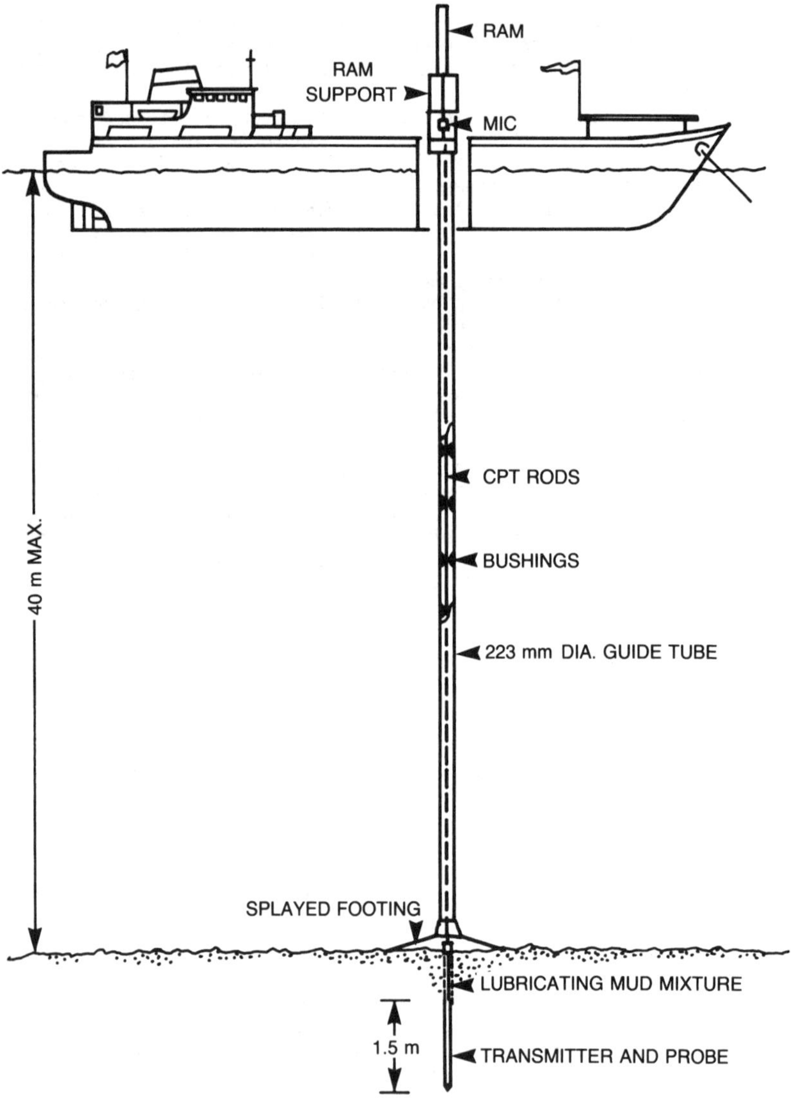

SCHEMATIC OF CPT DEPLOYMENT

FIGURE 7

the guide pipe. These bushings were, therefore, removed. Buckling of the cone rods in the guidepipe was prevented by leaving the 100 mm (4 in.) drillpipe hanging in the guidepipe and hanging a string of BQ rods in the 100 mm drillpipe. The cone rods ran down the centre of the BQ. This technique also increased the available reaction weight at the expense of an additional half hour in casing set up time.

The average single push in seabed sands was about 15 m (49 ft.); the greatest 31 m (102 ft.). The limitation on the push length was either insufficient reaction to get through the very dense layer or, after this layer had been drilled through, signal attentuation at about 30 m of seabed penetration.

Setting or retrieving the casing system typically took 1 1/2 to 2 hours. Each push usually took about 1 1/2 hours for 30 m of penetration. The reliability of the cone tip transducers could have been improved. The 1 1/2 hours allows for the occasional recovery of rods, replacement of tip and resuming the push.

The drillship did not always retrieve and reset anchors when more than one CPT or borehole was carried out; rather the ship was winched within the anchor spread to the next position. About 150 m (500 ft.) of travel were available in any direction from the nominal centre point of the anchor grid. The casing system did not have to be retrieved onto the deck of the ship when moving within an anchor set. It was found sufficient to lift the casing 6 m (20 ft.) above the sea floor using the drill rig while the ship was winched to a new position. The casing was then set down. This technique allowed greater productivity than the raw times given would imply on those sites where multiple pushes were carried out.

In anything but a flat calm, the drillship is continuously moving up and down while the casing system is stationary. Generally, it was safe and reasonably convenient to work the CPT system with as much as 0.6 m (2 ft.) of total movement. The greatest movement worked to date has been about 1 m (3 ft.); this much movement is distinctly hazardous. CPT operations were usually put on standby at 0.6 m of movement. Larger movements were accepted while the equipment was recovered at the onset of the bad weather.

No difficulty was encountered with the mud injection technique. The digital electronics were reliable although intermittent signal loss was experienced on a number of occasions. The signal loss was eventually traced to insufficient low temperature stability in the surface decoding/recording unit of the system. Further circuit design and temperature compensation of this unit is presently being carried out to prevent an occurrence of intermittent signal loss in future. The signal loss situation did emphasize the need for signal level meters on the receiving unit so that one can readily evaluate whether a problem is downhole or surface.

RESULTS, CORRELATIONS AND OTHER REMARKS

Sand Foundation Sites

The cone penetrometer was used as both a stratigraphic profiling tool and as a means of inferring relative densities of sands. The cone tip used was the conventional end bearing/friction combination, the end bearing transducer being rated at 50 MPa (522 tsf) and the friction sleeve 0.5 MPA (5.22 tsf).

Foundation sites primarily underlain by sand typically had a metre or so of soft clay on the sand. The 50 MPa tip was relatively insensitive to this upper clay and the inferred clay strength could not be relied on for design. The underlying sand had a uniform medium to fine texture with a silt content varying between 2 and 13 percent, averaging 5% or so.

Density was calculated from the cone end bearing stress using the relative density correlations reported by Baldi et al (1). These relative densities were transformed into absolute density using maximum and minimum densities obtained from tests on representative samples. The ASTM D2049 standard was used in the maximum and minimum density testing; three separate testing laboratories were employed and average values used in calculations.

Density was also calculated from moisture content measurements on 63 mm (2.5 in) tube samples obtained in boreholes adjacent to the cone pushes. These tubes were obtained either by rapid push (looser sands) or driven with a downhole hammer (dense sand). The end of the sample tube was capped immediately after the sample reached the top of the borehole to prevent moisture loss during sample handling. Tube samples obtained in this manner were usually abut 500 mm (20 in) long; the moisture content was normally measured on a sub-sample about 75 to 100 mm (3 to 4 in) long from the centre of the sample. Any water bleeding from the sample during handling in the laboratory was collected, weighed, and then distributed uniformly to the whole sample when calculating moisture contents. Density was calculated from the moisture content by assuming full saturation and measuring the specific gravity of the sand.

The densities calculated by the two methods are compared in Figure 8. There is a clear correlation between the two methods with the cone derived density being, on average, 0.4 KN/m^3 (2.5 pcf) greater than the density derived from moisture content measurements. The standard deviation between the two densities is about 0.2 KN/m^3 (1.3 pcf). For comparison the range of dry density for the sand is 13.9 to 17.9 KN/m^3 (88.5 to 114 pcf).

Some comments are in order regarding the errors involved in the two methods of density measurement. It can be noted that:

- Penetration of the sample into the sand will tend to compact the sample leading to a lower moisture content. Further, some moisture might be lost during the time the sample leaves the mud/water in the borehole to the time the bottom cap is in place. No moisture-content derived density has ever been calculated to be greater than the cone-derived density. It follows that these two effects on sample moisture content are probably small, at least with the procedures reported herein.
- The Baldi et al (1) correlation employs vertical effective stress. Although no measurements were made of the in-situ horizontal stress field in the eastern Beaufort, measurements made in the west indicate horizontal stresses are remarkably large. It appears this may not be uncommon (2). Large horizontal stresses are anticipated in the eastern Beaufort. If they are present, then using the first invariant of effective stress will be more valid theoretically. It will also produce somewhat lower estimates of density from the cone data. It follows that improved interpretation of cone data will only come

about if horizontal stresses are routinely measured. The 'step-taper-vane' (4) shows much promise in this regard as it is compatible with cone penetration hardware.

- The plot on Figure 8 also includes random errors due to boreholes being 15 m (49 ft.) or so away from the associated cone push and depth mismatch of data. These errors should be more or less normally distributed.
- One single set of maximum and minimum densities have been used in the calculation. No account has been taken of the slight variations in in-situ silt content, largely because the silt content appears to have no effect on the cone friction ratio and thus silt content is not detectable "a priori" with the cone.

Figure 8 cannot be taken as clear demonstration that the cone penetrometer can be used to determine absolute density since the errors outlined above might be compensating. But, Figure 8 is clearly encouraging and in any event is sufficient for practical engineering purposes since we can employ a partial material factor in design. The authors know of no better approach.

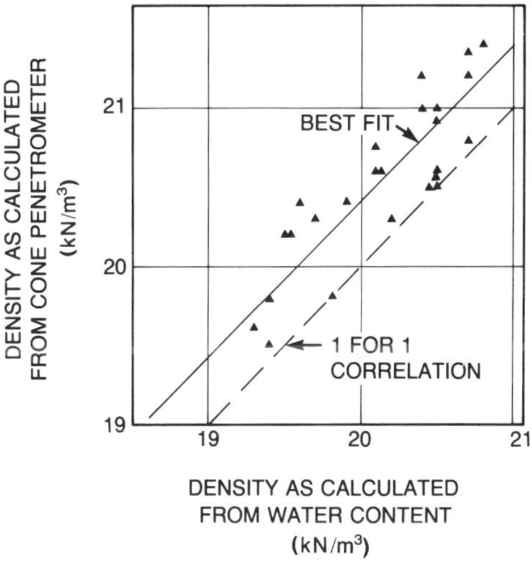

COMPARISON OF CALCULATED DENSITIES

FIGURE 8

Clay Foundation Sites

The cone penetrometer was initially used just as a stratigraphic profiling tool at sites primarily underlain by silt and clay. However, sufficient experience was accumulated in comparison to in-situ vane tests that the cone is now regarded as a suitable substitute for vane tests.

Both end bearing/friction and end bearing/pore pressure tips have been used at sites primarily underlain by clays and silts. The end bearing/pore pressure tip was used most of the time and found to be more useful of the two tips for subsequent data interpretation. The end bearing/pore pressure tip has also been used as a push in piezometer to measure the excess pressure in gas trapped by or in the clays.

Part of a cone trace is shown in Figure 9. It can be seen that a design minimum strength profile may easily be drawn from the trace by connecting the low q_c readings. The very low and variable strengths in the upper 2m or so are caused by ice scour. The seabed of the entire GCRI acreage is ice scoured to a greater or lesser extent. Strengths in this zone depend very much on the degree of remoulding or ploughing caused by ice. Strengths can vary substantially over horizontal distances as short as a metre. These low strength soils in the scoured zone are excavated before placement of the MAC berm. The only question is how deep the excavation need go.

The low strengths in the 0-3 m zone are not the low strengths the cone was used primarily to detect. The low strength shown at about 9.8 m is noteworthy. The strength was small but not zero as shown on the trace. What happened was the cone rods dropped 300 mm (1 ft.) under their own weight in the field with consequent signal loss, which the computer presently interprets as a zero strength. The interesting part of the event was that after the cone had indicated the low strength, the location was probed a second time holding the cone stationary for several minutes every 300 mm. In this mode, the end bearing/pore water pressure tip was used simply as a piezometer. The zone of very low strength was found to have a static excess pore water pressure equivalent to nearly 8 m of water, which is attributed to gas pressure. The site where the trace shown in Figure 9 was obtained is not suitable for the MAC.

A second example of a cone trace is shown on Figure 10. This is an end bearing/piezometric trace. Inspection of the raw data indicates two soils. But interpretation in terms of Dynamic Pore Pressure Ratio (DPPR) shows four soil types also shown on Figure 10. What is the DPPR? It is defined as:

$$DPPR = \frac{U - P}{q_c - \sigma_v}$$

where U = pore pressure measured by cone tip during penetration at 2 cm/sec (0.8 m/sec.).

P = static pore pressure (either assumed hydrostatic or measured with the cone stationary and peizometer fully equalized).

q_c = cone end bearing stress.

σ_v = vertical total stress (estimated).

Note that the DPPR, unlike the ratio U/q_c, is not affected by the water depth over the soil. This is important for offshore work and is theoretically correct. Also note that vertical total stress is used rather than the first invariant of total stress. Undoubtedly the invariant would lead to a more consistent interpretation. Unfortunately, a regional picture of the horizontal geostatic stress field has not yet been defined. The DPPR is an approximate measure of the pore pressure response of the soil to shear and can be thought of as similar in concept to Skempton's 'A' parameter.

Some comments are in order regarding the DPPR trace in Figure 10 and its stratigraphic interpretation.

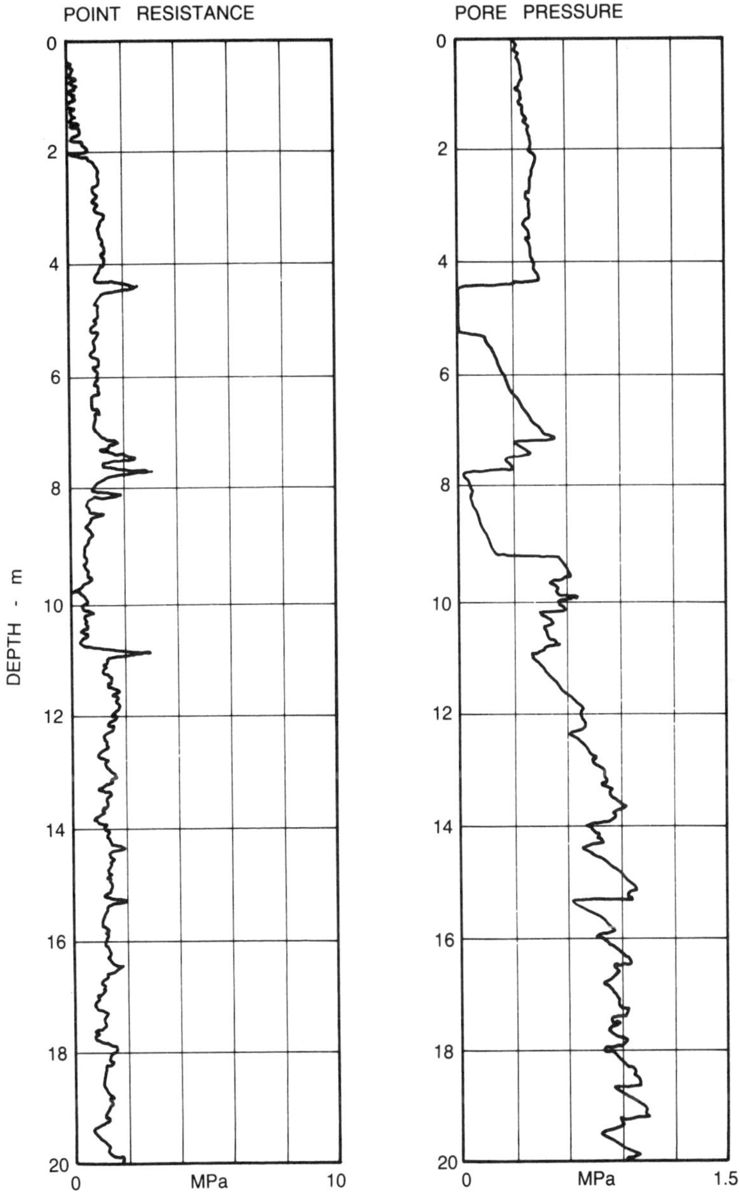

TYPICAL CONE RECORD WITH WEAK LAYERS

FIGURE 9

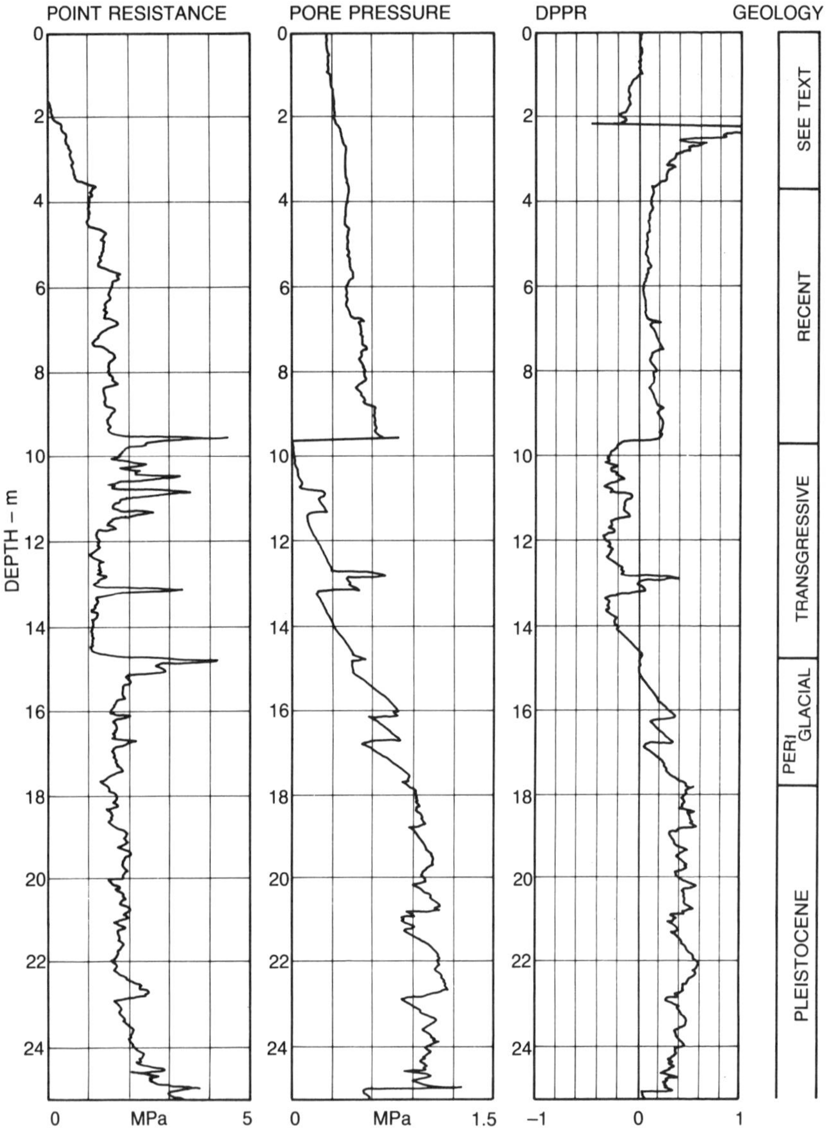

TYPICAL CONE RECORD – WESTERN BEAUFORT SHELF

FIGURE 10

The entire 25 m stratigraphic sequence whose CPT trace is shown in Figure 10 comprises silty clays and clayey silts, with a range of plasticity indices between approximately 12 and 30. Generally the soils are more or less horizontally bedded. The stratigraphic logging of undisturbed samples reflects changes in soil structure and interbedding frequency rather than changes in soil texture.

The 0-1.5 m portion of the CPT trace shows basically zero strength. Again this is caused by ice scour; strengths as low as 2.5 kPa 52.2 psf) have been measured on undisturbed samples taken from this zone. The trace showing near zero is more a reflection that all transducers have a lower detection limit and possibly that the zero offset voltage was not perfectly set. The slight oscillation of the pore pressure trace confirmed the tip was in soil at the time of testing; the DPPR shows a response which is much clearer and easier to interpret than a slight flutter in the pore pressure trace.

The DPPR shows very curious behaviour between 2.0 and 3.5 m on Figure 10. Undisturbed samples taken from this zone showed an apparently remoulded clay with seemingly undisturbed bedding. This zone is presently thought to be caused by strain softening during ice scour, without the soil being actually ploughed up. In this interpretation, the horizon at just below 2 m is predicted as the depth to which ice penetrated during scour at the cone test location.

The DPPR shows a near constant value from 3.5 to 9.5 m, jumps to a different, near constant value over the interval 9.5 to 13.6 m and is near constant again below about 17.8 m. The value of the DPPR in each near constant zone is quite different. What is interesting is that the logging of an adjacent borehole initially detected stratigraphic changes at 9.5 and about 16.5 m. The geophysical survey detected a reflector at 14.5 m. Undisturbed soil samples have been extruded subsequent to the DPPR intepretation; extremely careful logging of these samples has yielded the geologic interpretation shown on Figure 10. The DPPR allowed identification of geologic units that otherwise might have been missed. Scatter in laboratory data that might previously have been attributed to "natural variation in soil" is now seen to be a function of a change in geologic units.

A total of 107 vane tests were carried out using the McClelland Remote Vane (3) adjacent to cone soundings. The correlation between vane shear strength and cone end bearing is shown on Figure 11 expressed in terms of the cone factor N_k for each of the four identified soil types. The cone factor is defined as:

$$N_k = \frac{q_c - \sigma_v}{S_u}$$

where q_c = cone end bearing stress
σ_v = vertical total stress
S_u = vane shear strength

It can be seen from Figure 11 that each soil type has a range of cone factors. The Recent sediments have calculated cone factors, N_k, varying between 8 and 20. But, the values greater than 14 were recorded in the zone that has been heavily disturbed by ice scour. There can be substantial strength variation over horizontal distances as short as several meters in the ice scour zone. Consequently, not too much reliance should be placed on these numerically large cone factors in the Recent sediments; a N_k of about 12 is appropriate for design purposes

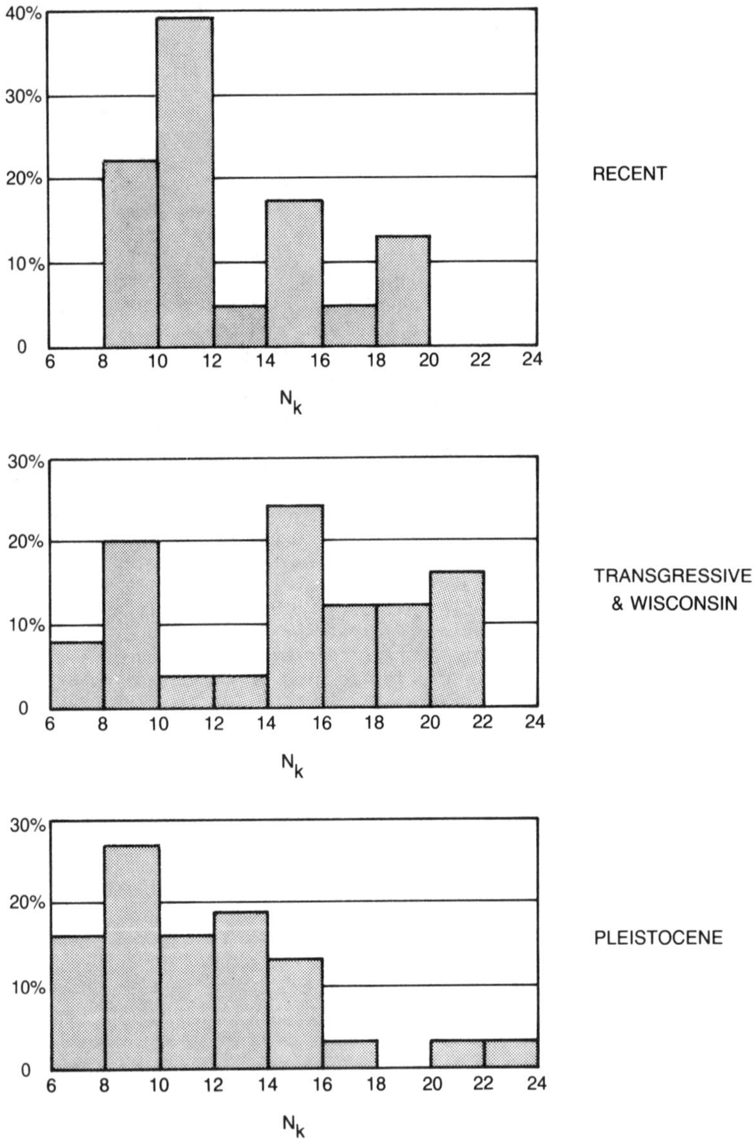

CONE FACTORS

FIGURE 11

at the 80% confidence level.

The Transgressive and Wisconsin sediments show an opposite trend in cone factors to the Recent sediments. The cone factor ranges from less than 8 to nearly 22. The negative DPPR values measured in much of the Transgressive and Wisconsin sediments suggest strongly dilatant material behavior. Undisturbed samples from adjacent boreholes showed many silt layers. Vane test and undrained interpretation of q_c cone values are not appropriate for silts. The large scatter in N_k would therefore seem to be an indication that neither test was undrained in many instances in these sediments. How could this be detected 'a priori'? On the present data it would appear that a large negative DPPR may well be an indication that undrained interpretation of q_c values should not be used.

The distribution of N_k in the Pleistocene sediments is more normal. Figure 11 illustrates a skewed distribution with a mean N_k of about 11. An N_k of 14 would be conservative at the 80% confidence level for design purposes.

Figure 11 illustrates that different distributions of N_k occur with different geologic strata. Attempts have been made to find correlations of individual N_k values with both plasticity index and DPPR. To date, no correlations have been found. The numerical values of N_k are within the range of values reported for the cone (5) in other geographic locations; the 80% confidence N_k are regarded as the local design correlations for the Beaufort shelf.

Berm Testing

GCRI constructed the first berm for deployment of the MAC during the 1982 arctic summer. The cone penetrometer was used as a quality assurance tool. Some 38 cone soundings were carried out in hydraulically placed sand fill.

An end bearing/piezometric cone was initially used, the idea being that the loose sand would tend to densify around the cone tip with consequent positive excess pore water pressure. No such piezometric response was measured. A conventional end bearing/friction tip was used for the bulk of the testing.

The interpretation of the cone data has followed the lines used for the foundation. End bearing stresses are transformed to relative density using the correlations of Baldi et al (1). Relative density is then translated to absolute density using laboratory maximum and minimum data. The berm testing differs from that of the foundation in that a full multiple regression analysis has been performed on the berm density data to define a 95% lower confidence density. Details on these procedures and typical data are presented elsewhere (6).

CONCLUSIONS

1. Cone penetration testing has been successfully carried out in the Beaufort Shelf. A new digitally encoded ultrasonic cone system was used. Some 1400 m (4600 ft.) of soundings have been carried out with the single greatest soundings as deep as 71 m (233 ft.) at one push.
2. The wireless feature of the digital cone allowed the use of mud injection to minimize rod friction. The reduction in rod friction

allowed a reduction in the reaction weight required to push the cone. This in turn allowed the cone to be deployed from a ship without special cranes, winches or other major structural modifications.

3. Cone data may be used quantitatively at sand sites for design of foundations for laterally loaded structures. However, the data to correlate and support the cone must be obtained carefully.
4. The end bearing/pore pressure cone data profiles clay sites well. The choice of a cone factor for design purposes should take into account the geologic unit in which the data was obtained. Different soil strata are observed most easily on the traces by calculating the dynamic pore pressure ratio (DPPR).
5. Caution should be used in an undrained interpretation of cone q_c data using N_k factors if a negative DPPR is measured.

ACKNOWLEDGEMENTS

The digital cone system was developed by Western Geosystems Inc. (Vancouver). Western Geosystems were also the cone testing contractor. Physical deployment of the cone system was carried out by the drilling contractor, Foundex Explorations (Vancouver), and much of the success of the cone is due to the wholehearted adoption of the system by the Foundex personnel. In particular, the authors are grateful for the enthusiasm of D. Diggle. Geotechnical consultants for the GCRI work were a joint venture of EBA Engineering (Edmonton) and McClelland Engineers (Houston); their efforts were much appreciated particularly with regard to remote vane testing. The authors would like to acknowledge the work of K. McKenzie (GCRI) who calculated many of the results presented and provided valuable assistance in production of the paper. Finally, the authors would like to thank the ASCE reviewers who made many helpful suggestions.

REFERENCES

(1) Baldi, G., Bellotti, R. Chiona, V., Jamiolkowski, M. and Pasqualini, E., "Design Parameters for Sands from CPT", Proceedings of the Second European Symposium on Penetration Testing, ESOPT II, Amsterdam, May 1982.

(2) Dalton, J., and Hawkins, P., "Fields of Stress - some measurements of the in-situ stress in a meadow in the Cambridgeshire countryside", Ground Engineering, May 1982.

(3) Doyle, E., McClelland, B. and Ferguson, G., "Wire-Line Vane Probe for Deep Penetration Measurements of Ocean Sediment Strength", Proceedings of Third Offshore Technical Conference, Vol. I, Houston, 1971.

(4) Fox, N., Handy, R., Trott, G., Remmes, B. and Moldt, S., "Determination of Horizontal Stress in Soils", Report No. FHWA/RD-81/118, U.S. Federal Highway Administration, 1981.

(5) Lunne, T., and Kleven, A., "Role of CPT in North Sea Foundation Engineering", Symposium on Cone Penetration Testing and Experience, Geotechnical Engineering Division, ASCE, October 1981.

(6) Stewart, H. and Jefferies, M., "Berm Construction for the Gulf Canada Mobile Arctic Caisson", to be published as paper 4552 1983 Offshore Tech. Conf. Houston.

IN SITU TESTS AND THEIR APPLICATION IN OFFSHORE DESIGN

Jean-Louis Briaud,[1] Member ASCE and Barry Meyer,[2] Member ASCE

INTRODUCTION

A platform is to be built in 700 ft of water; samples have been obtained from the site but the scatter of undrained shear strength is large; fortunately cone penetrometer profiles are available, these profiles help the engineer to reduce the scatter and to choose with less conservatism the shear strength design profile. Upstream from the platform a mud nose exists in gassy sediments; laboratory shear strengths from samples indicate that the slope is on the verge of failure; however the engineer suspects that decompression of the gassy samples resulted in a significant loss of strength; in situ vane tests results are available showing a strength twice as large as the miniature vane tests on the decompressed samples; the slope is much safer than would have been concluded from samples alone.

Here are two examples among many others where the client and the engineer have benefited from the use of in situ testing in offshore engineering. At the present time no engineer is ready to design an offshore structure on the basis of in situ tests alone without having seen, felt and tested in the laboratory, the soil on which the structure will be built but <u>the trend is here: in situ tests are becoming more and more useful to the engineer as water depths increase and as the need for less conservatism becomes essential to the client.</u>

In the first part of this paper the equipment available to perform in situ tests offshore is discussed; in the second part the use of the results in current practice is presented; in the third part recent advances that are not yet used commonly in design practice are described, while in the last part potential applications are suggested.

METHODS AND EQUIPMENTS

Detailed surveys and descriptions of offshore in situ testing equipment have been done [1,7,30,35]. The purpose of this section is not to describe all the pieces of equipment that have been or will be used offshore, but rather to describe the major pieces of equipment which are now available for routine investigation, as well as their

[1] Associate Professor, Center for Marine Geotechnical Engineering, Civil Engineering Department, Texas A&M University, College Station, Texas, 77843, U.S.A.

[2] McClelland Engineers, Inc., 6100 Hillcroft, Houston, Texas, 77081, U.S.A.

practicality and their range of operation. The vane, the cone penetrometer, the pressuremeter and the natural gamma logging equipment will be reviewed.

Equipment Used to Place the In Situ Testing Tools.

There are essentially two main systems to place the in situ testing tool: through the drill string and from a tethered sea bottom platform [Figure 1]. With a drill string, four components can be used: the drill string tensioner, the drill collars, the seabottom reaction frame, the drill string anchor. Combinations of these components are used to make suitable insertion systems.

The motion uncompensated drill string is the simplest of the systems. In this case the drill string moves with the boat and the drill bit moves up and down in the hole an amount equal to the boat vertical amplitude. The simplicity of this system makes it attractive for offshore operations but only in situ tools that do not require a controlled vertical push can be used. The use of motion uncompensated drill string is limited to 5 to 10 ft vessel heaves.

A stabilized drill string is required to operate the cone penetrometer from a floating vessel. Stabilization consists of stopping vertical motion of the drill string with respect to the bottom of the borehole [motion conpensation] and supplying a reaction force to resist the vertical thrust that may be as large as 10 tons.

The drill string tensioner is located on the vessel and is used to impose at the top of the drill string a tension force which decreases the compression between the drill bit and the bottom of the borehole. The applied tension is not constant, it is maximum when the vessel is at the top of the wave and minimum at the bottom of the wave. Because of the weight of the drill string the bit is in compression against the soil; this compression force called the drill bit load, is minimum when the vessel is at the top of the wave and maximum at the bottom of the wave. The drill string tensioner, often called motion compensator, only indirectly compensate for vertical motion of the vessel.

The weight of drill collars can be used to increase the drill bit load; however their use is limited.

A drill string anchor is a packer used to stabilize the drill string and to provide additional downward thrust. When the in situ testing tool is in position at the bottom of the drill string, the drill string is sealed, inflation ports leading to the packer on the outside of the drill string are opened and the drilling mud is pressurized; the packer inflates into the borehole wall [Figure 1]. The drill string anchor can provide up to 5 tons of additional vertical reaction. Wison III drillstring anchor [Fugro] is an example.

Seabottom reaction frames used to further stabilize the drill string are of two types. The first type clamps the drill pipe at the mudline but does not provide vertical downward thrust. Seaclam [Fugro] and Syminex One Function-Reaction Frame [Syminex] are two examples. If downward thrust is necessary, it must be provided by the in situ testing tool itself. The second type clamps the drill pipe and has downward jacking capabilities up to 10 tons of reaction and 1 meter of stroke. Stingray [McClelland] is an example. These units generally require a 10ft x 10ft moonpool.

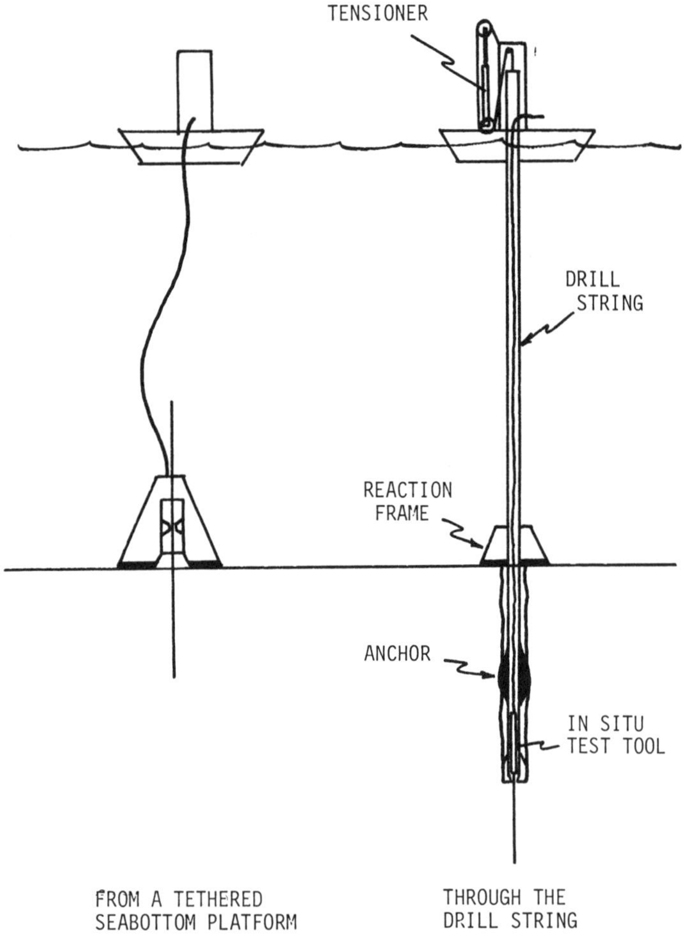

Figure 1 Systems to Place In Situ Testing Tools.

Tethered Seabottom Platforms [Figure 1] are frames resting on the seabottom and connected by flexible cables to the vessel. They do not provide drilling capability but provide up to 20 tons of thrust. Stingray in tethered mode [McClelland] and Seacalf [Fugro] are two examples of heavyweight platforms. Lightweight platforms such as MITS [Woodward- Clyde], LUT [Lehigh University] and Halibut [McClelland] are easier to handle but are limited to shallow testing [10 to 20 ft] and/or soft materials.

Equipment Used to Perform the In Situ Test

. The Vane Shear Test [VST]: is performed mainly to obtain a value of the undrained shear strength of cohesive soils. The tools available to perform a VST offshore are the Remote Vane and Halibut [McClelland] [15] MITS [Woodward-Clyde] [7], Fugro Vane [Fugro] and Lehigh Underwater Tower [Lehigh University] [34]. Figure 2 summarizes the data on these pieces of equipment. The use of the VST is limited to cohesive soils but has the advantage of being the simplest offshore in situ test: it can be used with motion uncompensated drill string and the tool itself is simple compared to the cone or the pressuremeter. A typical VST testing sequence would be to push a wireline vane at least 3 ft below the bottom of the borehole, perform a VST, drill 5 or 10 ft of borehole and repeat the above sequence. Under these conditions 2 to 3 tests can be performed in one hour, depending on water depth.

. The Cone Penetrometer Test [QS-CPT]: allows to obtain a profile of point resistance [q(max-CPT)], skin friction [f(max-CPT)] and more recently pore pressure [u]. The tools available to perform a QS-CPT offshore are Stingray in drill string or tethered mode [7] and Swordfish [29] [McClelland], Wison III in drill string anchor or seabottom reaction frame mode [36] and Seacalf [7] [Fugro], LUT [34] [Lehigh University], MITS [7] [Woodward-Clyde], XSP-40 [6] Naval Civil Engineering Laboratory]. Figure 3 summarizes the data on these pieces of equipment. The measurement of q[max-CPT] and u includes the hydrostatic pressure and the soil response; for deep water testing conpensation for hydrostatic pressure is a must in order to obtain accurate values for the soil response itself. The QS-CPT can only be used with a well stabilized drill string or a tethered seabottom platform; this makes its use more cumbersome than the VST. A typical QS-CPT testing sequence would be to take a 2 ft sample at the bottom of the borehole and then push the cone 10 ft below the bottom of the borehole, drill those 10 ft and repeat the sequence. Under these conditions 10 to 20 ft of QS-CPT sounding can be performed in one hour depending on water depth. From a tethered seabottom platform, the cone is pushed until the following is reached: system's thrust, capacity of the cone or length of the rods.

. The Pressuremeter Test [PMT]: allows to obtain an in situ stress strain curve of the soil by expanding a cylindrical probe into the borehole wall. During the test the pressure necessary to expand the probe and the increase in probe volume or diameter are measured. The PMT has not been used routinely in offshore investigations, however three tools are now available: the Pushed In Pressuremeter or P.I.P. [33] [Stressprobe], the WIreLine Expansometer or WILE [25] [Syminex],

Name	Years In Operation	Water Depth [ft] Design	Water Depth [ft] Actual	Depth Below Seafloor [ft] Design	Depth Below Seafloor [ft] Actual	Operational Method[2]	Vessel Type
Remote Vane	11	5000	1500	[1]	1500	MUDS/ SDS	Any drilling vessel
Halibut	6	5000	800	25	25	TSP	Any oceanographic or drilling vessel
Lehigh Underwater Tower	15	12000	12000	10	10	TSP	Any oceanographic or drilling vessel
MITS	4	1600	500	20	20	TSP	Any oceanographic or drilling vessel
Fugro Vane	3	1000	400	[1]	85	MUDS/ SDS	Any drilling vessel

Figure 2. Currently Available Vane Equipment

Name	Years In Operation	Water Depth [ft] Design	Water Depth [ft] Actual	Depth Below Seafloor [ft] Design	Depth Below Seafloor [ft] Actual	Operational Method[2]	Vessel Type
PIP	3	1000	550	[1]	400	SDS	Vessel with moonpool [3]
WILE	3	1000	650	[1]	15	SDS	Vessel with moonpool
PAM	2	3300	870	200	100	TSP	Supply boat or large oceanographic vessel

Figure 4. Currently Available Pressuremeter Equipment

NOTES: [1] included in design water depth.
[2] MUDS - Motion Uncompensated Drill String
SDS - Stabilized Drill String
TSP - Tethered Seabottom Platform
[3] Can be used on any drilling vessel if seafloor reaction frame is not required.

Name	Years In Operation	Water Depth [ft] Design	Water Depth [ft] Actual	Depth Below Seafloor [ft] Design	Depth Below Seafloor [ft] Actual	Operational Method	Vessel Type
Stingray, Downhole QS-CPT	6	2500	1200	600^{+}	600	SDS	Vessel with moonpool
Stingray, Shallow QS-CPT	4	2500	1200	100	100 [4]	TSP	Vessel with moonpool
Wison III w/Drill String Anchor	4	2100 [1]	1000	[1]	800	SDS	Any drilling vessel
Wison III w/Seaclam	2	2100	650	[1]	330	SDS	Vessel with moonpool
Swordfish	1.5	1800 [1]	100	[3]	300	SDS	Any drilling vessel
Seacalf	10	2100	740	100	100 [4]	TSP	Vessel with moonpool
MITS	4	1600	500	20	20	TSP	Any oceanographic or drilling vessel
XSP-40	1.5	200	100	20-40	38	TSP	Any oceanographic or drilling vessel

NOTES: [1] Combined water depth and penetration is dependent on umbilical length.

[2] SDS - Stabilized Drill String
TSP - Tethered Seabottom Platform

[3] Dependent on drilling capability.

[4] Depends on soil type and strength.

Figure 3 Currently Available QS-CPT Equipment

the Pressiometre Autoforeur Marin or PAM [12] [Institut Français du Petrole]. Figure 4 summarizes the data on these pieces of equipment. Pressuremeter probes are complicated pieces of equipment; also the pressure in the probe includes the hydrostatic pressure in the borehole and the soil resistance; as in the case of QS-CPT, the PMT probe must be pressure compensated for an accurate measurement of the soil resistance when testing in deep waters. The tools presently available must be used with a stabilized drill string or a tethered platform; however a pressuremeter which could operate with a motion uncompensated drill string is feasible and is being studied at Texas A&M University in cooperation with operators, oil companies and a manufacturer. It is estimated that one to two tests per hour can be expected when pressuremeter testing offshore.

Recently a presio penetrometer has been built by the Laboratoires des Ponts et Chaussees in cooperation with the Institut Français du Petrole [22]. The tool is driven in place, the dynamic q[max-CPT] is recorded during driving and the pressuremeter test is performed after driving.

. <u>Natural gamma logging and shear wave velocity</u> tests can also be performed [7]. The natural gamma logging test is gaining popularity as it is easy and quick to perform. Once a borehole is completed, [e.g. 350 ft deep B.M.L. in 500 ft of water] the probe is lowered to the bottom of the drill string. Logging takes place while raising the probe at a constant velocity. Under these conditions it takes approximately 2 hours to run 2 uphole logs of the same hole.

USE OF THE RESULTS: STATE OF PRACTICE

Vane Test Results

The main use of the VST consists of obtaining an undrained shear strength profile which will help the engineer in choosing a design S_u profile. Unconsolidated undrained [UU] triaxial tests on driven samples and miniature vane tests on driven samples usually give S_u value equal to 3/4 of the in situ vane S_u values [32]. As a result, the trend has been in the past to use S_u values corresponding to 3/4 of the in situ VST S_u values for design purposes. However it has been shown recently that S_u values from UU tests and miniature VST on 3 in. pushed samples are equal to the in situ VST S_u values [32]. This has warranted the use of the full in situ VST S_u value for design purposes. It is important to note that these results apply to the McClelland Remote Vane and cannot yet be generalized until further proof is obtained with other VST systems.

Generally the VST S_u values will present less scatter in the strength profile than driven samples S_u values and therefore, a VST S_u profile can be selected with more confidence.

In gassy sediment the decompression of the samples leads to very conservative S_u values. A 40% loss in S_u value has been shown to exist due to the decompression of gassy samples [14]. In such cases the in situ VST S_u values are less conservative and more reliable than the decompressed sample S_u values.

Cone Penetration Test Results.

One essential use of the CPT results is to accurately define stratigraphy. For example the exact thickness of a sand layer at depth can be obtained with accuracy from the q[max- CPT] profile; the strength and thickness of that layer may be critical for pile capacity and pile driveability. Also, the friction ratio, 100 x f[max-CPT/ qmax-CPT], and pore pressure response in the case of piezocones can be used to classify the soil penetrated [8,27,38] and provides a complement to laboratory classification.

Samples are commonly obtained every 10 ft. As a result, soil parameters are available only every 10 ft. The CPT profile allows to complete the soil profile by showing the relative variation of the soil strength within the 10 ft spacing. CPT results also help in reducing the scatter that may exist when the same soil property is measured by using different tests. Continuous CPT profiles have shown that the zone of disturbed soil can extend as much as 1 to 2 feet below the bottom of the borehole.

Undrained shear strength is generally computed using the following equation [8, 27, 38].

$$S_u = \frac{q[max\text{-}CPT]}{N_k} \qquad [1]$$

where q[max-CPT] = net cone tip resistance [26], and
N_k = 10 to 20, as determined from site specific correlations or other available correlations.

The same problem arises when selecting an N_k factor as occurs when trying to apply an empirical adjustment factor to an S_u profile. Soils are inherently different and there is no single N_k factor applying to all soils.

Dynamic pore pressure profiles are not commonly used; however they generally indicate stratigraphy with even greater precision than the q[max-CPT] profile; also the magnitude of the excess pore pressure is an indication of the permeability and density of the soil and may be negative for dense-dilatant cohesionless soils.

The q[max-CPT] value can be used to estimate the relative density D_r of a sand [38]. In many cases it is difficult to obtain undisturbed sand samples for laboratory testing. Obtaining D_r from q[max- CPT] may be the only way of estimating the relative density which is a key parameter in liquefaction studies.

Another essential use of the CPT results is the evaluation of the ultimate capacity of piles. Here, two methods are used in practice: the de Ruiter and Beringen method and the Schmertmann method. The procedure presented by de Ruiter and Beringen [37] and recommended by Lunne and Kleven [26] is summarized in Figure 5 and is detailed in references 8 and 37. This method allows for a maximum skin friction of 1.2 tsf and a maximum point bearing of 150 tsf.

Schmertmann's method is summarized in Figure 6 and detailed in references 8 and 38. The two above methods are used in practice but may lead to different ultimate pile capacity. In one example [8], the de Ruiter and Beringen method gave 142 tons for a pile in sand and 53.5 tons for a pile in clay, while Schmertmann's method for the same

	SAND	CLAY
Ultimate Skin Friction of Pile	Minumum of f[pile] = 2400 psf f[pile] = f[CPT] f[pile] = $\frac{q[CPT]}{300}$ [Compression] f[pile] = $\frac{q[CPT]}{400}$ [Tension]	f[pile] = $\alpha \frac{q[CPT]}{N_k}$ with $N_k \simeq 20$ α = 1 in N.C. clay α = 0.5 in O.C. clay
Ultimate Point Resistance of Pile	Minimum of q[pile] = average of q[CPT] within zone of influence of pile point q[pile] = limiting value, function of q[CPT] and OCR	q[pile] = $N_c \frac{q[CPT]}{N_k}$ with N_c = 9 $N_k \simeq 20$

Figure 5 The de Ruiter and Beringen Method [after de Ruiter and Beringen (37)]

	ALL SOILS
Ultimate Skin Friction for Pile	f[pile] = α f[CPT] Charts of α versus D/B for sands Charts of α versus f[CPT] for clays
Point Resistance for Pile	q[pile] = average of q[CPT] within zone of influence of pile point

Figure 6 Schmertmann Method [after Schmertmann (38)]

cases gave 182 tons and 111 tons, respectively. The pile was 1.5 ft in diameter and 36 ft long.

Pressuremeter Tests Results

The pressuremeter test is not used routinely in offshore investigations. When it is used it is generally for estimating the behavior of piles subjected to large horizontal loads, for the determination of a modulus of deformation for the soil or for the determination of a value of the undrained shear strength. The pressuremeter test has also been used for the prediction of jack up rig leg penetration and as a sensing probe below large North Sea gravity structures. A lot more can be done with pressuremeter results as will be seen in later sections.

Natural Gamma Logging.

Recording the level of natural radioactivity of the soil is used to give another continuous record of stratigraphy and soil type since clays have higher content of natural radioactive elements than sands. This test which is run through the drill string also provides the following pieces of information: a check on water depth and on the length of the drill string since a lower radioactive count will be recorded everytime the probe passes by the thicker wall of a drill string connector.

USE OF THE RESULTS: NEW DESIGN METHODS

In recent years several new design methods have been published; they are not used routinely in practice but are the results of extensive research and have been checked against numerous case histories. These methods are summarized in this section.

Cone Penetrometer f-w and q-w Curves.

The prediction of the pile head load-deflection curve for a pile loaded vertically makes use of two types of curves: the pile-soil friction f vs vertical pile movement w and the point pressure q vs w at the point [Figure 7]. Bustamante and Gianeselli [13, 8], developed a method to obtain f[max-pile] and q[max-pile] for the pile from q[max-CPT] of the cone penetrometer. This method consists of taking;

$$q[\text{max-pile}] = k_p\ q[\text{max-CPT}] \quad [2]$$

$$f[\text{max-pile}] = k_s\ q[\text{max-CPT}] \quad [3]$$

$$f[\text{max-pile}] \leq \text{limiting } f_{Lim} \text{ value} \quad [4]$$

where k_p, k_s and f_{Lim} depend on the pile installation procedure and the type of soil.

This method is based on approximately 50 onshore pile load tests. For the example reported previously, Bustamante and Gianeselli method gave a pile capacity of 119 tons in the sand and 39 tons in the clay.

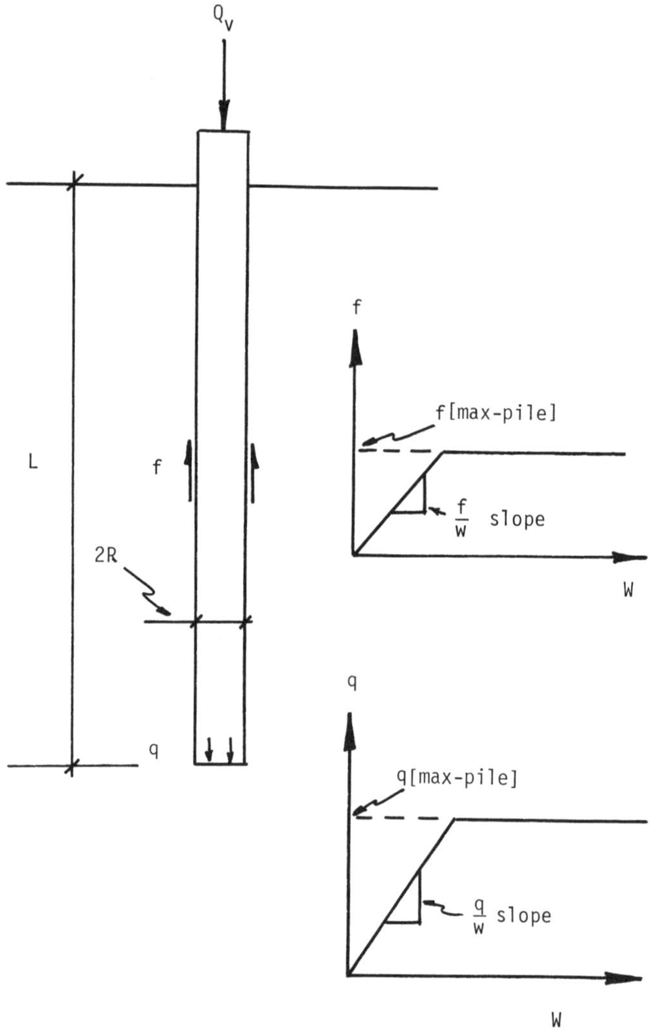

Figure 7 Elastic-Plastic Load Transfer Curves For Vertically Loaded Piles.

Verbrugge [39] has proposed a method to model the f-w and q-w curves by an elastic-plastic model [Figure 7]. The slopes of the linear portions of the two curves are obtained from linear elasticity tempered with experience from full scale load tests; these slopes are given by:

$$\frac{f}{w} = 0.11 \frac{E}{R} \qquad [5]$$

$$\frac{q}{w} = 1.56 \frac{E}{R} \qquad [6]$$

where R is the radius of the pile

For bored piles, the modulus E in kPa is obtained by correlation to q[max-CPT] in kpa through:

$$E = 3200 + 2.2\, q[\text{max-CPT}] \qquad [7]$$

For driven piles the modulus E given by equation 7 is multiplied by 3 before using Equations 5 and 6. The value of f[max-pile] is correlated to the value of q[max-CPT]. Values of q[max-pile] are not given.

Pressuremeter f-w and q-w Curves

Three methods exist to obtain the f-w and q-w curves from pressuremeter tests; two of them use preboring [Menard type] pressuremeter results [18,16], one of them uses selfboring pressuremeter results [3]. The Menard-Gambin method [18,8] models the f-w and q-w curves as elastic-plastic curves [Figure 7] where the slopes, for pile radii R less than 0.30 m, are given by:

$$\frac{f}{w} = \frac{E}{CR} \qquad [8]$$

$$\frac{q}{w} = 2 \frac{E}{R} \qquad [9]$$

where C depends on the length to diameter ratio of the pile.

For bored piles, E is the modulus obtained from a preboring pressuremeter test and for driven piles, E is the reload modulus obtained from the slope of an unload-reload cycle performed during a preboring pressuremeter test. The above two equations are modified for R larger than 0.30 m. Both f[max-pile] and q[max-pile] are obtained from the limit pressure p_L [28].

The Frank-Bustamante method [16,13] models the f-w and q-w curves as bilinear elastic-plastic curves; the first slope in the elastic range is given by [16]:

$$\frac{f}{w} = \alpha \frac{E}{R}\text{;with } \alpha = 2.5R \text{ and R in meters} \qquad [10]$$

$$\frac{q}{w} = 5.5 \frac{E}{R} \qquad [11]$$

Here again E is the first loading modulus from a preboring pressuremeter test. The second slope in the elastic range is 5 times softer than the first one and the change in slope occurs at one half of f[max-pile] and q[max-pile] which are obtained from the net limit pressure p_L^* [13]:

$$f[\text{max-pile}] = f[p_L^*, \text{ soil type, pile material, construction and method}] \qquad [12]$$

$$q[\text{max-pile}] = kp_L^* + q_{ov} \qquad [13]$$

where k is the bearing capacity factor which depends on the type of soil and the pile installation procedure, and q_{ov} is the total vertical overburden pressure at the pile point. Design curves have been developed to obtain f[max-pile] and q[max-pile] for known values of p_L^* and q_{ov}. These curves are based on the results of over 200 pile load tests performed by the Laboratoires des Ponts et Chaussees within the last 10 years. The value of α in Eq. 10 and the constant 5.5 in Eq. 11 are based on a limited number of case histories on bored piles.

Baguelin also developed a method which makes use of selfboring pressuremeter results [3,4] and uses an elastic-plastic model for the f-w curve. The slope is taken as:

$$\frac{f}{w} = \frac{G}{kR} = \frac{E}{2k[1+\nu]R} \qquad [14]$$

where $k = 1 + \text{Ln}\ [L/2R]$
and L = embeded pile length

Equation 14 is based on linear elasticity theory where the shear modulus G is the selfboring pressuremeter initial modulus. The f[max-pile] value is obtained from the pressure at 20% volumetric expansion p_{20} [4]. Although Baguelin does not make recommendations for the q/wslope, it seems to be consistent with his approach to take from elasticity:

$$\frac{q}{w} = \frac{8G}{\pi[1-\nu]B} = \frac{2}{\pi[1-\nu^2]}\ \frac{E}{R} \qquad [15]$$

Values of q[max-pile] are then obtained from the bearing capacity equation [4]:

$$q[\text{max-pile}] = \psi\ [p_{20} - p_o] + q_{ov} \qquad [16]$$

where p_0 is the total horizontal pressure at rest, and ψ is the bearing capacity factor [4]. If no selfboring pressuremeter results are available a pseudo-Baguelin method can be used if preboring pressuremeter results are available; in this case G is taken as the reload shear modulus and f[max-pile] and q[max-pile] are obtained from reference 5.

As a basis for comparison with equations 9, 11 and 15, it might be added that if the q-w response is considered to be similar to the expansion of a sphere in a linear elastic full space, the q-w slope is:

$$\frac{q}{w} = \frac{2E}{[1+\nu]R} \qquad [17]$$

and that for cylindrical expansion the equation is:

$$\frac{q}{w} = \frac{E}{[1+\nu]R} \qquad [18]$$

Although it bears no resemblance with an offshore pile, an example of the results obtained by using the above pressuremeter methods on a 2.5 ft diameter, 10 ft long drilled shaft in sand is shown in Figure 8. This example as well as the comparison of equations 8, 10 and 14 on one hand and equations 9, 11 and 15 on the other hand show that there is a need for unifying the pressuremeter design methods.

Pressuremeter p-y Curves.

Several methods exist to obtain p-y curves from pressuremeter data [4,5,9,17,19,21]. Two methods will be presented in this paper. The first method, developed by Baguelin and based on selfboring pressuremeter results consists of considering that [4]:

$$p_{pile} = p_{PMT} \times B_{pile} \quad [19]$$

where p_{pile} is the soil resistance per unit length of pile [lb/ft] and p_{PMT} is the pressure exerted by the pressuremeter on the borehole and B_{pile} is the pile diameter. It also considers that:

$$y_{pile} = \Delta R_{PMT} \times \frac{R_{pile}}{R_{PMT}} \quad [20]$$

where y_{pile} and R_{pile} are the lateral deflection and the radius of a pile element and ΔR_{PMT} and R_{PMT} are the increase in radius and the initial radius of the pressuremeter borehole, respectively.

The second method, proposed by Briaud et.al. [9,10] makes use of preboring pressuremeter results. The mechanism of soil response to the lateral movement of the pile is broken into two components: the frontal reaction [q-y curve] and the friction reaction [f-y curve] [Fig. 9]. This approach is consistent with the f-w/q-w approach for vertically loaded piles. The pressuremeter gives the q-y curve readily but theory must be used to obtain the f-y curve from the pressuremeter curve [10]. The q-y and f-y curves are then added according to the areas overwhich they apply. The resulting p-y curves are multiplied by a reduction factor close to the surface to recognize the fact that the critical depth is larger for the pile than for the pressuremeter; the reduction factor is not only a function of relative depth but also a function of displacement and soil strength. The details of this method are presented in reference 10.

All the new design methods presented in this section have been checked against onshore pile load tests only.

POTENTIAL USE OF THE RESULTS

Influence of Rate of Loading on Pile Behavior.

There is increasing evidence to show that in cohesive soils, pile capacities will be larger at fast loading rates than at slow loading rates. Advantage should be taken of this rate dependent phenomenon since waves load the offshore piles in seconds while design capacities are based on load tests which last hours. The problem is to find reliable methods which quantify the increase in soil resistance with

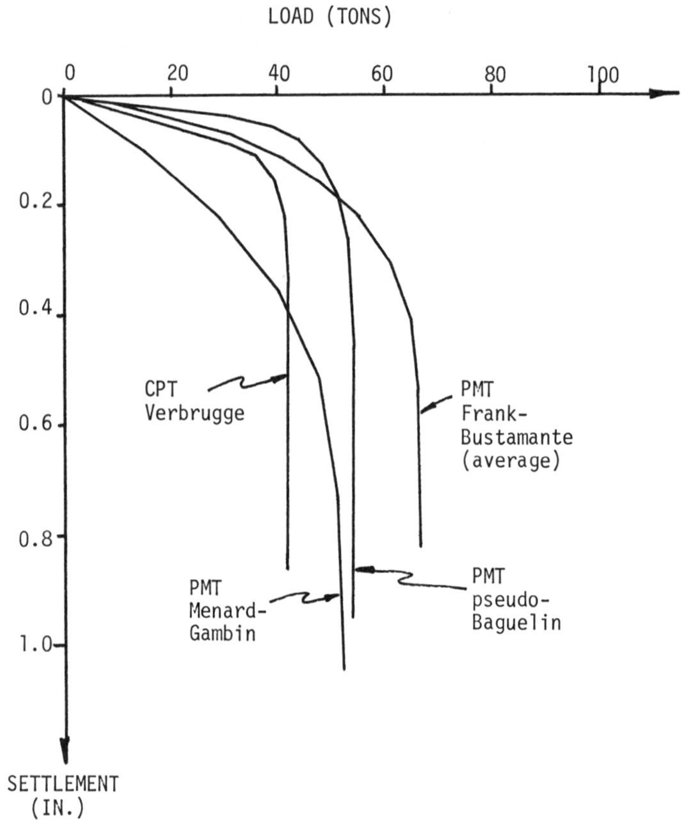

Figure 8 Example Comparison of PMT and CPT Top Load-Top Movement Predictions For a Vertically Loaded Pile

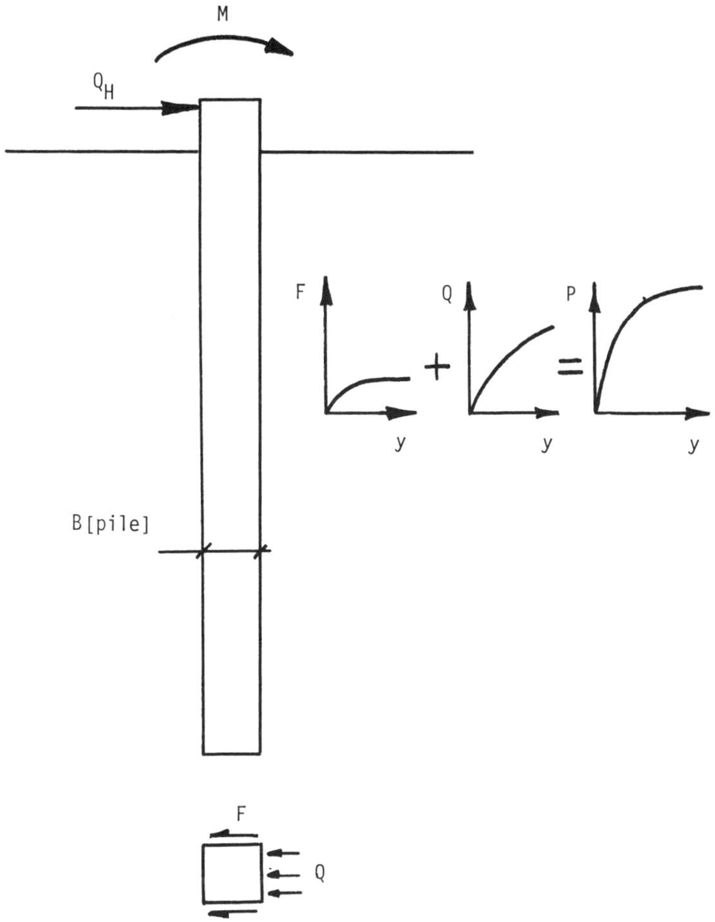

Figure 9 Load Transfer Curves for Horizontally Loaded Piles [after Briaud et.al. [10]]

increase in loading rate. The in situ test can help in this respect: the in situ vane test could be performed at higher rotation rates, the cone penetrometer test could be pushed faster than 2 cm/s, and the pressuremeter could be performed in less than 10 minutes. As far as the CPT is concerned however, a penetration rate of 2 cm/sec is already extremely high and this may be the reason why reduction factors are needed on both friction and point CPT values when matching hour long load test data. Another problem is the location of the friction sleeve. Most of the friction capacity of long piles is developed outside the zone of influence of the point, yet the CPT measures the friction within that zone. If this zone goes up 8 diameters along the pile as used in Schmertmann's method [38], and if this zone has an influence on the friction value, then the sleeve friction should be located at least one foot behind the point for the standard cone.

The increase in pile capacity with increase in loading rate is usually quantified in terms of percent increase per log cycle of loading rate. Numbers varying from 10 to 15 percent are commonly quoted; however penetrometer test data [20] and two pressuremeter tests recently performed in a soft plastic marine clay [Sabine, Texas] [Figure 10], tend to indicate percentages significantly higher. This discrepancy is probably due to the compressibility of the piles which leads to higher loading rates at the top of the pile and lower rates at the bottom. Indeed the load-displacement at the pile top represents an average soil response over the depth of the pile while CPT and PMT data represent local soil responses [2].

Influence of Cyclic Loading on Pile Behavior.

Cyclic loading of piles by waves will lead to a decrease in soil stiffness and soil strength which may offset the corresponding increase due to rate of loading effects. Cyclic loading is a more complex phenomenon and it appears that the pressuremeter can help in quantifying cyclic response [23]. An example is given in Figure 11. Similar plots could be obtained for the limit pressure versus number of cycles. A load transfer curve method similar to those described earlier, but applicable to cyclic loading is being developed [31].

CONCLUSIONS AND RECOMMENDATIONS

Laboratory tests and in situ tests are complementary; because of their definite advantages in situations where laboratory tests are of limited help in situ tests are growing more popular. For the continued development and design application of in situ tests it is necessary to consider the following points:

1. Measure pore pressure systematically - Although this type of measurement is difficult, it is evident that the use of effective stress and pore pressure rather than total stress alone leads to a better description of soil behavior. Therefore it is recommended that pore pressure be measured during in situ tests.

2. Develop theoretical interpretation - Due to the difficult boundary and drainage conditions, the theoretical interpretation of in situ test results is complicated. Development of the theoretical interpretation will lead to more complete use of the results.

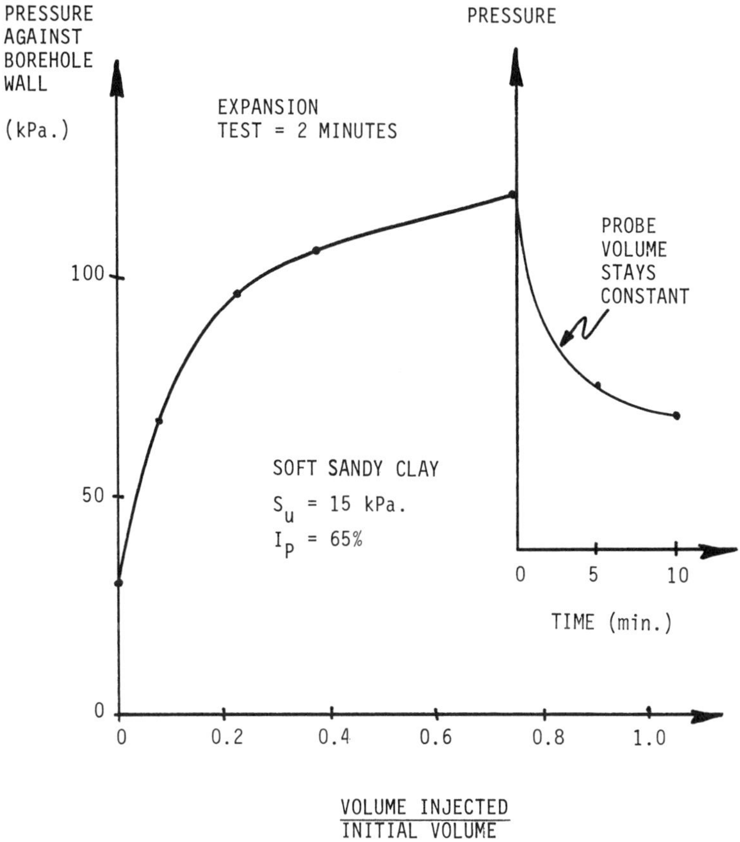

Figure 10 Results of a Pressuremeter Test Brought to the Limit Pressure in Two Minutes.

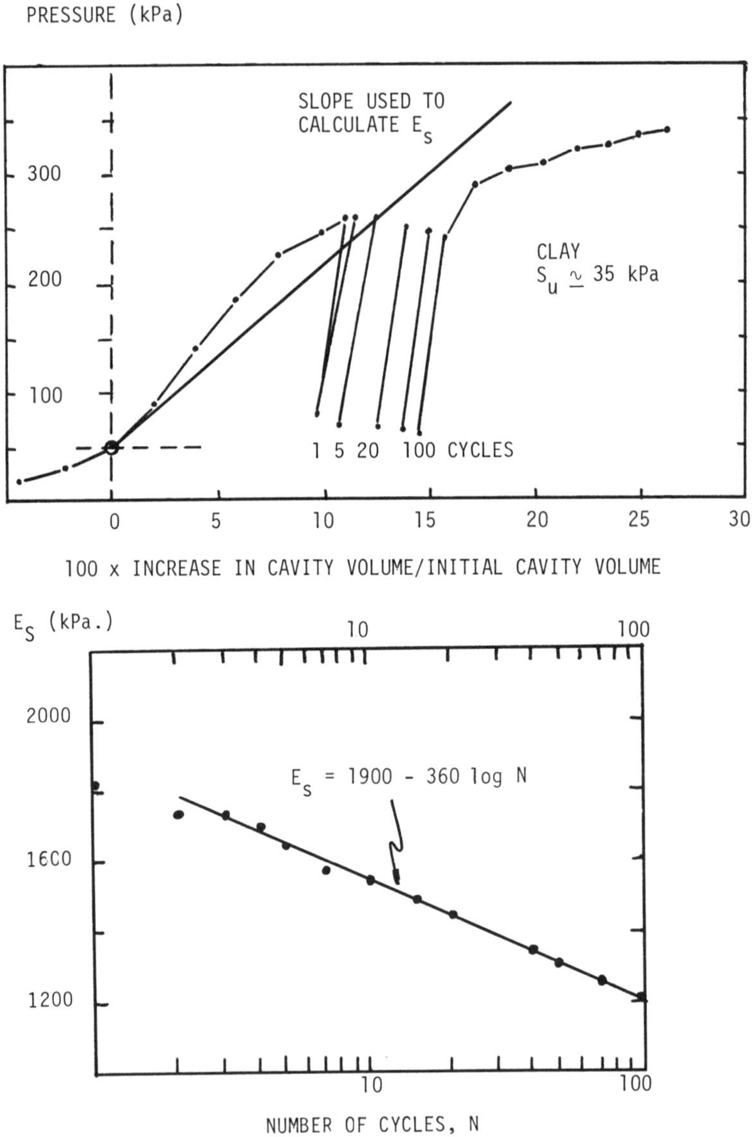

Figure 11 Cyclic Pressuremeter Test Results [after Briaud et al. (11)]

3. Improve understanding of scale effect - Some in situ tests are a small scale representation of the full scale engineering problem; this is the case of the cone penetrometer compared to a pile. Scale effect raises questions as to the direct application of the small scale test results to the full scale engineering problem.

4. Strive for standardization - Many direct comparisons have been made between the results of in situ tests and the actual be havior of the structure. However the use of these valuable comparisons is often hindered by the fact that from one case history to the next the in situ tests were not done according to the same procedure.

5. Choose the test that best simulates the problem - Since theoretical interpretation of the in situ tests is presently limited it appears necessary to use the test that best simulates the full scale problem.

6. Equate disturbance to the soil created by the in situ test and disturbance to the soil created by the installation of the foundation - For example when predicting the load carrying capacity of a deep foundation by the pressuremeter method, it seems appropriate to drive the pressuremeter for driven piles and insert the pressuremeter in a prebored hole for a drilled shaft. For a slope stability problem a selfboring test would be recommended. This approach should be taken while awaiting further development of methods that include the simulation of the insertion procedure for the pile.

7. Rely on comparison with full scale test - While pursuing the development of steps 2 and 3 above, it appears essential to rely mostly on direct comparison of in situ test results with full scale results. Small scale modeling of the full scale problem might be useful for the study of the basic phenomenon but should not be used for in situ tests calibration.

8. Measure dynamic resistance of the soil - In the offshore environment most forces acting on the structure and on the soil are of a dynamic nature (winds, waves, currents, earthquakes, ice), therefore the development of dynamic in situ testing should be strongly encouraged.

9. Develop in situ testing equipment to work in 2000 m of water and up to 100 m of penetration - Oil and gas production facilities are planned for those water depths. Due to the high cost of soil investigations and the time consuming aspect of wireline techniques in these water depths, multifunction tethered seabottom platforms should be developed.

ACKNOWLEDGEMENTS

Part of the information presented in this article was gathered during a research project sponsored by the United States Geological Survey and monitored by Dr. L. Garrison. The first author also wishes to acknowledge his team of graduate students.

REFERENCES

1. Andresen, A., Berre T., Kleven A., Lumme, T., "Procedures Used to Obtain Soil Parameters for Foundation Engineering in the orth Sea," Marine Geotechnology, Vol. 3 No. 3, 1979.

2. Audibert, J.M.E., Personnal Communication, October 1982.
3. Baguelin, F., Frank, R., Jezequel, J.F., "Parameters for Friction Piles in Marine Soils," Proceedings 2nd International Conference in Numerical Methods for Offshore Piling, University of Texas, Austin, April 1982.
4. Baguelin, F., "Rules for the Structural Design of Foundations Based on the Self-Boring Pressuremeter Test,"Symposium on the Pressuremeter and its Marine Applications," April 1982.
5. Baguelin, F., Jezequel, J.F., Shields, D.H., "Pressuremeter and Foundation Engineering", Trans Tech Publications, Rockport, Mass., 1978.
6. Beard, R., Lee, H.J., "A 40 ft Static Cone Penetrometer," Offshore Technology Conference, Paper No. 4300, Houston, 1982.
7. Briaud, J.L., "In situ Tests to Measure Soil Strength and Soil Deformability for Offshore Engineering," Report 1980-1 Center for Marine Geotechnical Engineering, Civil Engineering Dept., Texas A&M University, College Station, TX, USA, October, 1980.
8. Briaud, J.L., Tucker, L., Coyle, H.M., "Pressuremeter, Cone Penetrometer and Foundation Design," Texas A&M University Short Course Notes, April 1982.
9. Briaud, J.L., Smith, T., Meyer, B., "Design of laterally loaded Piles Using Pressuremeter Test Results," Symposium on the Pressuremeter and its Marine Applications, April 1982.
10. Briaud, J.L., Smith, T., Meyer, B., "Pressuremeter Gives Elementary Model for Laterally Loaded Piles," International Symposium on Soil and Rock Investigations by In Situ Testing," Paris, May 1983.
11. Briaud, J.L., Lytton, R.L., Hung, J-T, "Getting More from a Pressuremeter Curve," ASCE Geotechnical Engineering Journal, 1983.
12. Brucy, F., Fay, J.B., Marignier, J., "Offshore Self-Boring Pressuremeter: the LPC/IFP probe," Symposium on the Pressuremeter and its Marine Applications, Paris, April 1982.
13. Bustamante M., Gianeselli, L., "Prevision de la CapacitéPortante des Pieux Isolés Sous Charge Verticale: Regles pressio metriques et penetrometriques." Bull. Liaison Labs, P et Ch. 133, May-June 1981.
14. Denk, W.E., Dunlap, W.A., Bryant, W.R., Milberger, L.J., Whelan, T.J., "A Pressurized Core Barrel for Sampling Gas Charged Marine Sediments," Offshore Technology Conference, Vol. 4, p. 43-52, 1981.
15. Doyle, E.H., McClelland, B., Fergusson, G.H., "Wireline Vane Probe for Deep Penetration Measurements of Ocean Sediments Strength," Offshore Technology Conference, Houston, Paper No. 1327, 1971.
16. Frank, R., Zhao, S.R., "Estimation par les Parametres Pressio metriques de L'Enfoncement Sous Charge Axiale de Pieux Fores dans des Sols Fins," Bull. Liaison Laboratoires des Ponts et Chaussees, No. 119, May-June 1982.

17. GAI Consultants, Inc., "Laterally Loaded Drilled Pier Research", Research Reports Vol 1 and 2, for the Electric Power Research Institute, January 1982.

18. Gambin, M., "Calcul du Tassement d'une Fondation Profonde en Fonction des Resultats Pressiometriques," Sols-Soils, No. 7, 1963.

19. Gambin, M., "Foundations Sugjected to Horizontal Forces Using Pressuremeter Data," Sols-Soils, No. 30/31, 1979.

20. Goncharov, B.V., "Soil Resistance When Jacking Piles," OSMOVANIYA FUNDAMENTY I MEKHANIKA GRUNTOV, No. 1, pp. 17-18, July 1966.

21. Hughes, J.M.O., Goldsmith, P.R., Fondall, H.D.W., "Predicted and Measured Behavior of Laterally Loaded Piles for the Westgate Freeway: Melbourne," A paper presented to the Victoria Geomechanics Society, August 1979.

22. Jezequel, J.F., Laney, J.L., Perrier, M., "Mise en Oeuvre du Pressio Penetrometre LPC a L'Aide du BSMVA de la Societe Techniques Louis Menard," Symposium on the Pressuremeter and its Marine Applications, Paris, April 1982.

23. Jezequel, J.R., LeMehaute, A., "Cyclic Selfboring Pressuremeter Tests," Symposium on the Pressuremeter and its Marine Applications," Paris, April 1982.

24. Kraft, L.M., Jr., Ahmad, N., Focht, J.A., Jr., "Application of Remote Vane Results to Offshore Geotechnical Problems," Offshore Technology Conference, Houston, Paper No. 2626, 1976.

25. Legier, A., "The Wireline Expansometer:, Symposium on the Pressuremeter and Its Marine Applications, Paris, April, 1982.

26. Lunne, T., Kleven, A., "Role of CPT in North Sea Foundation Engineering," ASCE St. Louis Convention Session 35, Cone Penetration Testing and Experience, 1981.

27. Marr, L.S., Endley, S.N., "Offshore Geotechnical Investigation Using Cone Penetrometer," Offshore Technology Conference, Paper No. 4298, Houston, 1982.

28. Menard, L., "The Menard Pressuremeter; Interpretation and Application of Pressuremeter Test Results to Foundation Design," Sols-Soils, No. 26, 1975.

29. Meyer, B.J., Harman, D.E., King, P.G., "Introduction of a New Offshore Cone Penetrometer [CPT] Device for the Gulf of Mexico., Offshore Technology Conference, Paper No. 4299, Houston, 1982.

30. Noorany, I., Bea, R.G., "Methods of Geotechnical Explorations for Large Offshore Structures," Proceedings of the 15th Conference on Marine Technology, Ocean Energy, the Marine Technology Society, Washington, 1979.

31. Puech, A., Brucy, F., "Offshore Foundation Design Using Selfboring Pressuremeter Test Results," Symposium on the Pressuremeter and Its Marine Applications, Paris, April 1982.

32. Quiros, G. W., Young, A. G., Pelletier, J., Chan, J.H.C., "Strength Interpretation in the Gulf of Mexico: State of the Art," Geotechnical Practice in Offshore Engineering, ASCE Specialty Conference, Austin, April 1983.

33. Reid, W.M., St. John, H.D., Fyffe, S., Rigden, W.J., "The Push-In Pressuremeter," Symposium on the Pressuremeter and its Marine Applications, Paris, April 1982.
34. Richards, A.F., McDonald, V.J., Olson, R.E., Keller, G.H., "In Place Measurement of Deep Sea Soil Shear Strength," Symposium on Underwater Soil Sampling, Testing and Construction Control, ASTM, Special Technical Pub. 501, 1972.
35. Ruiter, J. de, "The Use of In Situ Testing for North Sea Soil Studies,", Offshore Europe 75 Conference, Aberdeen Proceedings, Paper OE-75219 London, Spearhead Publications, 1975.
36. Ruiter, J. de, "Site Investigations for North Sea Soil Studies,: Offshore Technology Conference, Paper No. 2246, Houston, 1975.
37. Ruiter, J. de, Beringen, F.L., "Pile Foundations for Large North Sea Structures," Marine Geotechnology, Vol 3, No. 3, pp. 267-314, 1979.
38. Schmertmann, J.H., "Guidelines for Cone Penetration Test: Performance and Design,: U.S. Department of Transportation, Federal Highway Administration, Report FHWA-TS-78-209, July 1978.
39. Verbrugge, J.C., "Evaluation du Tassement des Pieux a Partir de L'Essai de Penetration Statique,"Revue Francaise de Geotechnique, No. 15, 1981.

STABILITY OF ARCTIC GRAVITY STRUCTURES

BY VITOON VIVATRAT[1], M. ASCE
BRIAN J. WATT[2], M. ASCE

INTRODUCTION

Hydrocarbon exploration and production activities in the arctic offshore are expected to intensify rapidly in the near future. These activities will be supported by new types of offshore structures. The controlling design considerations for these structures include (1) large and concentrated design ice loads; (2) extremely severe temperature regime, and (3) remoteness of the area of operation.

Among the arctic offshore structures currently under development is a reusable drilling structure which can be removed from an operating location, towed to and reinstalled at a new operating location. This structure type will perform a function similar to that of conventional mobile drilling units, such as jack-up barges, in the shallower regions of the Gulf of Mexico. Contrary to a site specific structure, a mobile unit must be designed to cope with a range of seafloor conditions which may range from competent to relatively weak soils.

The structure schemes currently under active development include a conical structure (Fig. 1a) and a caisson (vertical-sided) structure made up of interconnected vertical cells (Fig. 1b). The choice of the structural scheme is largely dictated by whether or not the structure must resist the large multiyear ice features encountered in the ice pack.

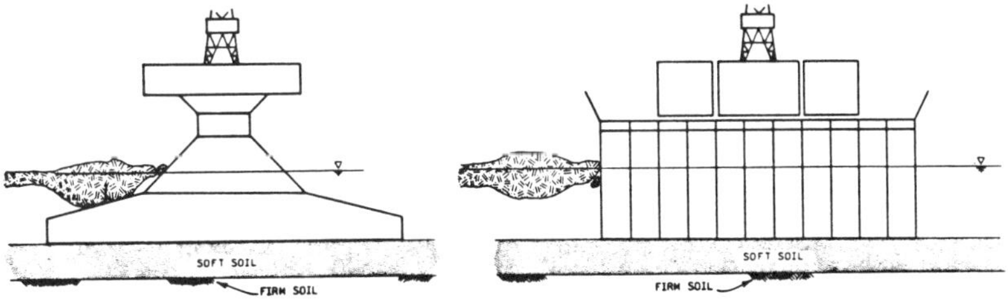

(a) Conical Structure (b) Caisson Structure

Fig. 1 Two structural schemes being considered for arctic offshore.

[1]Associate, Brian Watt Associates, Inc., 2350 E. North Belt Drive, Suite 450, Houston, Texas.

[2]President, Brian Watt Associates, Inc., 2350 E. North Belt Drive, Suite 450, Houston, Texas.

This paper discusses stability predictions for mobile or fixed gravity structures for the arctic offshore. The discussion will include (1) the selection of strength profiles for stability prediction, including the effects of consolidation under the structure weight; (2) the effects of finite soil layer thickness on stability at setdown; (3) the effects of structure configuration on soil failure mode, and (4) the effects of nonhomogeneous soil strength profile on sliding resistance.

Although the long-term (drained) behavior of the foundation is also of concern, the short-term (undrained) behavior under ice loading is believed to be more critical. This paper concentrates on undrained behavior under loads, with allowance for strength gain due to consolidation before loading.

It is recognized that the final design of structures of the type considered in this paper will rely heavily on sophisticated finite element analyses in which realistic soil constitutive models are used. Simple stability prediction will, however, still be needed to (1) establish technical feasibility; (2) perform parametric studies of the effects of various factors, and (3) provide a basis from which a rational and cost-effective numerical analysis program can be selected.

STRENGTH PROFILE SELECTION

Geological History

The arctic offshore area of interest in this paper is the region between water depths of 30 and 200 ft in the Alaskan Beaufort Sea. The subsea soil profile in this area generally consists of alternating layers of silts or clayey silts and silty sands. This layering is attributable to sea level fluctuation in the Pleistocene Period. As a result of this sea level fluctation, parts of the continental shelf were exposed to a cold climate for long periods of time, and permafrost was formed. Relic permafrost still exists at below the present seafloor (NRC, 15).

The soil layers which have the most impact on gravity structure design are within a depth of 50 ft below the mudline. Frequently, these comprise a thin veneer of silty sands overlying a 30 to 60 ft thick layer of relatively soft silts or clayey silts. Relic permafrost can sometimes be found within the silt layer. The silt layer tends to show a highly variable degree of "apparent" overconsolidation most likely due to the freezing and thawing action that it has undergone in the past. Wang, Vivatrat and Ruser (21) presented a compilation of typical index and engineering characteristics of this silty soil.

Factors Affecting Undrained Shear Strength

Strength Anisotropy The undrained shearing behavior of fine-grained soil samples in laboratory testing has been found to be highly anisotropic, i.e., dependent on the directions of the applied stress system (Ladd et al, 11). This anistropic behavior is attributed to the non-isotropic initial consolidation stress system in the ground, as well as the existence of preferred orientations in the soil fabric. Several analytical models have been proposed to represent this behavior, including the rigid-plasticity model by Davis and Christian (4), and the incremental plasticity model by Prevost (17).

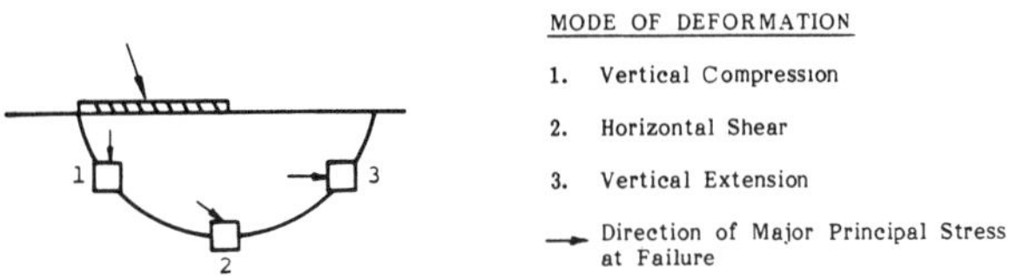

	TYPE OF TEST	UNDRAINED SHEAR STRENGTH RATIO c/p		
		OCR = 1	OCR = 2	OCR = 4
1.	Triaxial Compression CK_oU TC	0.46	0.77	1.30
2.	Simple Shear CK_oU SS	0.32	0.54	0.91
3.	Triaxial Extension CK_oU TE	0.12	0.20	0.34

Fig. 2 Estimated Undrained Shear Strength Parameters for Santa Barbara Silt.

The diagram at the top of Fig. 2 illustrates three different types of applied stress systems along an assumed failure surface in a bearing capacity analysis. These different modes of deformation include (1) vertical compression, (2) horizontal shear, and (3) vertical extension, and can be tested in the laboratory by various special testing devices and procedures, such as plane-strain (or triaxial) compression and extension, and the direct simple shear device.

Stress History In addition to the applied stress system, the undrained shearing behavior of a fine-grained soil sample is strongly influenced by its "Stress History" (Ladd et al, 11). The concepts of the "apparent" maximum past pressure (σ'_{vm}) and overconsolidation ratio (OCR) have been used to represent the cumulative effects of aging, as well as other physio-chemical processes occurring to the soil sample since the time of its deposition. Highly variable undrained shear strengths have been observed in fine-grained soil deposits subject to desiccation, weathering and leaching. This is not surprising since these processes can hardly be expected to produce uniform effect with depth or from location to location. Ladd, Moh and Gifford (10) successfully represented the strength ranges of the desiccated Bangkok clay deposit through bracketing the values of the "apparent" maximum past pressure.

As mentioned earlier, the Beaufort Sea soils are suspected to have undergone significant freezing and thawing in the past. This has resulted in a deposit with highly variable strength properties. The "apparent" maximum past pressure probably ranges from 2 to 12 ksf. For the purpose of discussion in this paper, the value of 4 ksf will be taken as the "design" condition.

Normalized Strength Parameters

A few soil exploration and testing programs in the Beaufort Sea have been conducted during the past several years. Most of the data, however, remains proprietary. For this paper, some laboratory test results for a similar silty soil (Santa Barbara Silt) from offshore Southern California will be used in lieu of actual soil data from the Beaufort Sea.

Results of K_0-consolidated, undrained triaxial compression and extension tests on normally-consolidated samples were presented by Prevost et al (18) together with the results from a simple shear test device. The stress-strain behavior from these tests is presented in Fig. 3. The undrained shear strength parameters from these tests are also summarized in the table in Fig. 2.

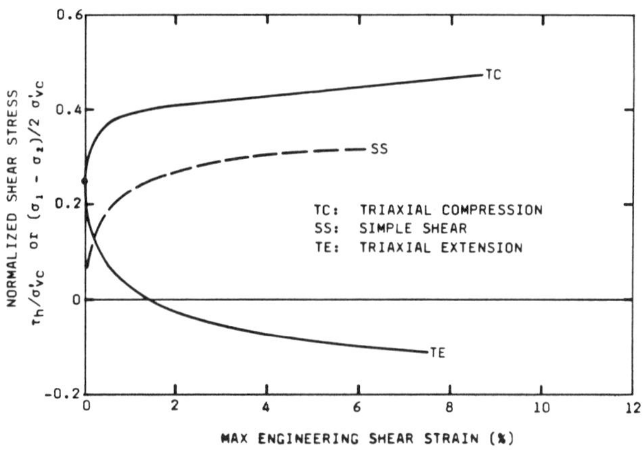

Fig. 3 Normalized Stress-Strain Behavior of Normally nsolidated Santa Barbara Silt.

The strength parameters for overconsolidated samples were not available. For this paper, the undrained shear strength ratio (c/p) for each mode of the deformation will be estimated with an empirical expression suggested by Ladd et al (11):

$$c/p = c/p\ (OCR = 1) \times OCR^m, \tag{1}$$

where m is 0.75. The strength parameters for OCR's of 2 and 4 are presented in Fig. 2. The degree of strength anisotropy was, in this approach, assumed to be constant with increasing OCR. Data from Ladd et al (11), however, suggests that the degree of strength anisotropy reduces as OCR increases.

Initial Strength Profiles

Fig. 4 shows the stength profiles for different deformation modes when the apparent maximum past pressure is 4 ksf. It is obvious that strength anisotropy can have a strong influence on the selection of the design strength and hence deserves serious consideration in the soil exploration and testing program.

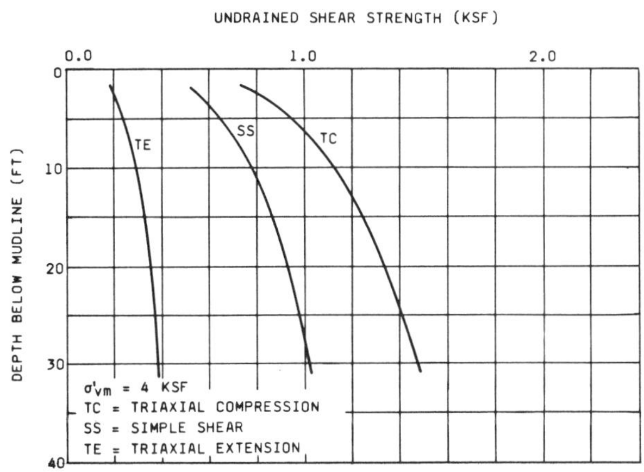

Fig. 4 Strength Profiles for Different Modes of Deformation.

For problems involving primarily horizontal sliding, the strength profile corresponding to the simple shear test is considered appropriate.

For bearing capacity problems, the strength selection can be based on some findings from rigid-plasticity solutions for anisotropic clays. These solutions were obtained from Slip-Line Theory (Hill, 8) using an anisotropic yield criterion proposed by Davis and Christian (4). Davis and Christian (4) also presented solutions for incipient footing failure, whereas solutions for steady state wedge penetration were presented by Vivatrat (20). These solutions show that a reasonable approximation (±10%) can be obtained by assuming the clays to be isotropic but with a strength equal to the average of the strength for vertical compression and extension.

The strength profiles in Fig. 4 indicate that the average of the vertical compression and extension strength is only slightly smaller than the simple shear strength. Therefore, the simple shear strength profile is also appropriate for bearing capacity problems.

Fig. 5 shows the simple-shear strength profiles corresponding to apparent maximum past pressures of 4 and 12 ksf. It can be seen that this much variability in the apparent maximum past pressure causes more than a factor of two variability in the undrained shear strength. This magnitude of variability in the apparent overconsolidation and the associated undrained shear strength is expected to be common for areas on the arctic continental shelf which were previously exposed. The modern arctic coastal lowland exhibits many morphological features which might have contributed to this variability, e.g., ice wedges, pingos, polygons, thaw lakes, etc.

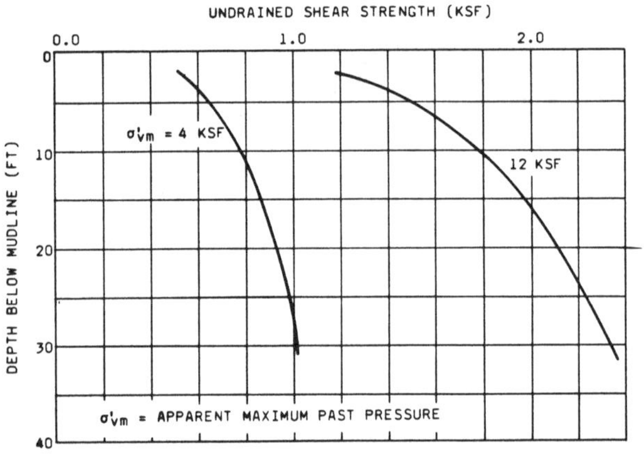

Fig. 5 Effects of Apparent Overconsolidation on Strength Profile.

The strength profiles in Fig. 5 were computed assuming the apparent maximum past pressure to be uniform with depth. This is a reasonable assumption for the area, and it provides a convenient basis for evaluating geotechnical conditions over a large area. Individual borings can, however, deviate substantially from this assumption and should be taken into account in a site-specific analysis.

Consolidation under the Structure Weight

For a gravity structure on a relatively weak deposit, the lowest resistance generally occurs right after the structure installation.

Subsequently, the foundation soils tend to gain in strength as consolidation under the dead weight of the structure progresses.

Traditionally, design load criteria are chosen to represent an acceptably low probability of failure during the design life of the structure. The maximum ice load on the other hand may not occur until some time after the structure installation. It is, therefore, necessary to evaluate the stability of the structure under the design ice load as a function of time after installation.

Fig. 6 illustrates the changes in the strength profile with time after structure installation. Curve 1 is the initial strength profile presented earlier in Fig. 5 (σ'_{vm} = 4 ksf).

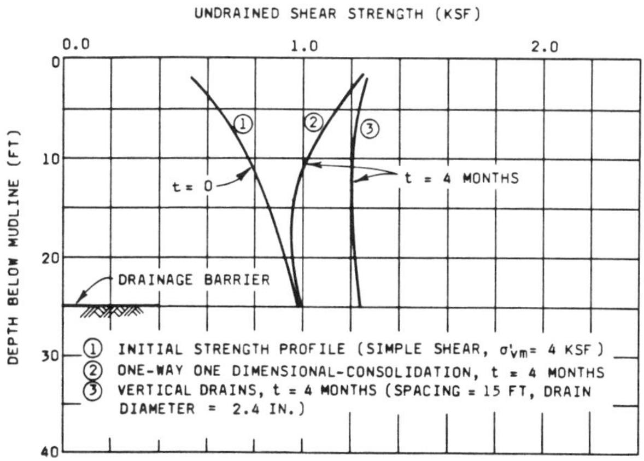

Fig. 6 Effects of Consolidation under Structure Weight on Strength Profile.

Curve 2 shows the strength profile at 4 months after the installation of a structure whose effective weight is equivalent to a surcharge pressure of 4 ksf. The curve was computed with the one-dimensional theory of consolidation, assuming one-way drainage with a drainage barrier at a 25-ft depth (representing perhaps an extensive layer of ice-bonded soils). The coefficient of consolidation, c_v, was taken to be $4x10^{-3}$ cm^2/sec. Fig. 6 shows that there is a substantial gain in strength, especially near the mudline. This will have a strong impact on the stability of the structure against horizontal sliding.

Curve 2 also applies for the case in which the layer thickness is 50 ft but two-way drainage exists, i.e., there is a sand layer below 50 ft depth. Note that Curve 2 shows the weakest soil to be at a depth of 17 ft after an elapsed time of 4 months. A number of these analyses were performed for different layer thicknesses, drainage conditions and surcharge pressures. Generally, it was found that, at 3 to 6 months after installation, the weakest soils tend to be located between 10 and 20 ft below mudline.

It should also be pointed out that the value selected for the coefficient of consolidation, c_v, is about 4 times the value considered appropriate for the normally-consolidated range. This is because the foundation soils will primarily be undergoing reloading in this case, and c_v for reloading is generally substantially higher than for loading in the virgin compression range. For cases involving both reloading and virgin compression, a variable c_v may have to be considered.

Comparison of Curves 1 and 2 in Fig. 6 shows that there is a substantial gain in the undrained shear strength due to consolidation under the structure weight after a period of 4 months. This points out the possibility that a cost-effective foundation design can be achieved if the rate of consolidation can be speeded up.

Artificial Drains

The use of artificial drains, such as sand drains and "wick" drains to speed up the consolidation process is quite common for embankments but has not been attempted offshore, to the authors' knowledge. On the other hand, gravity structures for soft soil sites are also relatively recent developments.

Curve 3 in Fig. 6 shows the strength profile at 4 months after set-down when artificial drains are used. The wick drains were assumed to have a spacing of 15 ft, and the equivalent drain diameter conservatively assumed to be 2.4 in. Based on field observations, Fellenius (6) and Hansbo (7) suggested an equivalent diameter for wick drains as high as 6 in.

With artificial drains, the consolidation of the foundation soils involves both vertical and radial drainage. The consolidation rate in this case can be estimated using a numerical method such as that presented by Olson, Daniel and Liu (16). Curve 3 in Fig. 6, however, was obtained simply by combining the independently estimated degrees of consolidation in the vertical and radial directions according to:

$$(1-U(z,t)) = (1-U_v(z,t))(1-U_h(t)) \qquad (2)$$

where $U(z,t)$ = the combined degree of consolidation at depth z and time t,

$U_v(z,t)$ = the degree of consolidation at depth z and time t according to 1-D theory,

$U_h(t)$ = the average degree of consolidation at time t for radial drainage.

The coefficient of consolidation for radial drainage, c_h, was taken to be $2c_v$. Existence of sand lenses could make higher values for c_h more appropriate.

EFFECTS OF FINITE SOIL LAYER THICKNESS ON SET-DOWN STABILITY

Problem Description

At the time of installation, a gravity structure will be floated in and ballasted down to the seafloor. Once the structure base is in contact with the seafloor, the ballast will be increased until the skirt system penetrates. (Short stiff skirts are more likely to be used on a mobile structure.) Differential ballasting, i.e., shifting of weight, may be needed to correct tilting. Finally, the ballast weight will be increased (or decreased) to the design value which is selected on the basis of resistance to ice and wave loads.

The first geotechnical concern is the bearing capacity at installation. Unless the ballasting procedure is spread out over a long period, the initial or in situ soil strength (Curve 1 in Fig. 6) should be used in stability calculations.

The traditional bearing capacity analysis based on Prandtl's solution is frequently inappropriate for analyzing set-down stability of an arctic gravity structure. This is because the base width (B) is significantly greater than the thickness of the most compressible soil layer (H).

The penetration resistance (or bearing capacity) of a nonhomogeneous cohesive soil layer has been investigated in the past. Davis and Booker (5) and Murff and Miller (14) treated problems in which the shear strength increases linearly with depth. Fig. 7 shows the slip-line and velocity fields for penetration of a rigid and smooth footing in this type of deposit. It can be seen that the failure zone depth is much more limited in this case than in the case of a uniform deposit.

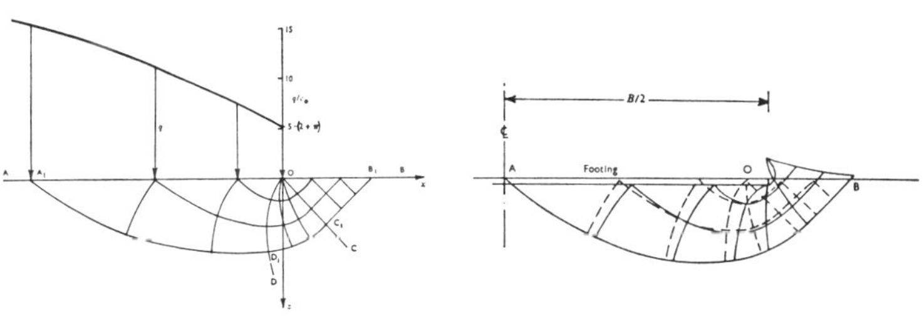

Fig. 7 Slip-Line and Velocity Fields for a Strip Load on a Soil Deposit with Increasing Shear Strength With Depth (after Davis and Booker, 5).

In the arctic offshore, site conditions can be frequently represented by a compressible layer near the mudline with a relatively low but approximately uniform shear strength, underlain by significantly more competent soils. This section concentrates on set-down stability under this condition.

Prediction Methods

Hill (8) presented the slip-line solution for the compression between two rough plates. The problem was further analyzed by Johnson and Kudo (9) and Collins (3), and the slip-line field for the case of a rough footing and a rough soil layer base is shown in Fig. 8.

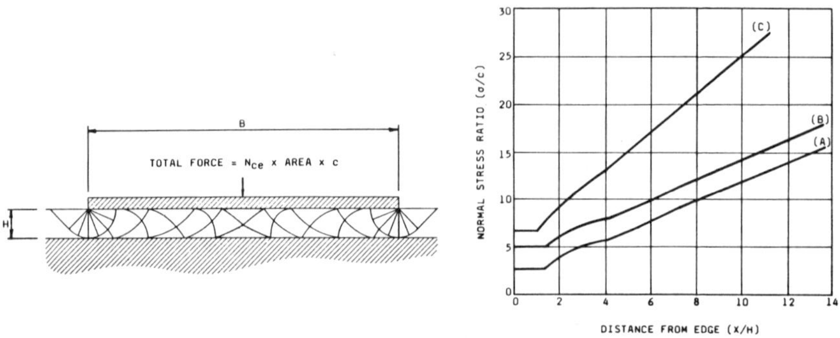

Fig. 8 Compression of a Finite Soil Layer by a Rigid Footing. (A) Rough Footing, Smooth Layer Base; (B) Smooth Footing, Rough Layer Base; (C) Rough Footing, Rough Layer Base. (Collins, 3)

Fig. 8 also shows the distribution of the normal pressure on the footing for three different cases: (A) rough plate and smooth layer base, (B) smooth plate and rough layer base, and (C) rough plate and rough layer base. Case (C) is considered appropriate for the case of a gravity structure with a properly designed skirt system.

The total load on the footing can be estimated by integrating the normal pressure over the footing area. Fig. 9(a) shows the load carrying capacity for the case of a rough strip footing in terms of an equivalent bearing capacity factor N_{ce} defined as

$$N_{ce} = \text{Average Footing Pressure/Shear Strength.} \qquad (3)$$

Fig. 9(b) presents the values of N_{ce} for circular footings. The curve was obtained by assuming that the stress field and pressure distribution for the plane strain case applies also in the axisymmetric case along a radial line. This assumption results in a 15 to 22% lower average footing pressure for a circular footing in comparison with a strip footing of the same width. This is approximately the same difference as predicted by Murff and Miller (14) with the upperbound method for a case of linearly increasing soil strength.

Based on these results and the initial shear strength profile of Fig. 6, the safety factor at set-down for a structure with a 500-ft diameter and a maximum ballast weight of 4 ksf is given by F.S. = 11.8 x 0.75/4.0 = 2.21.

Meyerhof and Chaplin (12) presented solutions for the above problems based on a similar approach using cycloidal failure surfaces and suggested the following:

strip: $N_{ce} = 4.14 + 0.50\ B/H$ (4)

circular: $N_{ce} = 5.05 + 0.66\ B/H$ (5)

which are presented as dashed lines in Figs. 9(a) and (b). These values are slightly lower than the values obtained above for strip footings but are essentially identical for the case of circular footings.

Vesic (19) interpolated between Prandtl's solution and Equation (4) and presented values for a modified bearing capacity factor for different ratios of the bottom and top layer shear strengths. His results do not predict a substantial increase in bearing capacity unless the bottom layer is much stronger than the top layer.

Comparison with Model Tests

Brown and Meyerhof (2) presented the results of a series of small scale model tests of strip and circular footings on finite layers of clays. The soil samples were prepared in the laboratory to have different strengths, and the ratio of the strengths of the bottom and top layers

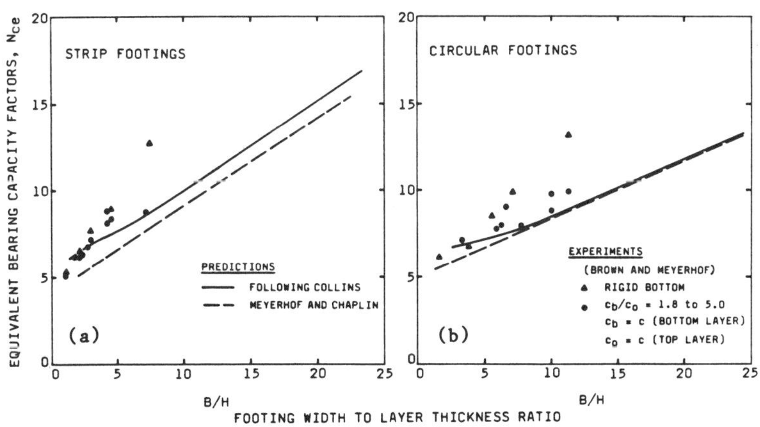

Fig. 9 Predictions and Experimental Values of Equivalent Bearing Capacity Factor for a Finite Soil Layer.

was varied between 1 and 5. Tests with a finite soil layer overlying a rigid base were also performed. Some of their test results are presented in Figs. 9(a) and 9(b) in comparison with the theoretically predicted values. At small B/H ratios, the experimental values of N_{ce} are close to the expected value of 5.14 (in case of strip footings). At larger values of B/H, the experimental values of N_{ce} increase rapidly, and the theoretically predicted curves become more or less the lower bound for the experimental results when the bottom clay layer is at least 1.8 times as strong as the top layer.

Although too much reliance should not be placed on this single set of data, the model test results by themselves show that an increasing B/H ratio can substantially increase the maximum bearing pressure of a rigid footing. Furthermore, these data seem to indicate, in contrast with Vesic's recommendation (19), that the solution for a finite soil layer overlying a rigid base applies as soon as the bottom soil layer has a strength which is two or more times the strength of the top layer.

STABILITY UNDER ICE LOADS

Soil Failure Modes

Ice loads will impose an inclined and eccentric load on the foundation of a gravity structure. Traditional bearing capacity analysis methods for this case rely on modifications to the Prandtl's solution to account for load inclination and eccentricity (API, 1). In addition, a separate check for sliding resistance is performed by assuming that a percentage of the soil shear strength, depending on the contact condition, is mobilized uniformly across the structure base (or at the level of the skirt tip). API (1) recommends the following safety factors for these stability considerations:

$$\begin{aligned} F.S. &= 2.0 \text{ for bearing;} \\ F.S. &= 1.5 \text{ for sliding.} \end{aligned} \qquad (6)$$

The traditional bearing capacity approach does not differentiate between positive and negative load eccentricity, i.e., the direction of the applied moment (about the centroid of the base) with respect to the direction of the applied horizontal force. Fig. 10 shows the deformation patterns predicted with a nonlinear finite element program by Prevost et al (18). It shows that the deformation mode changes significantly as the load eccentricity moves from one side of the centerline to another. The cone and the caisson configurations shown in Fig. 1 generally will be subject to negative and positive load eccentricity, respectively.

The conical structure will be subject to an inclined and eccentric load. The load inclination and eccentricity depends on the cone angle, the location of ice impact, structure weight, etc. For a shallow cone with a large base in shallow water, the load eccentricity can be rather large. The load inclination and eccentricity combination for a conical structure corresponds to the case illustrated in Fig. 10(a).

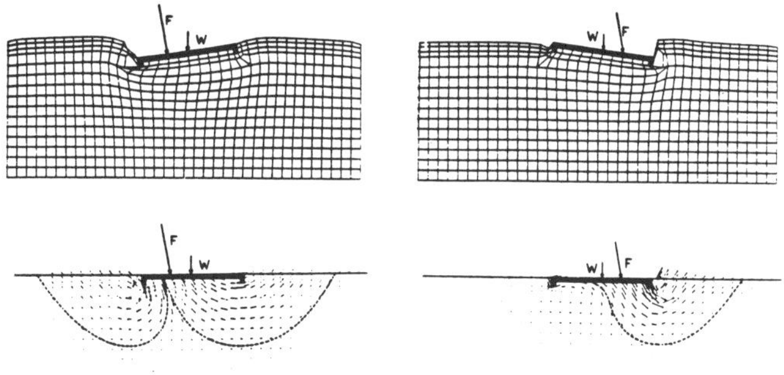

Fig. 10 Deformation Patterns for Strip Loads with Negative and Positive Eccentricities (Prevost et al, 18).

A caisson structure will be subject primarily to horizontal loads. The moment due to the horizontal loads results in a load inclination and eccentricity combination corresponding to Fig. 10(b). For water depths less than 100 ft, however, the load eccentricity resulting from the design ice loads is likely to be small, and the structure will be subject primarily to a sliding mode of deformation.

Positive Load Eccentricity - Caisson Structures

Fig. 11 shows the bearing capacity factor under an inclined and eccentric strip load for the case of a uniform soil profile. This result was obtained with a limit analysis method by Murff and Miller (13). The assumed collapse mechanisms are shown in the inset of Fig. 11. These collapse mechanisms correspond to the load situation in Fig. 10(b) - a positive load eccentricity (i.e., if the horizontal load is toward the right, the moment is clockwise).

For simplicity, the remainder of this section will consider only strip loads. For a given load inclination and eccentricity combination, the maximum vertical load (per unit length) is given by the bearing capacity factor N_{cv}, and the maximum horizontal load by $N_{cv} \tan\alpha$, provided that the horizontal load does not exceed Bc.

When computing the safety factor, it should be recognized that, in some cases, when a live load component (say, the horizontal load) exceeds the design value, the other live load components (i.e., the vertical load and moment) may exceed their design values as well. Consequently, the load inclination and eccentricity, at load levels greater than the design level, may be greater than the values determined at the design live load level. This problem can be dealt with by applying a load factor to all of the live load components and equating the safety factor to the load factor at failure. This approach generally requires an iterative process in the stability calculation.

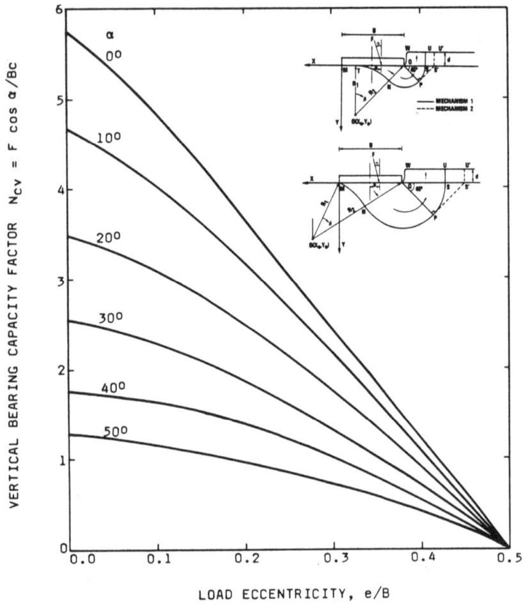

Fig. 11 Bearing Capacity Factor for Positive Eccentricity.

Negative Load Eccentricity - Conical Structures

For the load situation in Fig. 10(a), i.e., a negative eccentricity, a slip-circle type analysis method will be used to estimate the bearing capacity factor. The inset in Fig. 12 illustrates the collapse mechanism assumed by this approach. The slip-circle is also appropriate for problems involving primarily sliding when a "weak" zone exists below the mudline. This second case will be discussed in the next section.

The slip-circle method is essentially an approximate method for obtaining an upperbound to the correct solution since it postulates a possible failure mode without evaluating a complete stress distribution. Its major advantage in the case of a nonfrictional soil (ϕ = 0) is that the soil resistance can be evaluated purely on the basis of the geometry of the slip-circle.

In the slip-circle procedure, a grid is laid out for the centers of the slip-circles. Each slip-circle is assumed to pass through one edge of the footing and through or under the other edge. The resisting moment due to the shearing resistance of the soil is evaluated and compared with the applied moment. The results in term of the bearing capacity factor are presented in Fig. 12.

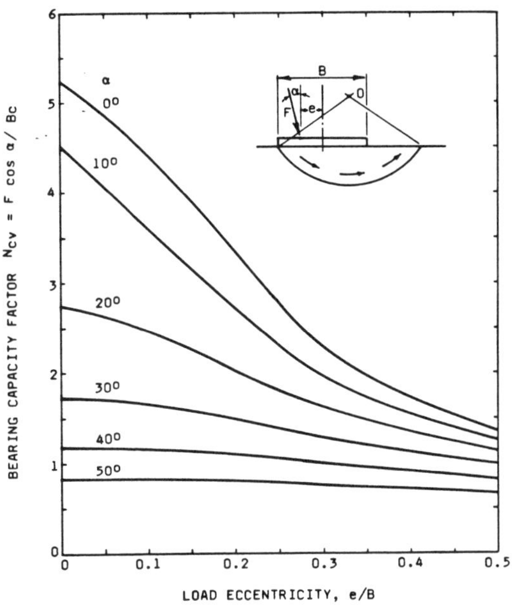

Fig. 12 Bearing Capacity Factor for Negative Eccentricity.

Comparison with Fig. 11 shows that the N_{cv} values are slightly lower than the positive eccentricity case (Fig. 11) when e/B is less than about 0.3. For greater values of e/B, the negative eccentricity case using the slip-circle method indicates larger values than in the case of positive eccentricity, and N_{cv} remains non-zero even at e/B of 0.5. This results from the fact that the failure mode in the slip-circle approach postulates a full contact between the footing and the soil, whereas, in Murff and Miller's approach, a less than full contact occurs at large eccentricity (see the top diagram of the inset in Fig. 11).

Interaction Diagrams Fig. 13(a) presents the results of Fig. 12 in terms of "interaction diagrams", which show the relationship between the vertical load, the horizontal load, and the moment about the centroid of the base at the time of failure. The vertical and horizontal axes show the normalized vertical and horizontal loads:

$$N_{cv} = F \cos\alpha / Bc; \qquad (7)$$
$$N_{ch} = F \sin\alpha / Bc,$$

where F = the total load at failure.

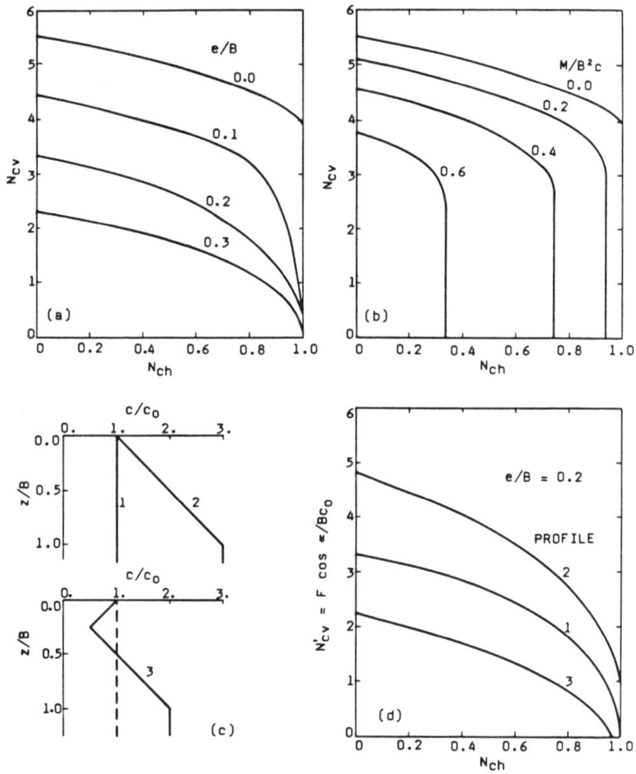

Fig. 13 Interaction Diagrams; (a) and (b) Uniform Soil Profiles, (c) and (d) Nonuniform Soil Profiles.

The moment about the centroid of the base is presented in terms of the eccentricity ratio $e/B = M/BF \cos \alpha$.

An interaction diagram illustrates how the resistance to one applied load component is reduced when another applied load component is increased. Take, for example, the case of an inclined but centrally located load ($e/B = 0$), Fig. 13(a) shows that when the horizontal applied load is increased from 0 to 0.8 Bc, the vertical load carrying capacity reduces from 5.5 Bc to 4.5 Bc.

The curves in Fig. 13(a) correspond to the case of constant load eccentricity, thus implying a decreasing moment as the vertical load decreases. To further illustrate the interaction between the vertical and horizontal loads and the moment, Fig. 13(b) shows the interaction diagrams when the moment is held constant and shows that, in the presence of an applied moment, the horizontal load carrying capacity is less than the traditionally computed "horizontal" sliding resistance (Bc). For example if M/B^2c is 0.4, the horizontal load carrying capacity is 0.75 Bc.

Nonhomogeneous Soil Profile Both the slip-circle approach and Murff's upperbound approach (13) are amenable to analysis of nonhomogeneous soil profiles. The following illustrates the effects of nonhomogeneous soil profiles as predicted by the slip-circle method.

Figs. 13(c) and 13(d) show the effects of nonhomogeneity in the undrained shear strength. Fig. 13(c) presents three normalized strength profiles 1, 2 and 3. Profile 1 is the case of a uniform soil. Profile 2 shows a soil whose strength increases with depth. Profile 3 represents a profile with a strong upper crust, and a weak zone at some depth below the mudline. Fig. 13(d) shows the interaction diagrams for these three profiles. It is evident that (1) when strength gains with depth, the load carrying capacities of the footing can be significantly increased, and (2) the existence of a weak zone below the mudline can significantly reduce the load carrying capacity.

Davis and Booker (5) made a comparison between the slip-circle approach and the slip-line solution for the bearing capacity of a soil deposit whose strength increases with depth and suggested that the slip-circle approach can significantly overestimate the bearing capacity when the strength gain with depth is significant. Although their conclusion was based on vertical, noneccentric loads for which the failure surfaces tend to go deeper than in the case of an inclined and eccentric load, caution should, however, still be exercised whenever this procedure is applied to a soil profile with a large strength gain with depth.

Sliding

The sliding mode of failure becomes predominant for both conical and caisson structures when the horizontal load is large relative to the horizontal sliding resistance Bc. For a nonhomogeneous strength profile, with the lowest strength at some depths below the mudline, the critical failure surface is not necessary the horizontal plane at the skirt tip but may resemble a very shallow slip circle passing through the toes of the structure. This failure pattern has also been predicted with a nonlinear finite element analysis using realistic constitutive relations and soil profiles.

This section considers the stability against sliding of a caisson-type structure (Fig. 1(b)). The strength profiles for this case are as shown in Fig. 6.

The problem will be idealized as a strip load. The ice load is 400 kips/ft horizontal and is applied at a point 50 ft above the mudline.

The footing width is 500 ft, and the structure dead weight is equivalent to a surcharge pressure of 4 ksf. The skirt tip is taken to be at 2 ft below mudline.

The applied loads correspond to a very small eccentricity, and the failure mode is, essentially, a sliding one. With a minimum undrained shear strength of 0.52 ksf at skirt tip, the safety factor against horizontal sliding immediately after set-down is 0.52 x 500/400 = 0.65.

The design ice load is, however, unlikely to occur until several months after set-down. During this time, consolidation under the structure dead weight will have progressed to some extent, as shown by the strength profile 2 in Fig. 6. After allowing consolidation for four months, the undrained shear strength at the skirt tip level will reach about 1.3 ksf and the safety factor against horizontal sliding at the mudline is therefore given by 500x1.3/400 = 1.63.

The strength profile 2 in Fig. 6 shows, however, that weaker soils exist below the skirt tip level. A slip-circle analysis using strength profile 2 indicates a safety factor of only 1.25 at 4 months. The critical slip-circle extends to a maximum depth of 22.5 ft. The relationship of safety factor with time is shown in Fig. 14. It shows that without artificially speeding up the consolidation rate, the safety factor does not reach the acceptable level of 1.5, until after at least 12 months.

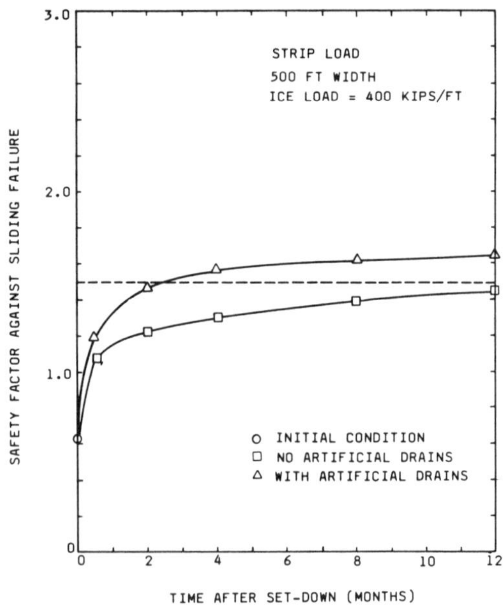

Fig. 14 Safety Factor against Sliding Versus Time.

Fig. 14 also presents the safety factor for the case in which artificial drains are used. These drains are the same as those discussed in connection with Fig. 6. The safety factor in this case increases rapidly with time after set-down and approaches 1.5 in 2.5 months. It approaches its long-term value (full consolidation) in about 12 months. For poorer initial soil strength conditions, the benefits of artificial drains will be more pronounced.

CONCLUSIONS

1. Arctic gravity structures must cope with large ice loads and soil conditions which can be relatively poor.

2. The large variability in the undrained shear strength can be represented by selecting upper and lower bounds of the "apparent" maximum past pressure.

3. Because of the large structure width, the thickness of the soft soil layer should be taken into account in stability prediction.

4. Conical and caisson structures will be subject to different modes of deformation under ice loads.

5. Sliding analysis of structures founded on nonhomogeneous soil strength profiles should include failure surfaces other than the horizontal plane at the skirt tip.

6. To achieve a cost-effective design, the strength gain due to consolidation under the weight of the structure can be taken into account. The magnitude and rate of strength gain depends on the consolidation characteristics of the site, as well as the structure ballast weight and configuration. The structural characteristics are factors which cannot be decided on the basis of foundation performance alone, since they influence the overall design, constructibility, transportation, floating draft, and cost.

7. Artificial drains can improve the foundation performance by speeding up the consolidation process.

REFERENCES

1. American Petroleum Institute, "Planning, Designing and Constructing Fixed Offshore Platforms," API RP 2A, 13th ed., 1982.

2. Brown, J.D. and Meyerhof, G.G, Experimental Study of Bearing Capacity in Layered Clays," Proceedings, Seventh International Conference on Soil Mechanics and Foundation Engineering, Mexico City, 1969, Vol. 2, pp. 45-51.

3. Collins, I.F., "An Optimum Loading Criterion for Rigid-Plastic Materials," Journal of the Mechanics and Physics of Solids, 1968, Vol. 16, pp. 73-80.

4. Davis, E.H. and Christian, J.T., "Bearing Capacity of Anisotropic Cohesive Soil," Journal of the Soil Mechanics and Foundations Division, ASCE, 1971, Vol. 97, No. SM5, pp. 753-769.

5. Davis, E.H. and Booker, J.R., "The Effect of Increasing Strength with Depth on the Bearing Capacity of Clays," Geotechnique, 1973, Vol. 23, No. 4, pp. 551-563.

6. Fellenius, B.H., "The Background and Theory of Vertical Drains with Particular Reference to the Alidrain," Notes for Lecture in Skelleftea, August 24, 1977, Montreal.

7. Hansbo, Sven, "Consolidation of Clay by Band-Shaped Prefabricated Drains," Ground Engineering, July, 1979, Vol. 12, No. 5.

8. Hill, R., The Mathematical Theory of Plasticity, Oxford University Press, 1950.

9. Johnson, W. and Kudo, H., "The Compression of Rigid-Perfectly Plastic Material between Rough Parallel Dies of Unequal Width," International Journal of Soil Mechanical Sciences, 1960, Vol. 1, pp. 336-341.

10. Ladd, C.C., Moh, Z.C. and Gifford, D., "Undrained Strength of Soft Bangkok Clay," Proceedings, 4th Asian Regional Conference on Soil Mechanics and Foundation Engineering, Bangkok, Thailand, 1971, Vol. 1, pp. 135-140.

11. Ladd, C.C., et al., "Stress-Deformation and Strength Characteristics," Proceedings, 9th International Conference on Soil Mechanics and Foundation Engineering, Tokyo, 1977.

12. Meyerhof, G.G. and Chaplin, T.K., "The Compression and Bearing Capacity of Cohesive Layers," British Journal of Applied Physics, 1953, Vol. 4, No. 20, pp. 20-26.

13. Murff, J.D. and Miller, T.W., "Foundation Stability on Nonhomogeneous Clays," Journal of the Geotechnical Engineering Division, ASCE, October, 1977, Vol. 103, No. GT10, pp. 1083-1095.

14. Murff, J.D. and Miller, T.W., "Shallow Penetration of Marine Foundations," Application of Plasticity and Generalized Stress-Strain in Geotechnical Engineering, Yong, R.N. and Selig, E.T., ed., ASCE, 1980, pp. 139-152.

15. National Research Council, Marine Board, Understanding the Arctic Sea Floor for Engineering Purposes, National Academy Press, 1982.

16. Olson, R.E., Daniel,D.E. and Liu, T.K., "Finite Difference Analyses for Sand Drain Problems," Proceedings, ASCE Specialty Conference on Analysis and Design in Geotechnical Engineering, 1974, pp. 85-110.

17. Prevost, J.H., "Mathematical Modeling of Monotonic and Cyclic Undrained Clay Behavior," International Journal for Numerical and Analytical Methods of Geomechanics, 1977, Vol. 1, pp. 195-216.

18. Prevost, J.H., Cuny, B., Hughes, T.J.R. and Scott, R.F., "Offshore Gravity Structures Analysis," Journal of the Geotechnical Engineering Division, ASCE, February, 1981, Vol. 107, No. GT2, Proceedings Paper 16045, pp. 143-165.

19. Vesic, A.S., "Bearing Capacity of Shallow Foundations," Foundation Engineering Handbook, Winterkorn, H.F. and Fang, H.Y., ed., Van Nostrand Reinhold Co., 1975.

20. Vivatrat, V., "Cone Penetration in Clays," Thesis Presented to the Massachusetts Institute of Technology in 1978, in Partial Fulfillment of the Requirements for the Degree of Doctor of Science.

21. Wang, J.L., Vivatrat, V. and Ruser, J.R., "Geotechnical Properties of Alaska OCS Marine Silts," Proceedings, Offshore Technology Conference, Houston, Texas, 1982, OTC Paper No. 4412.

BREAKOUT THEORY FOR OFFSHORE STRUCTURES SEATED ON SEA-BED

By Mostafa A. Foda[1]

INTRODUCTION

The spectrum of engineering problems in association with sea bed-structure interaction is widespread, and it represents a major challenge to the fast growing engineering activities in the offshore region. One important aspect of the interaction of offshore structures with sea bed, that has not received much attention in the past, is the phenomenon of detachment or breaking out of a structure from sea bottom. The phenomenon is important for a number of ocean and offshore engineering practices. In some cases, the objective is to achieve body breakout from sea floor, like for example, in salvaging sunken ships, submarine take-off, or repositioning a gravity structure. In others, the objective is to guarantee stability against body breakout, as for example, in the stability of submarine pipelines against floatation or sea floor breakout during severe storm conditions.

There has been very little work done on the mechanics of sea-floor breakout, with most of the work being related to salvage engineering practices (4). It is well known that actual salvage operations may last for several days in trials, and that sometimes the object never even breaks out due to the engineers' lack of understanding of the breakout phenomenon. As in the problem of detachment of two plates initially in contact, the required release force might be very large (infinite in the limiting case of two rigid impervious plates with no initial gap between them). Here, one of the bodies in contact is the deformable and porous sea bed, and it has been observed in all experiments on the breakout phenomenon that properties of the porous soil (e.g. permeability and stiffness) affect the breakout process significantly (3, 6). Therefore, analysis of the problem should account adequately for the response of the saturated porous sea bed.

THE MECHANISM OF BODY BREAKOUT FROM SEA FLOOR

The process of body breakout from sea floor is known to be a time-dependent phenomenon where complete detachment, or breaking out of body can only be achieved after the lifting force has been applied for a certain period of time, normally called 'breakout time'. The process is usually very slow in the beginning, concluded by a rather sharp release at breakout time. During the time of breaking out, negative

[1] E. & E. S. Div., Kuwait Institute for Scientific Research, P. O. Box 24885, Safat, Kuwait.

pressure, or 'mud suction' as it is sometimes referred to, will be developing at body-soil interface, giving rise to a certain sea-bed holding strength against body breakout. The body will eventually break loose from the bottom mud under the persistent application of lifting forces, as such mud suction resistance dissipates with time.

A breakout theory was worked out in (4), describing the development of mud suction beneath a body during a salvage lift-off operation. The theory, which was developed initially for unburied, flat-bottomed bodies resting on sea floor, is extended here for embedded objects with finite indentation to the sea floor. A simple and general relationship is arrived at for estimating the required breakout time as a function of the applied lifting force on the object.

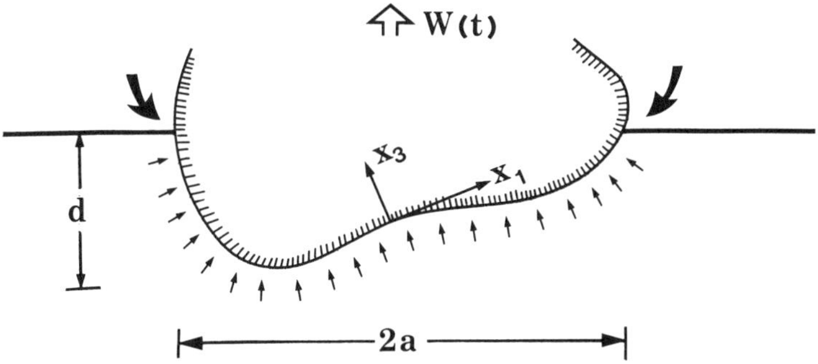

Fig.(1) Definition Sketch

Figure (1) depicts the general setting of a three dimensional rigid body seated on sea bed, causing finite indentation, of say depth d to the plane sea floor. The body-soil interface is assumed, in general, to be a smooth surface, characterized by some tangential length scale, say a, and that no soil exist above it. The last requirement is to eliminate the possibility of having to remove soil during body breakout, which would require additional lifting force. Here, we concentrate on breakout resistance due to mud suction forces.

Under the action of lifting force, the body will begin to move upward away from the sea floor at some velocity W(t), so that over the body-soil interface, a tiny gap filled with water will begin to develop. The gap-filling water can be supplied from two possible sources. One is the saturated pores in the sea bed below, and the second is the ambient water above sea bed, through the gap periphery (Figure 1). The governing equations will then consist of statements of conservation of mass and momentum in the gap as well as in the saturated porous bed.

Neglecting inertia effects, the lubrication theory may be used to describe the water motion in the gap (4), by assuming the gap thickness Δ to be much smaller than the tangential length scale of the gap, a.

As for the response of the saturated porous bed, Biot's (1) poro-elasticity equations are employed, where the main assumptions are that the porous solid skeleton is linearly elastic, and that the pore fluid obeys Darcy's law. Biot's equations have been used in many studies on sea bed dynamics and stability under sea wave action (e.g. 2, 7, 8, 9, 12). The governing poro-elasticity equations, which couple the pore pressure distribution in the sea bed with the porous solid displacements, are solved following the boundary layer approximation of Mei and Foda (9, 10). Biot equations in the sea bed and lubrication equations in the gap are coupled through conditions of velocities and stresses continuity across the gap-soil interface, and their solution would give a description to mud suction development during body breakout from sea floor. The solution procedure is given in detail in Reference (4) for unburied, flat bottomed bodies. Here we outline the main steps for extending the results to finite sea floor indentation situations, as the one shown in Figure (1).

BREAKOUT OF OBJECTS SEATED ON SEA FLOOR

Local orthogonal coordinates (x_1, x_2, x_3) are used to describe the water flow in the gap, with the x_1, x_2 plane being aligned locally to be tangent to the body-soil initial interface, with x_3 -coordinate pointing to the body (Figure 1). The lubrication equations for the water flow in the gap are given by, (e.g. 5),

$$\frac{\partial p}{\partial x_i} = \mu \frac{\partial^2 u_i}{\partial x_3^2} \quad (i = 1, 2), \quad \frac{\partial p}{\partial x_3} = 0 \qquad \text{(1a, b)}$$

$$\frac{\partial u_i}{\partial x_i} = 0 \qquad (i = 1, 2, 3) \qquad \text{(2)}$$

where u_i are the water velocity components, p is the excess pressure over its hydrostatic value, and μ is the water viscosity. As for tangential velocities, a simple order of magnitude argument may be used to show that tangential velocities at the gap boundaries, which are in general of the order of magnitude of the body velocity W, may be neglected as compared to typical pressure-driven tangential velocity within the gap, so that zero boundary values at the gap's upper and lower boundaries may be imposed (4). This approximation is consistent with the lubrication theory assumption of $\Delta/a<<1$ so that solving (1a) for the tangential water velocities in the gap would give,

$$u_i = \frac{-1}{2\mu} \frac{\partial p_o}{\partial x_i} (\Delta + \eta - x_3)(x_3 - \eta) \qquad \text{(3)}$$

where,

$$x_3 = \eta\ (t) \tag{4}$$

is the position of the soil surface (initially at $x_3 = 0$), deflecting under the flow-induced stresses in the gap. Here p_o is the unknown excess water pressure in the gap, and $\Delta(t)$ is the gap thickness which is in general a function of position and time. Now, we may invoke the conservation of mass, or continuity requirement in the gap by integrating (2) across the gap thickness Δ and using (3) to get

$$\frac{\partial \Delta}{\partial t} = n(\bar{u}_3 - \bar{v}_3) + \frac{1}{12\mu} \frac{\partial}{\partial x_i} (\Delta^3 \frac{\partial p_o}{\partial x_i}) \tag{5}$$

which simply gives the water budget in the gap during its breakout expansion. The result is that the expansion rate ($\partial\Delta/\partial t$) is equated to the sum of the rate of change of the tangential water flux in the gap (the last term in 5) plus the additional water flux into the gap through the porous soil surface: $n(\bar{u}_3-\bar{v}_3)$, where n is the soil porosity, $\bar{u}_3$ and $\bar{v}_3$ are, respectively, the water and the solid normal velocities at the porous soil surface. Such additional flux can only be accounted for by solving for the sea bed response during breakout.

Sea bed Response During Breakout:

It will be assumed that water drainage from the porous sea bed during body breakout (to supply water to the expanding body-soil gap) will be taking place over a thin boundary layer adjacent to the gap surface, having a thickness δ much smaller than the gap length scale, a. The rest of the sea bed, away from the gap surface, will be responding in an undrained fashion, having both solid and fluid velocities essentially the same. Such boundary layer structure may be anticipated, intuitively, to happen for low permeability soils when the duration of body breakout from the soil is short enough.

The main reason for such boundary layer behaviour is that soils of low permeabilities will be highly resistant to relative fluid-solid motion, so that both fluid and solid will be moving essentially in unison. The exception to this flow pattern will be near soil free surface, where water is more free to squeeze in and out of the free surface, so that drainage effects will be important there. The short time-scale requirement will guarantee that such drainage effects near the free surface will not have enough time to diffuse deep into the porous soil, and hence, the boundary layer thickness δ will remain much smaller than the problem length scale (i.e. a, for example).

Such boundary layer structure for saturated porous media response has been deduced and verified by Mei and Foda (9, 10) in general terms from Biot's poro-elasticity equations, under the action of general wave forcings, where they obtained from Biot equations an estimate of the boundary layer thickness δ as a function of soil's physical parameters (e.g. permeability, stiffness coefficients) and of the frequency of the forcing wave. For non-cyclic loading (4), the boundary layer thickness (or the influence of drainage) will increase, deeper in the ground, with time, so that the boundary layer structure will only be valid for short enough time. It was shown in Reference (4) that the boundary layer theory can be applied to describe the sea bed response during body breakout.

Following the boundary layer theory, the response of sea bed is represented by the summation of two solutions. One is the boundary layer solution (which may be denoted by superscript b for example), and the second is the outer, undrained, elastic solution (denoted by superscript o).

In the boundary layer, where drainage in the normal to the free surface direction is important, it was shown (8, 9) that the boundary layer solution for the pore pressure p^b is governed by Terzaghi's equation:

$$\frac{\partial p^b}{\partial t} = k\left(\frac{n}{\beta} + \frac{1-2\nu}{2G(1-\nu)}\right)^{-1} \frac{\partial^2 p^b}{\partial x_3^2} \tag{6}$$

where k is the permeability coefficient, G and ν are, respectively, the porous solid's shear modulus and Poisson's ratio, and β is the bulk modulus of the pore fluid. It is evident from the above equation (which is a one-dimensional heat equation) that the boundary layer thickness δ is given by,

$$\delta^2 = kt\left(\frac{n}{\beta} + \frac{1-2\nu}{2G(1-\nu)}\right)^{-1} \tag{7}$$

Darcy's law is also shown to be in effect in the boundary layer,

$$\frac{\partial p^b}{\partial x_i} = -\frac{n}{k}(u_i - v_i) \tag{8}$$

As for the outer, undrained region, the problem is reduced to a conventional elastostatic problem, where the solid skeleton elastic parameters G and ν are replaced with equivalent parameters, depending on both the fluid and the solid properties.

Now, we return to our problem (Figure 1). We start by solving the boundary layer equation (6), subject to the boundary condition from (5) and (8),

$$\frac{\partial p^b}{\partial x_3} = \frac{-1}{k} \left(\frac{\partial \Delta}{\partial t} - \frac{1}{12\mu} \frac{\partial}{\partial x_i} (\Delta^3 \frac{\partial p_o}{x_i}) \right) \quad \text{at } x_3 = \eta \tag{9}$$

Using Laplace's transform, we get at the gap-soil interface,

$$p_o^b = \frac{-\delta}{k\sqrt{\pi t}} \int_0^t \frac{d\tau}{\sqrt{t-\tau}} \left(\frac{\partial \Delta}{\partial t} - \frac{1}{12\mu} \frac{\partial}{\partial x_i} (\Delta^3 \frac{\partial p_o}{\partial x_i}) \right) \tag{10}$$

Next, we turn to the outer, elastostatic problem, where the forcings are the unknown flow-induced stresses at the soil-gap interface. From lubrication theory, it can be shown that the resulting shearing forces on the soil surface can be neglected as compared to the associated normal pressure forces. Furthermore, a general result was arrived at in (4) for the elastic response of a complete half-space under general pressure forcing on its plane surface. It was shown from elasticity equations that the trace of the resulting stress tensor $\tau^o_{kk} = \tau^o_{11} + \tau^o_{22} + \tau^o_{33}$ (where τ^o_{ij} is the two dimensional stress tensor in the saturated soil) is related to the applied normal (pressure) stress τ^o_{33} by the following relation,

$$\tau^o_{kk} = 2(1-\nu_e)\tau^o_{33} \tag{11}$$

for any τ^o_{33} distribution over the plane surface of the half-space. Here, ν_e is the equivalent Poisson's ratio for the composite solid-fluid undrained elastic system, and is given by, (see, e.g. 9).

$$\nu_e = \frac{1}{2} \left(\frac{2\nu}{1-2\nu} + \frac{\beta}{nG}\right) \left(\frac{1}{1-2\nu} + \frac{\beta}{nG}\right)^{-1} \tag{11a}$$

Now, for the case of non-plane half-space surface, as is the case here (Figure 1), the above result can be shown to be applicable just as well. This is because through conformal mapping (e.g. 11), non-plane surfaces may be mapped into plane ones, where forcings which were normal to the non-plane surface in the physical space will remain also normal to the plane one in the mapped space. The distribution of the pressure stresses will in general be different in the mapped and unmapped domains. However, the above result (11) will still apply since it is valid for general stress distribution. Of course, for undrained response, the outer pore pressure solution p^o is related to the trace of the stress tensor τ_{kk}. Such relation can be shown to be given by, (e.g. 9)

$$p^o = - \frac{(1-2\nu_e)\beta}{2(1+\nu_e)nG} \tau^o_{kk} \tag{12}$$

so that from (11) and (12), we get the important result for the outer part of the pore pressure solution at the free surface, namely,

$$p_o^o = \frac{1}{1+m} p_o, \quad m = \frac{nG}{(1-2\nu)\beta} \tag{13}$$

where,

$$p_o = p_o^o + p_o^b \tag{13a}$$

Thus, combining (10) and (13), we get the following integral - differential equation for the pressure p_o in the gap,

$$p_o = -\alpha \int_0^t \frac{d\tau}{\sqrt{t-\tau}} \left(\frac{\partial \Delta}{\partial \tau} - \frac{1}{12\mu} \frac{\partial}{\partial x_i} (\Delta^3 \frac{\partial p_o}{\partial x_i}) \right) \tag{14}$$

where,

$$\alpha = \frac{1+m}{m} \left(\frac{G/k}{\pi(1-2\nu)m + \frac{\pi(1-2\nu)}{2(1-\nu)}} \right)^{\frac{1}{2}} \tag{14a}$$

BREAKOUT FORCE-TIME RELATION

The above equation for the gap pressure p_o is formally the same as the one derived for horizontal gaps (4). However, differences will arise when integrating for the total mud suction F, which from body equilibrium is equal to the vertical lifting force. For non-plane gaps, we have

$$F = \int_{A_o} p_o \cos \theta \, dA_o$$

where θ is the gap surface angle with the horizontal and A_o is the gap area. In (4), Equation (14) was integrated to get a relation between the lifting force F and the spatially averaged gap thickness $\bar{\Delta}$ (t). After a good deal of manipulation, relations were obtained numerically for two examples of flat-bottomed unburied objects. One is for the breakout of two dimensional objects, e.g. a ship, and the other is for axisymetric objects, e.g. a caisson or an oil storage tank. The obtained results can be applied as well to objects slightly indenting sea floor such that cos θ is close to unity over the gap area. Sample results of calculation for two dimensional flat-bottomed objects are shown in Figure (2). Each curve in the figure gives the gap expansion history $\bar{\Delta}(t)$ under a constant applied lifting force F, where the breakout time t_b is defined by the time when the expansion rate becomes unbounded numerically. Figure (3) shows the resulting breakout force-time relation for the same body, with different soil properties. The obtained breakout time ranges between a few seconds to a few hours depending on the applied lifting force. These numerically obtained relations shows approximately the same functional relation,

$$t_b = A\ F^{-B} \tag{15}$$

with the exponent B approximately equal to 1.43 (4).

It will be shown now that the above breakout force-time relation is of general nature, and is not limited to flat bottomed, unburied objects. Furthermore, estimates for the coefficients A and B will be obtained for general embedement situation. This will be obtained from the general integral-differential equation (14) for the pressure p_o in the gap, following a simple order of magnitude argument. From (5), it is seen that the terms between brackets in (14) are equal exactly to sea bed drainage flow into the gap. Thus, for the dissipation of mud suction pressure p_o, such sea bed drainage has to vanish, with tangential water flux through the gap periphery compensating for it. Thus, both terms between brackets in (14) must be of comparable orders of magnitude for breakout to occur. This balance, therefore, will be taking place at time $t \sim t_b$, so that from (14) we have

$$\frac{\Delta_b}{t_b} \sim \frac{\Delta_b^3\ p_o}{12\mu\ a^2} \tag{16}$$

Furthermore, the balance between the right and left hand sides of (14) requires

$$p_o \sim \alpha\ t_b^{-\frac{1}{2}}\ \Delta_b \tag{17}$$

Also, if F is the lifting force per unit length for 2D objects, or the total lifting force for 3D objects, then we have

$$p_o \sim F\ a^{-r} \qquad r = \begin{cases} 1 \text{ for 2D objects} \\ 2 \text{ for 3D objects} \end{cases} \tag{18}$$

Combining (16), (17), and (18) we get an estimate of the breakout time t_b in terms of the breakout force F

$$t_b \sim C\ F^{-1.5}; \qquad C = (12\mu\alpha^2\ a^{2+3r})^{\frac{1}{2}} \tag{19}$$

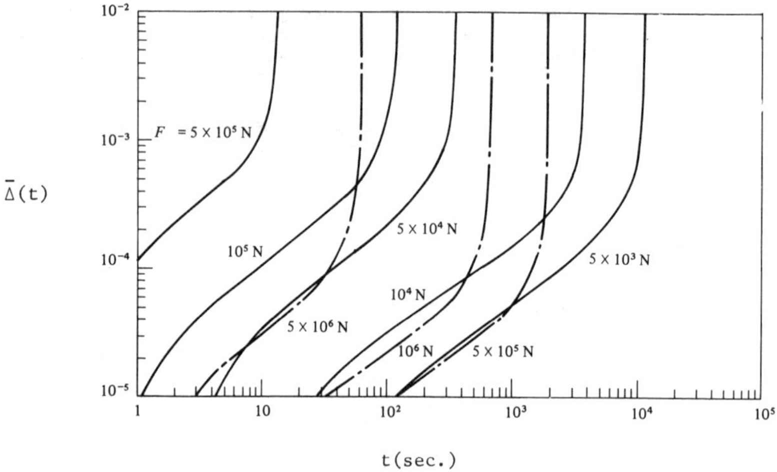

Figure (2): The gap-expansion history $\bar{\Delta}(t)$ under a two-dimensional body, where body-soil contact surface has a width of 2a = 10 m under constant applied lifting force F. The sea bed is sandy, with $k = 10^{-9}$ $m^3 s/kg$ and $G = 10^7 N/m$; —·— complete saturation, —— saturation slightly less than unity.

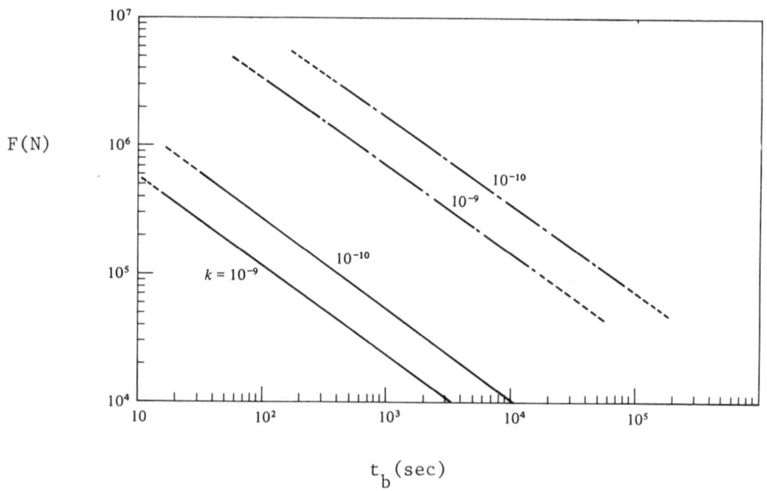

Figure (3): The breakout force-time relation for a two-dimensional body, with body-soil contact surface of width 2a = 10 m, for two different soil permeabilities $k = 10^{-9}$ and $10^{-10} m^3 s/kg$, with G fixed at $10^7 N/m^2$. .—. complete saturation, — saturation slightly less than unity

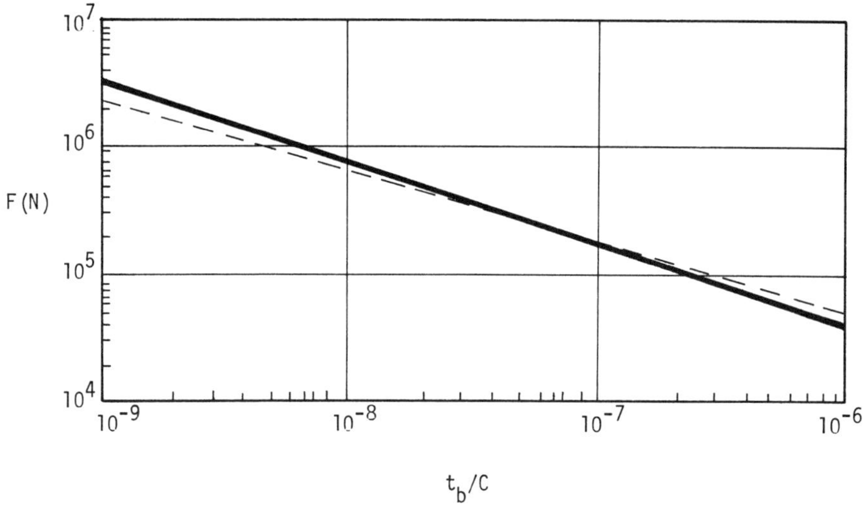

Fig (4): Comparison between the numerical curves in Figure (3); (solid line), and Equation (20); (dashed line) with a=10m and γ=3.7.

or,

$$t_b = \gamma \, C \, F^{-1.5} \tag{20}$$

where γ is some free coefficient which depends on the shape of the gap. Figure (4) shows comparison between the above relation and the numerically obtained breakout force-time relation (15) for horizontal gaps. First, it is seen that the exponents in (15) and (20) are approximately the same. Also it is noticed that when time is scaled with the coefficient C of (20) all four curves of Figure (3) are now falling into one single curve, meaning that C is the correct time scale for t_b. The best fit value for γ is obtained from Figure (4) to be 3.7. Thus, for flat-bottomed unburied objects of general geometry, equation (20) may be used as the general breakout force-time relation. On the other hand, for non-plane gaps, (19) might be used to get an order of magnitude estimate of the required breakout time for a given lifting force.

CONCLUSION

The present theory offers analytical expressions for the estimation of the time required to pull an object from a poroelastic sea bed. Porosity and soil deformation were shown to be important in body breakout. The applied lifting force is spent partly to pull up the soil skeleton and partly in pumping fluid into the expanding gap. Separation will eventually take place at breakout time. A simple and general power law is derived to relate breakout time with corresponding lifting force, to a free constant. The exact relation is obtained for flat bottomed objects, or objects slightly indenting the sea floor through the numerical integration of the gap pressure equation.

In the general breakout force-time relation, all the relevant physical parameters of our poroelastic model are lumped into one coefficient C (c.f. Equations 19, 20) which may be considered as a scale of the breakout time t_b. It is seen that for a given breakout force F, separation will be enhanced with increasing soil stiffness G, increasing soil permeability k, or decreasing contact length a.

The theory, however, has a number of limiting assumptions in modeling the two-phase response of the sea bed. In reality, the soil strains may exceed the elastic limit, leading to plastic yielding or fluidization. Controlled experiments are desirable to compare observed breakout behaviour against the present theory.

REFERENCES

1. Biot, M.A., "General Theory of Three-Dimensional Consolidation," Journal of Applied Physics, 12, 1941, pp. 155-164.
2. Bobby, W., Arockiasamy, M., Haaldar, A.K. and Reddy, D.V., "Finite Element Analysis of Pipe-Soil-Wave Interaction," Proc. Second Int'l Conf. on the Behaviour of Off-Shore Structures, 1979, London, England.

3. DeHart, R.S. and Ursell, C.R., Soutwest Research Institute Rep. on Contract N-ONR-336300, 1967.
4. Foda, M.A., "On The Extrication of Large Objects From the Ocean Bottom (The Breakout Phenomenon)," Journal of Fluid Mechanics, Vol. 117, 1982, pp. 211-231.
5. Lamb, H., Hydrodynamics, Cambridge University Press, 1932.
6. Liu, C.L., "Ocean Sediments Holding Strength Against Body Breakout," U.S. Naval Civ. Engng. Lab. Tech. Rep. R635, 1969.
7. Madsen, O.S., "Wave-induced Pore Pressures and Effective Stresses in a Porous Bed," Geotechnique, 28, 1978, pp. 377-393.
8. Mei, C.C. and Foda, M.A., "Wave-Induced Stresses Around a Pipe Laid on a Poro-Elastic Sea-Bed," Geotechnique, 31, 1981, pp. 509-517.
9. Mei, C.C. and Foda, M.A., "Wave-Induced Responses in a Fluid Filled Poro-Elastic Solid with a Free Surface- A Boundary Layer Theory," Geophysical Journal of the Royal Astronomical Society, 66, 1981, pp 597-631.
10. Mei, C.C. and Foda, M.A., "Boundary Layer Theory of Waves in a Poroelastic Sea Bed," Chapter 2 of Soil Mechanics, Ed. G.N. Pande and O.C. Zienkiewicz, John Wiley, 1982, pp 17-35.
11. Muskhelishvilli, V.I. "Some Basic Problems of the Mathematical Theory of Elasticity," Nordhoff, 1954.
12. Yamamoto, T., Koning, H.L., Sellmeiher, H. and Hijum, E.V., "On The Response of a Poro-elastic Bed to Water Waves," Journal of Fluid Mechanics, 87, 1978, pp 193-206.

Group Effect on Piles in a Liquefying Sand Deposit

By Pedro A. De Alba, M.ASCE[1]

INTRODUCTION

It is obvious that when liquefaction and cyclic mobility are induced by an earthquake in a saturated sand deposit, major structural damage will result to piled structures supported in the deposit. However, the consequences of lower levels of earthquake shaking for these foundation conditions are more difficult to predict. A basic question then becomes: how much pile settlement will occur under an earthquake-induced pore pressure increase short of actual liquefaction?

The work that will be described is the second part of a model study intended to examine pile behavior under these conditions. The first part of this study (Refs. 2,3) concentrated on modelling the problem of a single end-bearing displacement pile penetrating through a soft layer of negligible resistance into a saturated sand deposit. This supporting layer was then subjected to simulated earthquake loading and the development of pile settlement was observed as pore pressure increased. Tests were carried out in sand layers of two different densities and with pile loads representing a range of static safety factors from about 1.5 to 3.5. The applied effective stresses were 20 or 40 psi (138 or 276 kPa) representing much higher stress levels than those which could be applied in typical sand table tests. Scale effects were studied by testing model piles of three different diameters; 0.5 in (1.27 cm), 1.0 in (2.54 cm) and 1.63 in (4.14 cm). It was basically concluded that, once boundary effects were accounted for, single piles of three different sizes displayed essentially the same settlement behavior as pore pressure increased, and that the results might therefore be a guide to the behavior of full-size piles.

This first phase of the work left unanswered the question of how the overlapping stress-transfer zones of contiguous pile tips might affect the observed single-pile behavior. It was possible that under these higher localized stresses, pile settlement would develop more slowly. Consequently, further tests were carried out with a group of four closely-spaced piles.

APPARATUS

The same simple shear-type apparatus was used for the single pile and pile group studies. Fig. 1 is a schematic cross-section of the

[1]Associate Professor, Department of Civil Engineering, University of New Hampshire, Durham, NH.

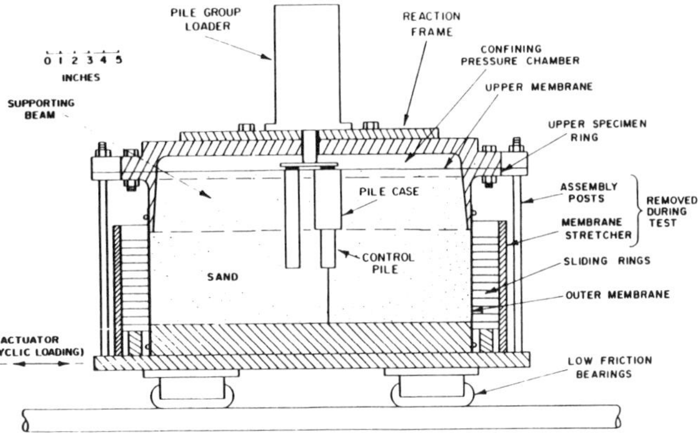

Fig. 1 Model Pile Group Testing Device

device. The bearing layer for the pile tips is represented by a 22-in (56 cm) diameter 6-in (15.2 cm) high sand layer. The diameter was chosen so that the lateral boundary would not interfere with the soil volume affected by pile insertion (about 2 to 4 diameters, ref. 4). For the 1-in (2.54 cm) diameter pile used in this work, the boundary distance was about 9 diameters for any pile in the group. For the group as a whole, the distance was about 3 times the group width.

The specimen height was limited by the height to diameter ratio of 1:4 chosen to avoid arching effects in the container and to approximate "free-field" conditions near the center of the specimen under cyclic loading. A consequence of the limited vertical dimension was that the rigid bottom of the sample chamber influenced pile behavior as will be discussed below.

The model pile group consisted of four, 1-in diameter brass piles. Settlement measurements were made for only one pile in the group. This control pile was designed so that it could be inserted after a certain level of confinement was applied to the soil specimen, thereby better approximating the soil volume change and stress conditions around a displacement pile. In order to permit vertical soil displacements to develop with pile insertion, a flexible upper boundary was required which precluded a simple rigid cap for the specimen chamber. A special two-part cap was therefore designed, as shown in Fig. 1.

The upper part of this cap forms the lid of a pressure chamber in which vertical confinement is applied to the specimen through a 0.015 in. (0.038 cm) thick rubber membrane. The lower part of the cap, below the membrane (upper specimen ring, Fig. 1), actually contains a passive part of the sand specimen itself. It consists of a 4-in (10.2 cm) high sand-filled steel ring to which a rigid supporting beam is attached across the diameter. This beam in turn supports a case from which the control pile is withdrawn after the specimen is confined.

The detail sketch of Fig. 2 shows that the pile is withdrawn from its case and loaded by a 0.063 in. (0.16 cm) - diameter stainless steel wire attached to the tip of the pile, which passes through a sliding seal in the base of the chamber and is attached to a loading piston. The inner sleeve of the control pile is only set into motion after a pile settlement in excess of one diameter has occurred, that is, after pile failure. The pile sleeve was designed to follow the pile motion after liquefaction. Pile and sleeve are sealed into the outer case with heavy 0.125-in. (0.318 cm) diameter O-rings. The clearance afforded by these seals permits rotation of the pile after it is inserted, thus allowing the pile to follow the motion of the specimen under cyclic load. Friction between the control pile and the O-ring seal was found to be less than 4 lb (17.8 N). To avoid creating a vacuum in the case or inner sleeve while withdrawing the control pile, the pile case is connected to atmosphere through a thin tube set into the support bar. Fig. 3 shows the upper specimen ring in place, with the control pile extended.

The other three "dummy" piles in the group were intended to provide the required overlapping stress zones underneath the pile tips. These piles were 1-in (2.54 cm) diameter brass shafts inserted from the top before the specimen was confined. They were spaced so as to form, with the control pile, a group of four piles separated 2.5 diameters center-to-center.

After the top cap was assembled and the specimen confined, the dummy pile group was loaded by means of another piston mounted on the chamber lid, Figs. 1, and 5. The piston shaft extended through a seal in the lid into the confining pressure chamber, loading a triangular aluminum cap which in turn loaded the dummy piles. Figs. 1 and 4 show that the load transfer was accomplished through hard rubber pads under the plate, which rested on the upper specimen membrane in line with the pile shafts.

All piles in the group penetrated 2.5 diameters into the lower 6 in (15.2 cm)-high active zone of the specimen, leaving the tips 3.5 diameters above the rigid base of the chamber. This active zone, upon which the cyclic stresses were imposed, was confined in a cylindrical rubber membrane supported by a stack of 0.5-in (1.27 cm) thick sliding steel rings. The sand specimen was thus isolated in a container formed by the base, the cylindrical membrane, the upper specimen ring, and the upper membrane. The specimen was loaded by fixing the cap to a rigid reaction frame and applying a cyclic load to the base, which moved on low-friction rollers, Fig. 1.

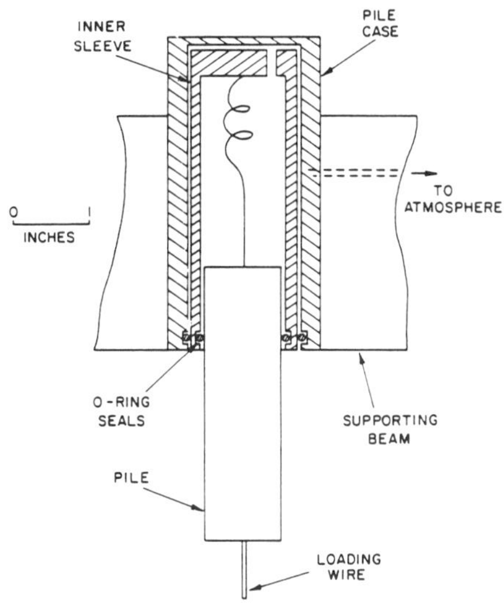

Fig. 2 Control Pile System

Fig. 3 Upper Specimen Ring in Place

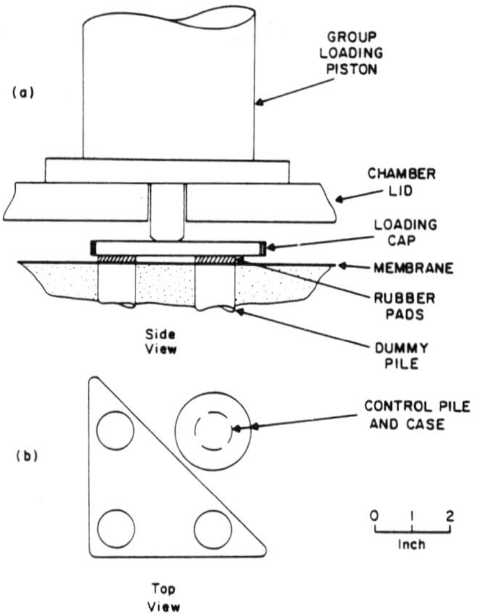

Fig. 4 Dummy Pile Loading System

The use of stacked sliding rings for lateral confinement was considered to provide a rough "stepped" surface under lateral deformation which would produce complementary cyclic shear stresses along the vertical perimeter of the specimen, balancing the imposed couple on the horizontal surfaces of the active zone. Earthquake loading was simulated by a constant-amplitude sinusoidal force applied to the base by a CKC electro-pneumatic loading system, at a frequency of 0.5 Hz.

MATERIAL

No attempt was made to scale the material, since it was felt that using a prototype sand would not affect the results as long as the pile diameter to sand size ratio was large. For the 1-in pile used in the pile group study, this ratio was about 64.

Consequently, a locally-available washed and screened material, Holliston 00 sand, was selected for the project. Holliston 00 is a clean uniform medium sand with subangular grains composed basically of quartz, feldspars, ferromagnesians and mica. Its properties may be summarized as follows:

Characteristic Diameters:

D_{50} = 0.40 mm
D_{60} = 0.43 mm
D_{10} = 0.24 mm

Coefficient of uniformity: Cu = 1.8

Percent passing 200 mesh: <1%

Maximum Density (ASTM-D2049-64): 107.4 lb/ft^3 (16.9 kN/m^3)

Minimum Density (Kolbuszewski, ref. 4): 88.2 lb/ft^3 (13.9 kN/m^3)

Mean Specific Gravity of Grains: 2.69

Friction Angle, ϕ, at D_r = 50%: 37°

SPECIMEN PREPARATION

Specimen preparation began by sealing the cylindrical outer membrane to the chamber base and stretching it against the stacked steel rings. Sand was then rained into this mold from a barrel with a perforated bottom, a technique similar to that described by Bieganousky and Marcuson (1). By experimenting with the drop height of the sand and the diameter and spacing of the perforations, a procedure was developed which consistently produced uniform specimens at a relative density, D_r, of 50%, within about 2 to 3%.

This density was chosen as representing the lower end of the range of D_r where a foundation might be considered feasible. The beneficial effects of group action might therefore be most pronounced. Extrapolation of the results to other densities is discussed in a subsequent section.

Once the lower, active, part of the specimen was poured into the ring stack, the upper specimen ring, supported by its assembly posts, was placed on the stack, Fig. 1. The control pile was then connected to its loading wire and retracted into its case. Next, the upper part of the specimen was poured. When the upper ring was filled, the sand layer was struck off at the level of the top of the ring. A positioning template was then placed on the sand layer and the three dummy piles were pushed into the specimen using an air piston. Fig. 5 shows the piles in position prior to insertion, connected by a temporary driving cap. Finally, the upper membrane was placed over the specimen, the dummy-pile loading cap was positioned over the membrane and the chamber lid was bolted down.

The specimen was saturated by first displacing the pore air with carbon dioxide gas, then flushing with deaired water and applying a back pressure of 20 to 30 psi. A degree of saturation ranging from 99.5 to 99.9% was obtained in the tests reported. The complex interior geometry of the specimen made higher degrees of saturation difficult to obtain. This was not considered to be a critical difficulty, since the object of the tests was to observe pile settlement with increasing porewater pressure, rather than to determine the liquefaction resistance of the sand specimen itself.

Fig. 5 Dummy Piles with Driving Cap in Positioning Template

With the specimen under backpressure, the chamber was rolled into the testing frame and bolted in. The membrane stretcher and assembly posts were removed, and the control pile was withdrawn from its case to a depth of 2.5 diameters into the active zone of the specimen, to the same level as the tips of the dummy piles. The vertical effective stress, σ', during the insertion of the 1-in (2.54 cm) diameter control pile was 2 psi (13.8 kPa). This value of σ' was fixed by the capacity of the control pile loading system; it should be noted that, during the tests on single piles, insertion effective stresses ranged from 20 psi (138 kPa) for the 0.5-in (1.27 cm) diameter pile to 0.5 psi (3.4 kPa) for the 1.63-in (4.14 cm) pile. Within this range, σ' during insertion did not seem to significantly affect subsequent pile behavior under dynamic load.

When the control pile had reached the required depth, σ' was increased to 20 psi (138 kPa) and the control pile was loaded to obtain the required initial static safety factor. Back pressure was also applied to the control pile case through the 'atmosphere' line of Fig. 2 to compensate for hydrostatic uplift. The ultimate capacity of the model piles had been obtained during the single-pile study. The failure load for each diameter was defined as that which would produce a pile settlement of 0.1 diameters, as proposed by Vésić (6). For the 1-in (2.54 cm) diameter control pile and an effective stress of 20 psi (138 kPa) the failure load was 355 lb (1580 N), corresponding to a tip stress of 452 psi (3119 kPa). It should be noted that this stress level approaches the range of tip stresses induced in full-size piles.

The final load was then applied to the dummy piles through the loading cap by activating the group loading piston. The total force on the cap was calculated to produce the same load on each of the dummy piles as on the control pile. In subsequent discussion, the group safety factor will be assumed equal to that of an individual pile under that pile's share of the load, i.e. a group efficiency of 100% is assumed.

TESTING

Once the pile group was in place and loaded, the specimen drains were closed and sinusoidal cyclic load was applied to the base of the chamber. The variables that were monitored during each test included the applied cyclic load, the dynamically-induced pore water pressure and the displacement of the control pile. Cyclic load was measured with a load cell, porewater pressure by a transducer set in the base of the chamber, and pile displacement by an LVDT attached to the control-pile loading piston, as indicated in Fig. 6. As previously noted, all the pile group tests were carried out with specimens at a relative density of 50%.

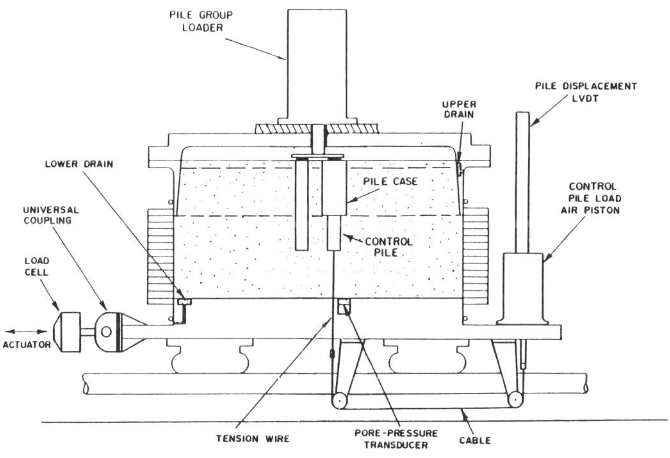

Fig. 6 Test Instrumentation

Results of four-pile group tests with safety factors, SF, ranging from 1.5 to 3.0 are shown in Fig. 7. Test data are summarized in terms of pile settlement ratio versus pore pressure ratio. Pile settlement ratio, $\Delta s/d$, is defined as pile settlement, Δs, divided by pile diameter, d. Pore pressure ratio, $\Delta u/\sigma_0'$ is dynamically-induced excess pore pressure, Δu, divided by initial effective stress, σ_0'.

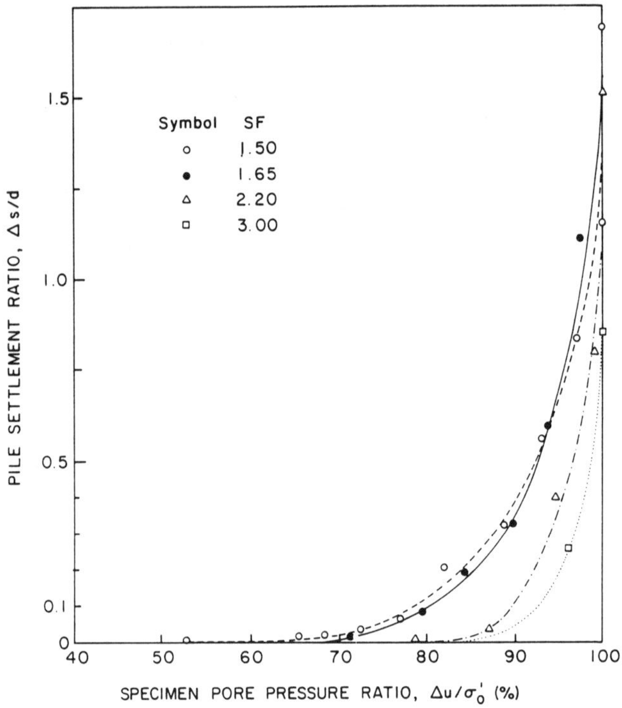

Fig. 7 Pile Group: Control Pile Settlement

The figure shows that essentially no settlement was observed, within the range of safety factors studied, until $\Delta u/\sigma_0'$ rose beyond a threshold level. Pile settlement then developed relatively slowly until $\Delta s/d$ reached about 0.05 to 0.1, and increased rapidly thereafter as residual pore pressure built up, with pile movement generally exceeding one diameter before liquefaction was reached.

Failure pore pressure ratio under dynamic loading was defined as that value of $\Delta u/\sigma_0'$ for which a $\Delta s/d$ value of 0.1 was observed. This definition is consistent with the definition of failure settlement accepted for static loads. Fig. 8 shows the variation of failure pore pressure ratio with initial static safety factor for the pile group.

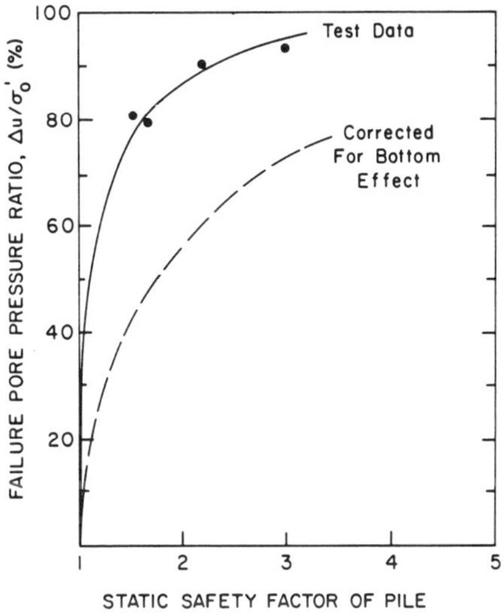

Fig. 8 Failure Conditions for Pile Group

DISCUSSION

The study of scale effects in the single pile tests has shown that differences in pile behavior between the three different sizes tested could not be attributed to differences in shaft friction which, by the Vesić (6) limiting friction criterion, amounted to less than about 5% of the total capacity in all cases. It was found, however, that these differences could be explained in terms of differences in relative depth, i.e. the ratio of sand depth between the pile tip and the chamber base, D, to the pile diameter, d. Fig. 9 shows that for the smaller values of D/d of the larger-diameter piles the pore pressure ratios at failure increased relative to those observed for the large relative depths of the 0.5-in pile. It was concluded that for values of D/d in excess of about 8 the results were unaffected by the presence of the rigid base.

For the pile group, it was calculated that shaft friction effects would represent about the same proportion of total ultimate capacity as for the single piles. In terms of relative depth, for the 2.5-diameter center-to-center spacing of the pile group, the exterior base width of the group was 3.5 diameters. An equivalent diameter for the base area would be about 3.95 pile diameters. Piles were inserted 2.5 diameters into the active zone of the specimen, giving a D-value of 3.5 diameters and, consequently, a relative depth D/d = 0.89. For this value of relative depth Fig. 9 would predict a $\Delta u/\sigma_0'$ at failure of about 0.8 for a single pile at a static safety factor, SF, of 1.6. By comparing this value with the pile group results of Fig. 8 it may be seen that this was indeed the failure pore pressure ratio observed. Good agreement is also obtained comparing the results of single pile and pile group tests at the same D/d for a safety factor of 2.8.

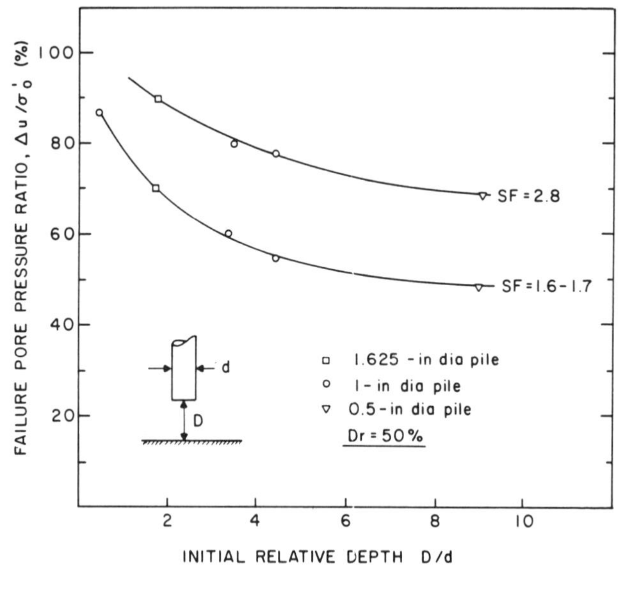

Fig. 9 Rigid Base Effect on Single Pile

A first conclusion from this result is that the piles behaved as a group; D/d for a single pile in this group would be 3.5, and the observed failure pore pressure ratio of this pile would not fit with the data from the single pile tests. However, if the four piles are considered to work as a group, the data fits well.

It may also be concluded that bottom effects for the pile group can be predicted from the single pile data of Fig. 9. Consequently, for a deep deposit (i.e. D/d > 8), the failure pore pressure ratio vs. safety factor relationship for the pile group will follow the dashed line in Fig. 8, which is the same curve that would be obtained for a single pile in a deep deposit.

Fig. 10, from the single pile study, shows evidence that this same failure pressure ratio vs. SF curve is also valid for a pile in a deposit with 68% relative density. This suggests that, in a deep deposit, the failure pore pressure ratio is uniquely related to safety factor for both isolated piles and pile groups.

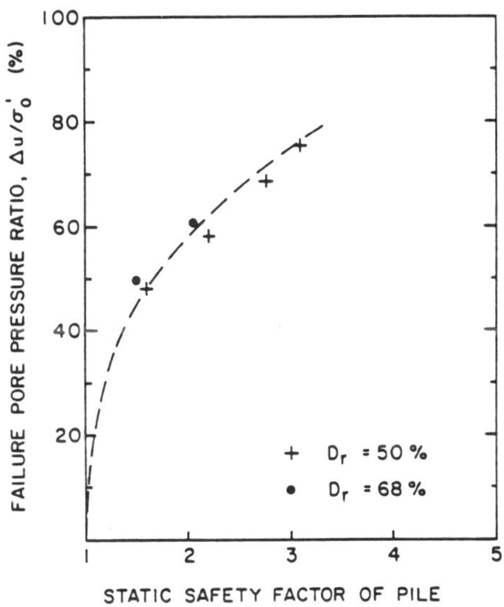

Fig. 10 Failure Conditions for Single Pile. Tests at D_r = 50 and 68%.

The data also suggest that the settlement ratio versus pore pressure ratio curves may also be the same at different densities if the static safety factors are the same. Fig. 11 shows $\Delta s/d$ versus $\Delta u/\sigma_o'$

data for a single pile in a deep deposit, i.e. D/d of about 9, at a relative density of 50%. These curves are analogous in shape to those of Fig. 7, obtained for a pile group with D/d = 1, but show a lower threshold $\Delta u/\sigma_0'$, as would be expected in the deep deposit. Based on the previous discussion, the Fig. 11 curves would also be representative of the settlement behavior of the pile group in a deep deposit.

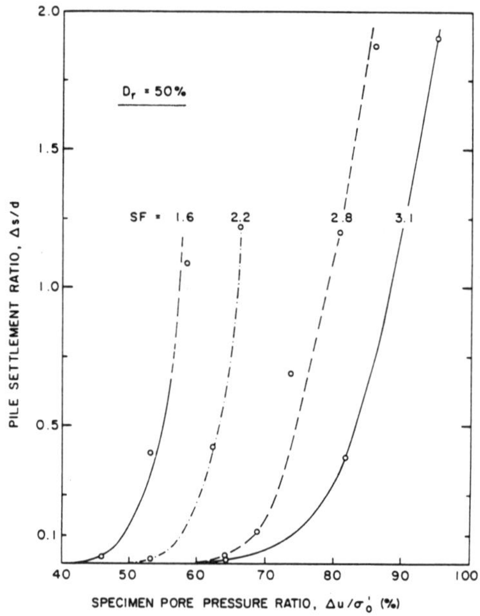

Fig. 11 Pile Settlement in Deep Deposits

It should finally be noted that the failure settlement of 0.1 diameters under dynamic load was chosen to be consistent with failure settlement under static load for driven piles, since it was considered that the pile insertion procedure most closely resembled a displacement pile. If a larger settlement at failure is accepted, i.e. 0.25 diameters as suggested for bored piles (6), the static load at failure will increase by about 25%. If failure settlement under dynamic load is then also considered to be 0.25 d, the pore pressure ratio at failure will also increase by about 6 or 7%. The net result is that the curve in Fig. 10 will drop by about 10%.

CONCLUSIONS

Comparison of model end-bearing pile behavior for pile groups and isolated piles has shown that for a thin liquefiable layer over a rigid base, a group of closely-spaced piles will settle more slowly with increasing residual pore pressure than an isolated pile. The difference in behavior is a function of the relative depth, D/d. For D/d values in excess of about 8, however, base effects become unimportant. Therefore, in deep deposits, piles in a group may be expected to follow the same settlement pattern as single piles.

Fig. 10 would thus represent earthquake-induced pore pressure at failure conditions for a pile group, failure being defined as an additional settlement of 0.1 diameters under dynamic loading. The single-pile test data suggest that the same dimensionless failure curve would apply for soil densities other than that tested, at least in the 50 to 70% relative density range. The individual piles in the group are expected to exhibit the settlement behavior shown in dimensionless form in Fig. 11.

Extrapolation of the data presented to another sand deposit with a stress history different from that of the virgin Holliston 00 sand specimens tested in this program is considered feasible, since the results are referred to the ultimate static bearing capacity values, which would be determined for the specific deposit in question.

Finally, it is considered that the consistency in behavior of pile groups and single piles of different sizes, when compared in dimensionless terms, indicates that the test results will provide insight into the behavior of full-size end-bearing pile groups. It should be noted, however, that the effects of shaft friction degradation under increasing pore pressure as well as the effects of significant superposed cyclic axial load on the piles remain to be investigated.

ACKNOWLEDGEMENTS

The work described in this paper was funded by the National Science Foundation as a Research Initiation Grant in the Earthquake Hazards Mitigation Program of the Division of Civil and Environmental Engineering. The testing program was greatly facilitated by the assistance of Arthur A. Nadeau and Stuart Hartford, undergraduate students in the Department of Civil Engineering at UNH.

REFERENCES

1. Bieganousky, W.A. and Marcuson, W.F., "Liquefaction Potential of Dams and Foundations, Report 1: Laboratory Standard Penetration Tests on Reid Bedford Model and Ottawa Sands," Research Report S-76-2. U.S. Army Engineer Waterways Experiment Station, Vicksburg, MI, Oct. 1976.

2. DeAlba, P., "A Model Study of Pile Settlement in Liquefying Sand Deposits Under Earthquake Loading," Civil Engineering Report No. 82-1, University of New Hampshire, Durham, NH, July 1982.

3. De Alba, P., "Pile Settlement in a Liquefying Sand Deposit", submitted to the Journal of the Geotechnical Engineering Division, ASCE, June 1982.

4. Kolbuszewski, J.J., "An Experimental Study of the Maximum and Minimum Porosities of Sands," Proceedings of the Second International Conference on Soil Mechanics and Foundation Engineering, Vol. 1, June 1948, pp. 158-165.

5. Meyerhof, G.G., "Compaction of Sands and Bearing Capacity of Piles," Journal of the Soil Mechanics and Foundations Division, ASCE. Vol. 85, No. SM6, Part 1, Proc. Paper 2294, December, 1959, pp. 1-29.

6. Vesić, A.S., "Design of Pile Foundations," National Cooperative Highway Research Program, Synthesis of Highway Practice No. 42, Transportation Research Board, National Research Council, Washington, D.C., 1977.

NOTATION

The following symbols are used in this paper:

d = pile diameter
D = distance from pile tip to bottom of chamber
D_r = relative density
SF = initial static safety factor
Δu = dynamically-induced excess pore water pressure
Δs = pile settlement
σ' = vertical effective stress
σ'_o = initial vertical effective stress

TORSIONAL BEHAVIOR OF MODEL PILES IN SAND

By R. N. Dutt[1], A. M. ASCE and M. W. O'Neill[2], M. ASCE

INTRODUCTION

Large structures such as deep water offshore platforms and tall bridge bents are subjected to loads of considerable magnitude due to wind and wave action. These loads can transfer significant torsional forces to piles by virtue of eccentric lateral loading. The need for understanding torsional behavior of piles has taken a renewed significance with the increased use of grouped piles in offshore structure design. Slender structures supported by one or more closely spaced groups can develop significant torques in individual piles. In the past, researchers have concentrated on axial and lateral interaction of piles, relegating the study of torsional pile-soil interaction to a secondary effort. Recently, however, rigorous analytical schemes have become available that consider torsion as a mode of loading (8, 9).

This paper describes the results of an experimental study of the torsional behavior of large model piles in sand. The primary objective was to formulate a criterion for describing soil-pile interaction in sand in the form of non-linear soil reaction curves, analogous to p-y and t-z curves. These curves were to be used in a discrete element numerical model to synthesize pile-head torque-twist behavior. Another important consideration was the study of the load transfer mechanism and its dependence on the relative density of sand and pile shape.

PREVIOUS WORK

O'Neill (6) established a closed form differential equation solution for the pile-head torque and twist relationship for the case in which both the pile and soil have linear torque-twist properties. Figure 1 illustrates a deformed increment representing a free body element from a torsionally linear cylindrical pile embedded in a torsionally linear soil. The solution of the differential equation for long piles results in equations relating torque distribution as a function of depth (z) and pile-head torque and twist in terms of soil and pile properties. These relationships may be expressed as:

---

1. Project Manager, Seafloor Engineers, Houston, Texas.

2. Associate Professor of Civil Engineering, University of Houston, Texas.

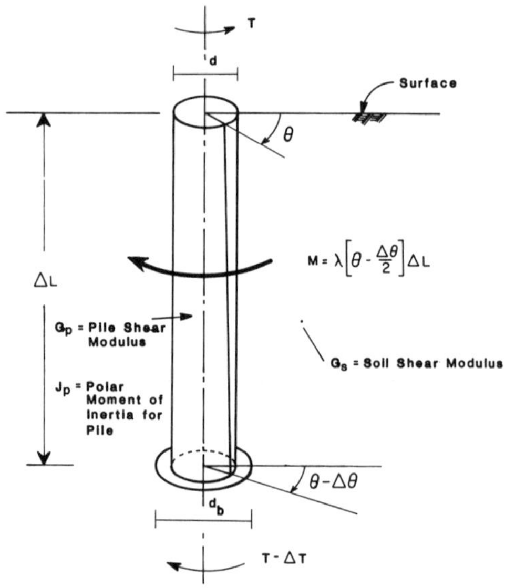

Fig. 1 Deformed Shape of a Torsionally Loaded Pile Segment

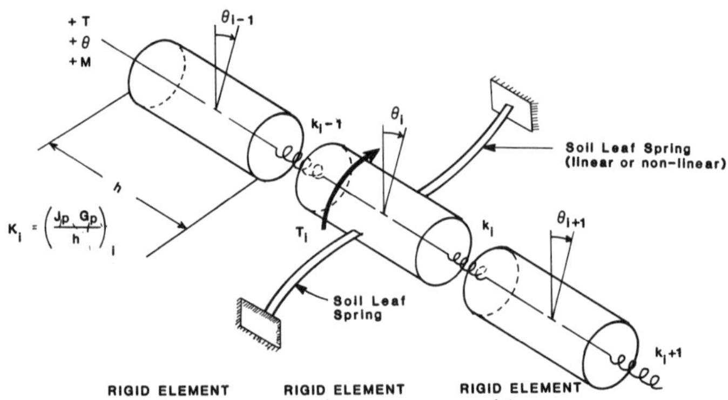

Fig. 2 Mechanical Discrete Element Model of Torsionally Loaded Pile (After O'Neill (6))

$$T(z) = T_0 e^{-z\sqrt{\beta\lambda}} \qquad (1)$$

$$\frac{T}{\theta \text{ pile head}} = \sqrt{\frac{\lambda}{\beta}} \qquad (2)$$

where $\lambda = \pi d^2 G_s$ and $\beta = (J_p G_p)^{-1}$.

Equations 1 and 2 yield reasonable results if the analysis is limited to small rotations. In order to include soil-pile non-linearities, O'Neill (6) also introduced a discrete element model that simulates torsional behavior of a circular, prismatic pile by a recursive solution of difference equations.

A three-element portion of the discrete element model is shown on Fig. 2. The mechanical model is composed of rigid elements connected by torsional springs. Each of these springs represents a concentrated torsional stiffness proportioned to the average $J_p G_p$ between points on the real pile. These springs can also simulate non-linear twist behavior. The spring constants can be varied to account for variable pile cross-section and/or shear modulus.

The pile is divided into n discrete elements of length h. For an elemental pile rotation θ_i, the restraining soil moment is equal to the sum of M_{0i}, a fixed moment, and $\theta_i S_i$, where S_i is the slope of the soil torsional resistance curve at θ_i. The moment in pile springs i and i-1 are:

$$M_i = K_i(\Theta_{i+1} - \Theta_i) \qquad (3)$$

$$M_{i-1} = K_{i-1}(\Theta_{i-1} - \Theta_i) \qquad (4)$$

Since will vary from element to element, the static equilibrium equation for rigid element i can be expressed as:

$$(T_i + M_{0i}) + K_{i-1}(\Theta_{i-1} - \Theta_i) + K_i(\Theta_{i+1} - \Theta_i) + S_i\Theta_i = 0 \qquad (5)$$

where T_i is an externally applied torque.

Equation 5 is a finite difference form of a second-order differential equation. This equation can be written in a matrix form in which the coefficients of the θ terms form a tridiagonally banded matrix. The resulting matrix equation is solved recursively. Strain softening behavior can be simulated with this model. In solving non-linear torque-twist problems, an iterative procedure is used by successively adjusting the S and M_0 values to insure compatibility between S and θ everywhere along the pile until closure within a specified tolerance is achieved.

Saul (10), in his general method of analysis of three-dimensional pile foundations, proposed a simple criterion for estimating the soil-pile interaction due to torsional loads by considering the pile as a semi-infinite shaft on an elastic foundation. The criterion representing the reaction of the pile may be expressed as:

$$\frac{T_z}{\Theta_z} = \frac{\delta}{L_e} K_T J_p G_p \quad \text{.................................... (6)}$$

Where δ and K_T are factors defined in the Notation. Equation 6 is limited to the linear case and does not consider soil conditions except indirectly through L_e, an equivalent length.

Poulos (8) developed an elastic method to determine the response of a single cylindrical pile subjected to torsion. He investigated the torsional relationships for both a uniform soil and a soil in which shear modulus and pile-soil adhesion increases linearly with depth. Figure 1 illustrates the problem considered in terms of loading, pile geometry and soil parameters. The method uses the elastic continuum theory for computing the soil rotation at the mid-point of each element in terms of the unknown interaction stresses. The corresponding pile rotations are then expressed in terms of these interaction stresses by considering the pile as a circular rod. Under conditions where the soil-pile interface remains elastic, the soil and pile rotation equations are solved in conjunction with a static equilibrium equation to obtain the interaction stresses and thus the pile rotation. Non-linear torque-twist relationships are simulated by considering elastic-perfectly plastic behavior of soil after slippage occurs between the pile and soil. Based on this quasi-elastic analysis, Poulos presented parametric solutions dependent on pile geometry and relative stiffness for the pile-head torque-twist relationships. For a uniform diameter pile, the pile-head rotation (twist), ϕ_0, of a pile subjected to torque T_0 may be expressed as follows:

For uniform soil

$$\phi_0 = \frac{T_0}{G_s d^3} \cdot \frac{I_\phi}{F_\phi} \quad \text{.................................... (7)}$$

For linearly increasing shear modulus and pile-soil adhesion

$$\phi_0 = \frac{T}{N_G d^4} \cdot \frac{I'_\phi}{F'_\phi} \quad \text{.................................... (8)}$$

The influence factors I and F are functions of pile length, L, pile diameter d, a relative rotational stiffness factor K_T, and torsional capacity of the pile. The factors can be evaluated using Fig. 3.

Randolph (10) presented closed form solutions for the torsional stiffness of a pile assuming the soil to be an elastic perfectly plastic medium. These closed form solutions agree well with the solutions of Poulos.

The solutions developed by both Poulos and Randolph are limited to the analysis of pile-head torque and twist behavior and do not permit computation of the dissipation of torsional stresses with depth.

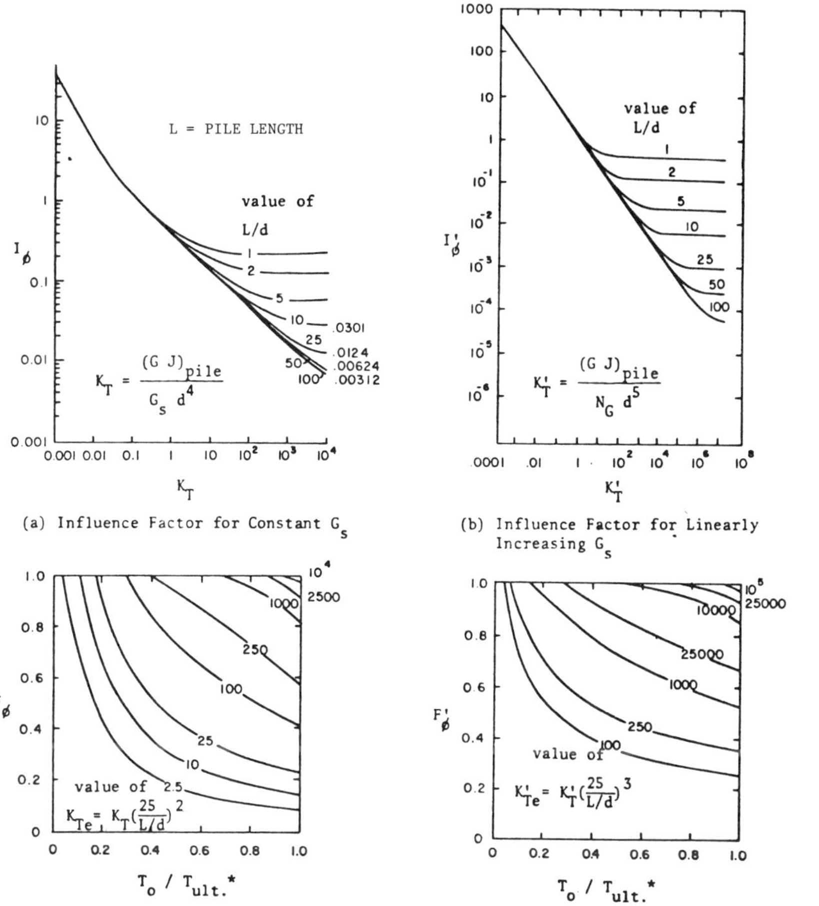

(a) Influence Factor for Constant G_s

(b) Influence Factor for Linearly Increasing G_s

(c) Yield Correction Factor for Constant G_s and Adhesion

(d) Yield Correction Factor for Linearly Increasing G_s and Adhesion

* $T_{ult.}$ = Ultimate torsional capacity of pile

Fig. 3 Functional Values of Factors for Torsional Behavior of Circular Pile (after Poulos,8)

Furthermore, the elastic solutions do not permit relaxation of shear at the pile-soil interface after slippage has occurred (strain-softening behavior), resulting in possible errors in the nonlinear range.

Because of the preceding limitations, the model proposed by O'Neill is given further consideration herein. Its solutions, however, are based on the premise that the individual torque-twist properties of soil and pile are known. Very little information is presently available which describe torsional load transfer from a pile to the surrounding soil mass. Cadling and Odenstad (1) first advanced a formula for the soil shear strain corresponding to the measured values of torque and twist by the vane shear device. In Fig. 4 are shown two views of a typical rigid cylinder embedded in a homogenous material enclosed by a cylindrical container and subjected to a rotational moment. For this condition the torsional restraint of the soil over a length L of a rigid cylinder in terms of soil shear stress-strain properties can be expressed by:

$$T_0 = 2\pi L \gamma_0^2 \tau_0 \quad \text{(9)}$$

$$\Theta = \frac{\gamma_0}{2} \quad \text{(10)}$$

For determining the torque-twist curve for the soil, Tucker (11) proposed the following representation of the soil stress-strain characteristics in torsional shear:

$$\gamma = A\tau^n \quad \text{(11)}$$

Where γ is the unit shearing strain due to a unit shearing stress, τ. The parameters A and n are properties of the material. Substituting Eq. 11 in the general solution of Cadling and Odenstad, the following expression for the angle of rotation at the surface of the cylinder is obtained:

$$\Theta = \frac{1}{2n} A\tau_0^n \quad \text{(12)}$$

The values of A and n are not directly obtainable for piles, but the torque-twist curves from rotational tests performed on soil samples in the laboratory, such as the vane shear test on clays, can be used.

Ha (5), based on the work by Tucker (11) and O'Neill (6), presented guidelines for determining soil torque-twist curves for clay. Criteria presented by Ha are primarily based on Eqs. 9 and 12 and the use of rotational shear tests performed on soil samples in the laboratory or in-situ.

TORSIONAL MODEL PILE TEST ARRANGEMENT

A 1.9-in. (48.3mm) OD circular pile and a square pile of 2-in. (50.8mm) outside dimensions were used in this investigation. The piles, made of soft aluminum, had smooth surfaces and were embedded and driven to a depth of 60-in. (1.52m) in air-dried loose and dense sands which

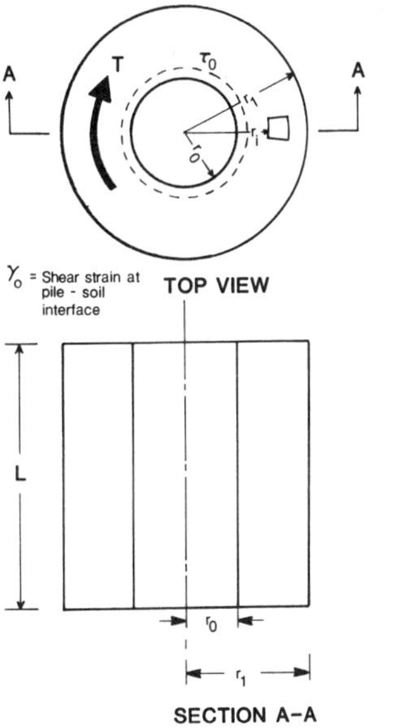

Fig. 4 Rigid Cylinder in Soil Subjected to Rotation

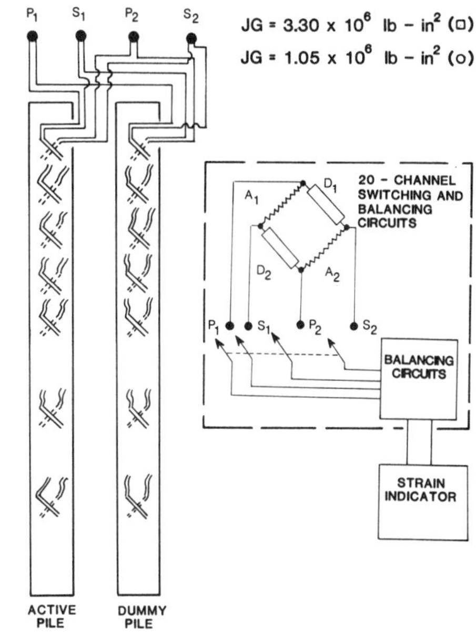

Fig. 5 Electrical Arrangement

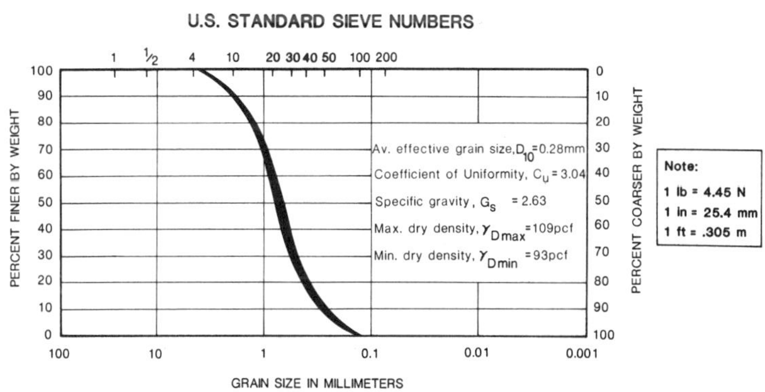

Fig. 6 Physical Properties of Sand

had been deposited under controlled density conditions. These piles were instrumented with electrical resistance strain gages. Fig. 5 illustrates the layout of the strain gages on the piles and the electrical arrangement for the measurement of torque.

The test facility was a rolled-steel cylindrical tank, five feet in diameter and six feet deep. The test pit was completely filled with air-dried sand under controlled conditions to achieve loose and dense state of compaction. Torsional load was applied to the torque shaft through a system of weights hanging over frictionless pulleys supported at the two ends of an A-frame assembly. The applied torque was transmitted to the pile head through a universal joint coupling located just above the pile head.

SOIL PROPERTIES

Sand used in this investigation was obtained from recent alluvium of the Colorado River near Columbus, Texas. On the basis of grain size distribution and microscopic examination, the sand is poorly graded medium sand and is primarily composed of subangular to rounded impure quartz particles with traces of plagioclase. Pertinent physical and strength characteristics of this material are summarized in Fig. 6 and Table 1. The average dry densities in the test pit under loose and dense conditions were 96 and 107 pcf, respectively. Further details of sand placement techniques and properties may be found in Ref. 4.

RESULTS

The test results for the two instrumented piles loaded torsionally in both loose and dense sand are presented in Figs. 7 through 10. These include the measured pile-head, torque-twist curves and the measured torque distribution along the pile length. Torque distribution was measured only in those piles that were embedded. Some of the salient features and characteristics of the torsional pile-head curves can be summarized as follows:

1. The initial portion of all the curves is linear, and the slope of this linear portion appears to be dependent on the shape and torsional rigidity of the piles and the relative density of the surrounding sand.

2. The decreasing rate of the increase in torque with rotation in the non-linear portions of the curves is an indication that the rotation, which produces soil failure, is achieved near the top of the pile at a level of torque well below the failure torque for the pile as a whole. Additional rotation may result in a decrease in interfacial pile-soil shear stress near the top while continuing to mobilize stresses farther down.

3. Influence of the shape and torsional rigidity of the piles is seen by comparing Curves (3) and (4).

TABLE 1. STRENGTH PROPERTIES OF TEST SAND

TRIAXIAL TEST		DIRECT SHEAR TEST (Soil-Soil)		DIRECT SHEAR TEST (Soil-Soft Aluminum)	
DRY DENSITY γ_D ,pcf	FRICTION ANGLE ϕ ,Degrees	DRY DENSITY γ_D ,pcf	FRICTION ANGLE ϕ ,Degrees	DRY DENSITY γ_D,pcf	ANGLE OF WALL FRICTION ϕ , Degrees
98.2	39.5	96.3	37.2	96.0	23.6
106.8	42.5	107.0	40.3	107.0	26.3

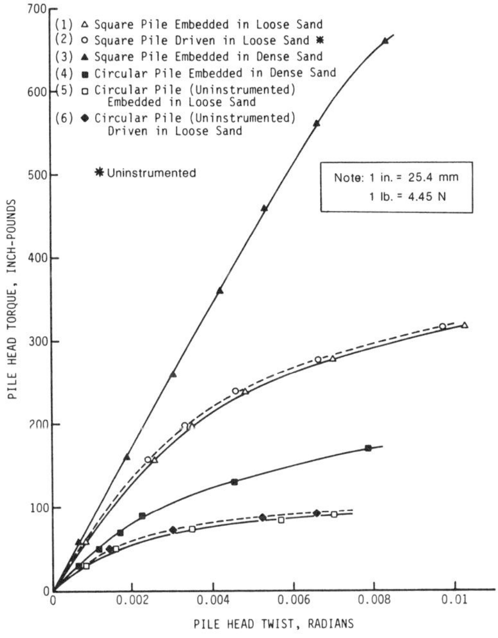

Fig. 7 Measured Pile-Head Torque-Twist Curves

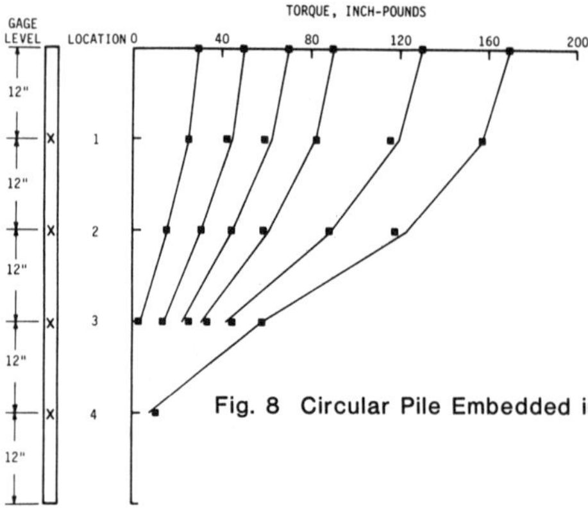

Fig. 8 Circular Pile Embedded in Dense Sand

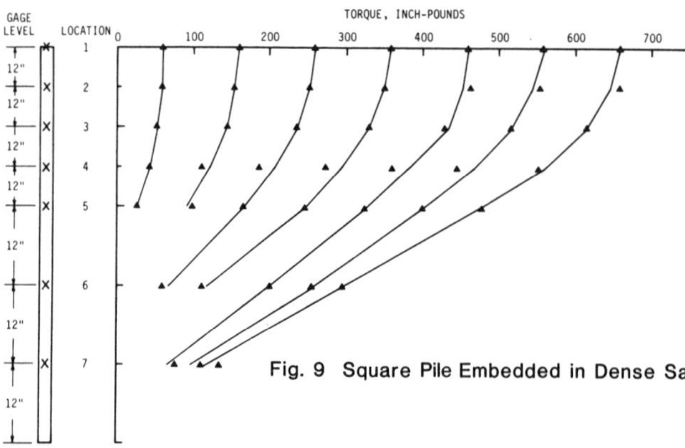

Fig. 9 Square Pile Embedded in Dense Sand

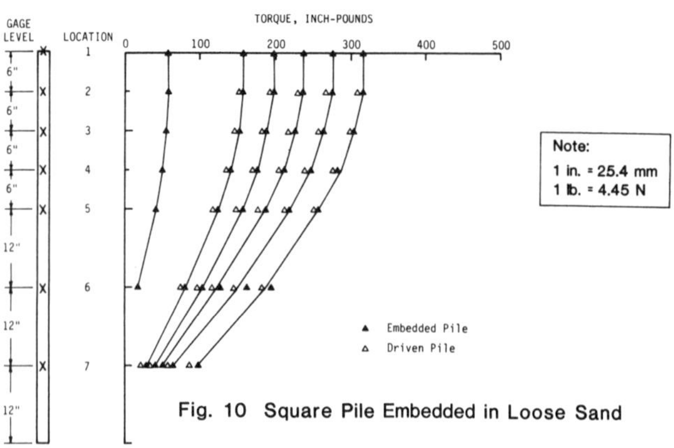

Fig. 10 Square Pile Embedded in Loose Sand

4. An increase of approximately four-fold in relative density from the loose to the dense state results in more than a two-fold increase in the apparent pile head failure torque, as evidenced by comparing Curves (1) and (3), although an accurate definition of failure was never actually achieved in any of the tests.

5. Driving resulted in a very slight increase in the torsional resistance of the square pile in loose sand compared to the embedded condition (Curves (1) and (2)). Curves (5) and (6) for the uninstrumented circular pile both embedded and driven in loose sand shows a similar trend in soil-pile response. These small increases could be attributed to possible densification of the mass of loose sand as a result of driving.

The torque distribution curves of Figs. 8, 9, and 10 were utilized to compute the unit soil torque-twist response relationship along the piles. The approach is similar to methods utilized by Coyle and Reese (2) and Coyle and Sulaiman (3) for computing the shear transfer to the soil versus pile movement for axially loaded piles. The torque-twist curves obtained using the above procedure were expressed as shear stress-strain mobilized at the soil-pile interface. Equations 9 and 10 were utilized for this purpose.

The computed shear stress-strain curves at various depths below the surface of the sand are shown in Figs. 11, 12, and 13. It is readily evident from these curves that the mobilized shear stresses in the soil increase with depth. Furthermore, the rate of increase in stress transfer shows a decreasing trend for the square pile in both the loose and dense conditions at greater depth. However, for the circular pile in dense sand, this rate of increase is almost constant.

In order to relate the ultimate shear stress transferred to the soil to the in-situ vertical stress, a knowledge of the coefficient of lateral earth pressure, K_z, is a prerequisite. The value of K_z depends on a number of factors such as pile shape, effective stresses in the soil, method of installation, and depth of embedment. Since all the proposed theoretical and semi-empirical expressions for K_z are for the axial mode of loading, either in compression or in tension, it was deemed advisable to evaluate K_z independently based on the tests described herein for the torsional mode of loading. In so doing, the dependence of K_z on embedment was also obtained. This was realized by considering the maximum (or failure) shear stress mobilized on the shear surface (a cylinder with d = diagonal distance between corners in the square pile) and the effective vertical stress, σ'_z, at depths where the shear stress-strain curves were generated. The results of this analysis are summarized for all embedded piles in Fig. 14. Some of the pertinent findings are:

1. K_z is dependent on the relative density of the sand. For the pile with square cross-section, a four-fold increase in the

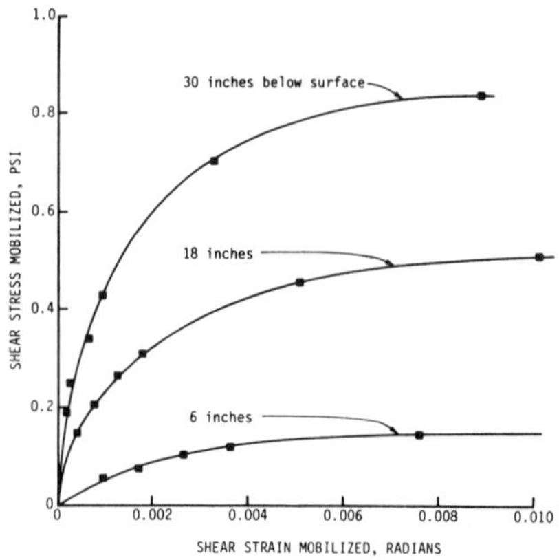

Fig. 11 Circular Pile, Dense Sand

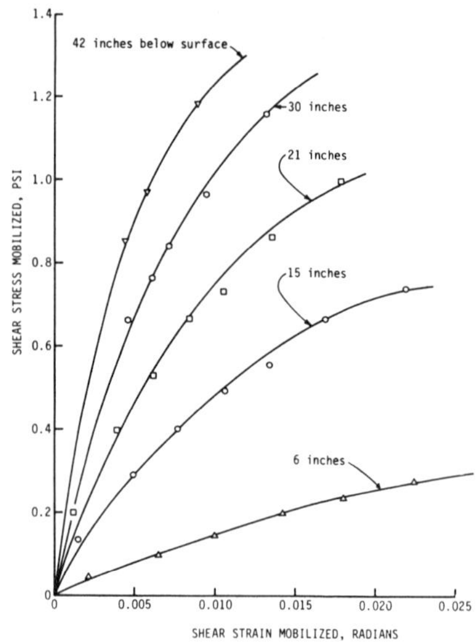

Fig. 12 Square Pile, Dense Sand

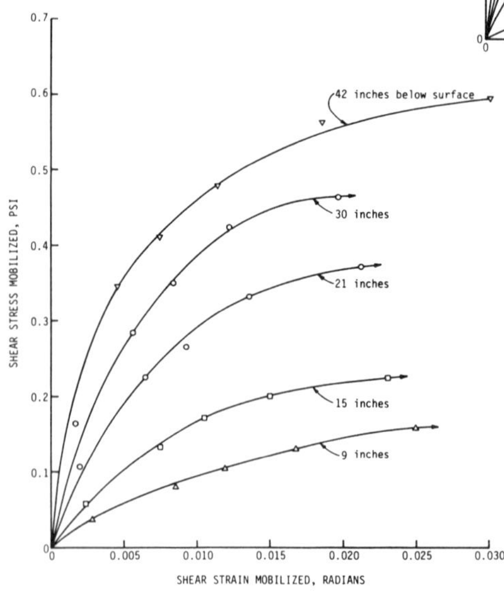

Note: 1 in. = 25.4 mm
1 psi = 6.89 kPa

Fig. 13 Square Pile, Loose Sand

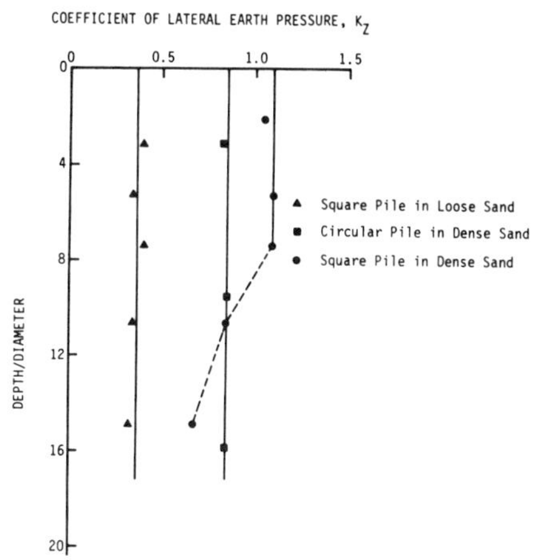

Fig. 14 K_Z vs. Depth For Embedded Piles

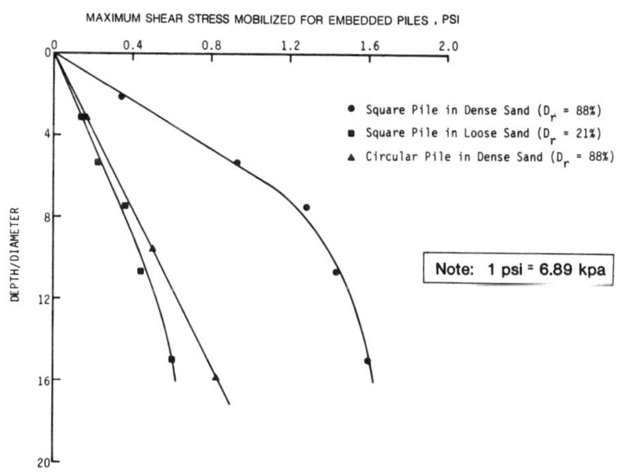

Fig. 15 Distribution of Maximum Shear Stress

relative density of the surrounding sand resulted in approximately a three-fold increase in K_Z.

2. For the circular pile in dense sand, K_Z appears to be independent of depth.

3. K_Z values in dense sand are in the range normally assumed for axially loaded piles.

4. The effect of the pile shape is also manifested in the increased values of K_Z in the dense sand environment for the square pile compared to that of the circular pile under the same circumstances.

5. For the square pile in dense sand below a depth of approximately 8 pile diameters, K_Z shows a decreasing trend. This could be attributed in part to the development of arches in the sand grains surrounding the pile, resulting in the mobilization of less shear stress in the denser soil at depths below the zone where arches are first formed. The arching phenomenon is possibly associated with the warping of the square pile in torsion.

Failure in torsion, as defined by a zero rate of shear stress increase with increasing rotation, was never observed, as can be seen in Figs. 12 and 13. For purposes of evaluating K_Z the maximum shear stress was taken to be that corresponding to a rotation of approximately 0.02 radians, which at large depths required some extrapolation. The resulting trend is shown in Fig. 15, in which a decreasing rate of load transfer with increasing depth can be seen in the square pile in dense sand. This decreased rate is a common phenomenon for axially loaded piles in sand. Experimental observations by Coyle and Sulaiman (3), and Vesić (12, 13), among others, indicate that for axially loaded piles below depths of 10-20 d, the rate of increase may decrease and may asymptotically approach a constant value. Residual stresses in the pile and grain crushing of the surrounding soil mass as a result of installation by driving are some of the major factors cited for such decrease.

Finally, the factors n and A were computed from the load transfer data and are presented in Figs. 16 and 17.

PROPOSED PROCEDURE FOR TORQUE-TWIST CURVES IN SAND

Based on observed characteristics and findings, specifically with regard to the parameters that define the shear stress-strain behavior of sand, the following tentative outline of a procedure for the development of the unit torque-twist curves for the soil is proposed:

1. Determine the properties of the sand, preferably the equivalent effective unit weight of the sand mass, angle of internal

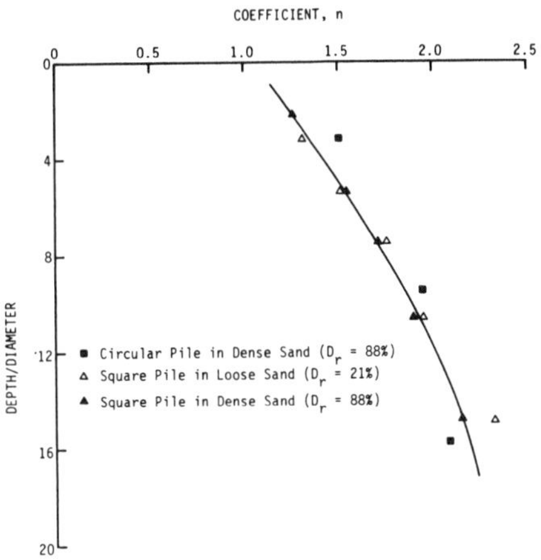

Fig. 16 Variation of n with Depth

Note: 1 psi = 6.89 kPa

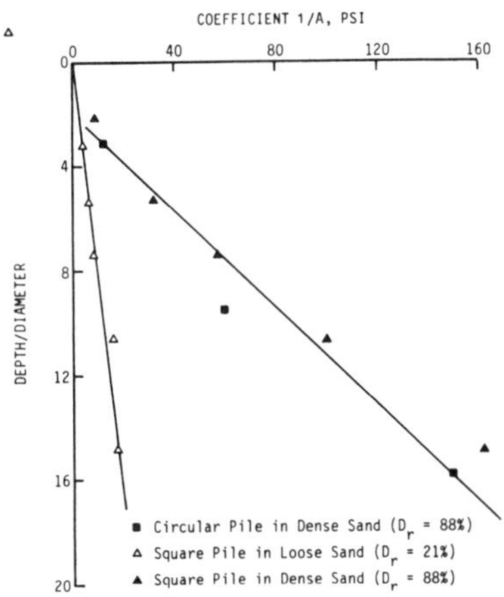

Fig. 17 Variation of 1/A with Depth

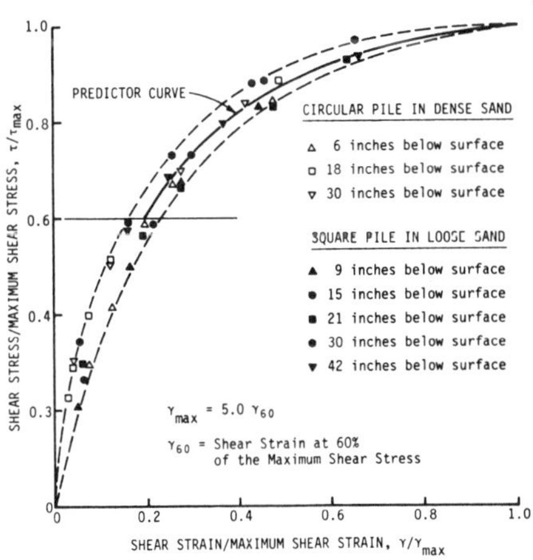

Fig. 18 Normalized Shear Stress-Strain Curves

friction, ϕ, and the apparent angle of friction, δ, between pile material and sand.

2. Determine the relevant coefficient of lateral earth pressure, K_Z, from Fig. 14.

3. Determine appropriate values of shear strength parameters n and 1/A from Figs. 16 and 17, interpolating where necessary.

4. Compute the maximum shear stress, τ_{max}, developed in the soil for each pile increment as follows:

$$\tau_{max} = K_Z \sigma'_v \tan\delta$$

5. Generate the soil shear stress-strain curves for each increment of the pile using the criteria:

(a) For $\tau < 60\%$ of τ_{max}

$$\gamma = A\tau^n$$

(b) For $\tau > 60\%$ of τ_{max}

(i) Compute $\gamma_{max} = 5A(\tau_{60})^n$

Where τ_{60} = 60 percent of the maximum shear stress.

(ii) Use the normalized predictor curve of Fig. 18 to generate the shear stress-strain curve in the nonlinear range between τ_{60} and τ_{max}. Figure 18 was developed directly from Figs. 11, 12 and 13.

6. Determine the unit soil torque-twist relationship for a length h of the pile from the soil shear stress-strain curves developed above and the following equations:

$$T = \frac{\pi D^2 h \tau}{2}, \qquad \Theta = \frac{\gamma}{2}$$

Discretion should be exercised when using values of n and 1/A from Figs. 16 and 17, since their variation with soil grain size distribution and mineralogical composition and method of pile installation, have not been investigated.

The procedures outlined above for the development of torque-twist curves in sand and the program developed by O'Neill were used to predict the measured pile-head torque-twist relationships for the square and circular piles embedded in loose and dense sand environments. For the circular pile procedures advanced by Poulos and Randolph were also utilized to predict the measured pile-head torque-twist relationship. The measured and predicted results are presented on Fig. 19.

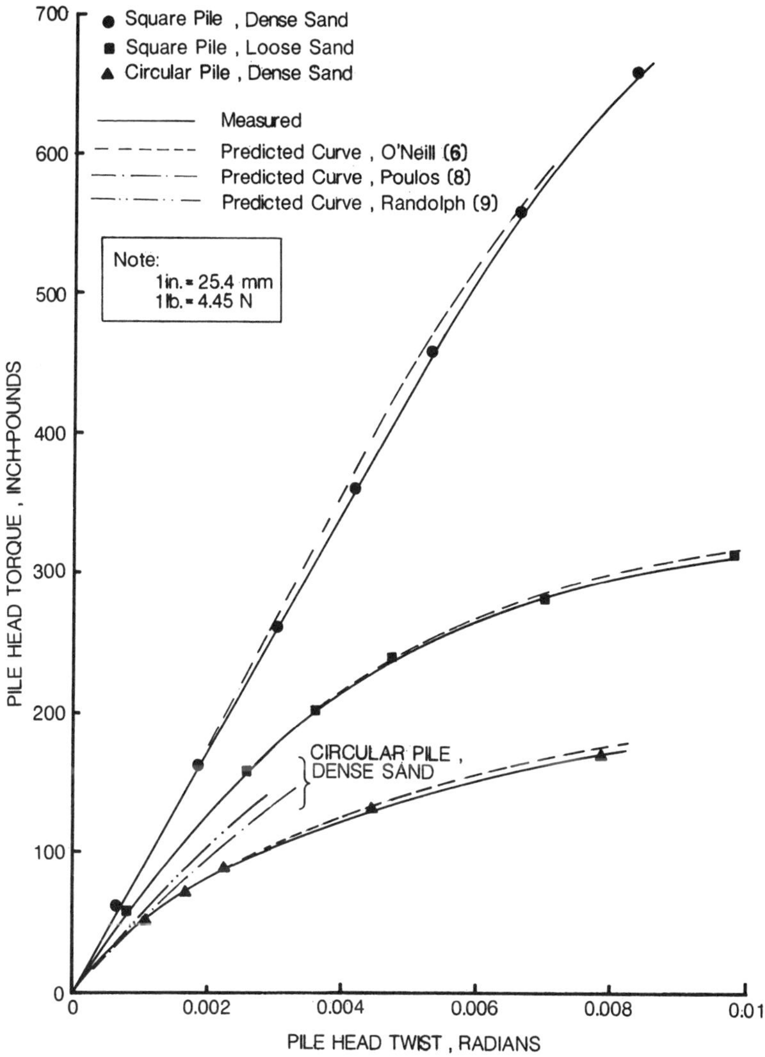

Fig. 19 Comparison of Measured and Predicted Pile-Head Torque-Twist Curves

Comparison of the measured and predicted pile-head torsional response indicates overall good correlation over the entire range of values with the computational procedure developed by O'Neill. For the circular pile, excellent agreement is observed up to about 40 percent of the torque using the criteria proposed by Poulos and Randolph. The deviation in the high torque range may be due to progressive failure beginning at the top and representation of soil by a simple elastic-plastic model (7).

CONCLUSIONS

Based on the results of this model pile study, the following conclusions are drawn:

1. Shear transfer may become essentially constant below a normalized depth that is dictated primarily by the shape and size of the pile and the relative density of the surrounding sand.

2. K_z determined from the torsional load tests is within the range of values reported for the axial mode of loading.

3. The proposed criteria simulating the torsional soil-pile interaction accurately represented the transfer of load from the pile to the soil for the model test piles.

4. A discrete element computational scheme for predicting pile-head torque-twist curves and the torque distribution along the pile developed by O'Neill yields good results provided representative soil response curves are used.

5. The pile-head torque-twist solutions advanced by both Poulos and Randolph provide good correlations in the linear range.

The authors can find no full-scale torsional load test data in sands that can be used to substantiate the proposed criteria. This warrants testing on full-scale instrumental piles to evaluate scale effects and the validity of the proposed criteria. Further, such tests should investigate the effects of installation methods and the influence of axial and lateral loads on torsional soil-pile interaction. Until such studies are performed, careful judgment should be exercised in applying the proposed criteria to full-scale pile design.

REFERENCES

1. Cadling, L. and Odenstad, S., "The Vane Borer," Proceedings, Royal Swedish Geotechnical Institute, 1950.

2. Coyle, H. M. And Reese, L. C., "Load Transfer for Axially Loaded Piles in Clay," Journal of the Soil Mechanics and Foundations Division, American Society of Civil Engineers, Vol. 92, No. SM2, March 1966.

3. Coyle, H. M. and Sulaiman, I. H., "Skin Friction for Steel Piles in Sands," Journal of the Soil Mechanics and Foundations Division, American Society of Civil Engineers, Vol. 93, No. SM6, November 1967.

4. Dutt, R. N., "Torsional Response of Piles in Sand," Ph.D. Dissertation, University of Houston, Houston, Texas, December 1976.

5. Ha, H. B., "Analysis of Generally Loaded Nonlinear Three-Dimensional Pile Groups Considering Group Effects," Ph.D. Dissertation, University of Houston, Houston, Texas, May 1976.

6. O'Neill, M. W., "Determination of the Pile-Head Torque-Twist Relationship for a Circular Pile Embedded in Clay Soil," Master's Thesis, The University of Texas at Austin, Austin, Texas, May 1964.

7. O'Neill, M. W. and Dutt, R. N., Discussion of Poulos (1975), "Torsional Response of Piles," Journal of the Geotechnical Engineering Division, American Society of Civil Engineers, Vol. 102, GT6, June 1976.

8. Poulos, H. G., "Torsional Response of Piles," Journal of the Geotechnical Engineering Division, American Society of Civil Engineers, Vol. 101, No. GT10, October 1975.

9. Randolph, M. F., "Piles subjected to Torsion," Journal of the Geotechnical Engineering Division, American Society of Civil Engineers, Vol. 107, No. GT8, August 1981.

10. Saul, W. E., "Static and Dynamic Analysis of Pile Foundations," Journal of the Structural Division, American Society of Civil Engineers, Vol. 94, No. ST5, May 1968.

11. Tucker, R. L., "An Investigation of a Rigid Cylinder Subjected to Rotational and Axial Loadings in a Clay Medium," Master's Thesis, The University of Texas at Austin, Austin, Texas, May 1960.

12. Vesić, A. S., "Investigations of Bearing Capacity of Piles in Sand," Proceedings, North American Conference on Deep Foundations, Mexico City, Vol. 1, 1964.

13. Vesić, A. S., "Tests on Instrumented Piles, Ogeechee River Site," Journal of the Soil Mechanics and Foundations Division, American Society of Civil Engineers, Vol. 96, No. SM2, March 1970.

NOTATION

The following symbols are used in this paper:

d = Diameter of pile;

F_ϕ = Correction factor for pile-soil slip in uniform soil;

F'_ϕ = Correction factor for pile-soil slip in soil with linearly increasing shear modulus;

G_p = Shear modulus of the pile material;

G_s = Soil shear modulus;

I_ϕ = Influence factor for elastic rotation in uniform soil;

I'_ϕ = Influence factor for elastic rotation in soil with linearly increasing shear modulus;

J_p = Polar moment of inertia of the pile;

K_T = Combined shape and participation factor
= 1, for end bearing pile,
= 2, for friction piles;

L_e = the effective pile length;

N_G = Rate of increase of soil shear modulus with depth;

r_o = Radius of pile;

T = Torvane applied at the top of the pile;

$T(z)$ = Torque present in the pile at any depth, Z.

β = Pile stiffness;

δ = End restraint condition

δ = 0, for hinged connection
δ = 1, for a fixed connection

λ = Soil stiffness;

$\Theta(z)$ = Rotation of the pile at any depth, Z;

γ = unit weight of soil;

σ'_v = Effective vertical stress

UPLIFT RESISTANCE OF MODEL SINGLE PILES

By K.P.R. Chaudhuri[1], M.ASCE; M.V. Symons[2]

SUMMARY

Fully-instrumented, straight shafted model single piles were embedded in dense and medium sand beds and subjected to uplift loading under controlled laboratory conditions. Tests were performed with piles of different diameters, lengths and surface roughnesses. The resulting mechanisms of uplift load transfer were determined by measurement of axial loadings within the piles and by calculation of shear stress distributions down the pile shafts. The limiting depths, beyond which the average frictional resistance attains constant final values due to arching effects, are defined for the various pile/sand systems. Comparisons are made between the present test results, previous model scale tests, field investigations and theoretical predictions.

INTRODUCTION

Single piles subjected to compressive loading have received considerable attention from previous investigators with the result that, for normal operating conditions, it is now possible to analyse their behaviour with some confidence. The same attention has not been devoted to single piles subjected to uplift loading, however, and the increasing use of straight shafted piles to resist both compressive and uplift loading has created a situation in which an accurate assessment of uplift resistance is a more critical requirement than it used to be.

In an effort to understand uplift resistance, previous investigators have carried out model scale laboratory and full size field tests and have attempted analytical approaches. Reported model scale (35,17,4,22,24,38,6,7,16,32,5) and full size (13,27,3,12,37,21,23) tests tend to be deficient in a number of important aspects: the majority were conducted at shallow embedment depths only; in many cases the uplift loading was applied after the pile had first been tested to failure in compression; most of the full size field tests provided limited information on the soil properties. The analytical approaches (13,29,22,11,18,43,25) all assume that the uplift resistance of straight shafted piles in sand is entirely dependent upon skin friction between the pile shaft and the surrounding soil and adopt a limiting friction approach when assessing the ultimate value of this uplift resistance. For this purpose an equation of the form

---

1. Geotechnical Engineer, Dantata and Sawoe, Consulting Engineers, Kano, Nigeria.
2. Lecturer, Dept. of Civil Engineering and Building Technology, University of Wales Institute of Science and Technology, Cardiff.

$$Q_u = (0.5)K_s \tan\delta . \gamma . \pi d H^2 \qquad (1)$$

where Q_u = ultimate uplift resistance of pile
K_s = earth pressure coefficient on pile shaft
δ = angle of friction between pile shaft and sand
γ = unit weight of sand
d = pile shaft diameter
H = embedded length of pile

is normally employed. A number of the investigators have tried to mathematically formulate the parameter $K_s \tan\delta$ but, as the limiting friction assumption is an oversimplification of the actual soil/structure interaction, their findings cannot be applied with confidence.

Understandably, in view of the deficiencies outlined above, straight shafted piles in uplift are invariably designed on the basis of empirical formulae and by conducting full size tests in the field. Further experimental data relating to the behaviour of single piles in uplift is therefore needed if reliable methods of analysis are to be established. As full size testing is expensive and difficult to carry out, model scale testing will undoubtedly play as important a role in the future understanding of piles in uplift as it has for piles in compression.

This paper reports the results of a testing programme into the behaviour of model scale straight shafted single piles in uplift.

APPARATUS AND TESTING

(a) General procedure

The testing arrangement is shown in Figure 1. The following procedure was adopted for the establishment and performance of a test:- the pile was placed in a 1.10m diameter by 0.765m or 1.530m high circular test bin and clamped into position. Sand was either poured to provide a homogeneous medium sand stratum or placed in compacted layers to form a dense sand stratum. Displacement transducers and dial gauges, fastened to fixed datum frames, were placed in position to measure movements of the pile and sand surface. The pile was unclamped and the instrumentation zeroed before commencement of the test, thus eliminating pile self-weight from subsequent load measurements. A continuous uplift pull was applied vertically at the centre of the pile cap by a 5mm diameter steel rope attached to a constant rate-of-strain machine operating at a rate of pull of 0.95mm per minute. The resulting loads and displacements were measured by means of the load cells within the piles and the linear transducers and dial gauges. The output from the load cells and transducers was monitored by a high speed data-logging and print-out system with readings taken at pre-set time intervals.

This procedure was carried out for the pile diameters, embedded depths, surface roughnesses and sand states shown:-

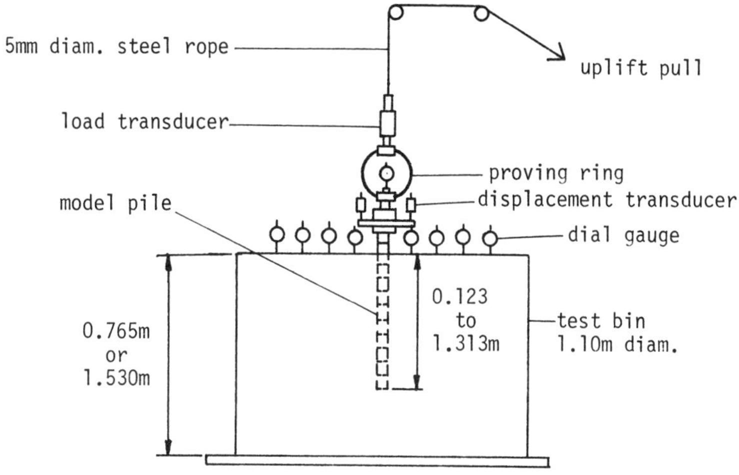

Fig. 1 General testing arrangement

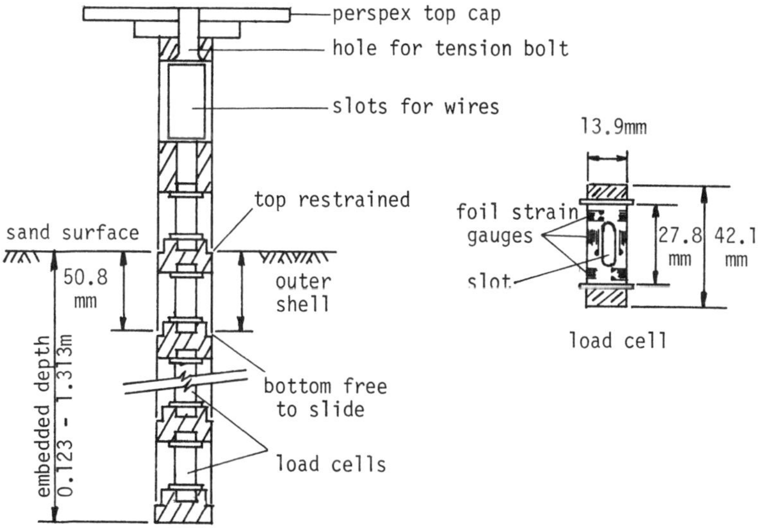

Fig. 2 Diagrammatic representation of model piles and load cell

Pile diam. (mm)	Embedded depth of pile (m)	Surface finish	Sand state
25.4	0.175; 0.336; 0.498; 0.661; 0.826; 0.989; 1.151; 1.313	Smooth	Medium
25.4	0.123; 0.229; 0.336; 0.445; 0.553; 0.661; 0.826; 0.989; 1.151; 1.313	Smooth	Dense
38.1	0.229; 0.336; 0.445; 0.553; 0.661	Smooth	Dense
39.2	0.229; 0.336; 0.445; 0.553; 0.661	Rough	Medium
39.2	0.229; 0.336; 0.445; 0.553; 0.661	Rough	Dense
50.0	0.229; 0.336; 0.445; 0.553; 0.661	Smooth	Dense

(b) Model piles and instrumentation

The outer body of the model piles consisted of a series of smooth, hollow brass sections, 50.8mm long, restrained at their top ends and free to slide at their bottom ends. Three different diameters of pile were formed with this smooth surface - 25.4, 38.1 and 50.0mm. The piles were made sufficiently long to allow a first testing programme with a maximum embedded depth of 0.661m. The 25.4mm diameter pile was later lengthened to allow a second testing programme with a maximum embedded depth of 1.313m in order that depth/diameter ratios of up to 50 could be examined. After completion of the smooth surfaced pile testing programmes, the 38.1mm pile was coated with adhesive and sand to simulate a rough surfaced pile for a third testing programme; the adhesive and sandcoating increased the pile diameter to 39.2mm.

The inner core of the piles was formed from a series of load cells which measured the distribution of axial load within the pile (Figure 2). The structure of the pile was such that it allowed each load cell to record the uplift load exerted at the top of the pile minus the frictional load developed down to the bottom of its outer shell. This also meant that the difference in recorded load between two successive cells, divided by the surface area of one outer shell, represented the average shear stress or skin friction developed on the lower of the two shells considered.

The load cells consisted of 42.1mm long by 13.9mm diameter hollow brass tubes provided with screw threads at their ends to make them effective in tension. Four slots were cut in the cell walls to increase sensitivity. Because small load variations had to be measured, 4 longitudinal and 4 lateral precision foil strain gauges were connected to the cell walls and wired in a configuration which minimised the effects of lateral strain and eccentricity. The output from the cells was monitored by a Solartron Compact 33 data-logger and printed through a Facit 4070 high speed paper tape punch. Full details of the piles, load cells and instrumentation are described elsewhere (33).

(c) Test media

A uniform medium/fine grained sand, which allowed a close control of density with the sand pouring and compaction techniques employed, was used for all tests. 97% of its particles by weight lay in the particle size range 0.15 to 0.40mm with effective size = 0.17 and uniformity coefficient = 1.70. Its limiting minimum and maximum dry densities = 1.45 and 1.72 g/ml. It possessed the following properties in its medium and dense states respectively:-
average unit weight = 1.55 and 1.67 g/ml; relative density = 39.4 and 83.2%; angle of shearing resistance = 35° and 43°; angle of friction between smooth brass surface and sand = 11.5^{o} and 12.0^{o}; angle of friction between rough surface and sand = 34° and 41°.

RESULTS AND ANALYSIS

(a) Load/displacement relationships

Typical load/displacement curves for embedment depths of up to 0.661m are given in Figure 3. It should be noted that the uplift loads do not include the pile self-weights, which were balanced to zero before the start of the tests. The curves for the 25.4, 38.1 and 50.0mm diameter smooth surfaced piles in dense sand all lacked the characteristic shape normally associated with piles in uplift, viz. a rapid fall after attainment of peak uplift load to a residual value. Instead, they tended to level off after peak uplift load for quite considerable displacements before gradually decreasing to residual values. This behavioural characteristic of smooth surfaced piles in dense sand has been observed by previous investigators (35). The curves for the 39.2mm diameter rough surfaced pile in dense and medium sand possessed the characteristic shape with a well-defined peak uplift load followed by failure to residual values at higher displacements. The 25.4mm diameter smooth surfaced pile in medium sand displayed an intermediate characteristic, with a well-defined peak followed by levelling-off before decreasing to a residual value.

(b) Peak uplift load relationships

The peak (or ultimate) uplift load is required for design purposes, with a factor of safety allocated to ensure that failure states are not approached. The relationship of this peak load to pile displacement, embedment depth and diameter is of primary interest to the engineer.

(i) Peak uplift load displacement/embedment depth - in order to compare the displacements required for development of peak uplift load, plots of dimensionless displacement Δ/H and Δ/d (where Δ = displacement at peak uplift load, H = embedded depth of pile, d = pile diameter) against H are given in Figure 4. For smooth surface piles in dense sand, the displacement at peak uplift load increases with embedment depth; no definite trend is apparent for smooth surfaced piles in medium sand. For a given embedment depth, higher displacements are required to develop peak uplift load for dense sand than for medium sand. For rough surfaced piles in both dense and medium sand, the displacements required to reach peak uplift load increase steadily with embedment

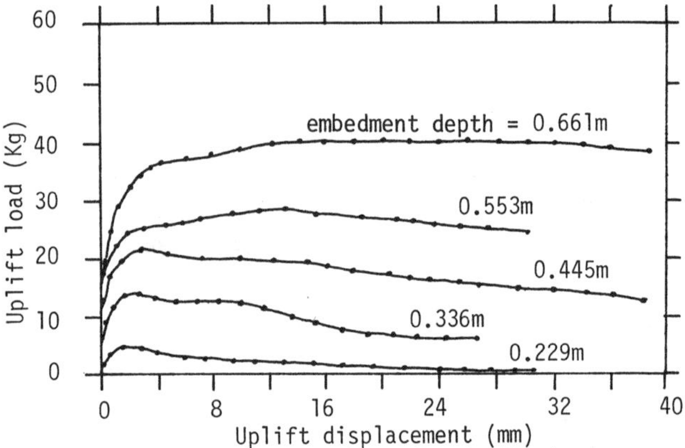

(a) 25.4mm diam. smooth surfaced pile in dense sand

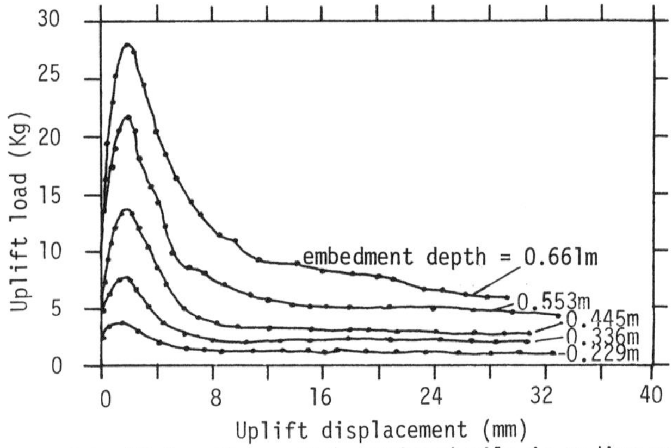

(b) 39.2mm diam. rough surfaced pile in medium sand

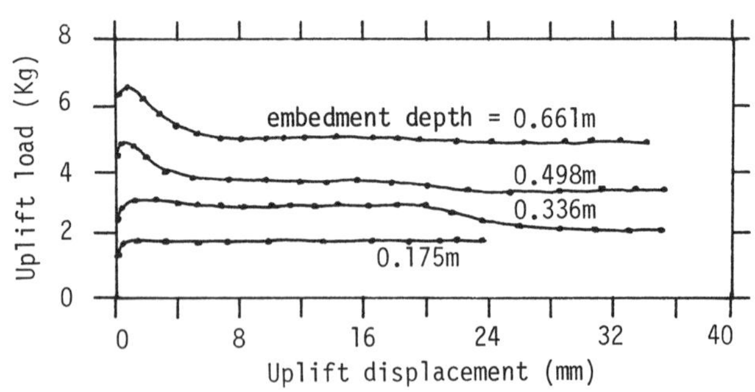

(c) 25.4mm diam. smooth surfaced pile in medium sand

Fig. 3 Load/displacement relationships

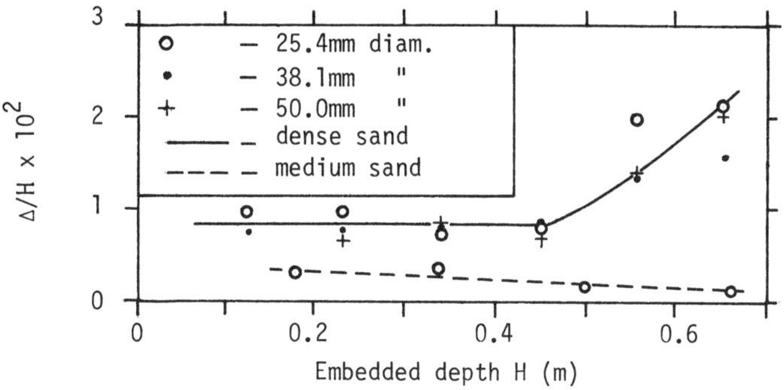

(a) Δ/H vs. H - smooth surfaced piles, dense and medium sand.

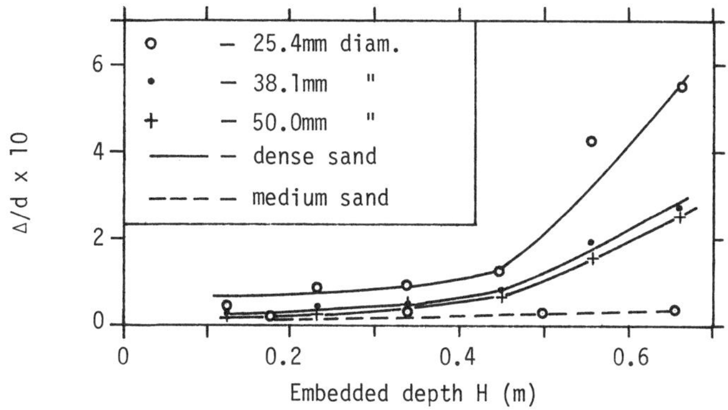

(b) Δ/d vs. H - smooth surfaced piles, dense and medium sand

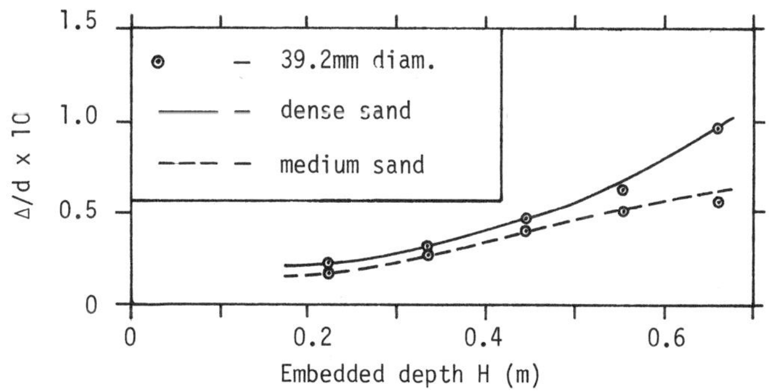

(c) Δ/d vs. H - rough surfaced pile, dense and medium sand

Fig. 4 Dimensionless displacement/embedded depth relationships depth: again, higher displacements are required in the case of dense sand. It may be generally concluded that the displacements required to reach peak uplift load in the case of smooth surfaced piles are appreciably higher then those required for rough surfaced piles.

(ii) Peak uplift load/embedment depth - the peak uplift load/embedment depth relationships (Figure 5) show a significant increase in load with depth for piles in dense sand: the increase being more pronounced in the case of rough surfaced piles. For both smooth and rough surfaced piles in medium sand, the increase of load with depth is small. The extra depth of embedment tests carried out with the 25.4mm smooth surfaced pile are included in Figure 5(c). For dense sand, the apparent steady rate of increase of peak uplift load alters after an embedment depth of 0.661m, the curve changing shape from concave upwards to concave downwards with a tendency towards a limiting value. For medium sand, the peak uplift load maintains its small, virtually linear increase at the higher embedment depths.

(iii) Peak uplift load/diameter - the peak uplift load/diameter relationships for smooth surfaced piles in dense sand (Figure 6) indicate a linear relationship for each embedment depth with higher peak uplift loads generated by the larger diameter piles. Theoretically, the peak uplift load should be directly proportional to pile diameter for a given embedment depth (See Equation (1)), but this is not the case for the results shown (i.e. $^{Qu}/d$ does not have the same value for the 25.4, 38.1 and 50.0mm diameter pile results for any of the embedment depths). This result is contrary to the theoretical proposition and to the findings of a previous investigation (21).

(c) Axial load and shear stress distributions

Typical axial load and shear stress (or skin friction) distribution diagrams are given in Figure 7. Limitations of space do not allow inclusion of all results. All unreported tests exhibited similar patterns of axial load and shear stress distribution. The shear stress distributions for the 25.4mm diameter pile are reported because the high embedded depth/diameter ratio tests yielded altered curve characteristics.

The main features of the results are:-

(i) axial load decreases with depth as skin friction erodes its value. The magnitude of axial load at all depths increases as the pile upward displacement increases, until peak uplift load is reached; thereafter the axial load decreases at all depths until residual values are achieved at high displacements.

(ii) shear stress (or skin friction) increases with depth, reaching its maximum value at a level varying between 50% and 70% of the pile's embedded depth. It then decreases rapidly to a small or zero value at the pile toe. This characteristic may be explained as follows:- when a pile is pulled upwards, the sand surrounding it will try to move upwards and outwards. Any resulting movement of the sand will increase its void ratio in the pile/soil interface region, due to removal of sand vertically upwards near the top of the pile and by the creation of a gap under the uplifted pile toe. This produces a ring of relatively loose sand around the pile surrounded by compacted sand, a situation ideal for the establishment of arching conditions (42). The arching causes a

(a) Rough surfaced pile, dense and medium sand

(b) Smooth surfaced piles, dense and medium sand

(c) 25.4mm diam. smooth surfaced pile, dense and medium sand

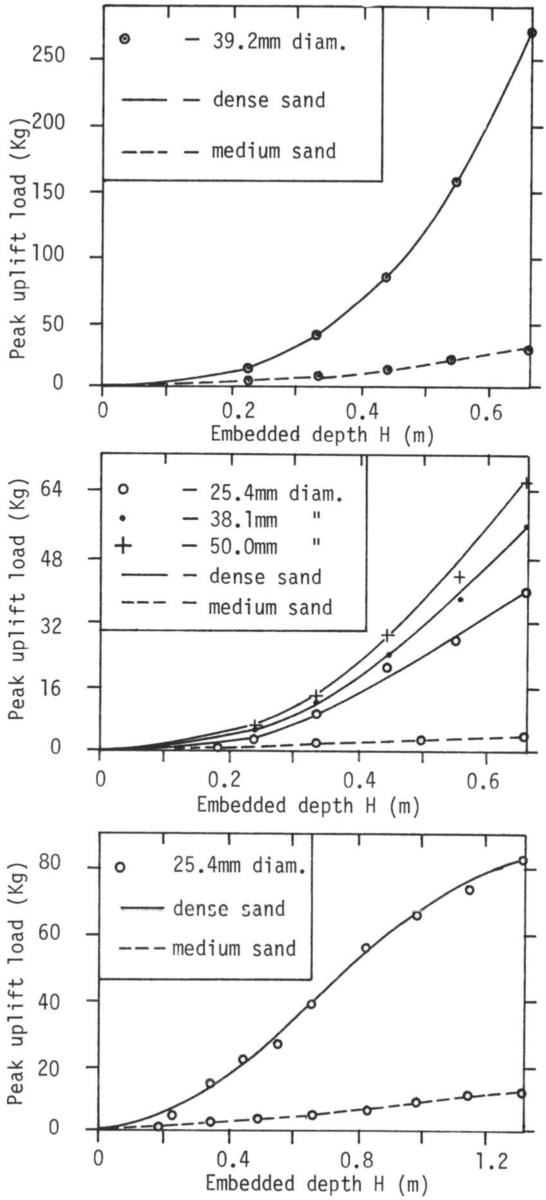

Fig. 5 Peak uplift load/embedded depth relationships

Fig. 6 Peak uplift load/ diameter relationships - smooth surfaced piles in dense sand

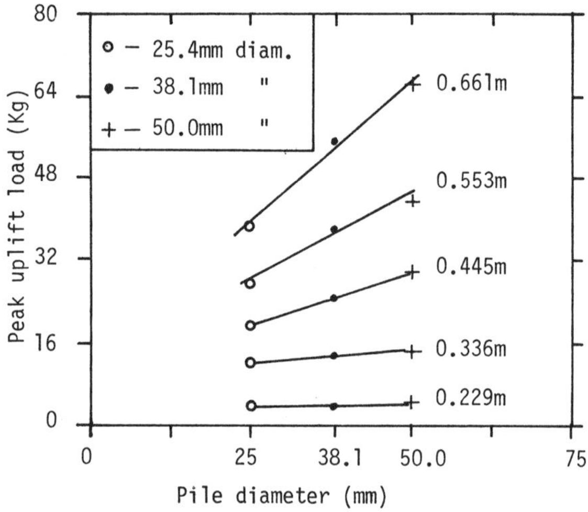

Values on curves denote percentage of peak uplift load.
Dashed lines represent residual values at high displacements.

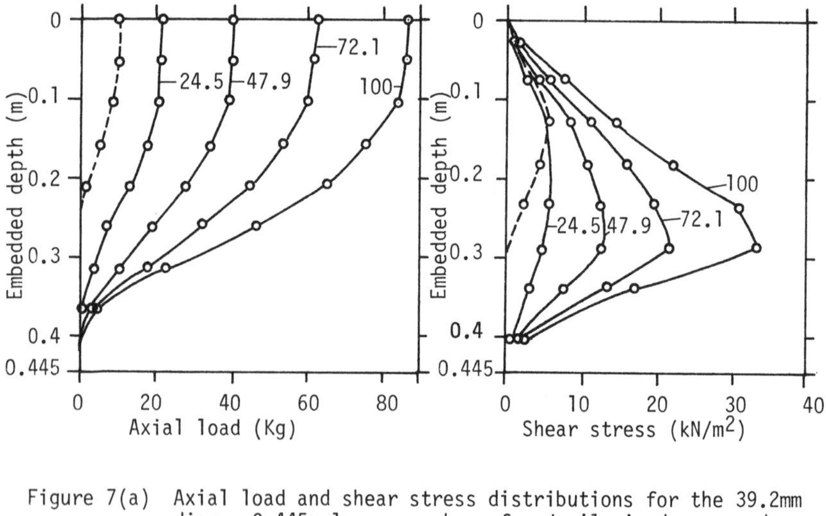

Figure 7(a) Axial load and shear stress distributions for the 39.2mm diam., 0.445m long, rough surfaced pile in dense sand.

Value on curve denotes embedded depth for that test result.

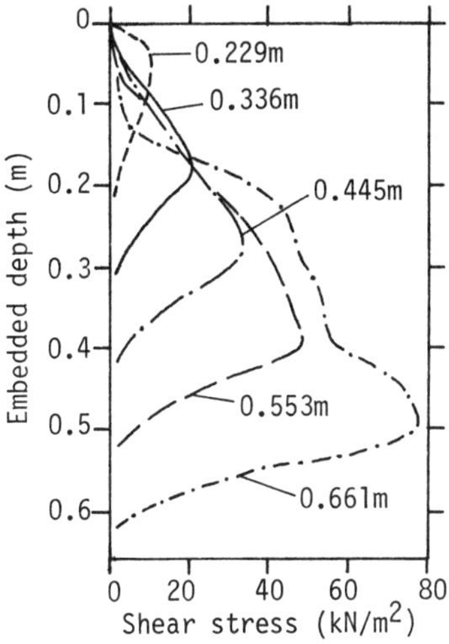

7(b) 39.2mm diam. rough surfaced piles in dense sand

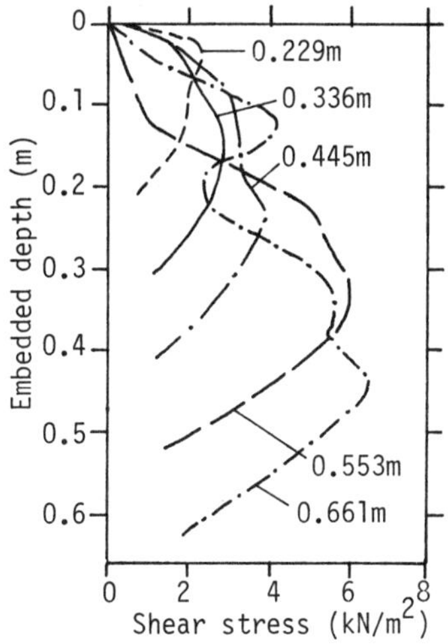

7(c) 39.2mm diam. rough surfaced piles in medium sand

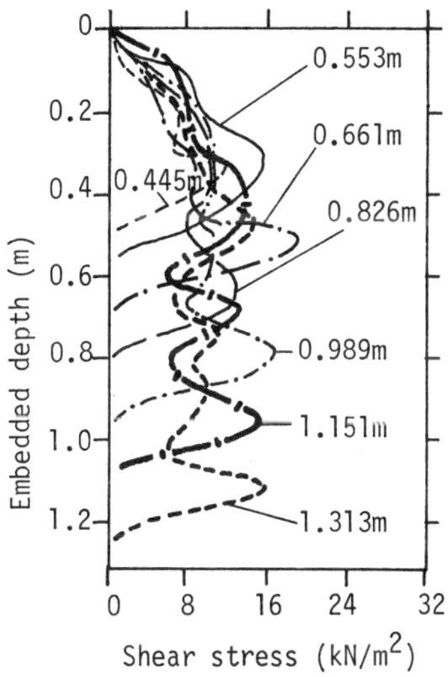

7(d) 25.4mm diam. smooth surfaced piles in dense sand

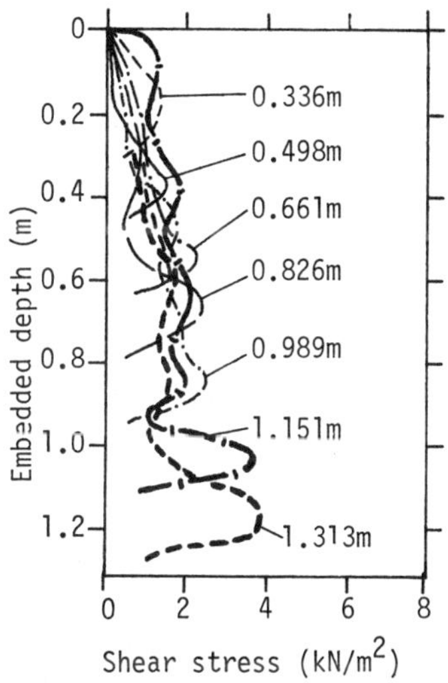

7(e) 25.4mm diam. smooth surfaced piles in medium sand

Figs. 7(b)-(e) Shear stress distributions for 39.2mm diam. rough surfaced and 25.4mm diam. smooth surfaced piles in dense and medium sand decrease in vertical stress in the lower regions of the pile leading to a reduction in the shear stress or skin friction developed. The result is the approximately parabolic shear stress/depth characteristic obtained.

(iii) the maximum values of shear stress developed vary significantly with the surface finish of the pile. Generally, the maximum shear stresses developed on rough surfaced piles are some 4 times (for dense sand) and some 3.5 times (for medium sand) as large as those developed on smooth surfaced piles.

(iv) the residual values of shear stress at high displacements indicate a significant decrease of uplift resistance and a complete absence of frictional resistance over the lower part of the pile. This absence is probably due to loosening of the sand in this region as particles fall into the cavity created under and around the pile toe as large uplift of the pile takes place. Even rough surfaced piles in dense sand exhibit these characteristics and there is no doubt that the sand mass around the pile is completely loosened at large pile displacements.

(v) the axial load and shear stress distributions described above are similar to results presented by previous investigators (28,34,45,21, 2, 10) despite significant differences in testing programmes and techniques.

(vi) the shear stress distribution curves for the 25.4mm diameter, high embedment depth piles do not follow the characteristic parabolic shape. Instead, they take the form of a wave curve in which the shear stress continually increases and decreases. This form of wave curve has been observed by previous investigators (34,39). An explanation of this change in mechanism is that, when the depth of embedment is great, there is more chance of unequal displacements occurring within the sand as it attempts to follow the pile upwards. This causes localised density changes within the sand adjacent to the pile shaft, leading to localised reductions in the shear stress or skin friction.

(d) Limiting depth considerations

Average frictional resistance (peak uplift load/embedded shaft area) against embedded depth/diameter ratio plots for the piles (Figure 8(a)) yield values for the limiting or critical depth (H_c) beyond which the average frictional resistance attains a constant value. This behavioural characteristic of long piles has been mainly observed in compressive loading and only occasionally in uplift (15,44, 40,10,38) and has been attributed to arching mechanisms in the sand preventing full overburden pressures from being effective along the lower reaches of long pile shafts. The limiting embedded depth/diameter ratios obtained are: approximately 30 for smooth surfaced piles in dense sand; approximately 11 for smooth surfaced piles in medium sand; approximately 15 for rough surfaced piles in medium sand. The limiting ratio for rough surfaced piles in dense sand cannot be reported as the maximum depth of embedment tested was insufficient to reach the limiting case.

The results of the present investigation are compared with the limiting ratios obtained by previous investigators (20,15, 36,44,40,10,

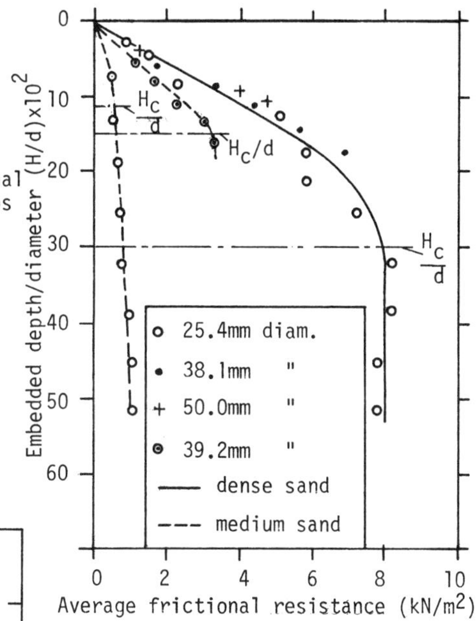

Fig. 8(a) Average frictional resistance/H/d relationships for piles in dense and medium sand.

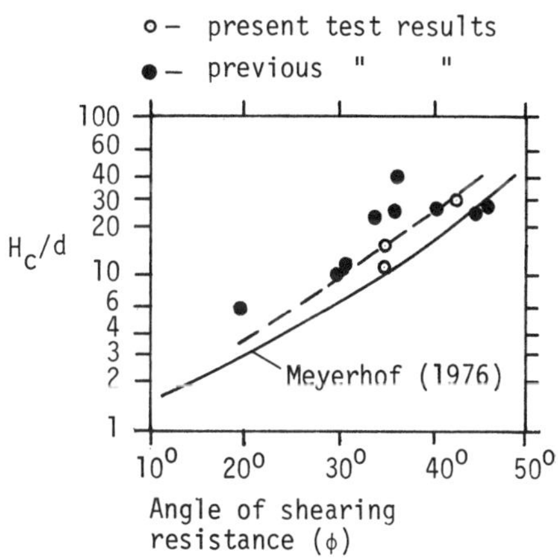

Fig. 8(b) Comparison of limiting depth results.

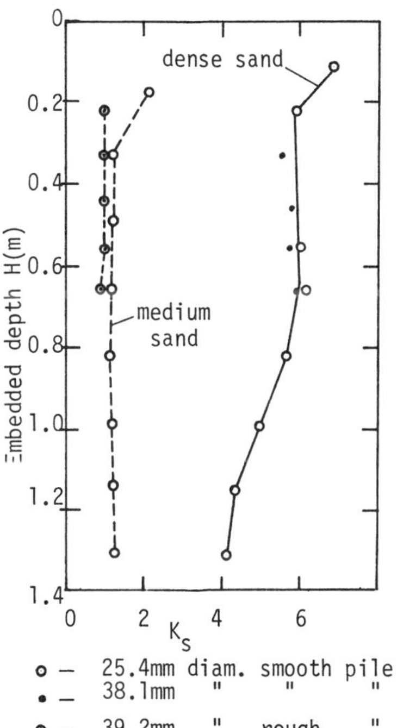

Fig. 11 K_s/embedded depth relationships for piles in dense and medium sand.

6) in Figure 8(b), in which limiting embedded depth/diameter (H_c/d) ratios are plotted against angle of shearing resistance (ϕ). Some compressive test results are included because of the shortage of uplift test data available. The relationship proposed by Meyerhof in 1976 (26) is shown as a full line and tends to yield lower H_c/d values than those obtained in the present and previous investigations.

Figures 8(a) and (b) indicate that the value of limiting depth is dependent on: the density of the sand; the pile surface finish; and probably on the method of pile installation.

(e) Comparisons with model scale and field investigations

In order to compare the present results with those of previous model scale (20,17,4,9,46,22,41,43,24,25,6,5) and full size field (13,27,31,35,8,1,14,21,37,40,19,2) investigations, values of the coefficient of earth pressure on the pile shaft (K_s) are plotted against angle of shearing resistance (ϕ) in Figures 9(a) and (b). Full line curves are given for the supposedly bounding active pressure and passive resistance values, K_a and K_p, and for the values proposed by Meyerhof in 1973 (25). The average values of K_s obtained in the present investigation for piles in medium sand lie between the K_a and K_p limits and compare well with the results of previous investigations. The average value of K_s for the piles in dense sand is close to, but greater than, the K_p value. This exceeding of the passive resistance limit in dense sand has been reported in previous model scale investigations (20,46,9,24,25,5), but not in full size field tests. The following physical explanation may account for the phenomenon:- as uplift loading is applied to a pile in dense sand, the sand adjacent to the pile surface is subjected to high shear stresses and tries to dilate. This volume expansion is suppressed by the surrounding dense soil mass leading to an increase in radial (normal) stress on the pile surface. This "locking-in" of the soil particles within the shear zone leads to high skin frictional resistances and to high K_s values. A previous investigation has measured radial stresses at soil/anchor interfaces of up to 10 times the effective overburden pressure in dense sand (30) and it has been observed that, when the soil in which a pile is embedded has been sufficiently compacted, a reliable estimate of uplift resistance can be obtained by assuming a passive resistance case (43). For piles in medium sand it is likely that local shear failure takes place around the pile shaft as uplift proceeds with re-orientation of the sand particles leading to a state approaching the active pressure condition.

The scatter of results apparent in Figures 9(a) and (b) is due to a number of factors but it may generally be concluded that the value of K_s depends upon: the angle of shearing resistance of the soil; the diameter and surface roughness of the pile; the effective unit weight of the soil; the method of installation of the pile.

(f) Comparison of measured and theoretical uplift resistances

Equation (1) can be re-written in the dimensionless form

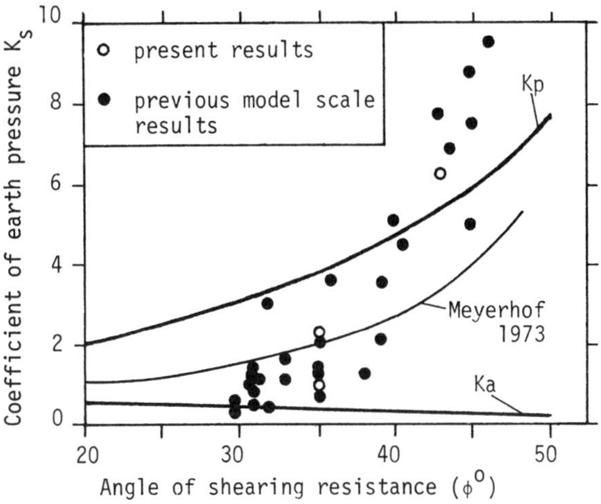

Fig. 9(a) Comparison of present results with previous model scale test results (for straight shafted piles above limiting depth)

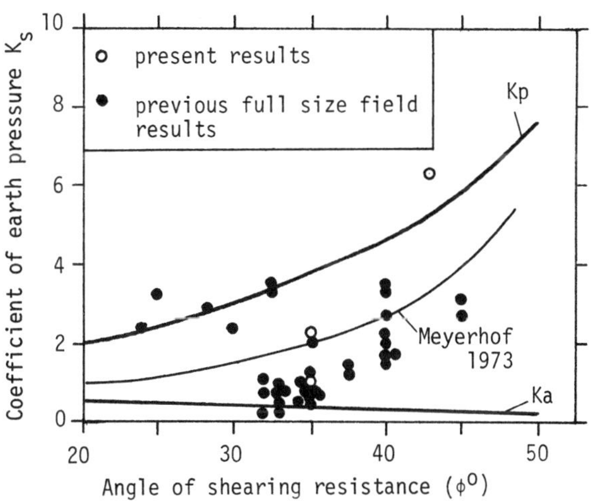

Fig. 9(b) Comparison of present results with previous full size field results (for straight shafted piles above limiting depth)

$$\frac{Qu}{{}^{\pi}/_{4}.d^{2}.H\gamma} = 2K_s \tan\delta({}^{H}/d)$$

$$\text{or } \frac{Qu}{AH\gamma} = 2K_s \tan\delta({}^{H}/d) \qquad (2)$$

$$\text{where } A = \frac{\pi d^2}{4}$$

Assuming $K_S \tan\delta$ remains constant with depth, then a plot of $(\frac{Qu}{AH\gamma})$ against $({}^{H}/d)$ will yield a straight line with a slope of $2K_S \tan\delta$. This relationship is valid down to the limiting depth only, because of the influence of arching mechanisms upon the K_S values below this level.

The results of the present investigation are compared, on this dimensionless basis, with theoretically predicted values based on the work of Meyerhof, Hansen and Lendi (25,11,18). The results obtained are given in Figures 10(a) - (d). Predicted results based on the assumption that $K_S = K_p$ are also included. The experimental results provide the expected straight line relationships, the only exception being the rough surfaced pile in dense sand which yields very high dimensionless load values. This is probably attributable to the failure plane passing through the soil mass instead of coinciding with the weakest plane of failure along the pile/sand interface. The dial gauges monitoring soil surface movement certainly recorded significant heave in the rough surfaced pile in dense sand tests, implying that the failure mechanism involved the lifting of a cone of soil around the pile.

The theoretical predictions are seen to be considerably in error and it may be concluded that none of the equations covers all the variables involved in the complex interaction between pile and soil. In general, predictions based on Meyerhof's suggested K_S/ϕ curve (Figure 9) provide the closest estimates although, for dense sand, the simple assumption that $K_S=K_p$ should be made.

Below the limiting depth, the average frictional intensity attains a constant value, becoming independent of the overburden pressure. Equation (2) no longer applies and a decrease of K_S values with depth is implied. Plots of K_S values (obtained by back-working with Equation (1)) against depth for the present testing are shown in Figure 11 and the decrease of K_S values below the limiting depth is confirmed for the 25.4mm diameter pile in dense sand.

The limitations of the theoretical predictions are due to many factors but, if they cannot correctly predict results for an ideal model condition in which soil properties and parameters are known exactly, it is unlikely that they can be used with confidence for the prediction of the uplift resistance of full size piles, particularly when they are driven. This is the reason why heavy reliance is still placed upon empirical approaches with experimentally adjusted constants, high factors of safety, and pile load testing.

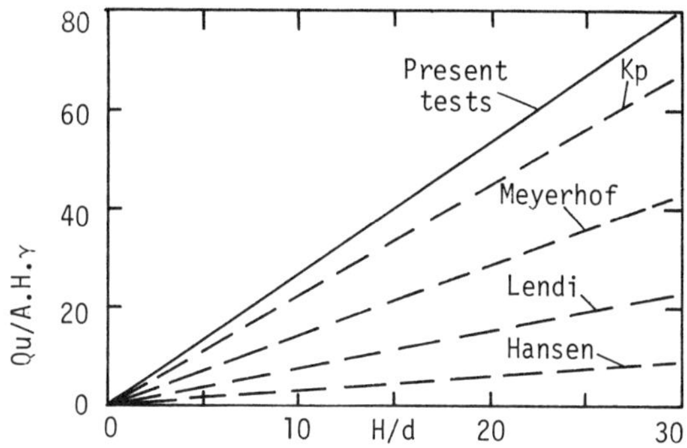

(a) Dense sand, smooth surfaced pile (ϕ=43°, δ=12°)

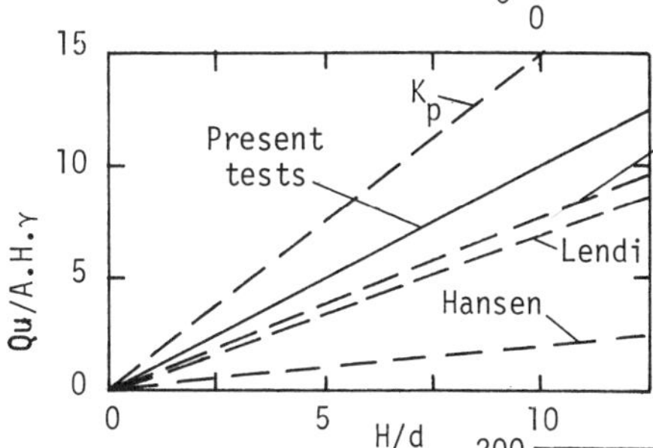

(b) Medium sand, smooth surfaced pile (ϕ=35°, δ=11.5°)

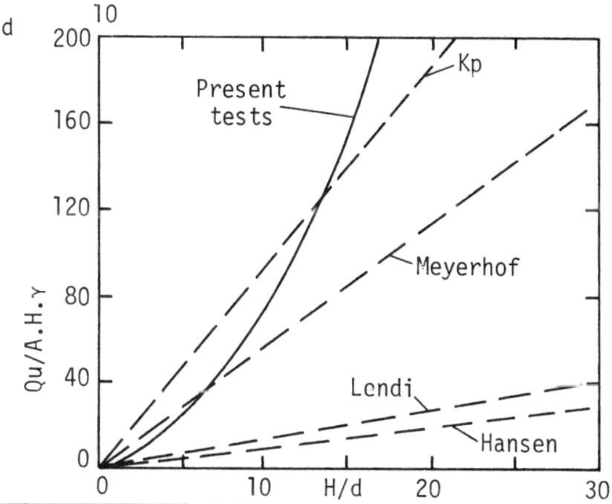

(c) Dense sand, rough surfaced pile (ϕ=43°, δ=41°)

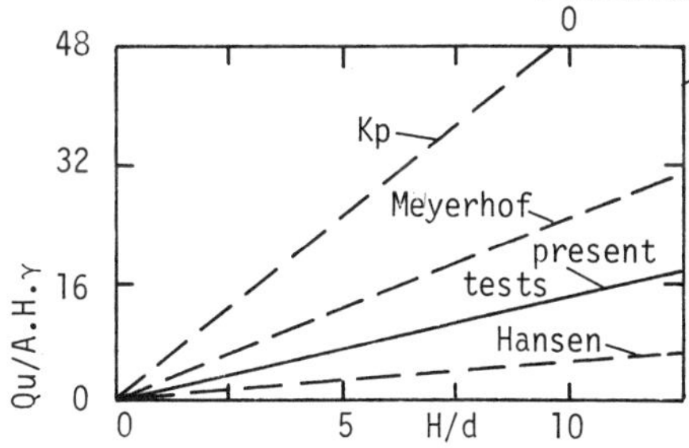

(d) Medium sand, rough surfaced pile (ϕ=35°, δ=34°)

Fig. 10 Comparison of present results with theoretical predictions (for straight shafted piles above limiting depth)

CONCLUSIONS

These conclusions relate to model studies involving straight shafted piles in medium and dense sand subjected to vertical uplift loading.

1. The load/displacement curves varied in characteristic shape, depending upon the surface texture of the pile and the density of the sand. The displacements required to attain peak uplift load were independent of pile diameter but generally increased with pile embedded depth.
2. Peak uplift load increased linearly with pile diameter but not in a directly proportional manner.
3. Uplift resistance was mobilised progressively from the sand surface downwards. The development of this resistance is dependent upon the relative displacement between pile and sand and, hence, on the deformation properties of the pile and sand.
4. Shear stress (or skin friction) distributions exhibited approximately parabolic shapes with maximum values generally attained at levels of between 50% and 70% of the pile's embedded depth. Piles with high embedded depths did not exhibit this characteristic distribution.
5. Limiting depths of embedment, beyond which the average frictional resistance attained a constant value, were encountered. The values of limiting embedded depth/diameter ratios obtained were:- 30 and 11 for smooth surfaced piles in dense and medium sand respectively; 15 for rough surfaced piles in medium sand.
6. Existing theoretical analyses were found to be significantly in error when compared with the experimental results. Meyerhof's K_s/ϕ relationship yielded the best estimates for pile uplift resistance in medium sand and a $K_s=K_p$ assumption provided quite accurate estimates for piles in dense sand.

REFERENCES

1. Adams, J.I., and Hayes, D.C., "The uplift capacity of shallow foundations," Ontario Hydro Research Quarterly, Vol. 19, No. 1, 1967, pp.1-13.
2. Adams, J.I., and Klym, T.W., "A study of anchorages for transmission tower foundations," Canadian Geotechnical Journal, No. 9, 1972, pp.89-104.
3. Begemann, H.K.S.P., "The maximum pulling force on a single tension pile calculated on the basis of results of the adhesion jacket cone," Proceedings of the 6th International Conference on Soil Mechanics and Foundation Engineering, Montreal, 1965, Vol. 2, pp.229-233.
4. Broms, B.B., and Silberman, J.O., "Skin friction resistance for piles in cohesionless soils," Soil Mechanics and Foundation Engineering, Vol. 10, September 1964.

5. Chaudhuri, K.P.R., Ghataora, G.S., and Symons, M.V., "Uplift resistance of model piles and pile groups," Proceedings of the 2nd International Conference on Numerical Methods in Offshore Piling, Austin, Texas, April 1982, pp.239-255.

6. Das, B.M., and Seeley, G.R., "Uplift capacity of buried model piles in sand," Journal of the Geotechnical Engineering Division, ASCE, Vol. 101, No. GT7, 1975, pp.711-715.

7. Das, B.M., Seeley, G.R., and Pfeifle, T.W., "Pullout resistance of rough rigid piles in granular soil," Soils and Foundations, Vol. 17, No. 3, 1977, pp.72-77.

8. Downs, D.I., and Chieurzzi, R., "Transmission Tower foundations," Journal of the Power Division, ASCE, Vol. 92, No.PO2, 1966, pp.91-114.

9. Graham, J., "Plastic failure in granular material; plane strain problems in soil mechanics," Ph.D Thesis, Queen's University, Belfast, 1966.

10. Hanna, T.H., and Tan, R.H.S., "The behaviour of long piles under compressive loads in sand," Canadian Geotechnical Journal, Vol. 10, No. 3, 1973, pp.311-340.

11. Hansen, J.B., "A theory for skin friction on piles," Danish Geotechnical Institute, Bulletin No. 25, 1968, pp.5-12.

12. Hunter, A.H., and Davisson, M.T., "Measurements of pile load transfer," Performance of deep foundations, ASTM STP 444, American Society for Testing Materials, 1969, pp.106-117.

13. Ireland, H.O., "Pulling tests on piles in sand," Proceedings of the 4th International Conference on Soil Mechanics and Foundation Engineering, 1957, Vol. 2, pp.43-45.

14. Jain, G.S., and Gupta, S.P., "A comparative study of multiple under-reamed pile with large diameter pile in sandy soil," 3rd Budapest Conference on Soil Mechanics and Foundation Engineering, 1968, pp.563-570.

15. Kerisel, J., "Deep foundations, basic experimental facts," Conference on Deep Foundations, Mexico, 1964, Vol. 1, pp.5-44.

16. Kulhawy, F.H., Kozera, P.W., and Withiam, J.L., "Uplift testing of model drilled shafts in sand," Journal of the Geotechnical Engineering Division, ASCE, Vol. 105, 1979, pp.31-47.

17. Lebegue, Y., "E'tude experimentale des efforts d'arrachage et de frottement negatif sur les pieux en milieu pulverulent," Annales ITBTP, No. 199-200, 1964, pp.808-822.

18. Lendi, P., "Beitrag zur erdstatischen Berechnung Von Verankerungen im Lockergestein," Institut fur bauwissenschaftliche Forsehung, No. 6, March 1969, Zurich, pp.1-75.

19. Littlejohn, G.S., "Recent developments in ground anchor construction," Ground Engineering, Vol. 1, No. 3, 1968, pp.32-36, 46.

20. Macdonald, H.F., "Uplift resistance of caisson piles in sand," M.Sc. Thesis, Nova Scotia Technical College, Canada, 1963.

21. Mansur, C.I., and Hunter, A.H., "Pile tests Arkansas River project," Journal of the Soil Mechanics and Foundation Division, ASCE, Vol. 96, No. SM5, 1970, pp.1545-1582.

22. Mazurkiewicz, B.K., "Skin friction on model piles in sand," Danish Geotechnical Institute, Bulletin No. 25, 1968, pp.13-48.

23. McClelland, B., "Design of deep penetration piles for ocean structures," Journal of the Geotechnical Division, ASCE, Vol. 100, No. GT7, 1974, pp.709-745.

24. Meyerhof, G.G., "The uplift capacity of foundations under oblique loads," Canadian Geotechnical Journal, No. 10, 1972, pp.64-70.

25. Meyerhof, G.G., "Uplift resistance of inclined anchors and piles," Proceedings of the 8th International Conference on Soil Mechanics and Foundation Engineering, Vol. 2.1, 1973, pp.167-172.

26. Meyerhof, G.G., "Bearing capacity and settlement of pile foundations," Journal of the Geotechnical Engineering Division, ASCE, Vol. 102, No. GT3, 1976, pp.197-227.

27. Milovic, D., "Pulling out resistance of foundations in loess soil," International Conference on Soil Mechanics and Foundation Engineering, Hungarian Academy of Science, Budapest, 1963, pp.281-287.

28. Mohan, D., Jain, G.S., and Kumar, V., "Load bearing capacity of piles," Geotechnique, Vol. 13, 1963, pp.76-86.

29. Norlund, R.L., "Bearing capacity of piles in cohesionless soils," Journal of the Soil Mechanics and Foundation Division, ASCE, Vol. 89, SM3, 1963, pp.1-35.

30. Ostermayer, H., "Construction, carrying behaviour and creep characteristics of ground anchors," Diaphragm walls and anchorages, Proceedings of the Institution of Civil Engineers, London, 1975, pp.141-151.

31. Paterson, G., and Urie, R.L., "Uplift resistance tests on full size transmission tower foundations," Conference Internationale des Grands Reseaux Electrique a Haute Tension, Paper 203, 1964, pp.1-24.

32. Raju, V.S., and Thomas, T.M., "Influence of installation methods on frictional characteristics of concrete piles in sand - a model study," Journal of the Institution of Engineers, India, Part C14, 1981, pp.187-191.

33. Ray Chaudhuri, K.P., "Uplift resistance of foundations," Ph.D. Thesis, University of Wales Institute of Science and Technology, Cardiff, 1977.

34. Robinsky, E.I., and Morrison, C.F., "Sand displacement and compaction around model friction piles," Canadian Geotechnical Journal, Vol. 1, No. 2, 1964, pp.81-93.

35. Schmidt, B., and Kristensen, J.P., "The pulling resistance of inclined anchor piles in sand," Danish Geotechnical Institute, Bulletin No. 18, 1964.

36. Shenton, M.J., "The stresses acting on a driven pile in a cohesionless material," Ph.D. Thesis, University of Nottingham, England, 1965.

37. Sowa, V.A., "Pulling capacity of concrete cast in-situ bored piles, Canadian Geotechnical Journal, Vol. 7, 1970, pp.482-493.

38. Tan, R.H.S., and Hanna, T.H., "Long piles under tensile loads in sand," Geotechnical Engineering, Vol. 5, Part 2, 1974, pp.109-124.

39. Tapasvi, V.T., "Stresses in granular soils induced by skin friction on pile shafts," Ph.D. Thesis, University of Leeds, England, 1966.

40. Tavenas, F.A., "Load test results on friction piles in sand," Canadian Geotechnical Journal, Vol. 8, 1971, pp.7-22.

41. Tejchman, A., "Skin resistance of pushed and pulled piles driven into non-cohesive soil medium," Archiwum Hydrotechniki, Vol. 17, 1970, pp.430-463.

42. Terzaghi, K., "Theoretical Soil Mechanics," Wiley and Co., New York, 1948.

43. Tran-Vo-Nhiem, "Ultimate uplift capacity of anchor piles," Proceedings of the 4th Budapest Conference on Soil Mechanics and Foundation Engineering, 1971, pp.829-836.

44. Vesic, A.S., "Ultimate loads and settlements of deep foundations in sand," Symposium on Bearing Capacity and Settlement of Foundations, Duke University, Durham N.C., 1965, pp.53-68.

45. Vesic, A.S., "Tests on instrumented piles, Ogeechee River site," Journal of the Soil Mechanics and Foundation Division, ASCE, Vol. 96, No. SM2, 1970, pp.561-584.

46. Wiseman, R.J., "Uplift resistance of groups of bulbous piles in sand," M.Sc. Thesis, Nova Scotia Technical College, Canada, 1966.

MODEL TESTS OF PILES IN CALCAREOUS SANDS

by J.F. NAUROY[1] and P. LE TIRANT[2]

INTRODUCTION

Seabed in a great many zones where petroleum activities are now in progress - Arabian Gulf, Australia, Northwest coast of India, Southeast Asia, Mediterranean, Brazil - are covered by sediments containing more than 40% carbonates.

Calcareous sands, which are accumulations of sand-sized pieces of carbonate materials, have very different characteristics depending on their origin, i.e. detrital, chemical or skeletal (shelly, coralline or algal debris). Skeletal calcareous sand grains are often very angular and are characterized by the presence of intraparticle voids. Such sands tend to crush more readily under stress than terrigenous sands. Calcareous sands may be found in cemented or uncemented states.

Recent research has shown that the mechanical behavior of calcareous sands is actually different from that of conventional non-calcareous sands only by the extent of their compressibility (8).

The installation of petroleum structures in this type of soil often entails difficulties.

In calcareous sands that are lightly cemented and highly compressible, it is not rare to observe the free fall of piles for several meters under the effect of their own weights or that of a few hammer blows. On the other hand, for highly cemented layers, pile driving may be ineffective, and drilling may have to be used.

It is difficult to evaluate the bearing capacity of piles driven in calcareous soils. The current practice in designing piles for calcareous sands is often limited to using reduced values of the limiting unit skin friction and the limiting unit end bearing in relation to the values used for piles in non-calcareous sands (6).

Since 1978, the "Association de Recherche en Mécanique des Sols Marins" (ARMSM) has undertaken a vast research program on the behavior of calcareous sediments and the bearing capacity of piles in such materials.

This paper describes a project aimed at linking the bearing-capacity parameters of piles, i.e. skin friction and end bearing capacity, to the compressibility of calcareous sands.

This study is confined to driven, open- or closed-ended cylindrical steel piles in uncemented sands.

---

1. Geotechnical Engr., Institut Français du Pétrole.

2. Geotechnical Project Manager, Institut Français du Pétrole.

1 - CALCAREOUS SAND COMPRESSIBILITY

The compressibility of a calcareous sand is the result of four mechanisms:

- elastic deformation of the material,
- rearrangement of the grains,
- crushing of the grains,
- breaking of cementation bonds (when they exist).

The first two mechanisms are common to all granular materials. The last two are very important for calcareous sands.

1.1 - Grain crushing

The brittleness of the sand is characterized by the changing of its grain size distribution under stress. Crushing magnitude can be expressed in terms of a crushing coefficient, CC, defined by Datta et al. (5) as :

$$CC = \frac{\text{Percentage of particles of the sand, after subjection to stress, finer than } D_{10} \text{ of the original sand}}{\text{Percentage of particles of the original sand finer than } D_{10} \text{ of the original sand}}$$

For skeletal sands, grain crushing appears progressively beginning with low stresses: it increases with intraparticle porosity and with the grain angularity.

Intraparticle porosity n_{intra} is evaluated by a vacuum saturation method of the material developed by IFP (to be published). This porosity is nil for detrital sands and goes up to 20% for some algal sands.

Grain angularity is a parameter that is estimated directly by measuring the characteristic friction angle ϕ'_{ca} of the sand (7). The angle ϕ_{ca} is an intrinsic factor characterizing the interlocking capacity of the soil. Indeed, for sands with highly angular grains, ϕ'_{ca} is higher (40 to 50°) than for sands with rounded grains (30 to 35°).

1.2 - Cementation

The existence of cementation bonds between grains may give calcareous sands a certain "cohesion" which is revealed, for exemple, by an "apparent" overconsolidation of the sand. Cementation is difficult to quantify because it is far from being homogeneous. It may be nodular. Weak cementation may be destroyed by core drilling, thus hiding its existence, although its influence on the installation and behavior of a pile may be real. In-situ measurements will certainly have to be developed.

1.3 - Compressibility index

The compressibility of a sand may be determined using the oedometer

	Sand	Location	% $CaCO_3$	Origin	n_{intra} %
o	CA	Carribean	98	algal	20
⊕	F	Florida Bay, USA	100	shelly	15
*	CH	English Channel	90	shelly	8
■	C2	Brittany, France	83	shelly and coralline	8
□	AG2	Arabian Gulf	96	coralline	-
Δ	C1	English Channel	90	shelly	6
X	AG1	Arabian Gulf	90	detrital	0
.	S	Fontainebleau, France	0	terrigenous	0

Table 1

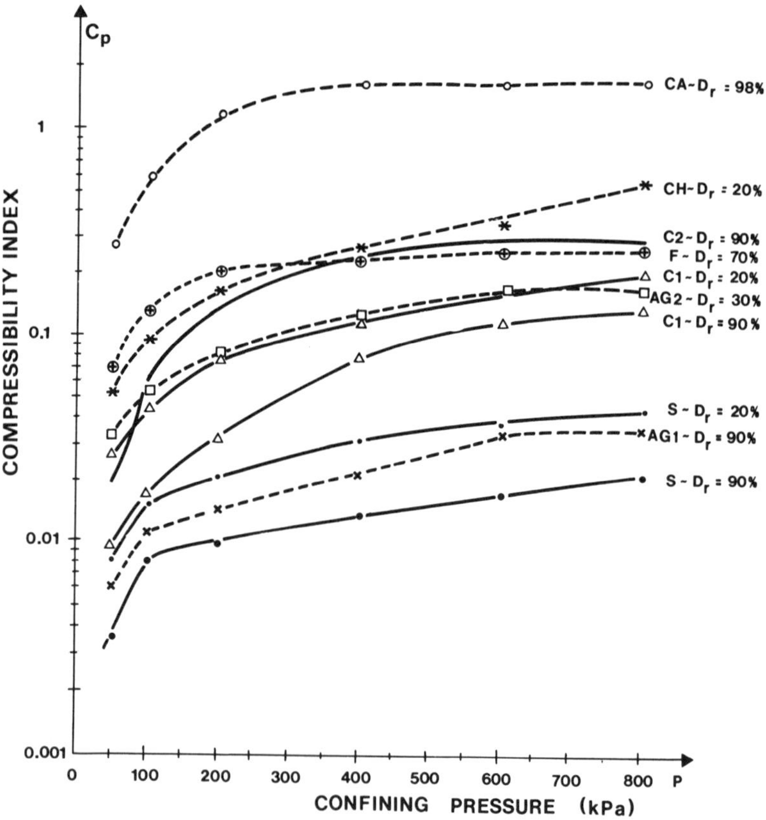

Fig. 1 - Effect of confining pressure on compressibility index for some uncemented calcareous sands.

test or the isotropic triaxial test.

The compressibility index C_p is defined as the slope of the e-log p curve.

C_p is extremely variable depending on the origin of the calcareous sands. Figure 1 illustrates this diversity for several uncemented sands.

Sands S, C_1 and C_2 (unbroken lines) are the sands used in our study. They are described in detail later on. The other sands (dotted lines) come from core samples from different sites throughout the world.

The compressibility index C_p tends toward a limit for high confining pressures. Detrital calcareous sands have a compressibility that is comparable to that of ordinary siliceous sands (S). Skeletal sands, which are generally more brittle, may be 10 to 100 times more compressible. The great influence of the intraparticle porosity can be seen.

2 - MODEL TESTS OF PILES

An important laboratory experiment was performed to determine and compare the bearing characteristics of piles in calcareous and siliceous sands.

2.1 - Testing device

Figure 2 illustrates the set-up of experimental equipment used. The test tank, 65 cm in diameter, was filled with saturated sand up to a height of 40 cm. A rubber membrane was placed on top of the sand pack to ensure isolation between the sand and the upper part of the tank underneath the cover. The sand could thus be confined by applying pressure p in the upper part of the tank: the stress applied to the sand was of oedometric type, just like overburden weight.

The model pile was passed through the upper part of the tank via a cylindrical tube attached to a circular plate with a hole through it. This plate was set in the middle of the sand pack and was linked to the rubber membrane (Fig. 2).

Tests were performed with two types of steel piles 85 cm long:

- a close-ended tubular pile of 2 cm in diameter (27 tests),
- an open-ended tubular pile with an outside diameter of 3 cm and a metal thickness of 1.5 mm.

The model piles were jacked or driven into the sand.

2.2 - Sands tested

Two calcareous sands were used, sands C1 and C2. Sand C1 is a coarse shelly sand with a uniform grain size distribution and very angular grains. It was sampled from the English Channel. Sand C2 is a coralline and shelly sand taken from a quarry in Brittany. It was deposited in the Miocene by a sea which covered western France at that

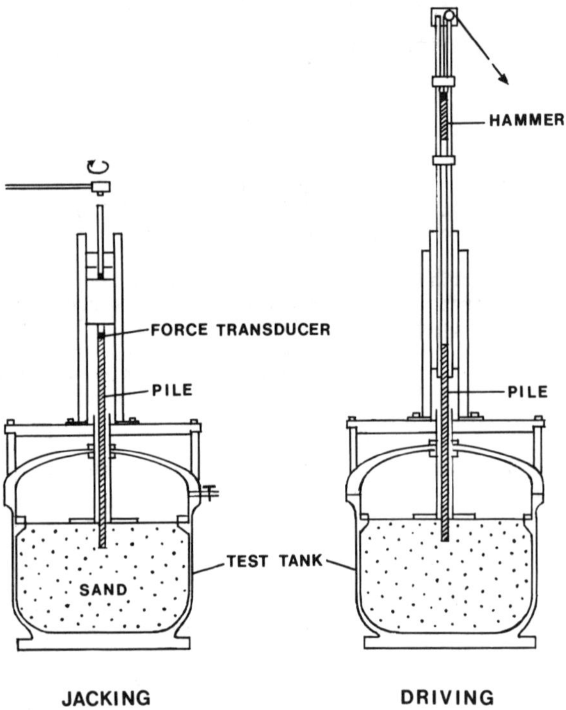

Fig. 2 - Testing Equipment

Sand	% $CaCO_3$	$\gamma_{d\,max}$ kN/m^3	$\gamma_{d\,min}$ kN/m^3	n_{intra} %	ϕ_{ca} (°)
S	0	17,2	14	0	33
C1	90	13,6	10,2	6	40
C2	83	13	9,4	8	39

Table 2

Characteristics of tested sands

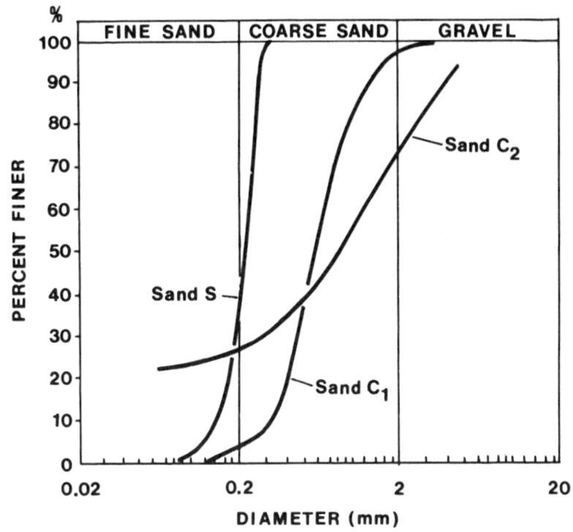

Fig. 3 - Grain size distribution curves of tested sands

time. It was a very brittle sand with a wide grain-size distribution.

The main characteristics of these sands are given in Table 2, and their grain-size distribution curves are plotted in Figure 3.

The compressibility of the sands was evaluated by isotropic compression tests using a triaxial device. The results are given in Figure 1. The extreme values of the compressibility index Cp of sands C1 and C2 are, respectively, about 5 and 15 times higher than those of sand S.

2.3 - Test procedure

Dry sand was poured into the tank so as to obtain a relative density of about 80 to 100%, i.e. 16 kN/m^3 for sand S, 13.1 kN/m^3 for sand C1, and 13 kN/m^3 for sand C2.

The sand pack was then saturated and subjected to a confining pressure of 50, 200 or 400 kPa, depending on the test.

The model piles were inserted either by down jacking (driving speed of about 0.5 mm/min) or by driving (falling weight from a height varying between 10 and 100 cm).

The piles were then pulled out at the rate of 0.5 mm/min. The pull-out procedure with controlled displacement is actually simpler in the laboratory than the incremental procedure used for standard pile tests. Comparative tests were performed.

3 - EXPERIMENTAL RESULTS

The experimental results obtained are perfectly homogeneous with sand characteristics and enable to bring out some important data concerning the bearing capacity of piles.

3.1 - Pile installation

Pile installation disturbs the surrounding soil and changes the stress state characterizing this soil in its virgin state. The normal stress on the pile wall evolve from σ_{h0} which is the horizontal stress in the soil before the pile is thrust in, to a value of σ_{hr} at the end of insertion. Figure 4 gives several recordings made with a total pressure sensor placed in the sand 5 cm away from the axis of the pile and at a depth of 15 cm. The readings of these sensors are solely qualitative because the shape and size of the grains are capable of influencing them.

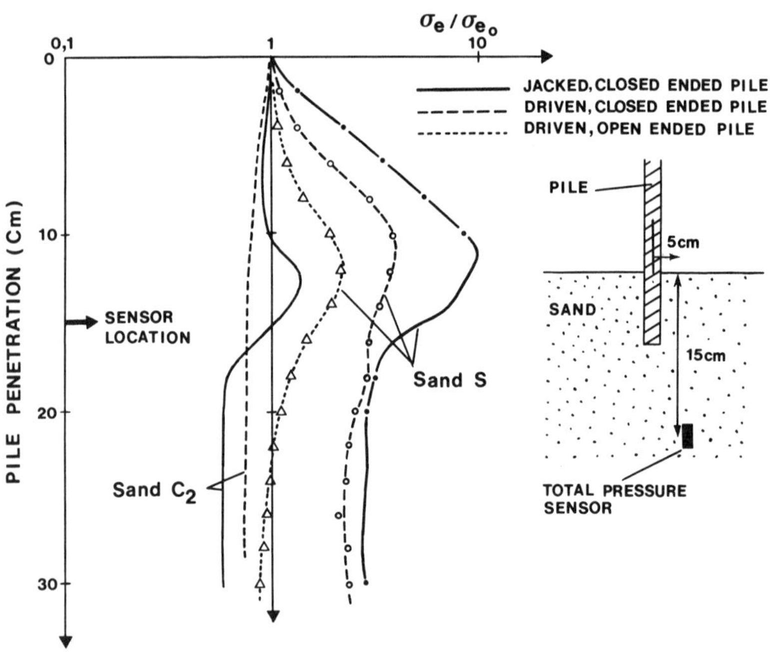

Fig. 4 - Variations of lateral pressure in sand during pile penetration (confining pressure 200 kPa)

3.1.1 - Closed-ended tubular piles

When a closed-ended tubular pile is jacked down into the tank, the pressure recorded σ_e can be seen to go via a maximum $\sigma_{e\ max}$ when the tip of the pile reaches a level a few centimeters above the sensor, and then it decreases and becomes stabilized at a value σ_{ef} different from the initial value σ_{eo} (Fig. 5). At equal confining pressures, the increase in the horizontal stress ($\sigma_{e\ max}/\sigma_{eo}$) is greater with sand S than with sands C1 and C2.

Likewise, it was seen that:

- σ_{ef} is higher than σ_{eo} for siliceous sand S,
- σ_{ef} is lower than σ_{eo} for calcareous sands C1 and C2.

When the closed-ended tubular pile is driven in, the pressure recorded by the sensor does not reach such a high level as with jacking. The maximum even disappears with sand C2.

These results, which confirm the measurements made by several authors with siliceous sands (9),(10),(12), lead to two important qualitative conclusions on the amplitude of stress σ_h along the shaft of the pile:

(i) σ_h is maximum for the tip of the pile and is higher with jacking than with driving.

(ii) σ_{h_r} along the pile depends on the compressibility of the sand.

- σ_{h_r} is higher than σ_{h_o} for the siliceous sand S,
- σ_{h_r} is lower than σ_{h_o} for the more compressible calcareous sands C1 and C2.

When the pile penetrates into the soil, this latter is compressed, deformed and pushed laterally aside. The amplitude of the lateral push obviously depends on the sand compressibility. It is willingly accepted that σ_H variation is linked to the pushing back of the sand, and hence to the compressibility. For the calcareous sands the crushing of the grains is a major phenomenon. Under the great stresses imparted by driving, this crushing is quite extensive. Such grain crushing is thus responsible for a drop in σ_h along the pile and therefore low skin-friction values.

3.1.2 - Open-ended tubular piles

During the driving in of open-ended tubular piles, the evolution of the pressure recorded by the sensor situated in the sand pack is the same as for a closed-ended tubular pile. However, a decrease is observed in the amplitude of the phenomena. And open-ended tubular pile effectively disturbs the soil much less than a closed-ended pile, although the existence of a plug may create conditions similar to those for a closed-ended pile.

The experimental study was limited to uncemented sands. The presence of in-situ cementation even to a slight degree, may help prevent lateral pressure from developing (4) or may even cancel it out.

3.2 - Skin friction

Skin friction during pull-out was measured by the controlled displacement of a pile jacked or driven to a depth of about 30 cm in the sand pack. Unit skin friction increased rapidly before reaching an appreciably constant value (Fig. 5). Pull-out tests performed according to the standard procedure of load increments gave unit skin friction values of the same order of magnitude. Table 3 groups the average values of the results obtained. Several important conclusions can be determined from these measurements.

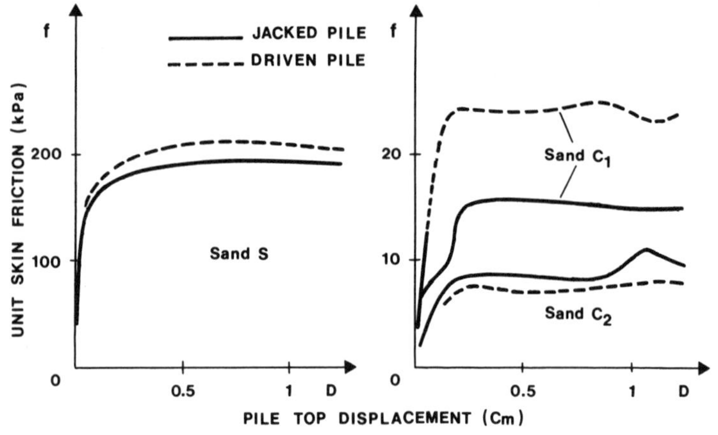

Fig. 5 - Results of pull-out tests (confining pressure = 200 kPa)

Model pile	Installation	Confining pressure	Unit skin friction (kPa) Sand S	Sand C1	Sand C2
Closed ended tubular	Jacking	50	41	16	
"	"	200	194	15	9
"	"	400	300	18	12
"	Driving	50	97	17	7,5
"	"	200	175	19	8
"	"	400	230	20,5	8,2
Open-ended tubular	"	200	108	19,3	2,4

Table 3 - Unit skin friction measured in model tests

3.2.1 - Influence of nature of sand

No matter how the pile is set in place (jacking or driving), the skin friction of the calcareous sands is always less than that of the siliceous sand S. The differences are all the greater as the compressibility of the sand increases.

3.2.2 - Influence of confinement (closed-ended tubular piles)

The skin friction of closed-ended piles generally increased with confining pressure but seems to have an extreme value f_ℓ (Fig. 6) of about 250 to 300 kPa for sand S, 20 kPa for sand C1 and 8.5 kPa for sand C2. Since sand compressibility also tends toward an extreme $C_{p\ell}$ when confining pressure increases (Fig. 1), an attempt can be made to connect f_ℓ and $C_{p\ell}$ (Fig. 7). Tests with other sands are obviously necessary to determine this relationship, but the information gathered is already considerable. Compressibility effectively seems to be the important parameter for designing driven piles.

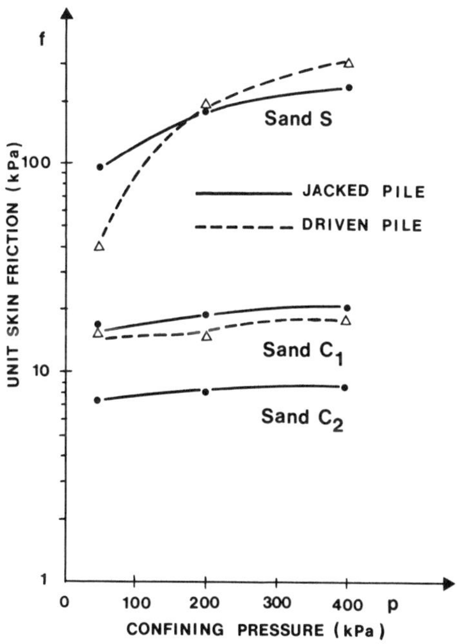

Fig. 6 - Effect of confining pressure on unit skin friction in model tests with closed-ended piles

3.2.3 - Influence of pile type

Tests show that the skin friction of an open-ended pile is less than that of a closed-ended one. The reduction even seems to increase with the compressibility of the sand – 2 times for sand S, and 6 times for sand C2. However, care must be taken not to scale up in-situ these values because the formation of a plug may transform an open-ended pile into a closed-ended one.

To try to quantify these phenomena and to specify the relationship between sand compressibility and skin friction of piles, ARMSM has been carrying out an important in-situ test project since 1982.

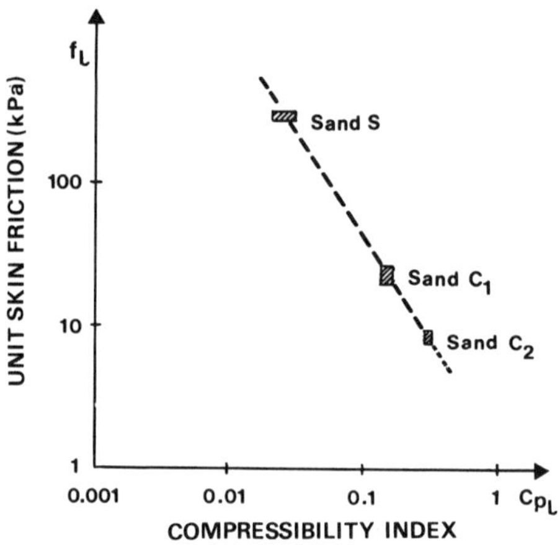

Fig. 7 - Limiting unit skin friction versus limiting compressibility index.

3.3 - End bearing capacity

The end bearing capacity is estimated by substracting the skin friction of the pile measured during pull-out from the total jacking force. This unit end bearing capacity quickly reaches a stabilized value q_s. Table 4 groups the results obtained.

Confining pressure	Unit end bearing capacity (MPa) Sand S	Sand C1	Sand C2
50	9.5	7.1	
200	23	12	3.5
400	37	16.7	

Table 4 - Unit end bearing capacity measured with closed-ended pile jacked in sands.

The end bearing capacity for the calcareous sands is always less than that for sand S. A particularly low end bearing capacity is found in sand C2. Here again we find the preponderant influence of compressibility.

The end bearing capacity also increases with confining pressure and likewise seems to end toward an extreme (Fig. 8).

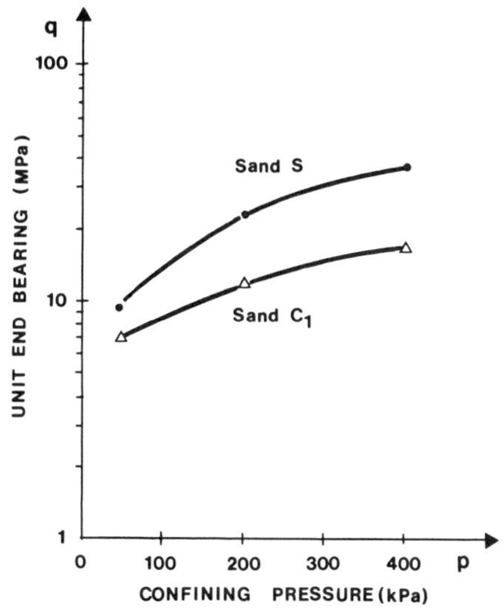

Fig. 8 - Effect of confining pressure on unit end bearing capacity in model tests

CONCLUSIONS

Calcareous sands have highly varying natures and behaviors. Compressibility is the essential parameter for characterizing them. It integrates different effects such as: (i) the physical characteristics of the grains (angularity, intraparticle porosity, brittleness), and (ii) the way the grains are assembled (grain-size distribution, density, cementation.)

Driving or jacking tests of piles in uncemented calcareous sands were performed in the laboratory. The results obtained served to bring out the phenomenology involved in installing piles and to provide the initial elements for determining bearing-capacity parameters from sand characteristics:

(i) Normal stress on the wall of the pile evolves from σ_{h_0}, which is the horizontal stress existing in the soil before penetration of the pile, to a value σ_{h_r} at the end of penetration. This value σ_{h_r}

depends on the compressibility of the sand. For a highly compressible sand, σh_r may be lower than σh_0, which explains the very low values of skin friction commonly observed with calcareous sands.

(ii) Skin friction and end bearing are all the less as the compressibility of the sand is high. There is a relationship between the limiting compressibility index $C_{p\ell}$ and the values of f_ℓ. More numerous tests are required to specify this relationship more accurately

The influence of sand cementation and the role of the plug in open ended tubular piles cannot be clearly evaluated by laboratory tests and thus imply having recourse to in situ pile experimenting. This is the aim of the research that has been undertaken since 1982 by IFP within the framework of "Association de Recherche en Mécanique des Sols Marins".

ACKNOWLEDGEMENTS

Most of the results presented in this paper were part of a research program carried out by IFP and sponsored by the "ASSOCIATION DE RECHERCHE EN MECANIQUE DES SOLS MARINS" (IFP, CNEXO, CFP; SNEA(P), CFEM, CG DORIS, ETPM, BOUYGUES OFFSHORE, SOLMARINE, SPIE OFFSHORE). Permission to release informations on these studies is gratefully acknowledged.

REFERENCES

1. Agarwal S.L., Malhotra A.K. and Banerjee R., "Engineering properties of calcareous soils affecting the design of deep penetration piles for offshore structures", Offshore Technology Conference, Paper 2792, Houston, May, 1977.

2. Agarwal S.L. and Rawat P.C., "Problems encountered in installation of offshore piles for fixed platforms", Offshore Technology Conference, Paper 3274, Houston, May, 1978.

3. Angemeer J., Carlson E.D. and Klick J.H., "Techniques and results of offshore pile load testing in calcareous soils", Offshore Technology Conference, Paper 1894, Houston, May, 1973.

4. Angemeer J., Carlson E.D., Stroud S. and Kurzeme M., "Pile load tests in calcareous soils conducted in 400 feet of water from a semi-submersible exploration rig", Offshore Technology Conference, Paper 2311, Houston, May, 1975.

5. Datta M., Gulhati S.K. and Rao G.V., "Crushing of calcareous sands during shear", Offshore Technology Conference, Paper 3525, Houston, May, 1979.

6. Datta M., Gulhati S.K. and Rao G.V., "An appraisal of the existing practice of determining the axial load capacity of deep penetration piles in calcareous sands", Offshore Technology Conference, Paper 3867, Houston, May, 1980.

7. Luong M.P. and Sidaner J.F., "Undrained behavior of cohesionless soils under cyclic and transient loading". International Conference on Recent Advances in Geotechnical Earthquake Engineering and Soil Dynamics, St Louis, April 1981.

8. Nauroy J.F. and Le Tirant P., "Comportement des sédiments marins carbonatés, Proc. 10th I.C.S.M.F.E, Vol. 3, 265-268, Stockholm, 1981.

9. Panet M., "Etude des fondations sur sable en modèle réduit", Annales I.T.B.T.P., Mars, 1972.

10. Puech A., Boulon M. and Meimon Y., "Tension piles : field data and numerical modelling", Second International Conference on Numerical Methods in Offshore Piling, Austin, April, 1982.

11. Settgast R.H., "Marine pile load testing in carbonate rocks", Offshore Technology Conference, Paper 3868, Houston, May, 1980.

12. Tcheng Y., "Fondations profondes en milieu pulvérulent à diverses compacités", Annales I.T.B.T.P., Mars 1966.

AXIAL CAPACITY OF STEEL PIPE PILES IN CLAY

By Norman D. Dennis, Jr.,[1] M. ASCE, and Roy E. Olson,[2] F. ASCE

INTRODUCTION

Development of theories for the prediction of the axial compressive and tensile load capacities of driven piles is made difficult by a variety of factors, e.g., disturbance of the soil and changes in the state of stress caused by pile driving, dragdown of soil from one layer into another, and pile/soil interaction during loading. In the absence of a well developed theory supported by field evidence, engineers have developed a series of empirical design rules utilizing coefficients which are evaluated using full scale pile load tests. The size of data bases used in these correlations has generally been small and the methods have been applied in cases where the supporting data have been inadequate.

In the terrestrial environment it is generally possible to overcome uncertainties in design by driving piles to refusal or by using pile load tests to verify design assumptions. In the case of piles driven offshore, load tests are prohibitively expensive and the use of excessively conservative design assumptions may lead to installation problems. There is, therefore, a continuing need for improvement in design methods.

The improvements are needed in the complementry areas of improving the fundamental theories and improving the supporting data base and empirical correlations. We have increased the size of the readily available data base, examined the accuracy of several existing design methods, and developed improved correlations which have led to a revised design procedure.

GENERAL EQUATIONS

For vertical piles subject to axial compressive and tensile loads, the pile capacity (Q) may be written as:

$$Q = Q_s + Q_p \pm W_p \qquad (1)$$

1. Department of Engineering, U.S. Military Academy, West Point, NY

2. Professor, Department of Civil Engineering, University of Texas Austin, Texas

where Q_s and Q_p are the loads transferred from pile to soil in side shear and end bearing, respectively, and W_p is the weight of the pile beneath the ground surface (use plus for tension tests and minus sign for compression tests). If Q_p is the net tip capacity then W_p becomes the weight of pile minus the weight of soil displaced, and thus becomes sufficiently small compared with $Q_s + Q_p$ that it is generally ignored. For tension loadings Q_p is assumed to be negligible.

Load transferred in side shear is taken as:

$$Q_s = \int f_s dA_s = \overline{f}_s A_s \tag{2}$$

where f_s and $\overline{f}_s$ represent local and average side shearing stresses and A_s is pile side area in contact with soil.

The tip capacity is taken as:

$$Q_p = q_p\ A_p = 9\ c_u A_p \tag{3}$$

where q_p is the net tip bearing capacity, A_p is the tip area, and c_u is the undrained shearing strength of the clay. For open ended pipe piles the tip capacity was taken as the smaller of: (1) the tip capacity of the wall of the steel pipe plus the side shear capacity of a full-length plug, (2) the tip capacity calculated as if the pile had a cover plate.

CALCULATION OF SIDE SHEAR CAPACITY USING EXISTING METHODS

Analysis of load transfer between a pile and soil, in side shear, involves consideration of soil-structure interaction as suggested by Coyle and Reese (2), with local transfer dependent on the relative movement of the pile and soil. Load transfer is also influenced by such effects as pile whipping during driving, drag-down of soil between layers, preboring and jetting, use of oversized cover plates, pile taper, and time-dependent reconsolidation of the disturbed soil. In the empirical methods of analysis considered here these effects must be included in the empirical factors. The methods considered in this study are summarized below. The names assigned to each method are ones we used in a computer program, and are used here only to simplify discussion.

ALP1 Method

In the ALP1 method (10), the average side shear, $\overline{f}_s$, is calculated using:

$$\overline{f}_s = \overline{c}_a = \alpha\overline{c}_u \tag{4}$$

where $\bar{c}_u$ is the average undrained shearing strength over the embedded length of the pile and α is an empirical coefficient, termed the "adhesion factor," which was made a function of $\bar{c}_u$ and pile material. The relationship between unit adhesion ($\bar{c}_a$) and undrained shearing strength used by Tomlinson (10) is illustrated in Fig. 1.

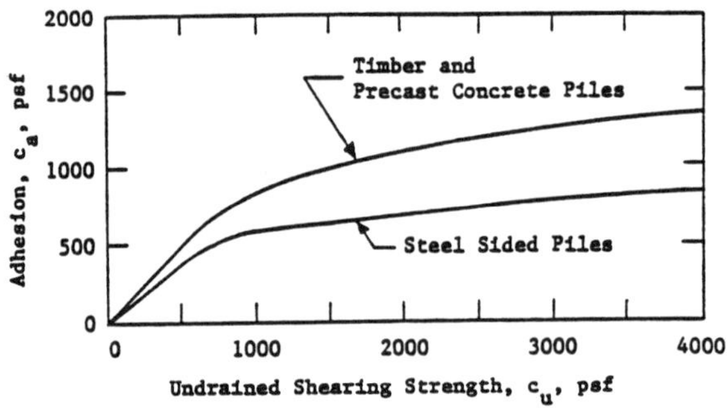

Fig. 1 Relationship between unit adhesion and undrained shearing strength after Tomlinson (10)

ALP2 Method

Based in part on field evidence obtained by excavating soil from around piles, Tomlinson (11) concluded that soil is dragged down from an overlying layer into the interface between the pile and soil in the next layer. He considered piles in three profiles: I, sand over "stiff clay:" II, soft clay over stiff clay, and III, stiff clay. He used Eq. 4 to calculate load transfer in the stiff clay, and made α a function of soil profile, average undrained strength in the stiff clay, and penetration of the pile into the stiff clay (Fig. 2). Tomlinson (11) calculated load transfer in an upper sand layer using Nordlund's (7) method and in an upper soft clay using $\alpha = 1$ and Eq. 4. He concluded that pile material exerted a comparatively minor effect on side shear and used data from 93 load tests on timber, precast concrete, cast-in-place concrete, and steel pipe piles in his correlations. In our application of the ALP2 method, clays were considered "soft" if c_u was less than 500 psf, and we used the API (1) method for overlying sand layers.

APIC Method

The APIC method is essentially the American Petroleum Institute (1) method. The side shear capacity in clay, Q_s, is calculated using:

$$Q_s = \sum_{\ell=1}^{N} \bar{f}_{s\ell} \, \Delta A_{s\ell} = \sum_{\ell=1}^{N} \alpha_\ell \, c_{u\ell} \, \Delta A_{s\ell} \qquad (5)$$

where $\overline{f}_{s\ell}$ is the average pile-soil shearing strength, $\Delta A_{s\ell}$ is the side area, α_{ℓ} is the adhesion factor, and $c_{u\ell}$ is the average undrained shearing strength of the clay, all in layer ℓ, and N is the number of layers penetrated by the pile.

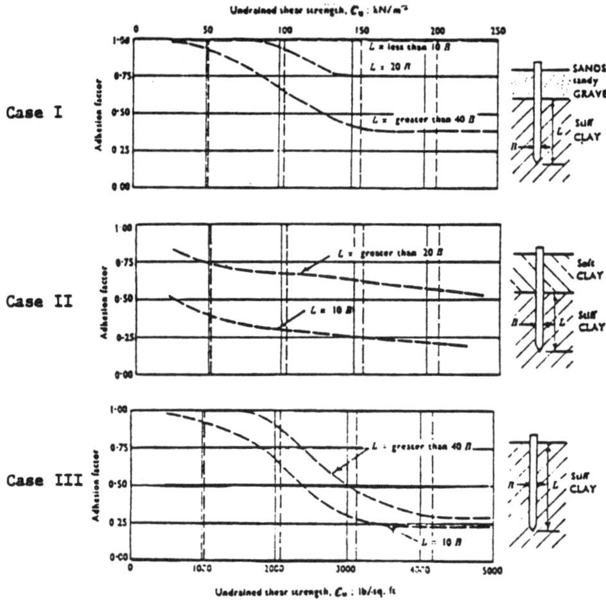

Fig. 2 Adhesion Factors Recommended by Tomlinson (11) and used in the ALP2 Method.

In the APIC method, clays are divided into two categories. Category 1 clays, here termed GULF clays (following the symbolism used in the computer program), are highly plastic clays, here including all clays with liquid limits and plasticity indices exceeding 50% and 35%, respectively. Category 2 clays are all other clays, here termed OTHR clays.

For normally consolidated and underconsolidated GULF clays, α is taken as one. For overconsolidated GULF clays α is taken as one but $\overline{f}_s$ is limited to the larger of 1000 psf or the c_u of a normally consolidated clay, whichever is larger.

For OTHR clays, α is 1.0 for $c_u \leq 500$ psf, 0.5 for $c_u \geq 1500$ psf, and varies linearly between those limits.

LAM1 Method

In the LAM1 method the side shear capacity is (12):

$$Q_s = \lambda(\sigma_m + 2c_m)A_s \tag{6}$$

where σ_m and c_m are the mean value of effective overburden pressure and undrained shearing strength, over the embedded length of the pile, and λ is an empirical factor which was found to depend on pile penetration (Fig. 3).

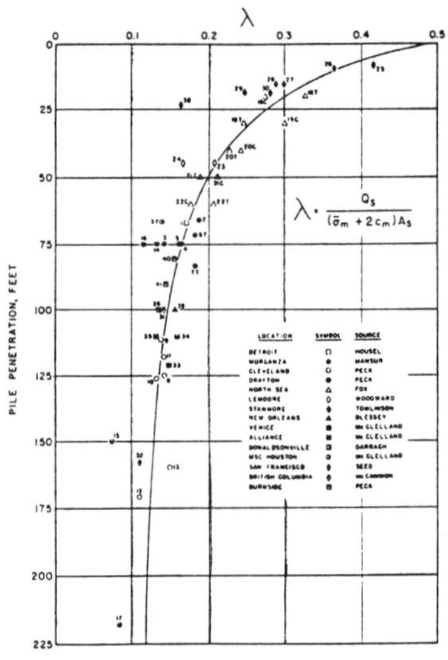

Fig. 3 Values of Lambda recommended by Vijayvergiya and Focht (12)

LAM2 Method

In the LAM2 method Eq. 6 is used but λ is given as (5):

$$\lambda = 0.178 - 0.016\ \ell n\ \pi_3 \tag{7a}$$

for normally consolidated soils and:

$$\lambda = 0.232 - 0.032\ \ell n\ \pi_3 \tag{7b}$$

for overconsolidated soils where:

$$\pi_3 = \pi B f_{max} L_e / AEu \tag{8}$$

and B = pile diameter, f_{max} is the peak pile-soil friction (taken as the mean undrained shearing strength), L_e is the embedded pile length, A is the cross-sectional area of pile material of modulus E, and u is the pile displacement needed to develop the side shear (taken as 0.1 inch). In the absense of consolidation data, Kraft et al. (5) considered a clay to be overconsolidated if $c/p \geq 0.4$. They set a lower limit of 0.14 on λ.

DATA BASE

A critical check on the accuracy of the predictive methods is a comparison of predicted (Q_c) and measured (Q_m) pile capacities using as large a data base as possible or, for a specific design application, a data base with conditions closely approximating those in the design problem. For this study, a data base of 1004 tests was developed using test results from the literature and unpublished data from government agencies, oil companies, and consulting firms.

Only data for tests carried essentially to plunging failure were used. Data used in this report were at a "defined" failure point, as opposed to plunging failure, because some piles appear not to reach plunging failure even at large displacements. Davisson's criterion (8, p. 215) was used to define the point of "failure."

The quality of collected data ranged widely. In the best cases, piles were fully instrumented, soil borings were near the test pile, soil properties were relatively uniform, soil properties were well defined (many measurements, reliable testing technique) and there seemed to be a reasonable consolidation period between pile driving and testing. For tests deemed good enough to be included in the data bank (about one test out of eight for which we had data), the quality of a test was denoted by an arbitrary "data quality factor" which ranged from 1 for the poorest acceptable test to 5 for the best tests. In the analyses reported here, only tests with data quality factors in the range of 2 to 5 were used.

The "strength" of the clay at a given site was measured, using a wide variety of methods, ranging from sophisticated laboratory and field tests to standard penetration tests. The most common measure was unconfined compression tests on 2-inch diameter thin-walled tube samples but Q-type and R-type triaxial compression tests and field vane tests were also often used. Lab vanes, Swedish fall cone, double ring shear, and other tests were also encountered. Reported strengths were used directly in analyses even though it is well known that they yield different values. In the case of the ALP1 and ALP2 methods the proper strength to use in the analysis would be the strength used in developing the original correlations (of the 93 tests used in the ALP2 correlations, 33 utilized unconfined compression, 11 field vane, and the technique was unknown for the other tests). Unconfined compression and miniature vane tests are specified in the API procedure and the engineer's interpreted strength is to be used in the LAM1 and LAM2 methods.

The data set used in this investigation was obtained by deleting from the master set of 1004 load tests all data for piles other than steel pipe piles, for piles with oversized cover plates or piles in prebored or jetted holes, tests with a data quality factor of one, tests where a load-settlement curve was unavailable, and tests where in situ vertical effective stresses could not be estimated with acceptable accuracy. The axial load capacities of the remaining piles were calculated using the methods discussed in this paper for clay layers and the API method (1) for sand layers, and tests were rejected if either the pile tip was in sand or a major part of the side capacity came from sand.

The final data set contained 84 load tests. Of these, 57 were entirely in clay, 11 derived 1-10% of their capacity in sand, 15 derived 10-20% of their capacity in sand, and one derived 29%, averaging 4% in compression tests and 3% in tension tests (based on analyses using the API methods in clay and sand). Of the 84 tests, 27 were loaded in tension and 57 were in compression. For the piles in compression, the calculated tip capacity was less than 10% of the calculated total capacity in 43 tests, 11-20% in 8 tests, 21-30% in 3 tests, and 31-40% in 3 tests, averaging 9%.

For the final data set, the average ratio of the peak reported load (often the plunging load) to the defined failure load was 1.06 with a range from 0.93 (3 tests had the load-settlement curve extrapolated to reach a defined failure) to 1.75.

It should be emphasized that calculated capacities are to be compared with "measured" capacities that are potentially in error. Measured capacities may be too high because of jack friction when loads are determined from hydraulic pressure in the jack and may be too low because of inadequate set-up time and because the long term static capacity may exceed the short term capacity measured in a load test, particulary for the pile tip. The measured capacity is influenced by the choice of failure criterion. Some of the measured capacities are potentially in error due to blunders, e.g., improper conversion of units, typographical errors, or use of incorrect calibration curves.

Q_c - Q_m COMPARISONS USING EXISTING METHODS

To accomodate load tests on piles of disparate sizes in a single diagram, Q_c and Q_m were plotted on log scales (Fig. 4). Lines were plotted with $Q_c/Q_m = \frac{1}{2}$ and 2 to help indicate scatter. The computer was programmed to prepare frequency diagrams for the ratio Q_c/Q_m on a log scale so that tests where say $Q_c = 2\ Q_m$ and $Q_c = \frac{1}{2}\ Q_m$ would plot the same distance from the $Q_c = Q_m$ point. For any test, a "factor of

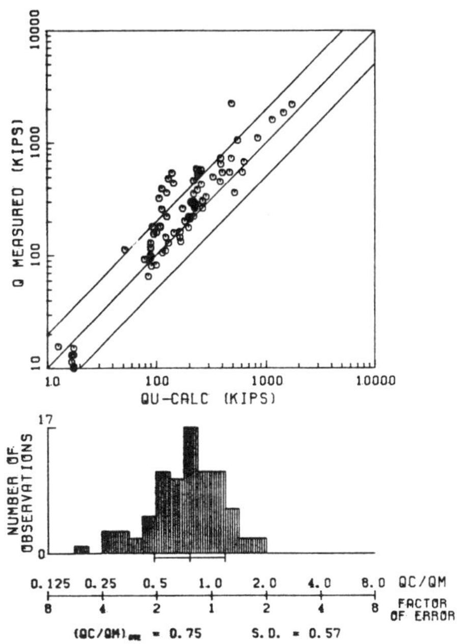

Fig. 4 Q_c - Q_m data, Method ALP1 (10)

error" was defined as Q_c/Q_m or Q_m/Q_c, whichever exceeded one. The mean value of Q_c/Q_m was calculated as:

$$(Q_c/Q_m)_{ave} = 10^{**}[\frac{1}{N} \sum_{j=1}^{N} \log (Q_c/Q_m)_j] \quad (9)$$

where ** denotes exponentiation, N is the number of tests, and $(Q_c/Q_m)_j$ is any one observation. The standard deviation, SD, was calculated as:

$$SD = 10^{**}\{ \frac{1}{N-1} \sum_{j=1}^{N} [\log (\frac{Q_c}{Q_m})_{ave} - \log (\frac{Q_c}{Q_m})_j]^2 \}^{\frac{1}{2}} - 1 \quad (10)$$

Comparison of Methods

Measured and computed capacities are compared in Figs. 4-8 for the ALP1, ALP2, LAM1, LAM2, and APIC methods, respectively. Statistical data are presented in Table 1.

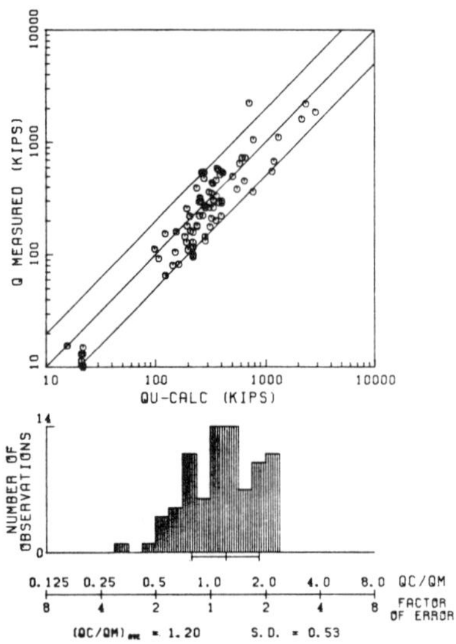

Fig. 5 Q_c - Q_m Data, ALP2 Method (10)

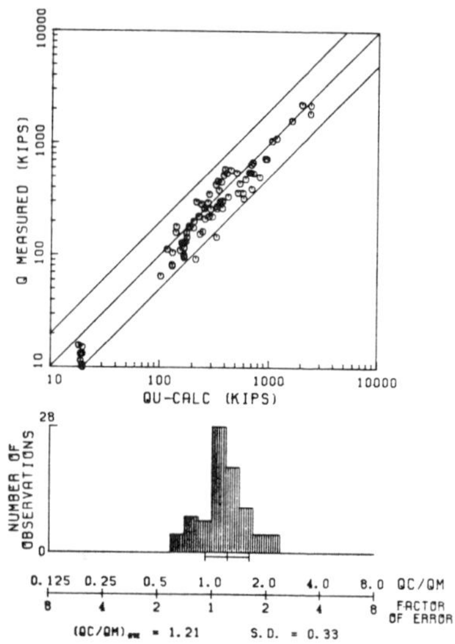

Fig. 6 Q_c - Q_m Data, LAM1 Method (12)

The ALP1 method leads to a significant underprediction of pile capacities for steel pipe piles and has the largest scatter. The ALP2 method leads to overprediction of capacities and a similar scatter. For both methods it was noticeable that piles with the largest values of Q_c/Q_m involved use of field vane shear tests or laboratory tests using samples of high quality (trimmed from 3-inch diameter, or larger, samples taken by pushing thin walled samplers) whereas tests with low values of Q_c/Q_m were often for cases where laboratory tests were performed on samples taken using smaller diameter, driven samplers. Data for tensile and compressive pile load tests were not significantly different and the existing differences were explainable in terms of sample and test quality.

Both lambda methods led to overprediction of pile capacities, with the LAM2 method being slightly better. The scatter in predicted capacities was much less for the lambda methods than for the two alpha methods. One reason for the reduced scatter in the lambda methods is that the term $(\sigma_m + 2c_m)$ in Eq. 6 is less effected by scatter in values of c_u than is the $\bar{c}_u$ term in the alpha methods (Eq. 4). Examination of the figures shows that the lambda methods are significantly more accurate for piles with capacities beyond 1000 kips, and none of those piles were in the original lambda

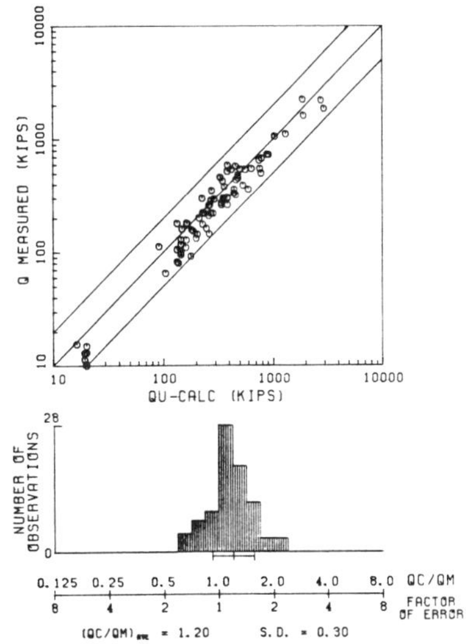

Fig. 7 Q_c - Q_m Data, LAM2 Method (5)

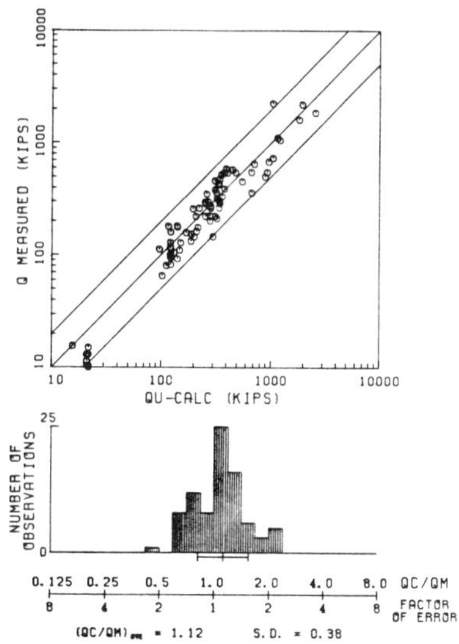

Fig. 8 Q_c - Q_m Data, APIC Method (1)

correlation so they constitute a valid test of the method.

The APIC method tended to overpredict capacities, but by less than did the lambda methods. However, there was more scatter in predictions, with the scatter again tending to correlate with sampling and testing quality.

DEVELOPMENT OF IMPROVED CORRELATIONS

Attempts could be made to improve correlations between measured and computed pile capacities using any of the existing formulations (alpha methods, beta methods, lambda methods) or new ones. Because of its simplicity, the alpha method was selected.

The data show an important effect of sampling and testing quality and thus indicate that a standard sampling and testing procedure should be used. We have chosen to define the standard as unconsolidated - undrained triaxial compression tests, performed using a confining pressure equal to the total overburden pressure, on samples taken using pushed thin-walled samplers, 3 inches in

Table 1 Statistical Summary of Analyses, using Existing Predictive Methods.

Method	Loading	N	Regression Line a	Regression Line b kips	Regression Line r	SEE kips	Q_c/Q_m Data Min	Q_c/Q_m Data Max	Q_c/Q_m Data Avg	Q_c/Q_m Data SD
ALP1	T/C	84	1.30	400	.88	209	0.22	1.86	0.75	0.57
	COMP	57	1.30	450	.96	128	0.26	1.75	0.74	0.48
	TENS	27	3.00	-160	.80	260	0.22	1.86	0.79	0.75
ALP2	C/T	84	0.78	62	.85	231	0.32	2.32	1.20	0.53
	COMP	57	0.72	76	.93	161	0.49	2.17	1.13	0.45
	TENS	27	2.10	-240	.81	255	0.32	2.32	1.36	0.66
APIC	C/T	84	0.90	25	.90	186	0.46	2.32	1.12	0.38
	COMP	57	0.84	43	.95	131	0.65	2.17	1.07	0.37
	TENS	27	1.20	-54	.81	254	0.46	2.32	1.22	0.40
LAM1	C/T	84	0.91	-8	.97	103	0.66	2.30	1.21	0.33
	COMP	57	0.87	27	.98	92	0.66	2.30	1.11	0.32
	TENS	27	1.00	-75	.97	108	0.89	2.10	1.45	0.22
LAM2	C/T	84	0.81	30	.96	123	0.65	2.14	1.20	0.30
	COMP	57	0.74	68	.97	102	0.65	2.00	1.12	0.29
	TENS	27	1.10	-79	.97	100	0.81	2.14	1.38	0.25

diameter or larger. No suggestion is made here that the standard procedure be followed. However, if a different procedure is used, then the strengths should be adjusted to the standard strength.

To begin the process of improving the correlations, values of α were obtained from load tests on steel pipe piles by subtracting the calculated tip capacity (Eq. 3) from the measured pile capacity and dividing by the side area and average uncorrected undrained shearing strength to obtain a value for α. Because the most common method of estimating shearing strengths was unconfined compression tests, only load tests where unconfined compression tests were available were used. The resulting (first iteration) values of α are plotted versus c_u in Fig. 9, where solid symbols denote cases where samplers were generally 2¼-inch diameter or smaller, were driven, and often had large area ratios (often wire line samplers), and the hollow symbols denote cases where a 3-inch diameter, pushed, thin walled sampler, or better, was used.

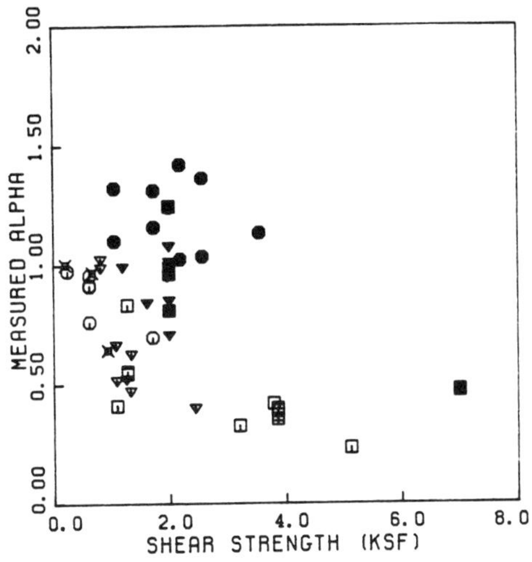

Fig. 9 Developed α Versus $\bar{c}_u$, with c_u Measured using Unconfined Compression Tests. The hollow symbols denote tests where the soil sampler was a 3-inch-diameter Shelby tube or better.

To increase the size of the data bank for tests with the better quality samplers, values of α were also determined for tests on untapered precast concrete piles. All values of α, for the better quality samples, were plotted versus c_u (Fig. 10) and a first iteration curve of α versus c_u was plotted. The present API curve for OTHR clays is shown for comparison.

The adhesion factors from Fig. 10 were then applied to the load tests on steel pipe piles for which strengths were measured using either laboratory tests (unconfined compression, triaxial compression) on high quality samples or in situ tests (vane, pressuremeter) and the ratio of Q_c/Q_m was plotted versus pile penetration (Fig. 11). Capacities seemed to be predicted tolerably well for piles of up to 100-foot penetration but the capacity of longer piles was underpredicted. The line in Fig. 11 could be used as a correction factor for pile length. The length effect is probably the result of interaction of a number of factors which cannot be separated using the data now available. Most of the short piles are in stiff clay where lateral movements of the piles during driving pushes the clay away and leads to reduced pile/clay contact stresses (10) and thus to reduced values of Q_m. All seven tests for piles with penetrations exceeding 250 feet were for piles in normally consolidated or lightly overconsolidated clays where good pile/clay contact is likely.

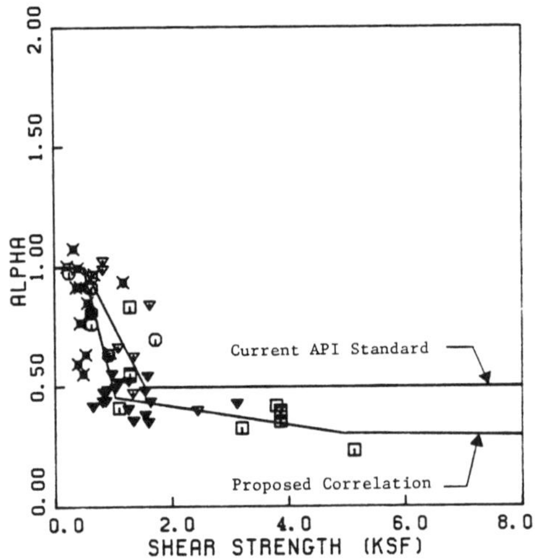

Fig. 10 Proposed Relationship Between α and c_u.

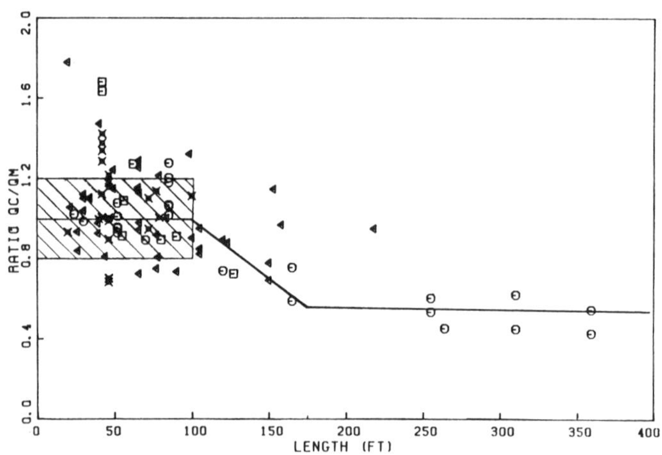

Fig. 11 Effect of Pile Penetration on the Ratio of Calculated to Measured Pile Capacity

Other effects that may have influenced the relationship include load shedding due to pile compressibility (2, 3, 6, 9), and dragdown of soil from one layer into the next (11). It is not clear, therefore, that the length correction shown in Fig. 11 has general applicability. It may also be noted that use of a dimensionless depth, e.g., the ratio of pile penetration to pile diameter, led to an appreciably poorer correlation.

Based on data shown in Fig. 9, and a point-by-point study of data in Fig. 11, it became apparent that the predictions would be improved by correcting the strengths to a standard value. It was found that on 13 sites where pile load tests were performed, the shearing strengths of the clays were measured using two or more techniques, thus allowing a limited amount of cross correlation. For these sites, the most often used measure of shearing strength was the unconfined compression test so other strengths were initially expressed in terms of the unconfined compressive strengths. On the average, the ratio of the unconfined compressive strength on high quality samples to another strength was 0.6 for field vane tests, 0.9 for Q-type triaxial compression tests on samples of high quality and 1.65 for laboratory unconfined compressive strengths of samples taken with driven samplers. Although the number of observations was small and the correction factors probably depend on various factors (plasticity, sensitivity, overconsolidation ratio, ...), the last factor is in agreement with a value reported by Emrich (4).

A revised method, called NCL1 (for new clay method 1), is now applied as follows. Calculate the tip capacity as:

$$Q_p = 9\, c_u F_c A_p \tag{11}$$

and the side capacity as:

$$Q_s = \alpha \bar{c}_u F_C F_L A_S \tag{12}$$

where c_u and $\bar{c}_u$ are the measured shearing strengths near the pile tip and averaged over the length of the pile, respectively, A_p and A_s are the areas of the pile tip and side, respectively, α is the adhesion factor, and F_c and F_L are correction factors for soil strength and pile penetration, respectively. The adhesion factor varies linearly as follows:

$\bar{c}_u F_c$ (psf)	0	600	1200	5000	∞
α	1.0	1.0	0.5	0.3	0.3

The correction factor for strength, F_c, is the ratio of the shearing strength determined using unconsolidated-undrained triaxial compression tests on samples of high quality to the shearing strength measured using some other technique. Values for F_c should be obtained from local experience. In the absence of such data, the engineer may use values found from the correlations reported earlier. These values would be 1.1 for unconfined compression tests on samples of high quality, 1.8 for unconfined compression tests on samples taken with typical driven samplers, and 0.7 for in situ vane shear tests. It should be noted that these values of F_c are based on a very small data set and do not show some well known effects, e.g., F_c should be larger for unconfined compression tests on stiff fissured and soft gaseous samples.

The correction factor for pile penetration, F_L, is:

L (feet)	0	100	175	∞
F_L	1.0	1.0	1.8	1.8

It should be re-emphasized that the data for long piles are for normally consolidated or lightly overconsolidated clays and may not be applicable for other soil conditions.

When the method was applied to open ended pipe piles, the side shear capacity was calculated as discussed above. The tip capacity was taken as the smaller of: (1) the tip capacity of a pile with a cover plate, and (2) the sum of the tip capacity of the base of the hollow pile and the side shear capacity of a full plug.

The NCL1 method was then applied to the data in the same data set used to study the existing methods. The method led to reasonably good correlations (Fig. 12) with a mean Q_c/Q_m of 1.0 and a scatter significantly smaller than found with the other methods. Statistical data are summarized in Table 2.

DISCUSSION

The recommended method is easy to apply and yields calculated capacities that compare well with measured capacities in the data bank used in this study. However, the method is empirical and is based on a relatively small number of tests. Developed pile capacities depend on a number of conflicting factors. The scatter in the values of Q_c/Q_m, found in this study, is unlikely to show the full scatter from all combinations of the conflicting errors. The corrections for method of measuring undrained shearing strength, and for pile length effects, are particularly suspect because of the small size of the relevant data bank. A long-term goal should be the development of rational methods of design.

Finally, it must be emphasized that the data set used in this study specifically excluded tapered piles, piles with oversized cover plates, and piles in prebored or jetted holes.

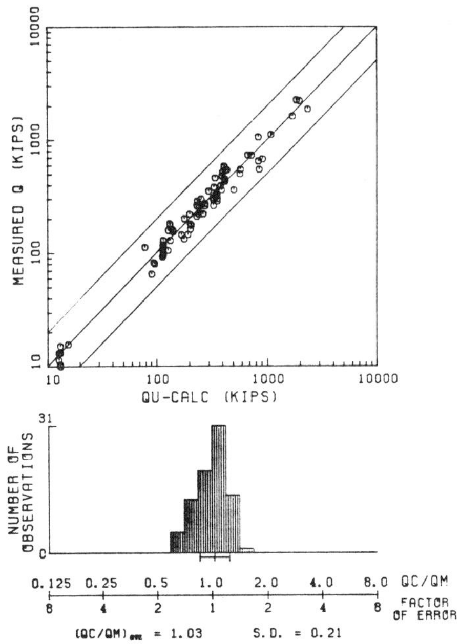

Fig. 12 Q_c - Q_m Data, Method NCL1

Table 2 Statistical Summary of Analyses using Method NCL1

Method	Loading	N	Regression Line a	Regression Line b kips	Regression Line r	SEE kips	Q_c/Q_m Data Min	Q_c/Q_m Data Max	Q_c/Q_m Data Avg	Q_c/Q_m Data SD
NCL1	Tens/Comp	84	0.98	6	0.97	104	0.70	1.55	1.03	0.21
NSA1	COMP	57	0.92	31	0.97	108	0.70	1.55	1.00	0.22
	TENS	27	1.20	-50	0.99	50	0.70	1.65	1.10	0.20

SUMMARY AND CONCLUSIONS

A data base of 84 pile load tests was developed for steel pipe piles, driven in clay (small amounts of sand near the surface for a few tests), where there was sufficient data to estimate effective overburden pressures with reasonable accuracy, where soil strengths were well defined, and where a load-settlement curve was available for the load test. Measured pile capacities were compared with values predicted using five methods that have been, are being, or potentially may be used in engineering practice. The two lambda methods had the least scatter but tended to overpredict capacities by about 20%. The current API method had slightly more scatter but overpredicted capacities by only about 13% on the average. The two alpha methods had the largest scatter and the greatest differences between measured and predicted capacities.

A major cause of scatter in prediction, for all five methods, was the different shearing strengths obtained using various methods of sampling and testing. To reduce the scatter, the analyses should utilize a standard strength which we have here taken as the unconsolidated-undrained triaxial compressive shearing strength measured on samples of high quality, preferably samples trimmed from ones taken using a pushed thin walled sampler of diameter 3 inches or more. An alpha-type method was developed using standardized strengths, a new α-c_u relationship, and a length correction factor. The resulting design method yielded an average ratio of calculated to measured capacities near one and the least scatter of any of the methods examined here.

ACKNOWLEDGMENTS

A major part of this research was performed with financial support from the American Petroleum Institute. The project advisory committee consisted of J.D. Allen (ARCO), D.D. Engle (Union Oil), L.M. Kraft, Jr. (McClelland Engineers), J.D. Murff (Exxon), B.S. Murphy (Amoco), J.R. Ruser (Chairman, Shell), G.E. Strickland (Chevron), and R.L. Thomas (Phillips). The guidance from the committee during the project and the permission from the American Petroleum Institute to publish the results are appreciated.

A major part of the literature survey and much of the early work on the computer program was carried out by D.G. Winter (Hart, Crowser, and Assoc.). Finally, we acknowledge the helpful critiques offered by the reviewers of the manuscript.

REFERENCES

1. American Petroleum Institute (1981), API Recommended Practice for Planning, Designing, and Constructing Fixed Offshore Platforms, API, Production Division, 211 North Ervay, Suite 1700, Dallas, Texas 75201.
2. Coyle, H. M., and L. C. Reese (1966), "Load Transfer for Axially Loaded Piles in Clay," Journ., Soil Mech. and Found. Div., ASCE, Vol. 92, SM2, pp. 1-26.
3. D'Appolonia, E., and I. P. Romualdi (1963), "Load Transfer in End-Bearing Steel H-Piles," Jour., Soil Mech. and Found. Div., ASCE, Vol. 89, No. SM2, pp. 1-25.

4. Emrich, W. J. (1970), "Performance Study of a Soil Sampler for Deep-Penetration Marine Borings, "Sampling of Soil and Rock, ASTM STP 483, pp. 30-50.
5. Kraft, L. M., J. A. Focht, Jr., and S. F. Amerasinghe (1981), "Friction Capacity of Piles Driven into Clay," Journ. Geot. Engr. Div., ASCE, Vol. 107, No. GT11, pp. 1521-1541.
6. Kraft, L. M., R. P. Ray, and T. Kagawa (1981), "Theoretical T-Z Curves," Jour., Geot. Engr. Div., ASCE, Vol. 107, No. GT11, pp. 1543-1561.
7. Nordlund, R. L. (1983), "Bearing Capacity of Piles in Cohesionless Soils," Journ. Soil Mech. and Found. Div., ASCE, Vol. 89, No. SM3, Paper 3506, pp. 1-35.
8. Peck, R. B., W. E. Hanson, and T. H. Thornburn (1974), Foundation Engineering, John Wiley and Sons, New York, NY.
9. Randolph, M. F., and C. P. Wroth (1978), "Analyses of Deformation of Vertically Loaded Piles," Jour., Geot. Engr. Div., ASCE, Vol. 104, No. GT12, pp. 1465-1488.
10. Tomlinson, M. J. (1957), "The Adhesion of Piles in Clay Soils," Proc. Fourth Intern. Conf. on Soil Mech. and Found. Engr., Vol. 2, pp. 66-71.
11. Tomlinson, M. J. (1971), "Some Effects of Pile Driving on Skin Friction," Behavior of Piles, Instn. of Civil Engrs., London, pp. 107-114. Response to Discussions on pp. 149-152.
12. Vijayvergiya, V. N., and J. A. Focht, Jr. (1972), "A New Way to Predict the Capacity of Piles in Clay," Offshore Technology Conference, Houston, Texas, Vol. 2, pp. 865-874.

LIST OF SYMBOLS

A	area (L^2)
a	slope of the linear regression line
A_p	total end area of the pile tip (L^2)
A_s	area of the pile side in contact with soil (L^2)
B	pile tip diameter (L)
$\underline{b}$	intercept of the regression line (F)
c_a	average pile soil adhesion along the pile side (F/L^2)
COMP	compressive load test
$\underline{c}_u$	local undrained shearing strength (F/L^2)
c_u	average undrained shearing strength along the pile side (F/L^2)
E	Youngs Modulus (F/L^2)
F_c	correction factor for shearing strength
f_{max}	peak pile soil friction (F/L^2)
F_L	correction factor for pile penetration
f_s	local unit shear at any point along the pile
L	total pile length (L)
Le	embedded pile length (L)
N	number of observations
q_p	unit tip resistance (F/L^2)
Q_c	calculated pile capacity (F)
Q_m	measured pile capacity (F)
Q_p	pile tip capacity (F)
Q_s	pile side capacity (F)

r	correlation coefficient
SD	standard deviation of Q_c/Q_m
SEE	standard error of the estimate (F)
T/C	tension and compression tests combined
TENS	tensile load test
u	displacement (L)
w_p	submerged pile weight (F)
α	adhesion factor, relating pile-soil adhesion to undrained shearing strength
λ	adhesion factor, relating pile-soil adhesion to undrained shearing strength and freefield effective overburden pressure.
π_3	pile-soil stiffness factor
σ_m	mean value of the free field effective overburden pressure (F/L^2)

AXIAL CAPACITY OF STEEL PIPE PILES IN SAND

By Norman D. Dennis, Jr.,[1] M. ASCE, and Roy E. Olson,[2] F. ASCE

INTRODUCTION

Prediction of the axial load capacity of driven piles in cohesionless soils is made difficult by a number of factors, including the problems associated with sampling and testing the soil, the large changes in stress caused by pile driving, and the nonhomogeneity of many deposits of cohesionless soil. As a result, engineers have resorted to relatively crude field tests, e.g., the standard penetration test (5), or simply to visual classification (1), in many cases. These methods contain factors that are evaluated from field load tests on full scale piles; their accuracy therefore hinges on the size and diversity of the data base. We have collected data for load tests on a variety of pile types in soil profiles ranging from soft clay to gravel. We are here concerned with pile load tests on steel pipe piles in cohesionless soils where design is based on visual classification only. The discussion is mainly applicable to offshore piling because the difficulties associated with laboratory and in situ testing of offshore cohesionless soils has led to the common use of a visual classification method.

APIS METHOD

The method in most widespread use for predicting pile capacities in cohesionless soils, based on visual classification of the soil, is the method recommended by the American Petroleum Institute (1). A slightly amplified version of this method is here referred to as the APIS method. The unit tip capacity (q_p) of a pile is calculated using:

$$q_p = \bar{\sigma}_{vo} N_q \tag{1}$$

where $\bar{\sigma}_{vo}$ is the free field vertical overburden pressure at the elevation of the pile tip, and N_q is a bearing capacity factor, which is assigned values of 8, 12, 20, and 40 for silt, sandy silt, silty sand,

1. Department of Engineering, U.S. Military Academy, West Point, NY

2. Professor, Department of Civil Engineering, University of Texas Austin, Texas

and sand, respectively. The above values are intended for use with medium dense to dense siliceous soils but no values are recommended for less dense, or carbonaceous, soils.

The local side shear (f_s) between the pile and soil at failure is taken as:

$$f_s = K \bar{\sigma}_{vo} \tan \delta \tag{2}$$

where K is the earth pressure coefficient and δ is the angle of friction between the pile and the soil. The value of K is specified (1) to lie between 0.5 and 1.0. Values of K used in this investigation, and specified for the APIS method, are shown in Table 1. The

Table 1 Values of K used in the APIS Method

Pile Type	Values of K Compression	Tension
Full Displacement	1.0	0.7
Partial Displacement	0.7	0.5

assigned values of δ (1) are 15, 20, 25, and 30 degrees for silt, sandy silt, silty sand, and sand, respectively. Again, these values apply to medium dense to dense, siliceous soils and no values are recommended for other cases.

Limiting values of q_p and f_s are often used in engineering practice. API (1) recommends use of limits set in accord with local experience, but does not recommend specific values. No limits were used in the APIS method.

DATA BASE

Through a survey of the literature, and correspondence with government agencies, oil companies, and consulting firms, data were collected for 1004 load tests where the load was carried essentially to plunging failure and soil properties were measured by some test, no matter how crude. Of the 1004 tests, 325 were on pipe piles but, of these, 142 were eliminated because the soil was at least partially cohesive, 40 were eliminated because the piles were installed using jetting or preboring or because the pile had an oversized cover plate, and 77 more were eliminated from the present discussion because of defects in the data (typically the nearest soil boring was deemed too far away, or the location of the water table was unknown), leaving 66 tests.

The load-settlement curves for piles in sand rarely define a plunging failure condition but, instead, the pile undergoes excessive settlements with only small increases in load. Data for two "failure" loads were used. Where load-settlement curves were available, Davisson's (6, p. 215) criterion was used to obtain a "defined" failure load. The peak applied load in the test, typically at rather large settlements, was also recorded. Data for both failure points will be given but the discussion will utilize data at the peak load because of the larger size of that data set. For the 58 tests for which a defined failure was known, the average ratio of peak applied load to defined failure load was 1.14, with a range from 1.00 to 1.52.

ANALYSES USING METHOD APIS

A computer program was written to perform analyses and prepare plots. The measured (Q_m) and computed (Q_c) pile capacities were plotted in a log-log diagram (Fig. 1) and, to aid in visual estimation

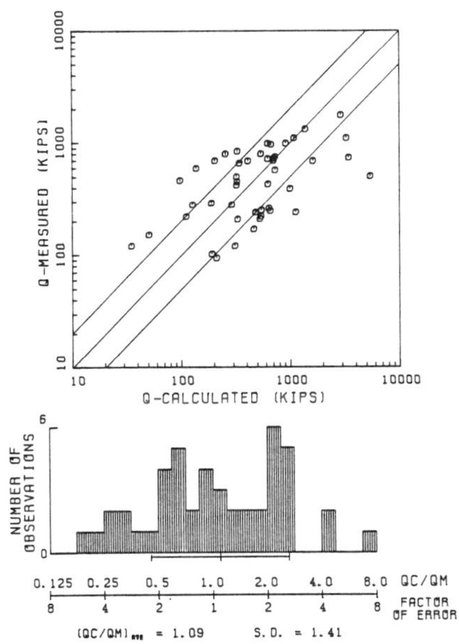

Fig. 1 Q_c - Q_m Data, Method APIS (Compression Tests Only)

of scatter, lines were drawn at $Q_c = \frac{1}{2} Q_m$ and $Q_c = 2Q_m$. The method used to evaluate the mean and the standard deviation are discussed in a companion paper (3).

For 46 piles tested in compression, Q_c/Q_m ranged from 0.21 to 10.67 and averaged 1.09 (Fig. 1). Many of the points with large values of Q_c/Q_m, including the one with $Q_c/Q_m = 10.67$, were for piles in calcareous sands where the API method was not intended to be used.

For 13 piles penetrating less than 50 feet, Q_c/Q_m averaged 0.42 and ranged from 0.21 to 0.69. For 24 piles in the 50-100 foot range, Q_c/Q_m averaged 1.77 (range 0.29-3.50), and for 7 piles in the 100-150 foot range, Q_c/Q_m averaged 2.74 (range 0.56-4.66). Thus, pile capacities tended to be underpredicted for short piles and overpredicted for long ones.

Data for 20 tension tests are shown in Fig. 2. The six tests for which the calculated capacities exceeded 1000 kips were all on long piles (penetrations of 148 to 271 feet) in calcareous sand. The other pile with a large Q_c/Q_m ratio was unique in that it was the only one that derived most of its capacity in gravel. For the 13 remaining tests in siliceous sands, Q_c/Q_m ranged from 0.40 to 1.14 and averaged 0.73. These piles penetrated 21 to 70 feet and thus had values of Q_c/Q_m similar to those for compression tests on piles having similar penetrations.

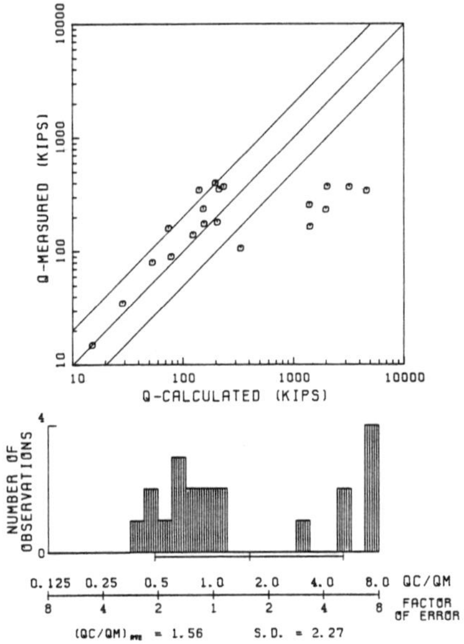

Fig. 2 $Q_c - Q_m$ Data, Method APIS (Tension Tests Only)

Taken collectively, the data indicate that the APIS method is conservative ($Q_c/Q_m < 1$) for short piles, possibly because these sands were dense, and unconservative for long piles, possibly indicating a need for a limit on q_p and f_s.

The data (66 tests) were reanalyzed with limits set on the average side shear, FSLIM, and the tip capacity, PTLIM. The side capacity applied to the outside of open piles and the tip capacity was the remaining capacity regardless of whether the pile appeared plugged or not. The results of the analyses are shown in Table 2.

Table 2 Effect of Setting Limits on Average Side Shear and End Bearing, Calcareous Sands Included

FSLIM ksf	PTLIM ksf	Average QC/QM	SD	Tests Effected Side	Tests Effected Tip
10	1000	1.22	1.74	0	0
2	200	1.18	1.59	6	11
2	100	1.11	1.51	6	25
1	200	1.04	1.17	27	11
1	100	0.97	1.09	27	25

If tests in calcareous sands are eliminated (59 tests remaining), the QC/QM ratios decrease significantly but the scatter remains high (Table 3).

Table 3 Effect of Setting Limits on Average Side Shear and End Bearing, Calcareous Sands Excluded

FSLIM ksf	PTLIM ksf	Average QC/QM	SD	Tests Effected Side	Tests Effected Tip
10	1000	0.97	1.16	0	0
2	200	0.95	1.10	3	10
2	100	0.89	1.01	3	24
1	200	0.89	0.95	20	10
1	100	0.82	0.88	20	24

If adequate data could be found, the APIS method could be improved by adjusting the values of K, δ, and N_q for density, grain hardness, and effects of varying confining pressures. An effort was made to use information from the data bank to develop an improved method, here termed method NSA1 (new sand, method 1).

METHOD NSA1

Source Data

The interpretation of load test data for piles in sand is made difficult by the lack of well defined soil properties and by the fact that both tip and side capacities are likely to be significant and uncertain in magnitude (in contrast, for friction piles in clay the tip capacity is typically small and reasonably easy to calculate so attention can be concentrated on the side shear).

The need to separate side shear and tip capacities restricts initial considerations to load tests on suitably instrumented piles, to tension tests where the tip capacity can be ignored, and to cyclic tension/compression tests. Even these data are ambiguous however. Pile driving typically causes large zero shifts in load cells, necessitating the taking of zero readings after driving and introducing errors due to residual loads left after pile driving (4). When cyclic tension/compression tests are used there is uncertainty as to the effect of one stage on the next. When only tension tests are used, there remains the question of the relationship of side loads in tension and compression.

Data were obtained for 21 tests on instrumented piles (11 steel pipe piles, 7 steel H-piles, 3 precast concrete piles) and 30 tests on non-instrumented piles in tension (20 steel pipe piles, 10 precast concrete piles). Six of the 21 instrumented piles were tested in both compression and tension. For these six tests, when tensile loads were applied, the instrumentation indicated tip tensile capacities of 20 to 42% (averaging 33%) of the total tensile capacity. On the assumption that the actual tip tensile capacity was zero and that a zero shift had occurred in the instrumentation, the indicated tip capacities were reduced to zero and the side shear capacity increased to equal the measured total capacity (3, 4). The corrected values for side shear in tension and compression were then within 10% of each other.

For the remaining 15 tests on instrumented piles, all of which were tested in compression only, there was no way to determine the zero shift. For these tests we assumed that the zero shift was the same as the average shift for piles tested in tension and thus we reduced the capacity attributed to side shear (by the load cells) by 33% and added that amount to the tip capacity.

Tip Capacities

The ratio of calculated (Q_{pc}) to corrected measured (Q_{pm}) tip capacities, for the 21 tests on instrumented piles is shown as a function of pile penetration in Fig. 3. For 19 of the 21 tests, the calculated tip capacities were smaller than the measured values. The two tests with higher calculated tip capacities were unique. One of these piles was the only one driven with a Bodine hammer. The other pile may have had its tip in a relatively loose layer because the driving resistance, measured in blows per foot, decreased by 50% in the last three feet of driving.

The previously calculated tip capacities could be corrected to bring them into conformance with "measured" values following several approaches, e.g., by setting a fixed upper limit on tip stress or by using the regression line in Fig. 3 to form a depth dependent correction factor. The latter approach was followed because it leads to a gradual change with depth (7) rather than a sudden change at the depth where the limiting capacity is first reached.

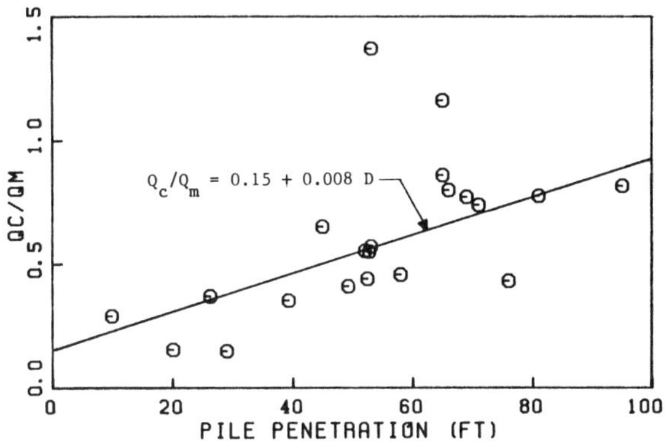

Fig. 3 Relationship Between the Ratio of Calculated to Measured Pile Tip Capacity and Pile Penetration

Accordingly, the previously calculated tip capacities may be corrected using the regression line of Fig. 3:

$$q_p = F_D \, \bar{\sigma}_{vo} N_q \tag{3}$$

where:

$$F_D = 1/(0.15 + 0.008\ D) \tag{4}$$

where D is the pile penetration in feet, and the other variables are as defined previously. From an analytical point of view it would have been preferable to use the dimensionless penetration ratio (D/B) in Eq. 4 but attempts to use D/B in Fig. 3 led to a substantial increase in the amount of scatter.

Side Capacities

In analysis of the side shear capacity, the variables subject to control are K and δ. No definitive data could be found to allow independent evaluation of these parameters using full scale field tests. Consequently, reasonable values are chosen for one and the other is adjusted to conform to field "measurements" of side shear capacity. In this application, the measured side shear capacity is defined as the measured total capacity minus the measured tip capacity corrected for zero shift in the load cells. Previous discussion indicated that measured side shear capacities were essentially the same in compression and tension so the same values would be used for K and δ. As a starting point, K was set at one for full displacement piles and δ was taken from the API (1) standard.

For steel pipe, and precast concrete (untapered) piles, in siliceous sands, the ratio of calculated (f_{sc}) to measured (f_{sm}) unit side shears at failure, for 6 compression and 23 tensile tests tended to increase with increasing pile penetration (Fig. 4). The observations are again in at least qualitative agreement with Vesic's (7) observation that the average developed side shears tend to approach

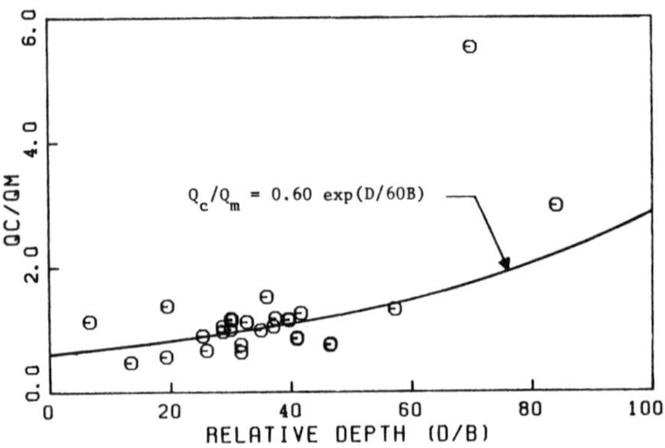

Fig. 4 The Relationship Between the Ratio of Calculated Unit Adhesion to Measured Unit Adhesion and Relative Depth

a limit as pile penetration increases. Again, instead of simply setting a limit on f_s, it seemed preferable to use the curve in Fig. 4 as a correction factor:

$$f_s = F_{SD}\, K\, \bar{\sigma}_{vo} \tan\delta \tag{5}$$

where:

$$F_{SD} = 1/[0.6 \exp(D/60B)] \tag{6}$$

where D and B are embedded length of pile and pile diameter, respectively. Separate analyses for steel H piles, where the side capacity was calculated using the area of the enclosing rectangle, indicated that K should be taken as 0.8 if the same values of δ are used (it would also be possible to use higher average values of δ to account for soil/soil shear around part of the area of the enclosing rectangle, and a lower value of K, but then two sets of δ values would be needed).

Effects of Density

When the correction factors, cited above, were incorporated in an analysis for piles driven into siliceous sands having a wide range in density it was clear that the effects of soil density, as indicated by standard penetration, or other tests, should be taken into account in assigning values for δ and N_q. On the assumption that only qualitative estimates of density would be available, the decision was made to classify soils using descriptive terms. The selected subdivisions and estimated values of δ and N_q are shown in Table 4. The numerical values selected, in general, conform to those proposed in the APIS method, however, the subdivisions of soil types has been expanded.

Calcareous Sands

Finally, data for piles in calcareous sands were studied. Data were found for six tension tests on open ended steel pipe piles. In accord with previous analyses K was chosen to be 0.8 and the values of δ shown in Table 3 were backcalculated from the measured capacities. Only a single compression test was available for a pile in calcareous sand. The values of N_q shown in Table 4 are based on that test and on the backfigured values of δ.

Table 4 Proposed Values of N_q and Delta for Visual Soil Characterization Methods

Description	Delta	N_q
Very Loose Siliceous Sand Medium Silt Loose-Medium Calcareous Sand* Medium Sandy Silt	15	8
Dense Silt Silty Sand Medium-Dense Calcareous Sand* Loose Siliceous Sand	20	12
Dense Sandy Silt Medium Siliceous Sand Medium Silty Sand	25	20
Dense Siliceous Sand Very Dense Silty Sand	30	40
Very Dense Siliceous Sand Dense Gravel	35	50

* Calcareous Sand is defined as consisting of at least 90% Calcium Carbonate.

Revised Methods

The collection of revisions to method APIS have led to a revised method, here called NSA1. The method is to be applied only to untapered piles in cohesionless soils. The total pile capacity under axial compressive or tensile loads, is taken as the sum of the tip and side shear capacities, these capacities being assumed to be independent of each other. The tip capacity is assumed to be zero for tensile loads. For compressive loads the tip capacity is calculated using Eqs. 3 and 4 with N_q taken from Table 4. The side shear capacity is calculated by inserting Eq. 6 into Eq. 5 and integrating over the surface area in contact with the soil. The earth pressure coefficient, K, is taken as 1.0 for closed ended piles and 0.8 for open ended pipe piles, and is the same for piles in tension and compression.

Comparison of Methods

The calculated and measured pile capacities are compared in Fig. 5. The revised method leads to a value of Q_c/Q_m closer to one than does the APIS method (1.10 vs. 1.22) and, more importantly, to a great reduction in scatter. Summaries of statistical data for analyses conducted using the APIS and NSA1 are presented in Tables 5 and 6.

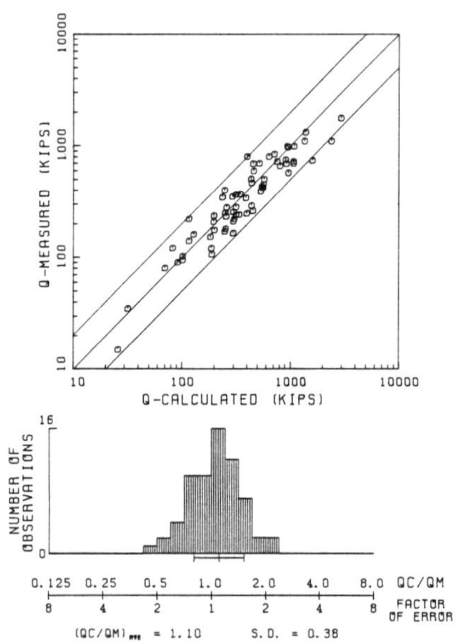

Fig. 5 $Q_c - Q_m$ Data, Method NSA1 (Combined Analysis using Tension and Compression Tests

Table 5 Summary of Statistical Data for Analyses Conducted using Methods APIS and NSA1 (Failure taken as ultimate reported load)

Method	Loading	N	Regression Line a	b kips	r	SEE kips	Q_c/Q_m Data Min	Max	Avg	SD
APIS	TENS/COMP	66	0.10	361	0.34	327	0.21	13.47	1.22	1.68
	COMP	46	0.15	426	0.42	333	0.21	10.67	1.09	1.41
	TENS	20	0.04	188	0.45	113	0.40	13.47	1.56	2.26
NSA1	TENS/COMP	66	0.60	137	0.89	159	0.49	2.17	1.10	0.38
	COMP	46	0.56	177	0.87	180	0.49	2.17	1.13	0.39
	TENS	20	0.98	10	0.85	67	0.62	1.79	1.02	0.35

Table 6 Summary of Statistical Data for Analyses Conducted using Methods APIS and NSA1 (Q_m taken at defined failure)

Method	Loading	N	Regression Line a	b kips	r	SEE kips	Q_c/Q_m Data Min	Max	Avg	SD
APIS	TENS/COMP	58	0.09	336	0.32	296	0.26	14.36	1.22	1.77
	COMP	42	0.12	391	0.40	299	0.26	10.93	1.26	1.47
	TENS	16	0.05	153	0.54	108	0.53	14.36	2.04	2.47
NSA1	TENS/COMP	53	0.49	144	0.86	160	0.50	2.56	1.25	0.42
	COMP	42	0.46	182	0.84	179	0.50	2.56	1.39	0.43
	TENS	16	0.91	5	0.86	66	0.68	2.18	1.13	0.39

Failure was defined as the maximum load applied in these tests. If the "defined" failure load had been used, then the average ratio of calculated to measured capacities (QC/QM) would have been about 1.25 for the NSA1 method. The average value of QC/QM could be lowered by setting limits on the calculated values of the average side shear and end bearing. The largest calculated side shear was 1.30 ksf and only 9 of the 66 tests had values over 1.00 ksf. If a limit of 1.00 ksf is set on side shear then the average QC/QM at "plunging" failure is 1.05 (SD = 0.39) if the limiting tip stress is 200 ksf and 1.00 (SD = 0.42) if the limiting tip stress is 160 ksf.

Limitations

The revised method, like its predecessor, is an empirical method which can be expected to yield reasonably accurate results only if the conditions under which it is applied resemble those in the data bank. The data bank used for preparation of Fig. 5 contained data only for untapered steel pipe piles driven into cohesionless soils without the use of jetting, preboring, or oversized cover plates. Only 7 tests on piles in calcareous sands were included, and only one of these was a compression test. Pile diameters ranged from 12 to 40 inches and penetrations from 10 to 168 feet for siliceous sands and 148 to 271 feet for calcareous sands. If method NSA1 is used, calculated and measured capacities may differ by up to about two times (Fig. 5) for cases in which pile/soil installation conditions are similar to those encountered in our data bank, and potentially by larger amounts for different cases. Designers should use any empirical method with caution.

SUMMARY AND CONCLUSIONS

Data were collected for 66 compressive or tensile axial loading tests, carried essentially to failure, on steel pipe piles in cohesionless soils where it was possible to estimate effective overburden pressures. All piles were driven without the aid of preboring or jetting and none had oversized cover plates.

Measured capacities were compared with capacities predicted using a method, here termed the APIS method, that closely resembles the API (1) method. The APIS method was applied to cohesionless soils with a wide range in density and no upper limits were set on either tip or side capacities. The method was applied to both siliceous and calcareous soils.

On the average, use of the APIS method led to about a 20% over-prediction of capacities but to individual ratios of calculated to measured capacities ranging from 0.2 to 11.0.

A revised method of analysis (NSA1) was developed which utilized a somewhat extended classification system that made provisions for soil density and grain hardness, and used modified values of δ and N_q. The same value of K was used for compression and tension tests. For displacement and non-displacement piles the value of K was set at 1.0 and 0.8, respectively. The equations used to calculate side and tip capacity were made a function of a pile penetration as well as the parameters above.

When the revised method was applied to the same data set used to evaluate method APIS it led to only a 10% overprediction of capacity on the average, and a large reduction in scatter as measured by the standard deviation of Q_c/Q_m.

ACKNOWLEDGMENTS

A major part of this research was performed with financial support from the American Petroleum Institute. The project advisory committee consisted of J.D. Allen (ARCO), D.D. Engle (Union Oil), L.M. Kraft, Jr. (McClelland Engineers), J.D. Murff (Exxon), B.S. Murphy (Amoco), J.R. Ruser (Chairman, Shell), G.E. Strickland (Chevron), and R.L. Thomas (Phillips). The guidance from the committee during the project and the permission from the American Petroleum Institute to publish the results are appreciated.

A major part of the literature survey and much of the early work on the computer program was carried out by D.G. Winter (Hart, Crowser, and Assoc.). Finally, we acknowledge the helpful critiques offered by the reviewers of the manuscript.

REFERENCES

1. American Petroleum Institute (1981), API Recommended Practice for Planning, Designing, and Constructing Fixed Offshore Platforms, API, Production Division, 211 North Ervay, Suite 1700, Dallas, Texas 75201.

2. Coyle, H.M., and R.R. Castello (1981), "New Design Correlations for Piles in Sand," Jour., Geot. Engr. Div., ASCE, Vol. 107, No. GT7, pp. 965-986.

3. Dennis, N.D., and R.E. Olson (1983), "Axial Capacity of Steel Pipe Piles in Clay," Am. Soc. of Civil Engrs., Geotechnical Practice in Offshore Engineering, Austin.

4. Hunter, A.H. and M.T. Davission (1968), "Measurement of Pile Load Transfer," Performance of Deep Foundations, ASTM, STP 444, pp. 106-117.

5. Meyerhof, G.G. (1956), "Penetration Tests and Bearing Capacity of Cohesionless Soils," Jour., Soil Mech. and Found. Div., ASCE, Vol. 82, No. SM1, Paper 866.

6. Peck, R.B., W.E. Hanson, and T.H. Thornburn (1974), Foundation Engineering, John Wiley and Sons, NYC.

7. Vesic, A.S. (1970), Tests on Instrumented Piles, Ogeechee River Site," Jour., Soil Mech. and Found. Div., ASCE, Vol. 96, No. SM2, pp. 561-584.

LIST OF SYMBOLS

B	pile diameter (L)
D	pile penetration (L)
F_D	correction factor for unit end bearing as a function of pile penetration
FSLIM	limit set on average side shear
F_{SP}	correction factor for unit side shear as a function of pile bearing-penetration
f_s	average unit side shear (F/L^2)
f_{sc}	calculated average unit side shear (F/L^2)
f_{sm}	measured average unit side shear (F/L^2)
K	earth pressure coefficient
N_q	bearing capacity factor
PTLIM	limit set on tip stress
Q_c	calculated pile capacity (F)
Q_m	measured pile capacity (F)
q_p	unit pile tip capacity (F/L^2)
Q_{pc}	calculated pile tip capacity (F)
Q_{pm}	measured pile tip capacity (F)
SD	standard deviation
δ	angle of pile-soil friction (DEG)
$\bar{\sigma}_{vo}$	effective overburden pressure (F/L^2)

CYCLIC AXIAL PILE RESPONSE - ALTERNATIVE ANALYSES

By Harry G. Poulos,[1] M.ASCE

SYNOPSIS

Two approaches to the analysis of cyclic axial pile response are described:

a) the "single step" method, which employs a single iterative analysis to determine the behaviour of a pile after a specified number of cycles, N

b) a "cycle-by-cycle" method which involves simulation of each cycle and hence involves a sequence of N cycles.

Within each approach, allowance can be made for the effects of cyclic loading on skin friction, ultimate base resistance and soil modulus. The paper presents, for a typical pile, a comparison between these two approaches and shows that the only major difference is in the computed mean displacement of the pile. A comparison is also made between two alternative representations of cyclic degradation, one based on cyclic displacement, and the other on the reversal of plastic stress. The influence on pile response of various factors such as loading rate, pile modulus and group effects is then examined for a hypothetical problem. It is shown that the theoretical solutions for pile load capacity are in broad agreement with values observed in field and model pile tests.

INTRODUCTION

The design of offshore piles to resist axial loading requires some assessment to be made of the influence of cyclic wave loading on the performance of a pile. Two counterbalancing phenomena need to be considered, the possible degradation of the pile due to cyclic loading, and the possible increase in pile resistance due to the relatively rapid rate of load application. Analyses incorporating one or both of these effects have been presented by Bea et al (1), Boulon et al (3), Matlock and Foo (9) and Poulos (11,14,15). The objectives of the present paper are as follows:

a) to compare two alternative methods of analysing cyclic pile response

b) to compare two alternative models of cyclic degradation

c) to examine factors which influence cyclic pile response.

[1]Professor, School of Civil and Mining Engineering, The University of Sydney, New South Wales, Australia.

For a typical offshore pile, the solutions from the two analysis methods are compared as are solutions obtained using the two alternative models of cyclic degradation. The influence on the pile response of such factors as loading rate, pile modulus and group effects is then examined, and finally, comparisons are made between theoretical combinations of mean and cyclic load to cause failure, and those measured in field and laboratory tests. Attention is confined to the case of cyclic loading of uniform amplitude; however, the analyses can be extended to consider variable or storm loading sequences, and it is also possible to utilize solutions for constant amplitude cyclic loading to consider storm loading sequences (15).

CHARACTERIZATION OF CYCLIC EFFECTS

In developing analyses of cyclic pile response, it is necessary to consider the influence of cyclic loading on ultimate skin friction, ultimate base resistance and soil modulus, and to also include consideration of the effects of loading rate on these properties. In order to quantify cyclic load effects, it is convenient to introduce the concept of degradation factors, the degradation factor being defined as:

$$D = \frac{\text{property after cyclic loading}}{\text{property for static loading}} \qquad (1)$$

The degradation factors for ultimate skin friction, ultimate base resistance and soil modulus are denoted as D_τ, D_b and D_E respectively. Some of the scant available data on degradation factors was reviewed by Poulos (15), and is summarized below.

Degradation of Skin Friction

The results of small-scale model pile tests in both clays and sands indicates that cyclic loading generally causes a reduction in the ultimate skin friction, and that this reduction is governed by the number of cycles and the amount of cyclic displacement ($\pm\rho_c$). For remoulded Hurstville clay (LL=43, PL=27), the relationship between skin friction degradation factor D_τ and normalizes cyclic displacement is shown in Fig. 1. ρ_c has been normalized both with respect to pile diameter d and also to the displacement required for full slip in a static test, ρ_{st}; the latter may be more generally applicable. It should be noted that static tests show no evidence of strain-softening interface behaviour. A normalized cyclic displacement ρ_c/ρ_{st} of 0.4 to 0.5 is required before degradation commences and D_τ becomes less than unity. Very large cyclic displacements ($\rho_c/\rho_{st} > |2|$) can cause reductions of skin friction in excess of 40% (i.e. $D_\tau < 0.6$) after 1000 cycles. The model of degradation involving relationships such as shown in Fig. 1 will be referred to as the 'cyclic displacement' model.

Model tests by Holmquist and Matlock (5) showed that "two-way" cyclic loading caused a much more severe reduction in skin friction than "one-way" cycling. Subsequently, Matlock and Foo (9) developed an alternative model in which cyclic degradation at a point on the pile only occurs if there is plastic reversal of stress at that point. This model

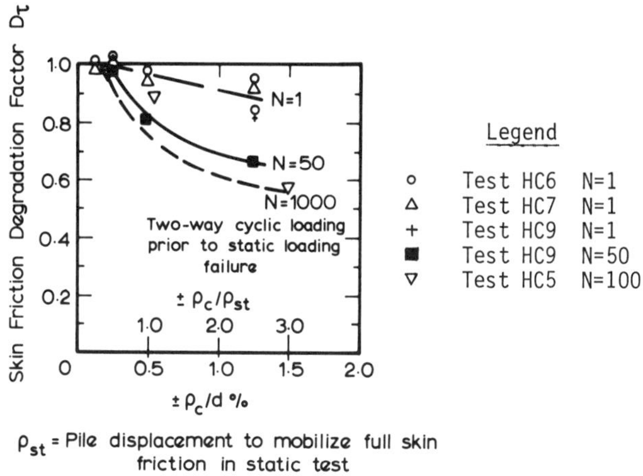

Fig. 1 Skin friction degradation factor for model piles in Hurstville clay (Ref. 13).

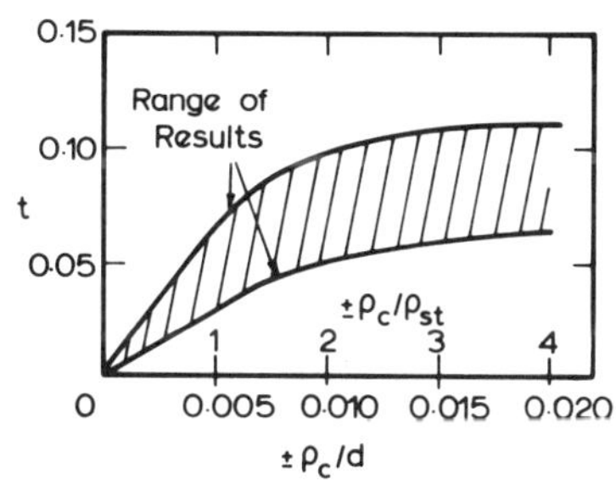

Fig. 2 Degradation parameter t derived from model pile tests in Hurstville clay (Ref. 13).

will be referred to as the 'reverse plastic stress' model, and in terms of the degradation factor, it can be expressed as follows:

$$D_\tau = (1-\lambda)(D'_\tau - D_{min}) + D_{min} \qquad (2)$$

where D_τ = degradation factor after plastic stress reversal

D'_τ = degradation factor existing before the last plastic stress reversal

D_{min} = minimum possible value of degradation factor

λ = parameter expressing extent of degradation during a plastic plastic stress reversal (typically in range 0.1 - 0.5).

The use of this model appears attractive at first since it only involves specification of two parameters, D_{min} and λ; however, there is relatively little data available on these parameters and, as will be demonstrated later, this approach may tend to underestimate the amount of cyclic degradation, as Poulos (13) has shown that "one-way" cycling without stress reversal can also cause degradation of skin friction.

Degradation of Ultimate Base Resistance

There appears to be no data on cyclic degradation of ultimate base resistance, with the exception of that by Van Weele (20) who deduced that in sand, the degradation of base resistance could be more severe than for skin friction. In the absence of other data, it has been assumed that the cyclic displacement degradation model can also apply to ultimate base resistance, except that, in normalizing the cyclic displacement in Fig. 1, the value of ρ_{st} is the displacement to cause full mobilization of base resistance; this will be significantly larger than the value to fully mobilize shaft friction.

The Matlock and Foo reverse plastic stress model (Eq. 2) can presumably be applied equally well to base resistance as to skin friction. With either model, it is necessary to allow for the possibility that the ultimate tensile capacity of the base is much less than the compressive capacity.

Degradation of Soil Modulus

Model pile tests suggest that an expression similar to that developed by Idriss et al (6) from cyclic triaxial tests on clay may be applied approximately to the degradation of soil modulus due to cycling. Thus, for the cyclic displacement model of degradation, the modulus degradation factor D_E may be approximated as:

$$D_E = N^{-t} \qquad (3)$$

where N = number of cycles

t = a degradation parameter which is a function of normalized cyclic displacement

An alternative form of Eq. 3 useful for incremental computations is (18):

$$D_E = D_E' - 0.5t(D_E')^{(1+t)/t} \qquad (4)$$

where D_E' = modulus degradation factor existing prior to the current cycle of loading

t = degradation parameter, as above.

The relationship between t and the normalized cyclic displacement, obtained from model tests in remoulded clay, is shown in Fig. 2; because of experimental scatter, a range of values of t is shown. It should be noted that, in the case of piles in sand, it is possible to obtain negative values of t at small cyclic displacements, thus indicating a stiffening of the soil with cyclic loading; at larger cyclic displacements, t may become positive, indicating a degradation of modulus.

For the reverse plastic stress model, the modulus degradation factor D_E is presumed to be the same as the skin friction factor D_τ in Eq. 2.

Loading Rate Effects

Both pile resistance and pile head stiffness tend to increase linearly with the logarithm of loading rate (1,4,7). A loading factor D_R may therefore be applied to the values of ultimate skin and base resistance and the soil modulus, in which

$$D_R = 1 + F_\rho \log_{10} (\zeta/\zeta_r) \qquad (5)$$

where F_ρ = rate coefficient (typically in the range 0.1 to 0.25)

ζ = actual loading rate

ζ_r = reference loading rate (e.g. for static load test)

In general, the rate of cyclic loading on offshore piles will be relatively high, so that the rate effects will tend to counteract the degradation due to cycling. In this paper, the rate factor D_R is applied in both the degradation models.

ANALYSIS METHODS

Two methods of analysis will be considered herein, both relying on essentially the same type of analysis, but differing in their treatment of the cyclic loading:

a) the "single-step" analysis, in which, for a specified number of cycles, N, a single analysis is carried out to determine the behaviour of the pile after N cycles. The analysis is iterative in that the soil parameters are adjusted to reflect the effects of cyclic loading after N cycles by obtaining compatibility between the cyclic deflections and the degradation factors at various points along the pile.

b) the "cycle-by-cycle" analysis, in which the application of each of the N cycles is modelled in sequence. After each cycle, the degradation factors are determined and used to adjust the ultimate skin friction, ultimate base resistance and soil modulus for the next cycle of load. Thus, this analysis requires a sequence of analyses, (for large values of N, this may involve a correspondingly large number of analyses); this contrasts with the single iterative analysis of the "single-step" analysis.

For both types of analysis, the pile is discretized into a series of elements, cylindrical for the shaft and annular elements for the base and any discontinuities in diameter along the shaft. The soil is considered initially to be a linearly elastic continuum, and by considering equilibrium and compatibility between the vertical movements of the pile and soil at each element, relationships may be derived between the increment in pile-soil stress, and the applied axial load increment. These relationships may be solved for the interaction stress increments and the incremental pile deflection, from which the over-all values can be obtained by addition to the existing values. Allowance can be made for pile-soil slip or soil yield at an element by specifying an upper limit to the value of the interaction stress at each element, and carrying out an iterative analysis. Nonhomogeneity of the soil along and beneath the pile is taken into account approximately (12).

As outlined by Poulos (15), the analyses are readily extended to the case of a pile group in which the piles are symmetrically arranged around the circumference of a circle. The soil displacement matrix is amended, using Mindlin's equations, to include components due to the surrounding piles in the group.

The single-step analysis is described in Ref. 14 and will not be repeated here. For the cycle-by-cycle analysis, the response of a pile after N cycles of uniform loading can be analysed as follows:

1) Starting with the static values of E_s, τ_a and p_{bu}, modified by the rate factor D_R, the pile is analysed for a load of P_{max}.

2) The load is then decreased to P_{min} and the analysis repeated. Finally the load is returned to zero, thus completing the first cycle of loading.

3) For the cyclic displacement model of degradation, the cyclic displacement at each element is calculated and the degradation factors D_τ, D_b and D_E determined. For the reverse plastic stress degradation model, D_τ, D_b and D_E are calculated from Eq. 2 if reversal of plastic stress has occurred at an element; otherwise, they are left unaltered for the next cycle.

4) The values of E_s, τ_a and p_{bu} for the next cycle are determined by multiplying the static values by the appropriate degradation

factor and the rate factor D_R; the next cycle of loading is then applied, as in steps 1) and 2), and the procedure repeated until the required number N of cycles have been simulated. A program, TAPCYC, has been written to carry out this analysis.

Because the analysis simulates each cycle of loading, the behaviour of the pile at all stages of the cycling process may be determined. Also, in contrast to the single-step analysis, it is possible to use the reverse plastic stress degradation model. A further advantage is that "parcels" of loading of different amplitude can readily be analysed; in this case, it is advantageous to use Eq. 4 to determine the modulus degradation factor D_E. On the other hand, this analysis generally involves considerably more computation than the single-step analysis, particularly for large values of N.

A comparison between the results of each type of analysis is presented below.

COMPARATIVE SOLUTIONS FOR TYPICAL PILE

In order to compare solutions from the alternative analyses and the alternative degradation models, a typical steel tube pile driven into a deep deposit of normally consolidated marine clay has been analysed. The pile is 90 m long, 1.8 m in diameter and has a 60 mm wall thickness; 15 shaft elements and one base element are used to divide the pile. The soil is assumed to have a static modulus of 125 times the in-situ effective overburden stress σ_{vo}', and an effective unit weight of 7 kN/m^3. The ultimate skin friction is 0.25 σ_{vo}' while the ultimate base resistance is 1.42 MPa in compression and 0.28 MPa in tension. For the cyclic degradation model, the degradation characteristics for both skin friction and base resistance are those shown in Figs. 1 and 2. For the reverse plastic strain degradation model, (Eq. 2), λ has been taken as 0.25 and D_{min} as 0.6. For both models, rate effects have been included via the factor D_R in Eq. 5, with a rate coefficient F_ρ of 0.1. The "standard" loading rate has been arbitrarily assumed to be 0.17 MN/min and would correspond to static failure in about 4 hours; the period of cyclic loading is 15 seconds.

Under static loading conditions, with no rate effects, the ultimate load capacities in compression and tension are 43.7 MN and 40.8 MN respectively, while the cyclic stiffness of the pile head (at cyclic load levels up to about ±20 MN) is 837.6 MN/m. If rate effects are included but no cyclic degradation occurs, the corresponding values are 55.44 MN, 51.7 MN and 930.7 MN/m.

In the analyses performed, N cycles of axial load of magnitude $\pm P_c$ about a mean load P_o have been applied. Attention is concentrated on the following quantities:

i) the ultimate compression load capacity after cycling, P_{uc}, and the combinations of mean and cyclic load which result in failure during cycling
ii) the cyclic stiffness of the pile head, K_c
iii) the mean vertical displacement of the pile head, ρ_m .

Comparison Between Analysis Methods

Fig. 3 shows a typical comparison between solutions from a single-step and cycle-by-cycle analyses, using the cyclic displacement degradation model. There is very close agreement between the two analyses for the ultimate load capacity after cycling and the cyclic stiffness of the pile head. In both cases, these quantities are independent of the mean load P_o, provided that the maximum load (P_o+P_c) does not exceed the load capacity. However, there is a distinct difference between the mean displacements, particularly at larger values of the cyclic load, with the value given by the cycle-by-cycle analysis being significantly larger than that from the single-step analysis, which predicts smaller maximum and minimum displacements. Examination of the variation of the minimum displacement with time shows that the single step analysis frequently predicts a decrease in the minimum displacement with increasing cycles, whereas the cycle-by-cycle analysis generally predicts an increase. This latter behaviour is in accord with observations from field and model tests, and suggests that the cycle-by-cycle analysis may be more satisfactory overall than the single step analysis. However, if only the cyclic stiffness or ultimate load capacity after a specific number of cycles is required, the single step analysis is a far quicker means of obtaining the solution.

All subsequent solutions presented in this paper have been obtained from the cycle-by-cycle analysis.

Comparison Between Degradation Models

Fig 4 compares the results of analyses from a cycle-by-cycle analysis incorporating the reverse plastic stress model and the cyclic displacement model. For the former model, values of $\lambda = 0.25$ and $D_{min} = 6$ have been chosen, in an attempt to obtain comparable degradation characteristics to those shown in Figs.1 and 2 for the cyclic displacement model. The latter predicts a more severe influence of cyclic degradation, with lower cyclic stiffness and higher mean displacement for a given value of P_c and N. This is not surprising since, in the reverse plastic stress model, relatively large cyclic displacements may occur without any resulting degradation, provided that reverse "slip" has not occurred, whereas, in the cyclic displacement model, these cyclic displacements may result in considerable degradation. This is clearly demonstrated in Fig.5, which compares the distribution of degradation factors given by each model for two values of cyclic load. The degradation indicated by the reverse plastic stress model is generally confined to the upper portion of the pile, whereas the cyclic displacement model shows (in this case) a more uniform degradation along the pile shaft.

An interesting finding is that, as with the cyclic displacement model, the reverse plastic strain model predicts values of the ultimate load capacity and cyclic stiffness of the pile which are independent of the mean load P_o (provided again that the maximum load $P_o + P_c$ does not exceed the load capacity). In the solutions which follow, the cyclic displacement model has been used in conjunction with the cycle-by-cycle analysis to examine some of the factors influencing the cyclic response of the pile.

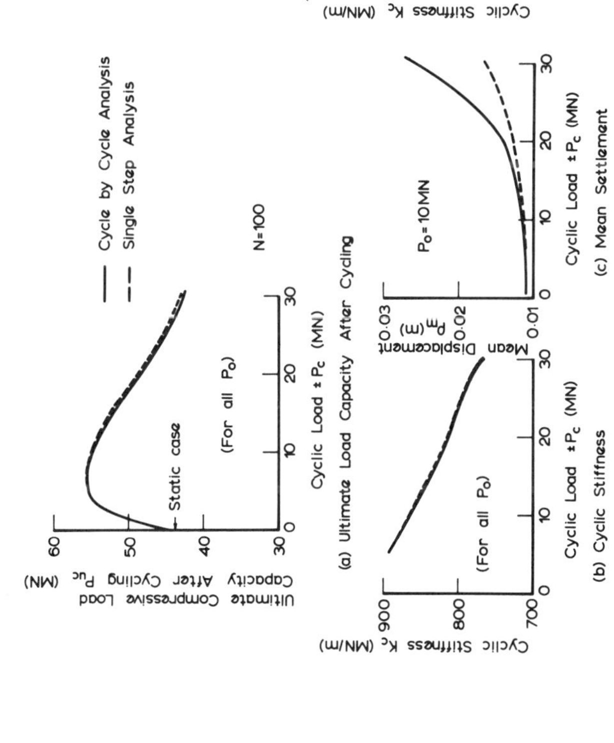

Fig. 3 Comparisons between solutions from alternative analyses.

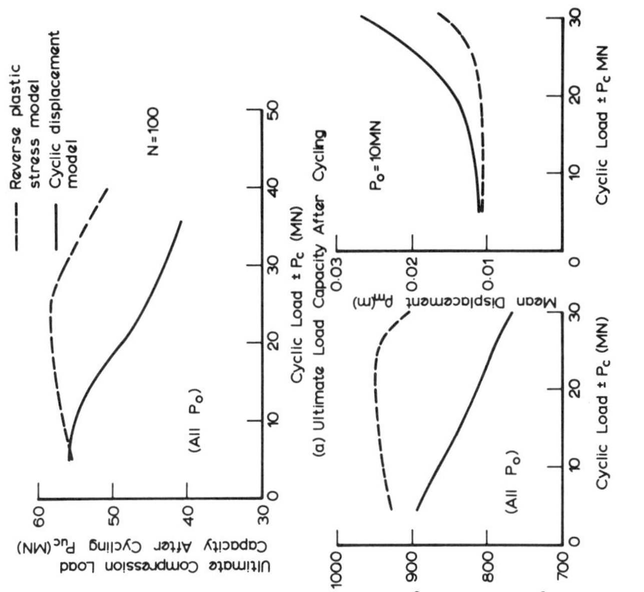

Fig. 4 Comparisons between solutions from alternative degradation models.

CYCLIC RESPONSE CHARACTERISTICS

Critical Load Combinations

A useful representation of the results of a cyclic response analysis is in terms of limiting (or critical) combinations of static and mean load which cause failure during cycling. Because the ultimate load capacity after cycling is independent of the mean load level, such limiting combinations can readily be deduced from the solutions for a zero mean level. Limiting load combinations for $N = 10$ and 100 are shown in Fig. 6. Also shown is the curve for static failure in compression, defined by $P_0 + P_c = P_{us} = 43.7$ MN. For small values of mean load P_0, the pile fails in tension, but for values of P_0 larger than about 3 MN, failure occurs in compression. For small values of cyclic load P_c, large values of P_0 can be tolerated because of the effects of loading rate. The effect of number of cycles is not great, although as expected, the ultimate load capacity after cycling decreases with increasing number of cycles.

If loading rate effects are excluded, significantly smaller load combinations are required to cause failure, and in all cases, (unless $P_c = 0$), the cyclic load which causes failure is considerably less than the value for static failure in compression (i.e. if no cyclic degradation occurred). Fig. 6 thus emphasizes the significance of loading rate effects in evaluating the cyclic response of piles.

Influence of Maximum Load

Examination of the results of analyses from the program TAPCYC reveals that the maximum pile displacement after a given number of cycles cycles is dependent primarily on the maximum load and not on mean load; thic conclusion applies to both degradation models considered. The relationship between the maximum load P_{max} and maximum displacement ρ_{max} is shown in Fig. 7 for $N = 100$ and analyses with and without loading rate effects. The effects of P_0 can be seen to be small. As would be expected, the inclusion of rate effects reduces the maximum displacement for a given maximum load.

Since the maximum displacement depends mainly on the maximum load, and cyclic displacement depends on the cyclic load (see Fig. 3b) it therefore follows that for any combination of mean and cyclic load, the maximum and cyclic displacements (and hence the mean and minimum displacements) may readily be estimated from the solutions for a single value of mean load (e.g. zero) and a series of values of cyclic load.

Influence of Number of Cycles

The solutions shown in Fig. 6 indicate a relatively minor dependence of ultimate load capacity on number of cycles N, (at least for values of N up to 100). The variation of cyclic pile head stiffness K_c with N is shown in Fig. 8. The influence of N increases as the amplitude of cyclic load increases. Fig. 9 shows the development of mean displacement with increasing N for $P_0 = 15$MN. Again, the influence of N becomes more marked as P_c increases. Also indicated in Fig. 9 is the ratio of the

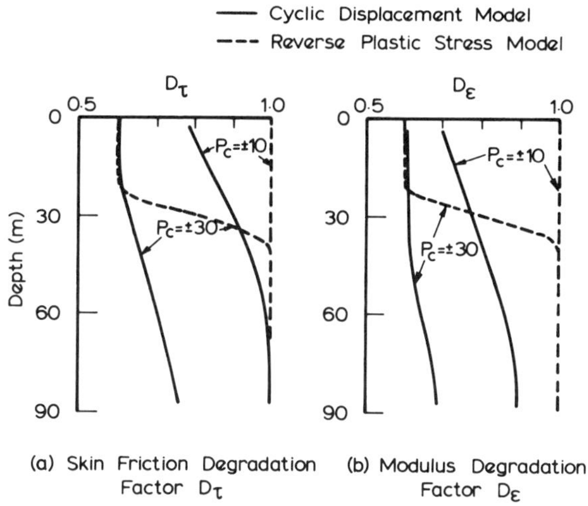

Fig. 5 Distribution of degradation factors for N = 100.

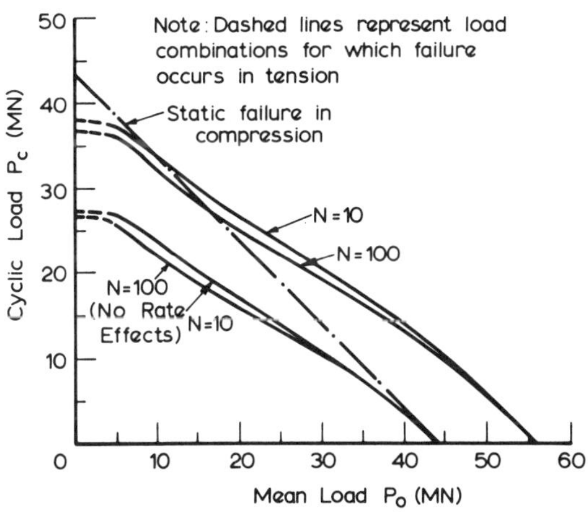

Fig. 6 Critical combinations of static and cyclic load to cause failure.

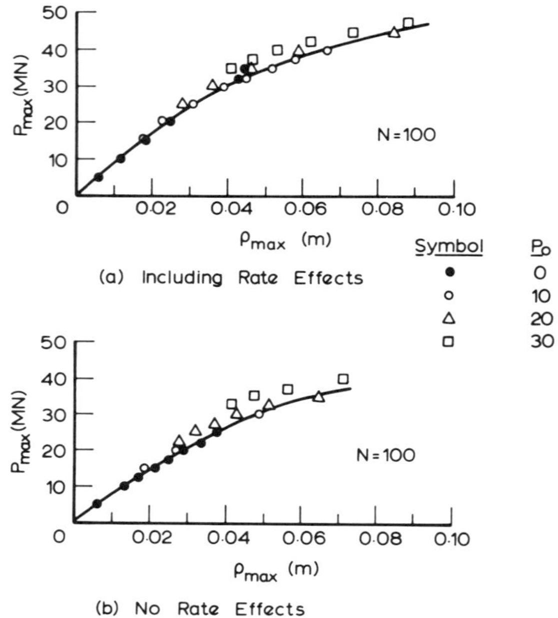

Fig. 7 Relationship between maximum load and maximum displacement - cyclic strain degradation model.

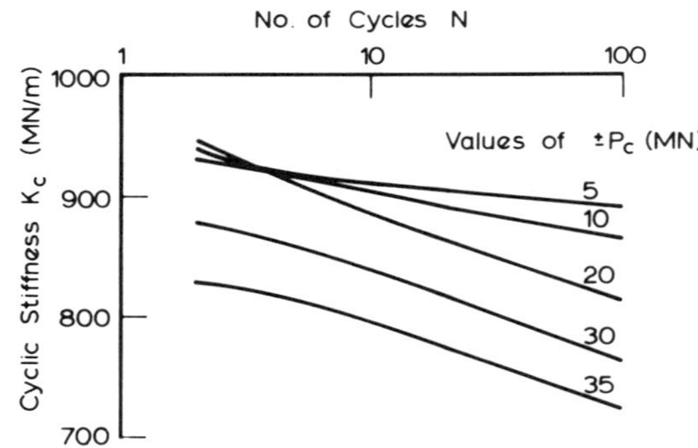

Fig. 8 Variation of cyclic stiffness with number of cycles.

mean displacement ρ_m to the static settlement ρ_s under a static load of 15 MN. For small values of P_c, ρ_m is less than ρ_s because of rate effects, but for larger values of P_c, ρ_m becomes significantly greater than ρ_s because of the effects of cyclic degradation on modulus and skin friction.

Influence of Pile Modulus

To examine the effects of pile modulus, calculations were carried out for piles having a modulus ten times and one-tenth that of the "standard" pile analysed previously. In terms of the pile stiffness factor K_b defined by Poulos (12). the corresponding values of K_b are 3170 and 31.7 respectively. The results of these calculations are shown in Fig. 10 in terms of the critical combinations of mean and cyclic load to cause failure for $N = 100$ cycles. The pile modulus has some influence on the critical load combinations, with the stiffer pile being capable of sustaining a larger cyclic load.

Fig. 11 shows the effect of pile modulus on the cyclic stiffness of the pile head, expressed dimensionlessly as the ratio of the pile head stiffness to the corresponding value for $P_c = 0$ (the latter value therefore does not involve loading rate effects). It can be clearly seen that the rate of reduction of cyclic pile head stiffness with increasing cyclic load amplitude becomes more pronounced as the pile modulus decreases.

It is therefore concluded that the cyclic response of a pile, and in particular the variation of cyclic pile stiffness, depends on the modulus of the pile, or more generally, on the pile stiffness factor i.e. ratio of the effective pile modulus to soil modulus. It follows that, for a given pile, the effects of cyclic loading if the pile is in a soft clay will tend to be less serious than in a stiff clay, since the pile stiffness factor in the former case will be higher. Furthermore, in stiff clays, the possibility of interface strain-softening could lead to additional degradation of pile performance.

Group Effects

As previously indicated (15), consideration of group effects will lead to a reduction in pile head stiffness and also in the ultimate load capacity after cycling because of the increase in cyclic deflection arising from interaction among the piles in the group. For a group of 8 piles equally spaced around the circumference of a circle of radius of 5.0 m, the critical load combinations to cause failure for $N = 100$ cycles are shown in Fig. 12, together with the corresponding curve for a single isolated pile. Loading rate effects are included in both cases. The group effects lead to a significant reduction in the level of cyclic load that can be sustained at any value of P_0. In addition, the cyclic stiffness of the pile load is reduced by a factor of about 2.7. Model tests reported by Matlock et al (10) suggest that these theoretical conclusions are valid, at least for small-scale model piles. It is therefore clear that the consideration of group effects is very important in the design of piles subjected to cyclic loading.

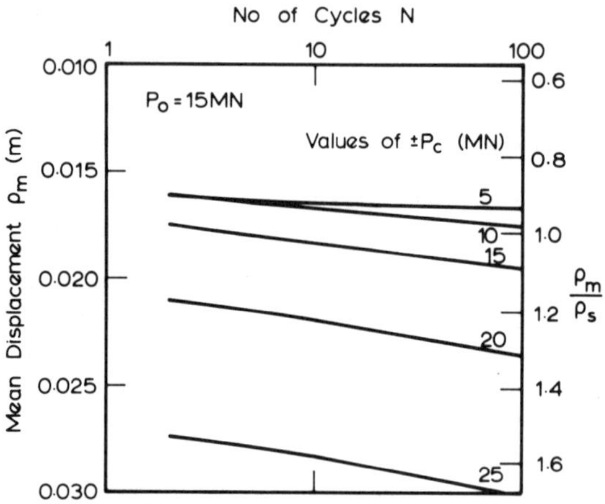

Fig. 9 Development of mean displacement with increasing cycles.

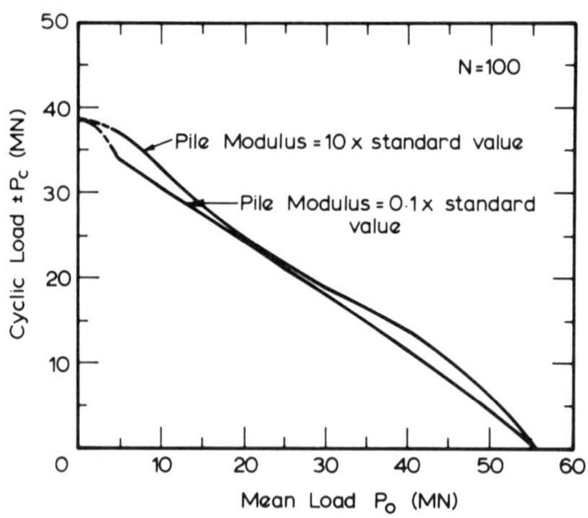

Fig. 10 Influence of pile modulus on critical load combinations.

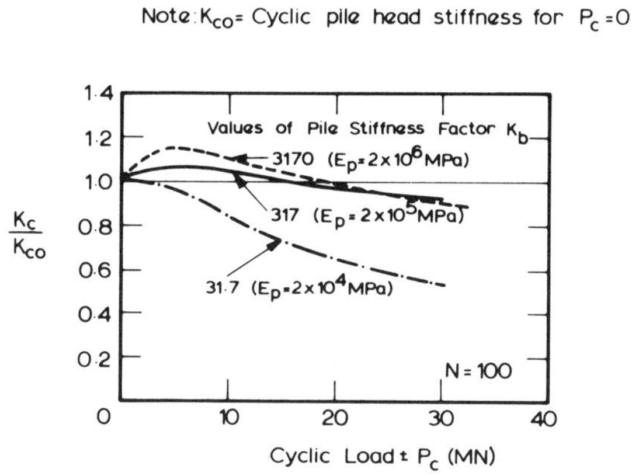

Fig. 11 Effect of pile modulus on cyclic pile head stiffness.

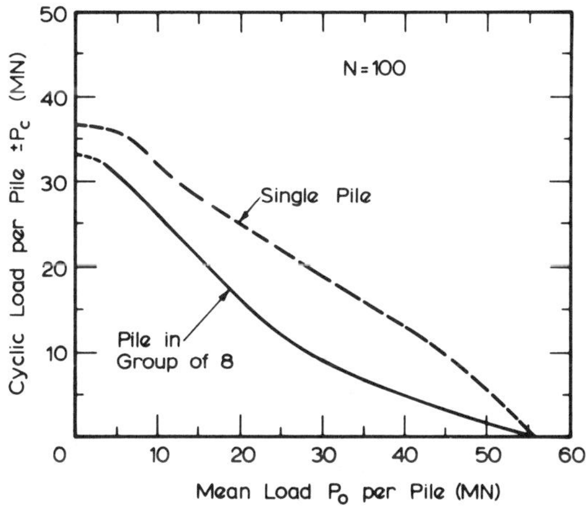

Fig. 12 Critical static and cyclic load combinations to cause failure for group and single pile.

Comparison with Available Field Data

Field data on the effects of cyclic load on pile load capacity have been obtained recently (8,16,17), while some earlier data has been summarised by Stevens (19) and Bea et al (1). Some data from small-scale laboratory tests has also been presented by Poulos (13). Detailed analysis of the field data is not yet feasible because of the wide range of soil types involved and the paucity of data on cyclic degradation effects in these soils; however, it would appear useful to examine the data in relation to the critical combinations of mean and cyclic load to cause failure. Such a comparison is shown in Fig. 13, with the cyclic and mean loads expressed in dimensionless form as proportions of the ultimate static load capacity P_{us}. Two theoretical curves are plotted, one in which rate effects are included and the other in which no allowance for loading rate is made. These theoretical curves are derived from the solutions obtained for the example pile studied herein, using the cyclic degradation model. It should be recognised that these solutions will not necessarily closely model the geometry or relative stiffness of the test piles, nor will the assumed degradation characteristics necessarily be representative of the real pile-soil behaviour in all cases.

Despite these possible limitations, it can be seen from Fig. 13 that the theoretical curve defining the critical load combinations without rate effects agrees well with observed data. The cases in which failure during cyclic loading have been observed lie close to this theoretical line while cases in which no failure has occurred lie below it. It should be noted that the theoretical solution for the reverse plastic stress degradation model lie well above those for the cyclic displacement model and would not be in good agreement with the data. It is also interesting to note that the cases reported by Stevens (19) in which 'no settlement' was observed have very low values of P_c/P_{us} (less than 0.1); the theoretical solutions (Fig.9) indicate that, in such cases, the vertical movement tends to stabilise, whereas at higher values of P_c/P_{us}, the mean vertical movement continues to increase with increasing cycles, corresponding to the region of 'continuing settlement' referred to by Stevens (it is presumed that this 'continuing settlement' does not imply failure of the pile, at least for the relatively small numbers of cycles considered herein).

Several of the theoretical characteristics of pile behaviour under cyclic loads are also observed in the field tests, for example, the progressive transfer of loads from the upper part of the pile towards the lower part during cycling, the dependence of cyclic pile head stiffness on the cyclic load amplitude, and the increasing displacement of the pile at relatively low cyclic load levels, even after 1500 cycles (16). On the other hand, there are some aspects of observed behaviour which conflict with the theoretical behaviour, notably, the observation by McAnoy et al (8) and Puesch (16) that the cyclic stiffness does not decrease significantly with increasing cycles. This may however be due to the nature of the soils in which the piles are located. For example, the stiffness of piles in sand tends to increase with increasing cycles, in contrast to piles in clay, for which the cyclic stiffness has been observed to decrease in model pile tests (13). It is therefore possible that in silty soils, the cyclic stiffness of piles (particularly at relatively low levels of cyclic loading) remains relatively constant

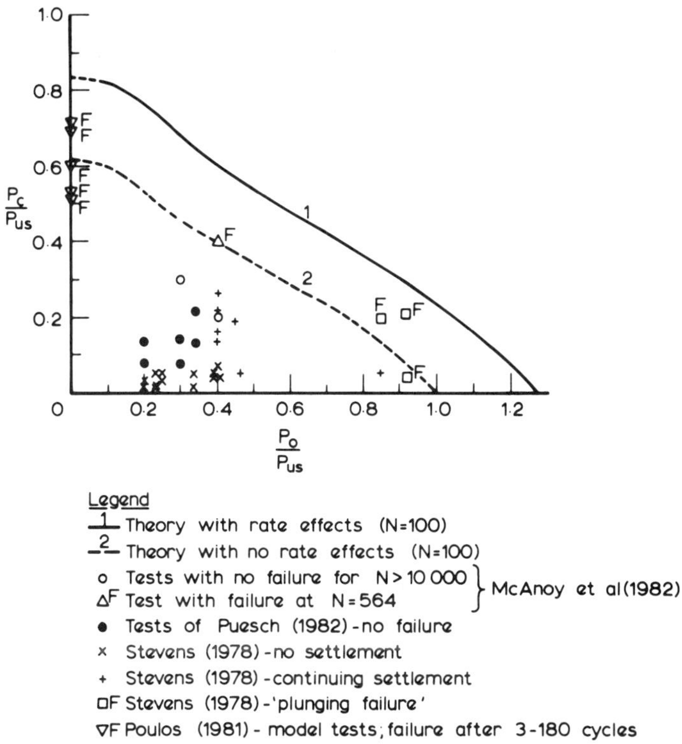

Fig. 13 Comparison between field tests and theory.

with increasing cycles. Clearly, further data on cyclic degradation characteristics (such as that shown in Figs. 1 and 2) is required for a variety of soil types.

CONCLUSIONS

The major objectives of this paper have been three-fold:

a) to compare two alternative methods of analysing cyclic pile response.

b) to compare two alternative models of cyclic degradation.

c) to examine some of the characteristics of cyclic pile response.

The cycle-by-cycle and the single-step analyses give almost identical solutions for the cyclic pile stiffness and the ultimate load capacity after cycling. However, the cycle-by-cycle analysis gives larger values of the mean displacement and is considered a more reliable indicator of real pile behaviour.

The cyclic displacement degradation model generally leads to a greater amount of degradation than does the reverse plastic stress model, although this evidence is by no means conclusive.

A useful method of representing the results of the analyses is to plot critical combinations of mean and cyclic load to cause failure in a specified number of cycles. Among the factors which influence this plot are the rate of load application, the ratio of pile modulus to soil modulus and group effects. It is found that, regardless of the degradation model used, the cyclic displacement depends on the amplitude of the cyclic load while the maximum displacement depends primarily on the magnitude of maximum load. Therefore, it is possible to determine the behaviour of a pile under all combinations of mean and cyclic load by carrying out analyses for a single value of mean load and a series of cyclic load amplitudes.

ACKNOWLEDGEMENTS

The work described in this paper forms part of a research project on Behaviour of Offshore Foundations being carried out at the University of Sydney. This project is supported by a grant from the Australian Research Grants Scheme.

REFERENCES

1. Bea, R.G., Audibert, J.M.E. and Dover, A.R. (1980). "Dynamic Response of Laterally Loaded and Axially Loaded Piles". Proc. 12th OTC Conf., Houston, Paper OTC 3749, pp.129-139.
2. Bea, R.G., Dover, A.R. and Audibert, J.M.E. (1982). "Pile Foundation Design Considerations for Deepwater Fixed Structures". 3rd BOSS Conf., Boston, Vol.1, pp.125-140.
3. Boulon, M., Desrues, J., Foray, P. and Forgue, M. (1980). "Numerical Model for Foundation Under Cyclic Loading, Application to Piles". Int. Symp. Soils Under Cyclic and Trans. Loading, Swansea. A.A. Balkema, Rotterdam, pp.681-694.

4. Gallagher, K.A. and St. John, H.D. (1980). "Field Scale Model Studies of Piles as Anchorages for Buoyant Platforms". Eur. Off. Pet. Conf. and Exhibition, London, U.K.
5. Holmquist, D.V. and Matlock, H. (1976). "Resistance-Displacement Relationships for Axially-Loaded Piles in Soft Clay". Proc. 8th OTC Conf., Houston, Paper OTC 2474, pp.554-569.
6. Idriss, I.M., Dobry, R. and Singh, R.D. (1978). "Nonlinear Behaviour of Soft Clays During Cyclic Loading". Jnl. Geot. Eng. Divn., ASCE, Vol.104, No.GT12, pp.1427-1447.
7. Kraft, L.M., Cox, W.R. and Verner, E.A. (1981). "Pile Load Tests: Cyclic Loads and Varying Load Rates". Jnl. Geot. Eng. Divn. ASCE, Vol.107, No.GT1, pp.1-19.
8. McAnoy, R.P.L., Cashman, A.C. and Purvis, D. (1982). "Cyclic Tensile Testing of a Pile in Glacial Till". Proc. 2nd Conf. Num. Meth. in Offshore Piling, Austin, Texas.
9. Matlock, H. and Foo, S.C. (1979). "Axial Analysis of Pile Using a Hysteretic and Degrading Soil Model". Proc. Conf. Num. Meth. in Offshore Piling, I.C.E., London, pp.165-185.
10. Matlock, H., Bogard, D. and Cheong, L. (1982). "A Laboratory Study of Axially Loaded Piles and Pile Groups Including Pore Pressure Measurements". Proc. 2nd BOSS Conf., Boston, Vol.1, pp.105-121.
11. Poulos, H.G. (1979). "Development of an Analysis for Cyclic Axial Loading of Piles". Proc. 3rd Int. Conf. Num. Meth. Geomech, Aachen, Vol.4, pp.1513-1530.
12. Poulos, H.G. (1979). "Settlement of Single Piles in Non-Homogeneous Soils". Jnl. Geot. Eng. Divn., ASCE, Vol.105, No.GT5, pp.627-641.
13. Poulos, H.G. (1981). "Some Aspects of Skin Friction of Piles in Clay Under Cyclic Loading". Geot. Eng. Vol.12, No.1, pp.1017.
14. Poulos, H.G. (1981). "Cyclic Axial Response of Single Pile". Jnl. Geot. Eng. Divn., ASCE, Vol.107, No.GT7, pp.41-58.
15. Poulos, H.G. (1982). "Influence of Cyclic Loading on Axial Pile Response". Proc. 2nd Conf. Num. Meth. in Offshore Piling, Austin, Texas.
16. Puesch, A.A. (1982). "Basic Data for the Design of Tension Piles in Silty Soils". Proc. 3rd BOSS Conf., Boston, Vol.1, pp. 141-157.
17. Puesch, A.A., Boulon, M. and Meimon, Y. (1982). "Behaviour of Tension Piles: Field Data and Numerical Modelling". Proc. 2nd Conf. Num. Meth. in Offshore Piling, Austin, Texas.
18. Singh, R.D., Kim, J.H. and Caldwell, S.R. (1978). "Properties of Clays Under Cyclic Loading". Proc. 6th Symp. on Earthquake Eng., Roorkee,Vol.1, pp.107-112.
19. Stevens, J.B. (1978). "Prediction of Pile Response to Vibratory Loads". Proc. 10th OTC Conf., Houston, Vol.3, pp.2213-2223.
20. Van Weele, A.F. (1979). "Pile Bearing Capacity Under Cyclic Loading Compared with that Under Static Loading". Proc. 2nd BOSS Conf., London, pp.475-488.

DESIGN CONSIDERATIONS FOR OFFSHORE PILES

By M. F. Randolph[1]

INTRODUCTION

The design of piled foundations for offshore structures is currently based largely on methods arising from experience with onshore piles. Differences in soil types, and the much greater pile penetrations encountered offshore, give rise to considerable uncertainty as to the reliability of these methods. For example, at high pile penetrations, essentially normally consolidated soils may be encountered with strengths greater than highly overconsolidated deposits found onshore. Current design procedures which relate skin friction directly to the strength of the soil, taking no account of the consolidation history, may prove unduly conservative and uneconomic in such circumstances. On the other hand, ratios of skin friction to the shear strength of the soil, derived from onshore test data, may prove too high, and thus unsafe, for piles in very heavily overconsolidated offshore deposits, where soil strengths can be in excess of 500 kPa.

The advent of tethered buoyant structures has brought the additional possibility of piles gradually pulling out under a mean tensile loading - an event made more likely by the large cyclic component of load. It is necessary, not only to estimate peak values of skin friction between the pile and the soil, but also to assess under what circumstances these values may reduce and what lower limit, or residual values of skin friction, may be assumed. It would be unduly conservative to design directly on residual values of skin friction, treating the additional available skin friction as providing a factor of safety. However, if peak values of skin friction are to be used in design, then it is necessary to ensure that degradation of the shear transfer between pile and soil will not occur during the working life of the pile. It thus becomes important to consider the structural response of the pile, and the effect of pile geometry and relative stiffness on the response to sustained and cyclic loading.

This paper concentrates on the shaft capacity of piles driven into clay, and outlines an approach for estimating peak and residual values of skin friction. Tentative design curves are presented for the parameters α (the ratio of skin friction, τ_s, to undrained shear strength, c_{uo}) and β (the ratio of skin friction to effective overburden pressure, σ'_{vo}). The dependence of these parameters on the consolidation history of the soil is represented by plotting them against the strength ratio, c_{uo}/σ'_{vo}. This ratio reflects the overconsolidation ratio (OCR) of the soil, but is more directly obtainable from standard site investigation data.

One of the key quantities, that affect the performance of a pile

---

[1]Assistant Lecturer, Cambridge University Engineering Department, Cambridge, England.

under monotonic or cyclic loading, is the relative compressibility or extensibility of a pile. The more compressible a pile is, the greater is the possibility of large relative movements occurring between the upper part of the pile and the soil, leading to degradation of the shear transfer. An approach is outlined for calculating a safe upper limit to the cyclic load amplitude, as a fraction of the ultimate shaft capacity of the pile, in terms of an appropriate stiffness ratio. The upper limit is chosen so as to avoid relative slip between pile and soil at ground level, thus reducing the risk of degradation of the soil immediately surrounding the pile.

The design approaches discussed in the paper are illustrated by results from two series of tests on model piles conducted at Cambridge University. The first series, carried out by Francescon (8) under a research contract with the Building Research Establishment, Department of the Environment, consisted of axial load tests on a full displacement pile jacked into clay. Tests were performed in clay which had been consolidated to known effective stress conditions, at different initial values of OCR. Pore pressure transducers, radial stress cells and axial load cells on the pile, enabled the stress changes occurring during pile installation, consolidation of the surrounding soil, and loading of the pile, to be carefully monitored. The second series of tests, performed by the author (15) and supported financially by McClelland Engineers, consisted of torsional load tests on piles of different flexibility, jacked into clay. The object of these tests was to study the effect of the flexibility of a pile on its performance under monotonic and cyclic loading.

UNIT VALUES OF SKIN FRICTION

Considerable research effort has been put into trying to understand the changes which occur in the soil due to the installation of a displacement pile. The strain path undergone by soil close to the advancing pile is complex, but the net displacement pattern may be divided into three main areas (see Fig. 1). Near ground level some surface heave may occur, while, at the advancing pile tip, a spherical bulb of movement takes place. Over much of the pile shaft, however, the major effect of pile installation is to cause radial movement of the soil away from the pile axis. The process is thus often idealised as the expansion of a cylindrical cavity under conditions of plane strain.

The high strain level in the soil around a driven pile will cause severe remoulding of the soil, and a resulting change in effective stresses. Forcing of the soil away from the advancing pile will lead to an increase in the total stress level in the soil. Adjacent to the pile, the combined effect of these two changes will be the generation of excess pore pressures, Δu. Randolph et al (16) suggest that the magnitude of the excess pore pressures may be estimated by

$$\Delta u \cong 4\, c_{uo} - \Delta p' \qquad (1)$$

where $\Delta p'$ is the change in mean effective stress due to shearing and remoulding of the soil. For lightly overconsolidated soil, the mean effective stress will decrease during shearing, and the above expression would indicate values of excess pore pressure greater than 4 c_{uo}. In heavily overconsolidated soil, $\Delta p'$ will be positive, as the soil attempts to dilate, and the excess pore pressures generated will be less than 4 c_{uo}. This trend of decreasing excess pore pressure with increasing

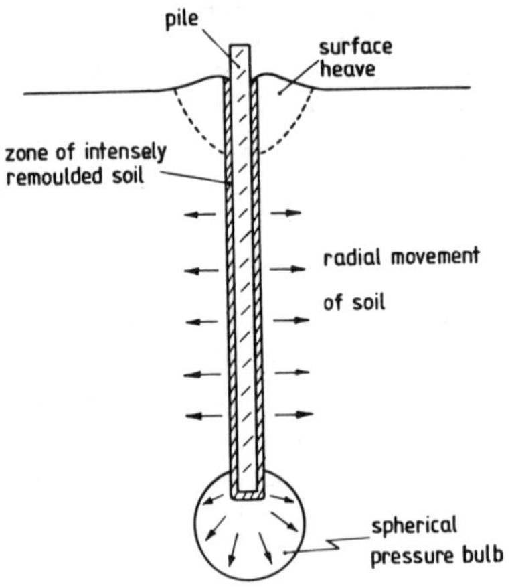

Fig. 1 Soil movements due to pile installation

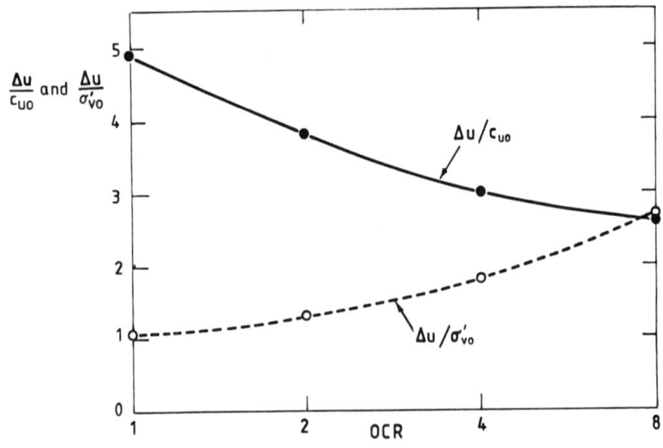

Fig. 2 Excess pore pressures due to installation of a full displacement pile (after Francescon (8))

OCR (for a given value of shear strength, c_{uo}) is reflected in results from model pile tests discussed by Francescon (8). Figure 2 shows his results, where the excess pore pressure generated by pile installation is plotted against the initial OCR of the soil. It may be also seen from this figure that excess pore pressures in excess of twice the effective overburden pressure, σ'_{vo}, are generated at high OCR.

Immediately after pile installation, high pore pressure gradients will exist in the radial direction. Typically, excess pore pressures will be neglibile beyond about 10 pile radii (4, 8). The resulting consolidation, as the excess pore pressures dissipate, will be mainly by outward radial flow of pore water and small inward recovery of the soil movements caused by installation. Finite element analyses of the consolidation process (5, 16) indicate that the changes in moisture content close to the pile, due to outward flow of pore water, are such as to increase the undrained shear strength of the soil by 40 - 60%. The corresponding effective stresses close to the pile at the end of consolidation were computed to be σ_r' (radial) $\sim$ 5 c_{uo} and σ_z' (vertical) $\sim$ 3 c_{uo}.

The above predictions of the final effective stress state and shear strength of the soil are reasonable for normally consolidated soil and are consistent with the limited available data (notably Seed and Reese (20) who took samples from close to piles driven into San Francisco bay mud). For more heavily overconsolidated soil, where the in situ shear strength, c_{uo}, may be as high as, or higher than, the effective overburden stress, σ_{vo}', the predicted final effective stresses appear unrealistically high. The predictions were obtained from finite element analyses where no strain was permitted in the vertical direction. In reality, as discussed by Randolph and Wroth (19), the soil is free to strain in this direction and a more realistic boundary condition might be to assume that the vertical stress, σ_z', returns eventually to close to the original effective overburden pressure, σ_{vo}'.

Meyerhof (12) has attempted to deduce the final radial effective stress acting on piles driven into overconsolidated clay. Starting from the limiting skin friction measured in a load test, and assuming that the friction angle mobilised at the pile-soil interface is the same as the internal angle of friction, ϕ', for the soil, he shows that the radial effective stress acting on the pile at failure lies between 1 and 2 times the best estimate of the original horizontal effective stress, σ_{ho}'. This back-analysis relies on key assumptions, such as assuming an interface friction angle of ϕ', and correct estimation of the value of σ_{ho}', but laboratory data from Francescon (8), shown in Fig. 3, support Meyerhof's findings. From Fig. 3, the final radial effective stress after consolidation of the soil around the pile, lies between 1.5 and 2 K_o σ_{vo}',

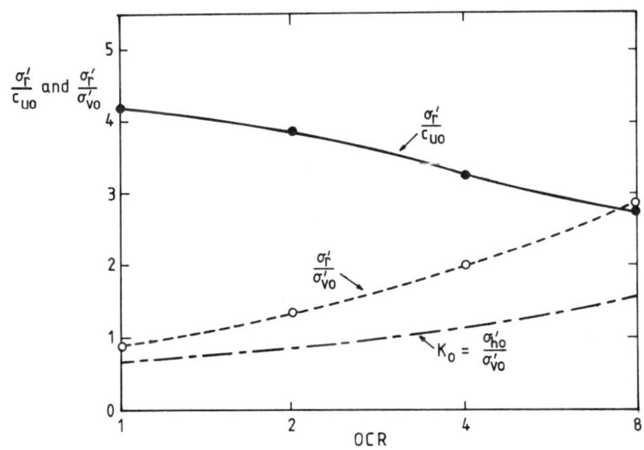

Fig. 3 Radial effective stresses after consolidation (after Francescon (8))

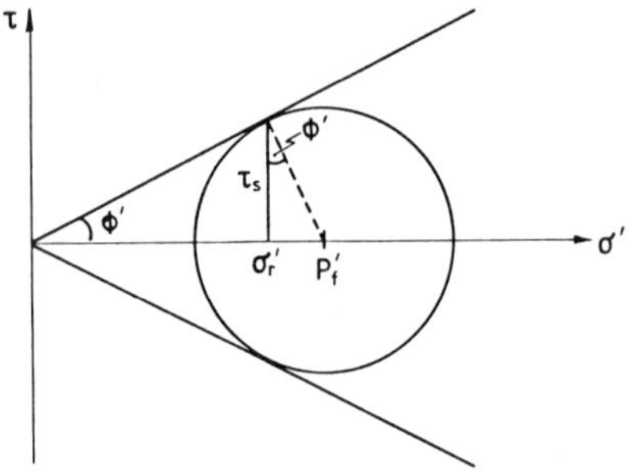

(a) Failure of soil continuum for normally consolidated soil

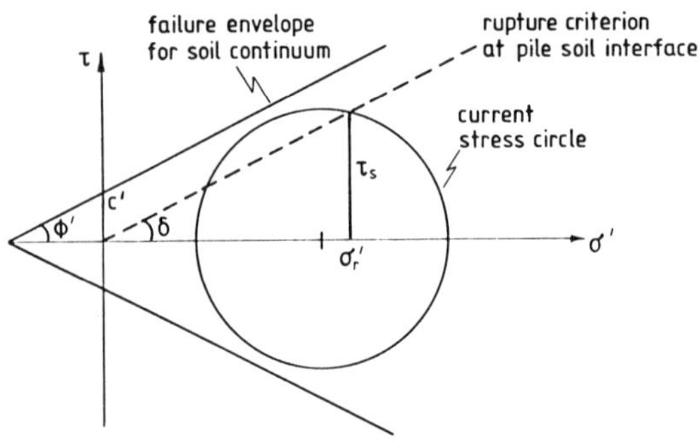

(b) Rupture at pile-soil interface prior to failure of soil continuum for heavily overconsolidated soil

Fig. 4 Possible modes of failure at pile-soil interface

where K_o is the at rest earth pressure coefficient.

When the pile is loaded, soil around the pile is deformed in a similar manner to soil in a simple shear test. For normally or lightly overconsolidated clay failure of the pile will correspond to failure of the soil adjacent to the pile (see Fig. 4(a)). As the pile plunges to failure, a vertical rupture surface is formed, and the directions of principal stress will rotate until the angle of friction mobilised on the pile shaft reaches its maximum value (taken equal to ϕ' in Fig. 4(a)). In order to estimate the peak skin friction, the relationship proposed by Randolph and Wroth (18) may be used

$$\tau_f = \frac{3 - \sin\phi'}{3} \cos\phi' \, (c_u)_{tc} \qquad (2)$$

where τ_f is the failure shear stress in a simple shear test and $(c_u)_{tc}$ is the shear strength measured in a triaxial compression test. The ratio $\tau_f/(c_u)_{tc}$ lies typically between 0.7 and 0.8 for common values of ϕ'.

Since the current shear strength of the clay will be some 50% greater than the in-situ strength (due to consolidation following pile installation), the peak skin friction, τ_s, may be expected to be just greater than the in situ shear strength, c_{uo}.

For heavily overconsolidated clay, which possesses some cohesion, c', the situation at failure of the pile may be rather different. Since the process of installing a pile will remould the soil close to the pile (reducing the cohesion to zero), it becomes possible for failure to occur at the pile-soil interface before the strength of the soil continuum is fully mobilised. Figure 4(b) shows this situation. Failure of the pile occurs, prior to mobilising the strength of the unremoulded soil, at a value of skin friction given by

$$\tau_s = \sigma_r' \tan\delta \tag{3}$$

where δ is the interface friction angle. As Meyerhof (12) has demonstrated, results from field pile tests are consistent with the combined assumptions, that $\sigma_r' \cong 1.5\ \sigma_{ho}' = 1.5\ K_o\ \sigma_{vo}'$ and that $\delta \cong \phi'$.

Any design approach should be consistent with established empirical design rules, irrespective of whether the skin friction is to be expressed as a fraction of the in situ shear strength, c_{uo}, or of the effective overburden pressure, σ_{vo}'. Consistency between the total and effective stress approaches may be aided by plotting the variation of $\alpha = \tau_s/c_{uo}$ and $\beta = \tau_s/\sigma_{vo}'$, against the in situ strength ratio c_{uo}/σ_{vo}' (noting that $\alpha\ c_{uo}/\sigma_{vo}' = \beta$). Figure 5 shows such a plot, derived from the considerations discussed previously, for values of ϕ' between 20° and 30°. For normally consolidated clay, the ratio α has been limited to unity, while values of β for overconsolidated clay are based on

$$\beta = 1.5\ K_o \tan\phi' \tag{4}$$

taking conservative values of K_o for given values of c_{uo}/σ_{vo}' and ϕ' (using relationships presented by Ladd et al (10)).

Data from model pile tests in Speswhite kaolin (with $\phi' \sim 23°$) reported by Francescon (8) show reasonably good agreement with the design curves for low and intermediate values of c_{uo}/σ_{vo}'. For high values of c_{uo}/σ_{vo}' (i.e. very heavily overconsolidated clay), considerable uncertainty exists in the appropriate choice of α and β. The curves in Fig. 5 have been shown by dashed lines for values of c_{uo}/σ_{vo}' greater than 1.6, emphasising this uncertainty. Values of β reported in the literature range as high as 3 or 4 (22), implying radial effective stresses, acting on the pile, which are greater than $K_p\ \sigma_{vo}'$, where K_p is the passive earth pressure coefficient. While it is possible that such high effective stresses may be locked in the ground during consolidation of the soil, it would seem an unsafe design philosophy to rely on them existing throughout the life of a pile. A more prudent approach would be to limit the value of β to $K_p \tan\phi'$ (shown in Fig. 5). In practice, soil with values of c_{uo}/σ_{vo}' greater than about 2 is usually only found in the upper few metres of a deposit. Shear transfer between offshore piles and the soil is often ignored in this region, due to post-holing effects arising from lateral loading of the piles.

It should be emphasised that the tentative design curves presented in Fig. 5 are based on theoretical models of the stress changes caused by full displacement piles. Similarly, most of the data studied by Meyerhof (12) were for solid, or closed ended, driven piles. In offshore

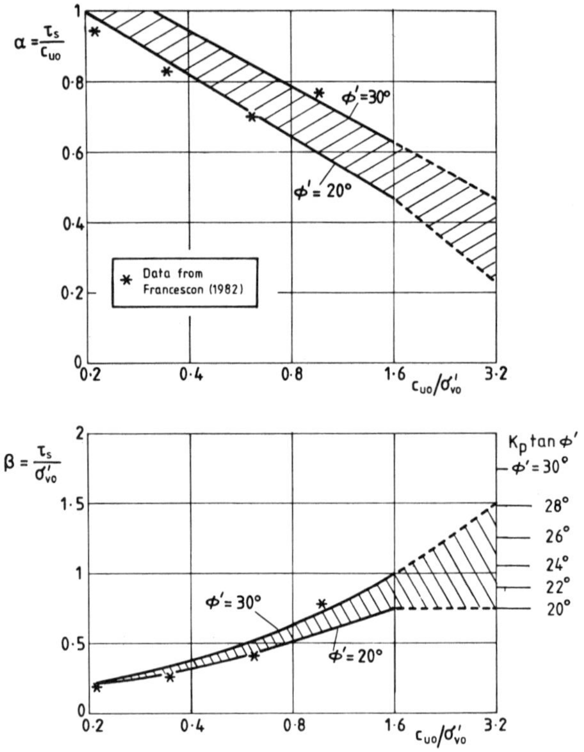

Fig. 5 Design curves for variation of α and β with c_{uo}/σ'_{vo}

applications, open ended steel piles are generally used. Such piles remain essentially 'unplugged' during installation, allowing soil to move up the inside of the piles. They are thus likely to cause smaller disturbance to the soil than full displacement piles. Carter et al (6) present the results of numerical studies of the effect of the area ratio of driven piles. For typical ratios of wall thickness to pile diameter for offshore piles, the ratio of net to gross pile area is around 10%. The results presented by Carter et al indicate that the effective stress level adjacent to such open ended pipe piles would be up to 20% lower than for full displacement piles. Consequently, it may be prudent to reduce the design parameters given in Fig. 5 by up to 20% when applying them to the design of open ended piles that remain unplugged during installation.

At large displacements of the pile relative to the soil, the shear transfer may reduce from a peak to a residual value, as a polished rupture plane develops. Model tests indicate that the effect of cumulative relative displacement under cyclic loading may be particularly severe (3, 21). Two separate effects contribute to the reduction in shear transfer. As the clay particles are re-aligned, some stress relief will occur, reducing the radial effective stress. It is not clear what lower limit to the value of σ_r' should be taken. A possible approach for driven piles would be to assume that the increase in horizontal effective stress due to pile installation, is subsequently lost. Thus the value of σ_r' reduces to σ_{ho}'. The second effect of relative displacement of the pile and soil, is a reduction in the friction angle on the rupture surface, to a residual value, ϕ_{res}. Ring shear tests (2, 11) have shown that the friction angle for the soil can reduce by a factor of 2 after as little as 30 - 50 mm of relative movement along a rupture surface.

The combined effects of reducing σ_r' and δ, lead to a residual value of skin friction which may be estimated as

$$(\tau_s)_{res} = \sigma_{ho}' \tan\phi_{res}' \tag{5}$$

The ratio between $(\tau_s)_{res}/(\tau_s)_{peak}$ may be obtained by comparing equations (4) and (5), giving

$$\frac{(\tau_s)_{res}}{(\tau_s)_{peak}} = \frac{\sigma_{ho}' \tan\phi_{res}'}{1.5\sigma_{ho}' \tan\phi'} = 0.7 \ \frac{\tan\phi_{res}'}{\tan\phi'} \ . \tag{6}$$

For $\tan\phi_{res}' \sim 0.5 \tan\phi'$, the ratio is 0.35. Model pile test data reported by Steenfelt et al (21) and by Francescon (8) show ratios of residual pile capacity under cyclic loading to peak capacity, varying between 0.3 and 0.54, thus lending some support to equation (6).

SETTLEMENT OF PILES

A solution for the ground level settlement, w_t, of piles under axial load, P, has been presented by Randolph and Wroth (17). Treating the soil as elastic, with a shear modulus, G, which varies linearly with depth as shown in Fig. 6, and Poisson's ratio, ν, the load settlement ratio for a solid pile of radius r_o, length ℓ and Young's modulus, E_p, may be written

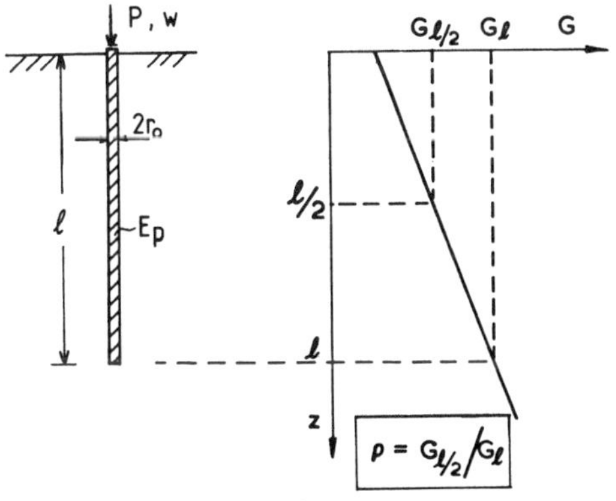

Fig. 6 Idealised variation of soil shear modulus with depth

$$\frac{P}{G_\ell r_o w_t} = \frac{\dfrac{4}{1-\nu} + \dfrac{2\pi}{\zeta}\,\rho\,\dfrac{\tanh(\mu\ell)}{\mu\ell}\,\dfrac{\ell}{r_o}}{1 + \dfrac{1}{\pi\lambda}\,\dfrac{4}{(1-\nu)}\,\dfrac{\tanh(\mu\ell)}{\mu\ell}\,\dfrac{\ell}{r_o}} \tag{7}$$

where G_ℓ is the shear modulus of the soil at a depth of $z = \ell$, λ is the stiffness ratio, E_p/G_ℓ, and ρ is the ratio of the shear modulus at mid-depth of the pile to the shear modulus at the pile base (see Fig. 6). The quantities ζ and $\mu\ell$ are defined by

$$\zeta = \ell n[2.5\,\rho(1 - \nu)\,\ell/r_o] \tag{8}$$

and $$\mu\ell = \left(\frac{2}{\zeta\lambda}\right)^{\frac{1}{2}} \frac{\ell}{r_o}\,. \tag{9}$$

The significance of the parameter, ζ, is that it gives the ratio of the local pile movement, w, normalised by the pile radius, r_o, to the local shear strain in the soil adjacent to the pile. Thus for mobilised shear stress at the pile shaft of τ, the local pile movement (or soil movement, assuming no slip) is

$$\frac{w}{r_o} = \zeta\,\frac{\tau}{G} \tag{10}$$

Typical values of ζ range between 3 and 5 (see also Baguelin and Frank (1)). The pile movement, w^*, necessary to generate full skin-friction may be calculated as

$$\frac{w^*}{r_o} = \zeta\,\frac{\tau_s}{G} = \zeta\,\alpha\,\frac{c_{uo}}{G} \tag{11}$$

For typical values of α and G/c_{uo}, the normalised movement to failure, w^*/r_o, will vary between about 1%, for piles in stiff heavily overconsolidated clay, up to about 4%, for piles in normally consolidated clay.

The degree of pile compression is governed by the quantity, $\mu\ell$, which represents the relative stiffness of the pile and the soil. For very stiff, or short, piles, $\mu\ell$ tends to zero and equation (7) reduces to

$$\frac{P}{G_\ell r_o w_t} = \frac{4}{(1-\nu)} + \frac{2\pi}{\zeta}\,\rho\,\frac{\ell}{r_o} \tag{12}$$

At the other extreme, for long compressible piles, $\mu\ell$ increases, and $\tanh(\mu\ell)$ tends to unity. The expression then reduces approximately (exactly for $\rho = 1$) to

$$\frac{P}{G_\ell r_o w_t} = \pi\rho\left(\frac{2\lambda}{\zeta}\right)^{\frac{1}{2}} \tag{13}$$

It may be noticed that this expression is now independent of the pile length, ℓ - implying that the effect of load applied to the top of the pile dies out before the base of the pile is reached. Equation (13) may be used in place of equation (7) for piles where $\mu\ell > 1.5$, or $\ell/r_o > 2\lambda^{\frac{1}{2}}$.

When applying equations (7) and (13) in design, it is necessary to check that the assumption of an elastic soil, and no slip between pile and soil, is reasonable. It is helpful in this context to calculate the pile load at which slip will first occur between pile and soil at ground level (i.e. where $w_t = w^*$). For long piles where equation (13) applies (most piles used offshore fall into this category) the load for which $w_t = w^*$ is given by

$$P^* = \pi\rho G_\ell r_o \left(\frac{2\lambda}{\zeta}\right)^{1/2} w^* \tag{14}$$

Substituting for w^*, and normalising by the ultimate shaft capacity $P_{ult} = 2\pi r_o \ell \bar{\tau}_s$, the limiting 'elastic' load level is given by

$$\frac{P^*}{P_{ult}} = \left(\frac{\lambda\zeta}{2}\right)^{1/2} \frac{r_o}{\ell} = \frac{1}{\mu\ell} \tag{15}$$

(where the assumption has been made that the ratio τ_s/G does not vary with depth).

At load levels above that given by equation (15), pile settlement may be calculated by treating separately the upper part of the pile, where slip between pile and soil has occurred, and the lower part of the pile. The settlement at the boundary between the two regions may be calculated using equation (7), with the load P reduced by the amount of the stress transfer in the upper part of the pile. The additional settlement due to compression of the pile in the upper region may be calculated from basic mechanics.

EFFECTS OF PILE COMPRESSIBILITY

It is clear from equation (15) of the previous section, that failure between pile and soil at ground level may occur at relatively low loads. Values of $\mu\ell$ for offshore piles lie typically between 3 and 4; thus failure starts to occur at 25 - 30% of the ultimate shaft capacity. The consequences of this are two fold. Firstly, under monotonic loading to failure, considerable slip between the upper part of the pile and the surrounding soil may occur before the lower part of the pile has reached failure. In a strain softening soil, the measured capacity of the pile will thus be less than that for an ideal (rigid) pile which mobilises the peak skin friction simultaneously down the whole pile shaft.

The second consequence relates to cyclic loading. The magnitude of the skin friction mobilised at the pile shaft has been shown to decrease significantly once relative slip occurs between pile and soil. This is particularly true when two-way shearing of the rupture surface occurs. For compressible piles, where failure is initiated at ground level at relatively low loads, it is possible for the pile-soil bond to be broken at each load cycle under quite moderate cyclic load amplitudes. For example, for $\mu\ell = 4$, equation (15) gives $P^*/P_{ult} = 0.25$. Thus a cyclic load range of 50% of the ultimate shaft capacity will cause two-way slip between pile and soil, at ground level, at each load cycle, regardless of the mean value of the applied load.

Equation (15) may be interpreted as a design guide-line for avoiding significant degradation of the soil around a pile under cyclic loading. In order to avoid relative slip between pile and soil occurring (in both directions) at each load cycle, the ratio of cyclic load amplitude, P_{cyc}, (i.e. half the cyclic load range) to the ultimate shaft capacity, P_{ult}, must be kept below $1/\mu\ell$. In practical terms, this may be achieved by designing piles of sufficient radius to ensure that the response to a given cyclic load amplitude is 'elastic'.

It is helpful to relate the quantity, $\mu\ell$, to the stiffness ratio, π_3, introduced by Murff (13), defined as

$$\pi_3 = \frac{\pi d \tau_s \ell^2}{(AE)_p w^*} \tag{16}$$

where d is the pile diameter and $(AE)_p$ is the cross-sectional stiffness of the pile section. (Note, the equivalent modulus E_p used in the previous section is related to $(AE)_p$ by $E_p = (AE)_p/\pi r_o^2$.) Substituting for w^*/τ_s from equation (11), it may be shown that

$$\pi_3 = \frac{2\pi r_o G\ell^2}{\pi r_o^2 E_p \zeta r_o} = \frac{2}{\lambda\zeta}\frac{\ell^2}{r_o^2} = (\mu\ell)^2 \ . \tag{17}$$

Thus the limiting cyclic load amplitude to avoid slip between pile and soil may be expressed as

$$\frac{P_{cyc}}{P_{ult}} \leq \frac{1}{\sqrt{\pi_3}} \tag{18}$$

The parameter, π_3, was originally used by Murff (13) to help quantify the effect of pile compressibility on the shaft capacity of piles in a strain-softening soil. Murff adopted a load transfer curve which strain-softened abruptly once the full skin friction was mobilised (see curve A in Fig. 7). Such a model of the soil response exaggerates the effect of pile compressibility, and a more appropriate model might be to assume that a further displacement, Δw_{res}, would be necessary to reach a residual

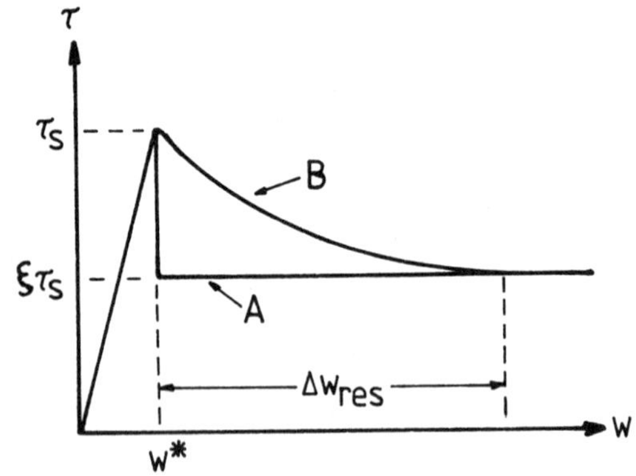

Fig. 7 Load transfer curves for pile-soil interface

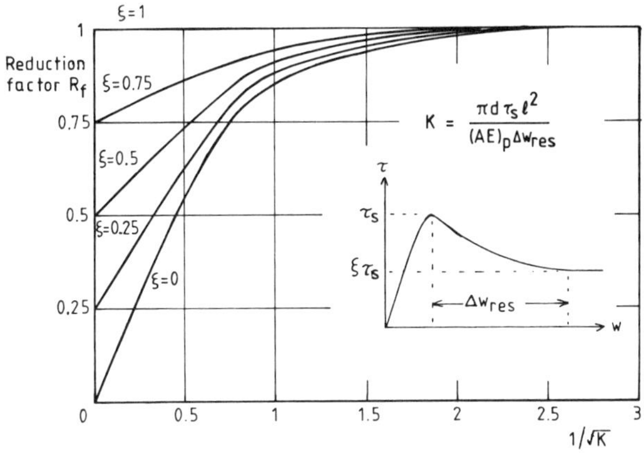

Fig. 8 Effect of stiffness ratio, K, on pile capacity

value of skin friction (curve B in Fig. 7). Following Randolph and Wroth (19), a new stiffness ratio, K, may be introduced, defined identically to π_3 except that w* is replaced by Δw_{res}, giving

$$K = \frac{\pi d \tau_s \ell^2}{(AE)_p \Delta w_{res}} . \tag{19}$$

Figure 8 shows the variation of reduction factor, R_f, (defined as the ratio of actual pile shaft capacity to that of an ideal, rigid, pile) with stiffness ratio, K, for different values of $\xi = (\tau_s)_{res}/(\tau_s)_{peak}$.

From ring shear tests (2, 11) significant degradation of the shear transfer across a rupture plane can occur after as little as 10 mm of relative movement. Typical plots of shear stress against displacement show values of ξ as low as 0.5 for Δw_{res} between 30 and 50 mm. The degradation appears particularly marked for clays of plasticity index greater than 30%. For a typical offshore pile, with ℓ = 100 m, d = 1.5 m, wall thickness = 50 mm, in soil with $(\tau_s)_{peak}$ = 150 kN/m^2, ξ = 0.5 and Δw_{res} = 30 mm, the value of K is 4.9. The reduction factor, R_f, from Fig. 8, is then 0.72, giving an actual capacity of 51 MN, compared with an ideal capacity for a rigid pile of 71 MN. For the same pile, taking the displacement, w*, necessary to mobilise full skin friction as 1% of the pile diameter (i.e. 15 mm) the stiffness ratio π_3 would be 9.8. Thus the safe cyclic load range is $2/\sqrt{9.8} \times 71$ = 45 MN.

In an attempt to substantiate the theoretical ideas discussed above, a series of torsional loading tests on model piles of varying flexibility was performed at Cambridge University (15). The mode of shearing of the soil around a pile loaded by torsion is different from that for axial loading. However, the equations governing the response of the pile are similar for the two forms of loading (14). Torsion tests have the advantage that piles of high flexibility (at model scale) are easily obtainable, whereas it is difficult to match the stiffness ratios of offshore piles at model scale under axial loading.

An initial series of tests on piles in coarse sand was conducted by

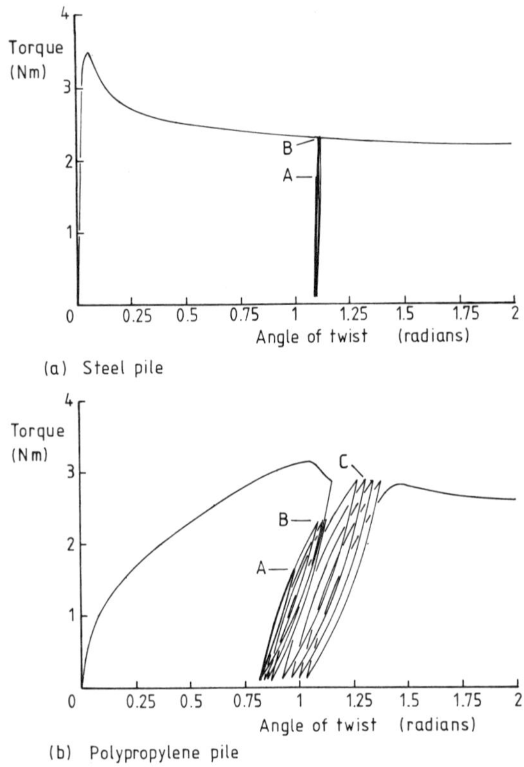

Fig. 9 Torsional response of steel and polypropylene piles

Dakin (7) as an undergraduate project, using steel, aluminium and perspex (acrylic) piles. Further tests in clay were then performed by the author, using piles made of steel, perspex, polypropylene and PTFE. The results to be presented are for a steel pile (diameter 10.6 mm) and a polypropylene pile (diameter 11.2 mm) jacked 300 mm into Speswhite kaolin clay, normally consolidated to $\sigma'_{vo} = 400$ kN/m^2. Both piles were coated with a thin layer of araldite and fine sand in order to achieve a consistent surface texture for the different piles. The shear modulus of the polypropylene pile varied from 440 MN/m^2 at low stress levels, down to 300 MN/m^2 at the upper range of torques applied during the tests.

The torque-twist relationships for the two piles are shown in Fig. 9. A monotonic loading test up to large angles of twist was performed on the steel pile in order to obtain the load transfer curve for the clay around the pile. Cyclic load tests were then performed between 2 and 50% (point A) and between 2 and 63% (point B) of the peak capacity. (Note that 63% of the peak capacity corresponded to the current capacity.) The polyproylene pile was loaded until failure was observed, followed by cyclic load tests between 3 and 53% (point A), 3 and 73% (point B) and

Pile type	Skin friction (kN/m^2) Peak	Residual	Stiffness ratios $(\pi_3)_{torsion}$	$K_{torsion}$	Reduction factor R_f Measured	Predicted
Steel	66.1	42.0	0.23	0.01	-	-
Polyprop-ylene	53.4	44.0	25 - 37	4.2	0.81	0.80

Table 1 Details from torsion tests on model piles

3 and 93% (point C) of the peak capacity. Ten load cycles were performed at each load level.

Table 1 summarises the salient points of each test. Equivalent stiffness ratios for torsional loading may be calculated as

$$(\pi_3)_{torsion} = \frac{1}{2}\frac{\pi d^2 \tau_s \ell^2}{(GJ)_p \psi^*} ; \qquad K_{torsion} = \frac{1}{2}\frac{\pi d^2 \tau_s \ell^2}{(GJ)_p \Delta\psi_{res}} \qquad (20)$$

where $(GJ)_p$ is the torsional rigidity of the pile, ψ^* is the angle of twist necessary to mobilise full skin friction (here taken as 0.05 radians) and $\Delta\psi_{res}$ is the additional angle of twist to achieve a residual condition ($\sim$ 0.6 radians). In calculating values of $K_{torsion}$ for the polypropylene pile, the range in modulus of the pile must be considered. Also, since the cyclic loading tests were conducted after failing the pile, τ_s has been taken as 49 kN/m^2 (the average skin friction available at the start of the cyclic loading tests).

The predicted reduction factor, R_f, for the polypropylene pile may be obtained by interpolation from Fig. 9 (taking ξ = 0.63) as R_f = 0.80. The average skin friction at failure of the polypropylene pile was 53.4 kN/m^2 or 81% of that for the steel pile. While such close agreement between prediction and measurement is fortuitous, the trend of decreasing pile capacity with increasing pile flexibility was observed throughout the test series. From Fig. 9(a), it may be seen that cyclic loading of the steel pile produced no degradation. Instead, a stable, and narrow, hysteresis loop was observed, even when the peak torque equalled the current pile capacity (point B). For the polypropylene pile tests (Fig. 9(b)), degradation of the pile response was apparent. For the lowest load cycle (3 - 57% of the current capacity), a stable hysteresis loop is observed. At the intermediate load cycle (3 - 80% of the current capacity) incremental collapse of the pile is observed, with each load cycle giving rise to additional permanent twist of the pile. At the highest load cycle (3 - 100% of the current capacity), incremental collapse occurs initially, followed by failure of the pile as further degradation of the soil takes place.

The estimated values of $(\pi_3)_{torsion}$ may be used, in conjunction with equation (18), to estimate the cyclic load range at which slip between pile and soil starts to occur. For $(\pi_3)_{torsion} = 25$, slip should start to occur for $P_{cyc} = 20\%$ of P_{ult}. The amplitude of the lowest load cycle is 27% of the current capacity. Although the hysteresis loop appears stable, the deformation range is about 0.1 radian, implying that two-way slip between pile and soil is just starting to occur at ground level (remembering that $\psi^* \sim 0.05$ radians). Thus the criterion given by equation (18) looks reasonable, if slightly conservative, as a design guide-line for avoiding slip between pile and soil at each load cycle, and the dangers of incremental collapse of the pile.

CONCLUSIONS

This paper has considered design aspects of the response of piles driven into clay, to axial loading. The major part of the capacity of such piles is derived from skin friction along the pile shafts. The design process for a particular application may then be considered in two stages:

(a) estimating profiles of peak and residual skin friction, over the depth of embedment of the pile, from the available soils data;

(b) assessing the response of the pile, as a structural member, to monotonic and cyclic loading.

The second of the two stages may lead to modification of the profiles of skin friction originally estimated, for example due to consideration of the effects of lateral loading of the pile.

Over the years, a better understanding is gradually being developed of the effective and total stress changes which occur in the soil adjacent to displacement piles. In due course, effective stress methods of design based on the stress path followed during pile installation, consolidation of the soil, and pile loading may emerge, similar to those reviewed by Kraft (9). At present, however, a precise quantitative understanding of the processes is still lacking, and tentative design curves such as those shown in Fig. 5 are based on a mixture of analytical studies and empirical observations. Change and refinement of such curves will no doubt occur as more reliable data from field and laboratory pile tests become available.

Pile load tests show evidence of a reduction in available skin friction with increasing relative displacement between pile and soil. This reduction may occur very rapidly (for example see Fig. 9(a)) and can allow progressive failure of the pile. Under cyclic loading, sucessive degradation of the soil at each load cycle - and the possibility of incremental collapse as shown in Fig. 9(b) - may be avoided by ensuring that the cyclic displacement amplitude is less than that necessary to cause slip between the pile and soil at each cycle. This implies a maximum cyclic load amplitude, which may be expressed in terms of the stifness ratio of the pile and the ultimate shaft capacity (see equation (18)).

There is still a paucity of reliable data from pile load tests in soil types and for stiffness ratios typical of offshore conditions. At the same time, new generations of offshore platforms, particularly those employing tension piles, present a continuing challenge for designers. In order to meet this challenge, theoretical models of pile and soil response must be supplemented by results from actual pile tests. The model pile tests discussed in the paper are being continued at Cambridge University by centrifuge tests on tension piles. Use of the centrifuge enables the stiffness ratio of prototype offshore piles to be matched corretly at model scale. Such tests, conducted in reconstituted clay, provide a means of assessing the effect of pile geometry and stiffness ratio, on the response to cyclic loading. However, there is still an urgent need for data from field scale pile tests, conducted in soil types relevant to offshore deposits.

REFERENCES

1. Baguelin, F. and Frank, R., "Theoretical Studies of Piles using the Finite Element Method", Proceedings 1st International Conference on Numerical Methods in Offshore Piling, Institution of Civil Engineers, London, 1979, pp. 83-92.
2. Bishop, A. W., Green, G. F., Garga, V. U., Andresen, A. and Brown, J. D., "A New Ring Shear Apparatus and its Application to the Measurement of Residual Strength", Geotechnique, Vol. 21, No. 4, 1971, pp. 273-328.
3. Bogard, D. and Matlock, H., "A Model Study of Axially Loaded Pile Segments', Report to American Petroleum Institute, 1978.
4. Burghignoli, A. and Carnana, R., "Indagine sperimentale sugi effetti dell'infissione di pali in terreni coesivi tenari", Rivista Italiana di Goetechnica, Vol. 13, No. 2, 1979, pp. 87-93.
5. Carter, J. P., Randolph, M. F. and Wroth, C. P., "Stress and Pore Pressure Changes in Clay During and After the Expansion of a Cylindrical Cavity", International Journal of Numerical and Analytical Methods in Geomechanics, Vol. 3, No. 4, 1979, pp.305-322.
6. Carter, J. P., Randolph, M. F. and Wroth, C. P., "Some Aspects of the Performance of Open and Closed Ended Piles", Proceedings 1st International Conference on Numerical Methods in Offshore Piling, Institution of Civil Engineers, London, 1979, pp. 165-170.
7. Dakin, M. R. A., "Torsion Tests to Study the Effect of Compressibility on the Loading Response of Piles", Final year project report, Cambridge University Engineering Department, 1982.
8. Francescon, M., Discussion in Proceedings 2nd International Conference on Numerical Methods in Offshore Piling, Austin, 1982, also forthcoming Ph.D. Thesis, "Model Pile Tests in Clay", Cambridge University.
9. Kraft, L. M., "Effective Stress Capacity Model for Piles in Clay", (to appear in) Journal Geotechnical Engineering Division, ASCE, Vol. 108, 1982.
10. Ladd, C. C., Foott, R., Ishihara, K., Schlosser, F. and Poulos, H. G., "Stress Deformation and Strength Characteristics", State-of-the-art Report, Proceedings 9th International Conference on Soil Mechanics and Foundation Engineering, Tokyo, 1977, Vol. 2, pp. 421-494.
11. Lupini, J. F., Skinner, A. E. and Vaughan, P. R., "The Drained Residual Strength of Cohesive Soils", Geotechnique, Vol. 31, No. 2, 1981, pp. 181-213.
12. Meyerhof, G. G., "Bearing Capacity and Settlement of Pile Foundations", Journal of Geotechnical Engineering Division, ASCE, Vol. 102, No. GT3, 1981, pp.197-228.
13. Murff, J. J., "Pile Capacity in a Softening Soil", International Journal Numerical and Analytical Methods in Geomechanics, Vol. 4, No. 2, 1980, pp. 185-189.
14. Randolph, M. F., "Piles Subjected to Torsion", Journal Geotechnical Engineering Division, ASCE, Vol. 107, No. GT8, 1981, pp. 1095-1111.

15. Randolph, M. F., "Torsion Tests of Model Piles", Report to McClelland Engineers, 1982.
16. Randolph, M. F., Carter, J. P and Wroth, C. P., "Driven Piles in Clay - The Effects of Installation and Subsequent Consolidation", Geotechnique, Vol. 29, No. 4, 1979, pp. 361-393.
17. Randolph, M. F. and Wroth, C. P., "Analysis of Deformation of Vertically Loaded Piles", Journal of Geotechnical Engineering Division, ASCE, Vol. 104, No. GT12, 1978, pp. 1465-1488.
18. Randolph, M. F. and Wroth, C. P., "Application of the Failure State in Undrained Simple Shear to the Shaft Capacity of Driven Piles", Geotechnique, Vol. 31, No. 1, 1981, pp. 143-157.
19. Randolph, M. F. and Wroth, C. P., "Recent Developments in Understanding the Axial Capacity of Piles in Clay", Ground Engineering, Vol. 15, No. , 1982, pp.
20. Seed, H. B. and Reese, L. C., "The Action of Soft Clay along Friction Piles", Proceedings American Society of Civil Engineers, Vol. 81, 1955, Paper 842.
21. Steenfelt, J. S., Randolph, M. F. and Wroth, C. P., "Instrumented Model Piles Jacked into Clay", Proceedings 10th International Conference on Soil Mechanics and Foundation Engineering, Stockholm, 1981, Vol. 2, pp. 857-864.
22. Tomlinson, M. J., "Adhesion of Piles in Stiff Clay", CIRIA Report, No. 26, 1970.

NOTATION

$(AE)_p$ - cross-sectional rigidity of pile
c_{uo} - in situ undrained shear strength
$(c_u)_{tc}$ - undrained shear strength in triaxial compression
d - pile diameter
E_p - Young's modulus of equivalent solid pile
G - soil shear modulus (subscript ℓ denoting level of pile base)
$(GJ)_p$ - torsional rigidity of pile
K - pile stiffness ratio (based on residual deformation)
K_o - at rest earth pressure coefficient
K_p - passive earth pressure coefficient
ℓ - pile length
P - load at pile head (subscripts 'ult' denoting ulimate shaft capacity, 'cyc' denoting cyclic load amplitude)
p' - mean effective stress
R_f - Reduction factor
r_o - pile radius
u - pore pressure
w - pile movement (subscript t denoting movement of pile head, superscript * denoting local pile movement to cause slip)
z - depth

α - ratio of skin friction to in situ undrained shear strength
β - ratio of skin friction to effective overburden pressure
Δ - change in quantity
δ - interface friction angle
ζ - parameter in vertical pile deformation
λ - stiffness ratio E_p/G_ℓ
μ - parameter in vertical pile deformation
ν - Poisson's ratio

ξ - ratio of residual to peak skin friction
π - mathematical constant
π_3 - pile stiffness ratio (based on deformation to first failure)
ρ - ratio of soil shear modulus at pile mid-depth to that at pile base
σ' - normal effective stress, subscripts ho - in situ horizontal
r - radial (at pile shaft)
vo - in situ vertical
z - vertical (at pile shaft)
τ - shear stress, subscripts f - failure in simple shear
s - failure at pile shaft (skin friction)
ϕ' - internal angle of friction of soil
ψ - angle of twist of pile (in torsion tests)

Subscript expressions:

cyc - cyclic amplitude
peak - peak value
res - residual value
torsion - torsional (as opposed to axial)
ult - ultimate value

DRIVEN PILES IN A GRANULAR CALCAREOUS DEPOSIT

By Javier Gómez Puyuelo,[1], Javier Sastre[1] and Antonio Soriano[2]

INTRODUCTION

This paper describes the experience gained in the use of driven steel pipe piles on calcareous sands of coralline origin. Six load tests to failure have been performed to investigate the load capacity of a deposit of soil of this type that forms the bottom on the southern coast of the island of Nonoc (Philippines), where a coal-unloading terminal for a Nickel and Copper Refinery is under construction.

Three load tests were carried out on the same pile at increasing driven lengths and three other load tests were run at a neighbouring pile also at three different depths within the coralline soil deposit.

At depths in the order of 45 m below sea level (water depths ranging from 10 to 20 m), a sound sedimentary rock exists that allowed the design of point bearing piles, since the main result of this load testing is that even if the sand and gravel deposit appeared as dense or medium dense under SPT testing, it would not develop the strength needed for an alternative floating pile foundation.

The results of load testing show that load capacities of piles within this deposit would be much lower, in fact several times lower, than with a similar deposit of siliceous sands and gravels that would give quite similar relative densities on a regular subsoil investigation, thus indicating the tremendous influence that the type of soil has on the capacity to withstand driven pile loads. However, a comparison of test results with pile driving resistance, through the use of pile driving formulae, confirms a better correlation, as would occur in other types of soil; hence, not only static resistance, but also dynamic resistance to penetration are, similarly, substantially reduced for this type of tropical formation, as compared to siliceous soil deposits.

[1] Ingeniero de Caminos, Empresarios Agrupados, S.A., Madrid
[2] Ingeniero de Caminos, Ph.D, Empresarios Agrupados, S.A., Madrid M.ASCE

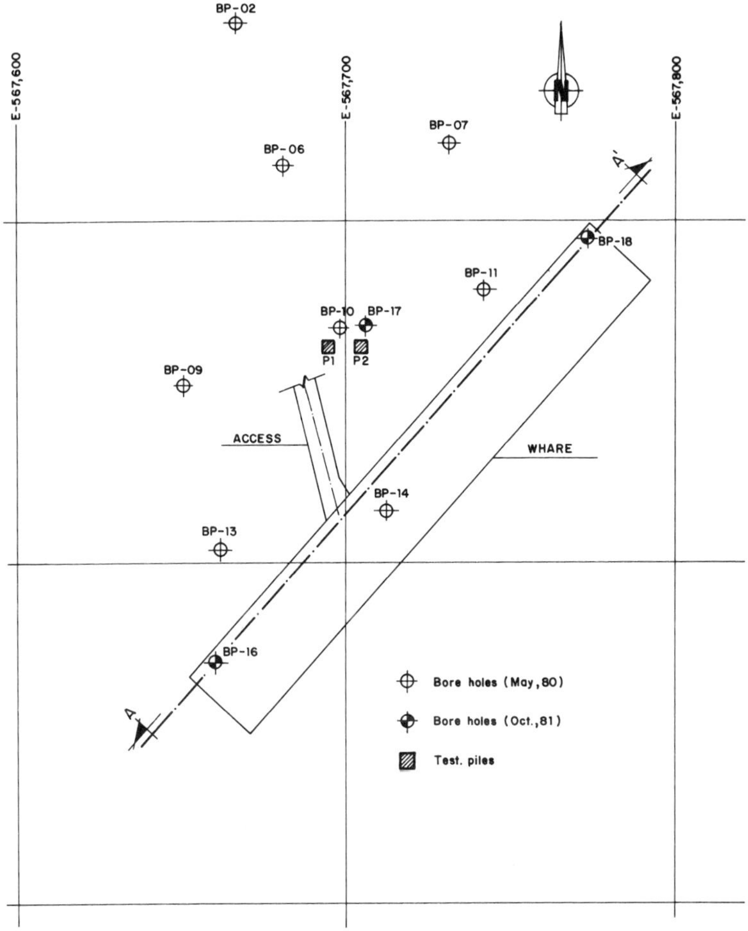

Fig. Nº 1 - PLAN VIEW OF THE SITE

SITE DESCRIPTION

Construction work in progress is partly on land and partly offshore. The pile foundation described here is off-shore, close to the coast, and with average water depths of 10 to 20 m..

The off-shore facility to be built is a coal unloading terminal 180 m long and 25 m wide, supported on 146 piles. Piles have been designed as steel pipes of 30 inches in diameter (0.762 m), with a wall thickness of 1/2 inch (12.7 mm), to be driven to bedrock through the more recent coralline formation.

At the initial stages of the project, some borings were performed to investigate the underground conditions at the wharf location as well as at other installations of this facility. Borings close to the wharf (BP-10, BP-11, BP-13 and BP-14) were driven up to depths that did not reach the underlying rock but were stopped at a depth where more cemented corals gave refusal to penetration testing. Along with the pile testing, more borings were performed along the future wharf line (borings BP-16, BP-17 and BP-18) that indicated the existence of the sedimentary rock at the somewhat constant depth of 45 m below sea level. Water depths were very different from place to place, giving rather steep slopes of the sea bed.

A sketch of the soil conditions is given in Fig. 2 and in Fig. 3 results of some "in situ" and laboratory tests are summarized.

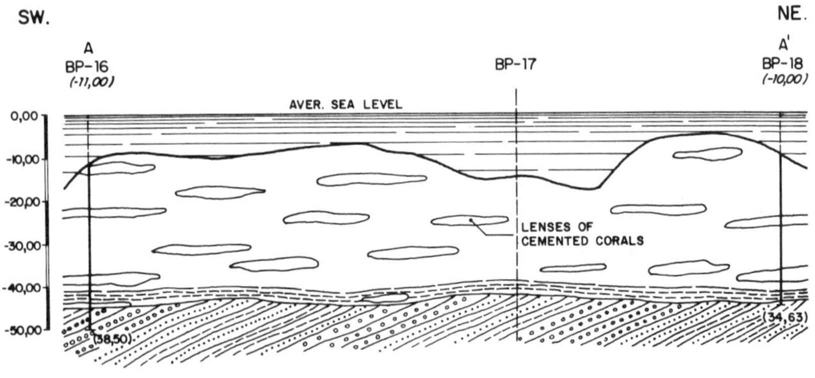

Fig. Nº 2 - SKETCH OF SOIL CONDITIONS

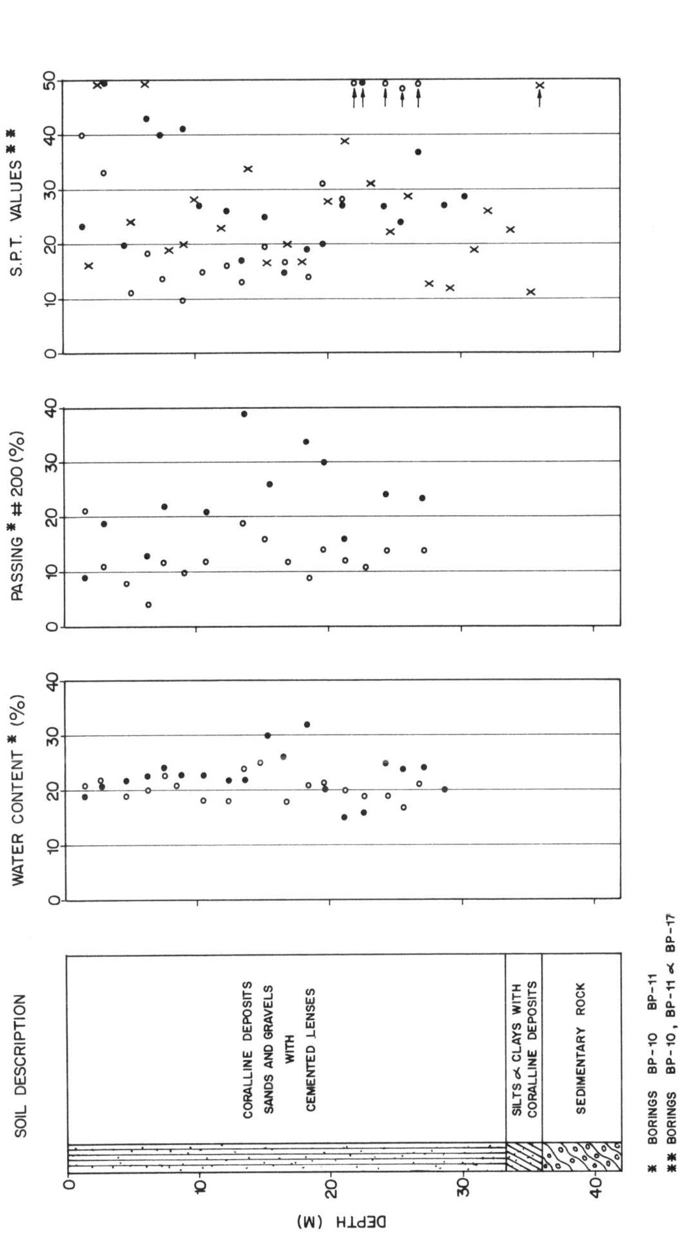

Fig. Nº 3 - TYPICAL SOIL PROFILE

The coralline deposit is mainly granular; the percentage of fines passing the ASTM N° 200 mesh is below 40% on all grain size analysis and these fines have no plasticity except for the lower part of the deposit, near the bedrock. No data on this lower part of the profile are available other than visual classification.

The natural water content varies within the deposit between the range of 15 to 30%.

SPT values are somewhat erratic and no correlation with depth or across the profile has been found. The test gave refusal at some cemented levels and the average value of the other test that did not give refusal was N(SPT) = 20, with a large scatter.

Due to the difficulties of obtaining undisturbed samples, no tests to estimate the shear strength were run other than unconfined compression of cemented corals and sedimentary rock. The unconfined compressive strength of the corals sampled at the cemented levels was as high as 5 MN/m^2 for the more resistant samples. The average value of the unconfined compressive strength of the sedimentary rock was 9.3 MN/m^2.

On the basis of these data, it was thought that two types of pile behaviour could be expected under driving. First, it was expected that most of the piles would encounter some of the cemented lenses and then further driving, without other means than the driving hammer, would not be possible. Secondly, it was expected that a few of the other piles could not find refusal. A KOBE-KB 60 hammer with a rated energy of 157 KN.m was to be used for the final part of the driving.

At the design stage, both conditions were analyzed and both were allowed to occur independently over the whole area of the wharf. Piles reaching refusal (less than 1 mm per blow) that was expected always after at least 17 m of driving (the shallowest cemented level found in the boring logs) were assigned a minimum failure load of 8.6 MN. Piles that would not reach refusal were designed to have at least 50 m of their length driven within the deposit and, for such length, the same vertical failure load of 8.6 MN was estimated. These estimates were based on an average value of the side friction of 42 KN/m^2 and point resistances in the order of 15 MN/m^2 for cemented levels and 8 MN/m^2 for the few cases where piles would not reach refusal at any depth when driving.

When designing the superstructure, provisions were made for differential settlements caused by different types of pile, short end bearing or long floating.

Finally, a load test was specified to check the accuracy of the predicted load capacities prior to the execution of the piling.

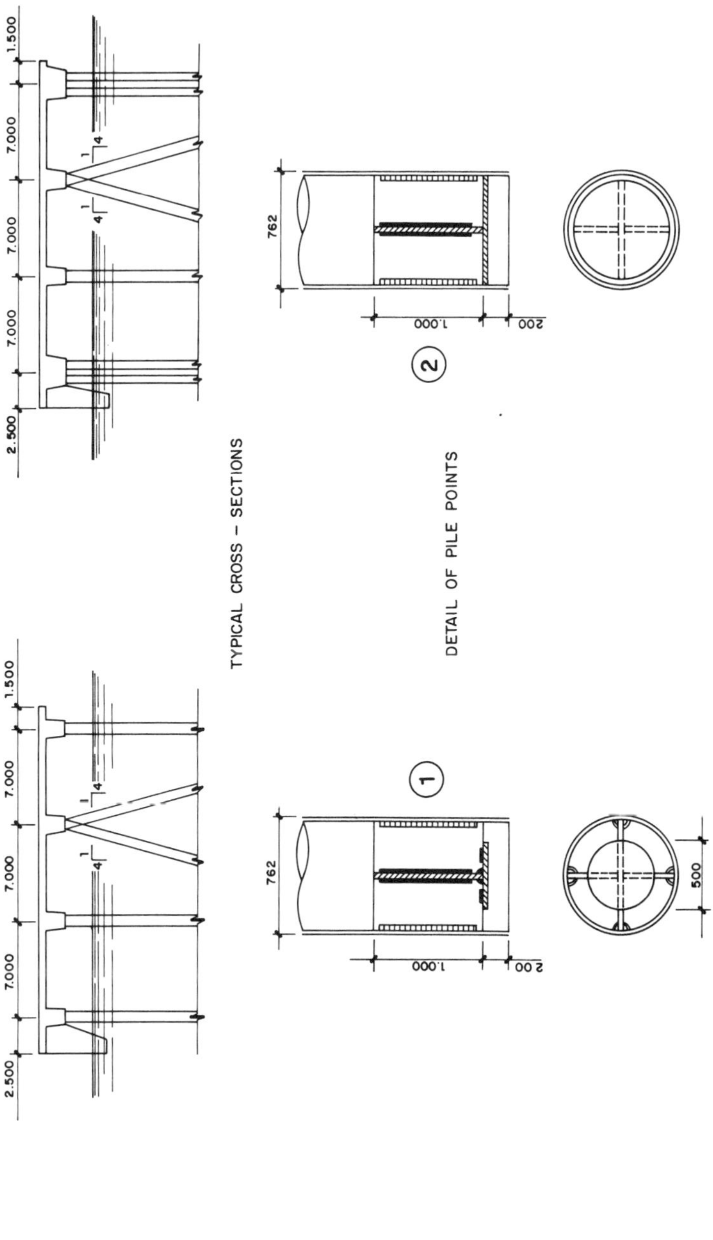

Fig. Nº 4 - DETAILS OF THE STRUCTURE

DESCRIPTION OF THE STRUCTURE

The pier is comprised of a mooring platform, an access gangway and a transition area, as indicated in Fig. 1. It is formed by an orthogonal 2m trapezoidal girder framework and 0.50 m mats embedded therein.

Two of the longitudinal girders, those which have to support the cranes, are supported in two groups of two battered or vertical piles, with a separation of 7.15 m between each group. The other longitudinal girders are supported by a single pile per node every 7.15 m, except for some which have to withstand greater loads. The separation between longitdinal girders is 7 m.

Figure 4 shows a detail of a typical section of the pier and also a detail of the systems used to seal the points of the piles. One of them, N° 1, was used to perform the load tests on the fist test pile. The second, with a greater seal, was used for the second test pile and then throughout the work.

Two mooring wharves, supported by 12 and 18 piles complete the facility.

The calculation of this structure, under the various loading conditions, gives a maximum vertical load of 300 t (2.94 MN), which each pile must be designed to withstand.

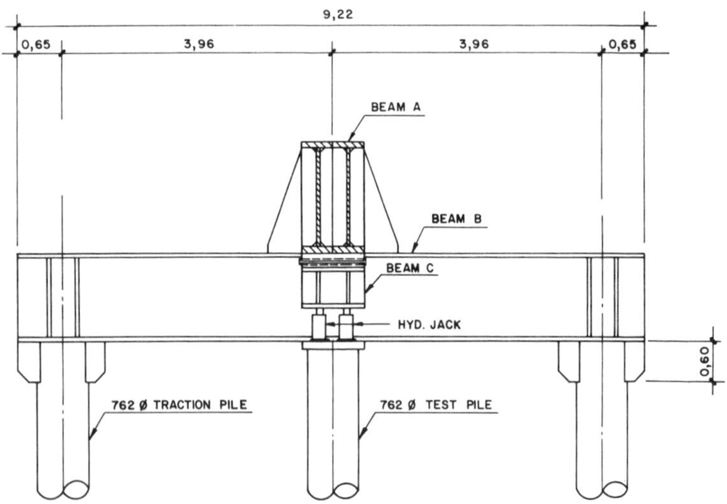

Fig. Nº 5 - LAYOUT OF LOAD TEST

TESTING OF PILE NUMBER 1

According to the specification of the project, a load test was to be carried out near the location of one of the borings where SPT was performed at 1.50 m intervals. At this point, the water depth was about 3 m, since the test site was chosen close to the coastline.

The layout of the test is shown in Fig. 5. Four traction piles and an H-shaped reaction frame were used to provide support for the maximum test load, which was specified as 5.9 MN.

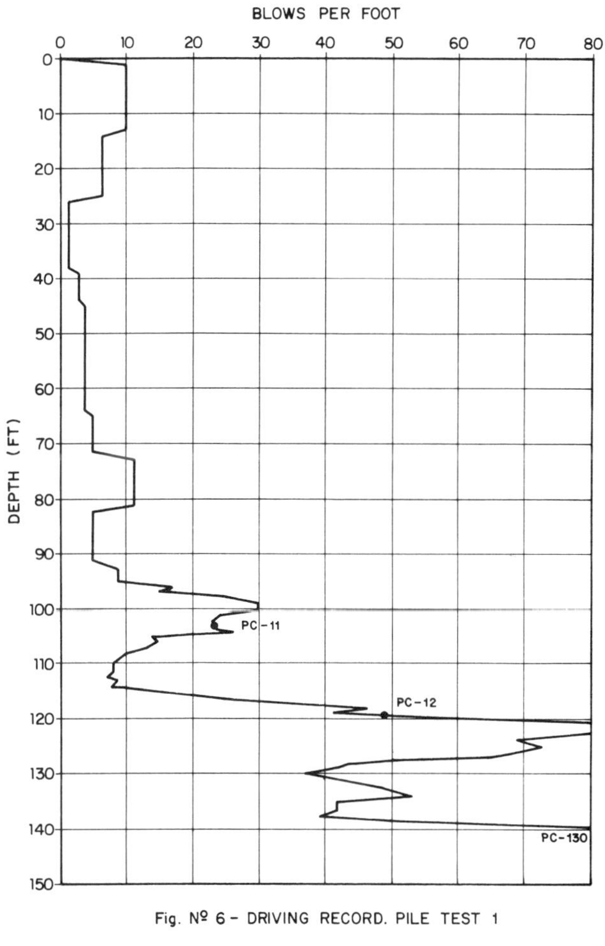

Fig. Nº 6 - DRIVING RECORD. PILE TEST 1

The load was applied by means of four hydraulic jacks of 2 MN each and with a maximum extension of 330 mm. Only displacements at the top of the piles were measured with standard equipment. Load tests were run between three and seven days after driving. Load was applied in steps of 200 to 500 KN (varied from test to test). Deflections were recorded at 30-minute intervals. When no more than 0.25 mm difference occurred between two consecutive readings, loading proceeded to next step. The duration of each load step in these conditions was in the order of three hours.

The four traction piles, as well as the tested pile, were driven with a Delmag-44 hammer (see Table 1) and driving records were obtained. The record of the blowcounts for the test pile is included in Fig. 6. Records of the four traction piles were quite similar.

When the driven length of the test pile was 31.4 m and the blowcount 23 blows per foot, as can be seen in the record of the driving log shown in Fig. 6 (point PC-11), the first load test was run and the load settlement curve given in Fig. 7 was obtained.

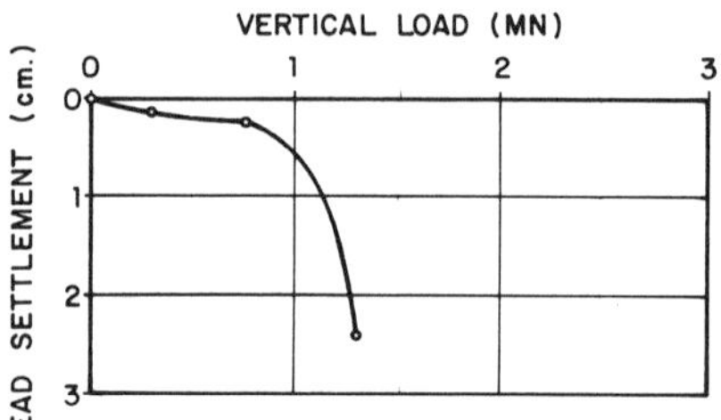

Fig. 7 - LOAD SETTLEMENT CURVE PILE N° 1. DEPTH: 31,4 m.

The failure load obtained from the test was interpreted to be 1.23 MN, i.e. seven times lower than the estimated failure load for end bearing piles and more than five times lower than that estimated for floating piles of this length, by use of the unit skin friction and point resistance, mentioned previously in this paper, that was used during the preliminary stages of the project.

The pile was driven deeper and the load test repeated at a depth of 36.3 m when the blowcount was 49 blows per foot. Fig. 8 shows the load settlement curve for this test. The failure load at this depth reached 2.35 MN, still well below the design service loads.

Finally, this test pile was driven to a depth of 42 m, with a blowcount higher than one hundred blows per foot with hammer D-44 and the third load test to failure was run. Failure load at this stage reached 4.4 MN, which is higher than the design service loads for the piles, but was still below the minimum to provide the required safety. Results are included in Fig. 9.

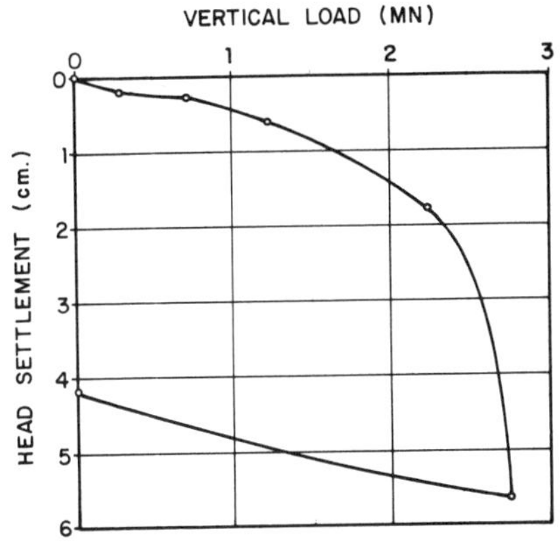

Fig. 8 - LOAD SETTLEMENT CURVE PILE N° 1. DEPTH: 36,3 m.

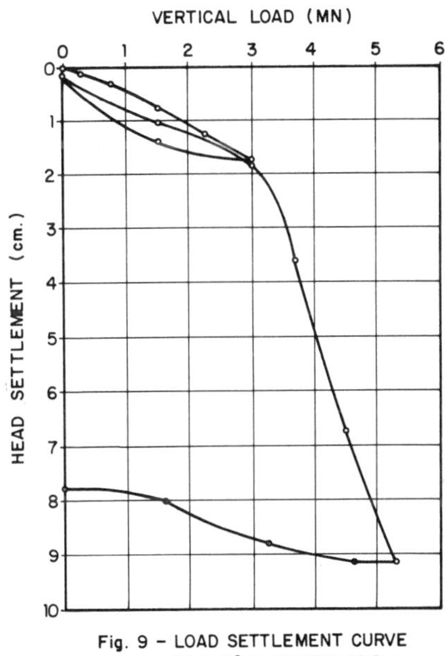

Fig. 9 - LOAD SETTLEMENT CURVE PILE N° 1. DEPTH: 42,0 m.

TESTING OF PILE NUMBER 2

Pile number 1 was driven with a somewhat close point, since a plate with a diameter of 0.50 m, which is smaller than the diameter of the pile (0.762 m) was attached at the point with a steel cross welded to the pile inner face, as shown in Fig. 4. During the load testing, soil inside the pile was detected as being several metres above the point, so a different point - a circular plate with a small central hole - was used to prevent a large soil column in the pile.

A heavier hammer, a KOBE-KB-60, was prepared for the last part of the driving.

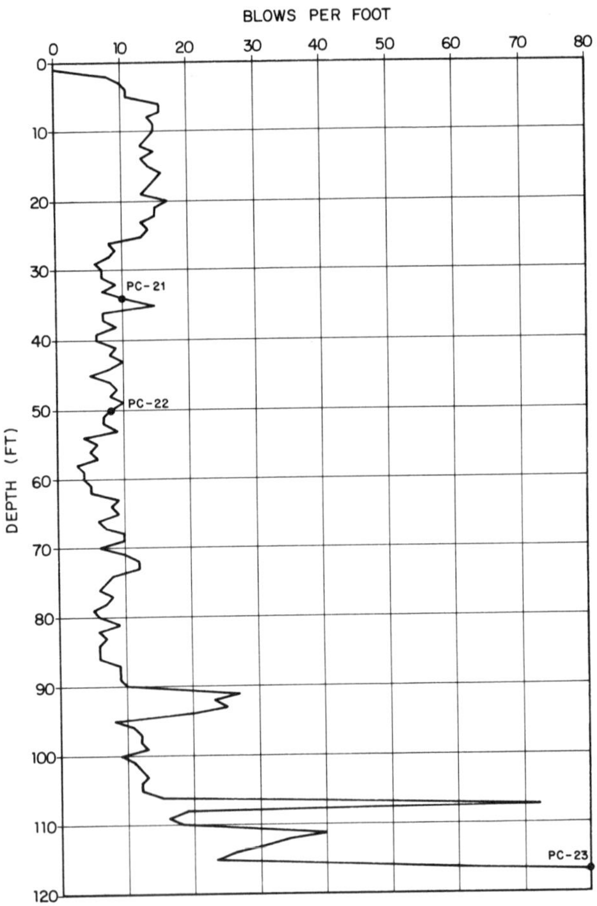

Fig. Nº 10- DRIVING RECORD. PILE TEST 2

Driving of test pile number 2 was done close to pile test number 1, in order to use two of the same traction piles for reaction. The driving record for this pile is given in Fig. 10. Since they are similar to those of the test pile, records of the blowcount for the last two piles are not included.

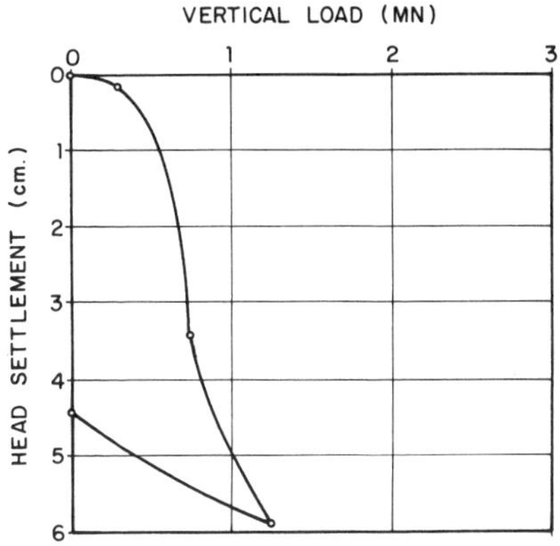

Fig. 11 - LOAD SETTLEMENT CURVE
PILE N° 2. DEPTH: 10,0 m.

The first load test was run at a depth of 10 m and a blowcount of 10 blows per foot, the ultimate load for this test being 1.23 MN. Results are given in Fig. 11.

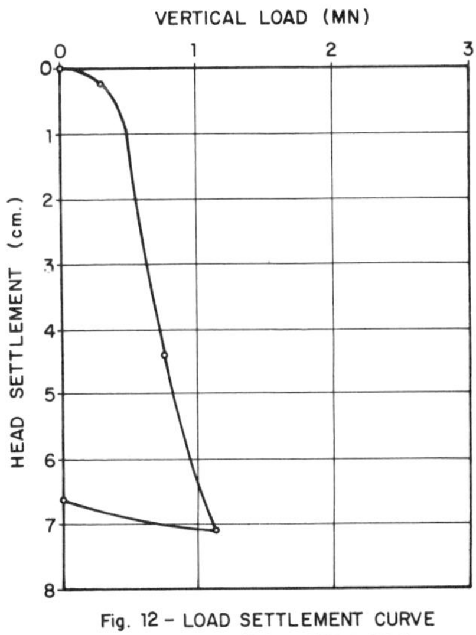

Fig. 12 - LOAD SETTLEMENT CURVE
PILE N° 2. DEPTH: 15,0 m.

The second load test was run after re-driving to a depth of 15 m and a blowcount of 8 blows per foot; in this case, the ultimate load was 1.0 MN. See Fig. 12. As can be seen from the driving log of this pile (Fig. 10) no substantial difference in the blow count between first driving and redriving after load testing was found, as was the case for the first test pile (Fig. 6).

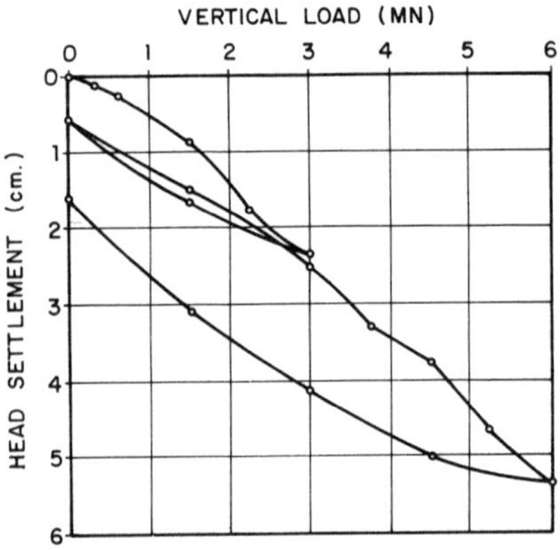

Fig. 13 - LOAD SETTLEMENT CURVE
PILE N° 2. DEPTH: 35,3 m.

Finally, the pile was driven to refusal (less than 3 mm per blow) using the larger hammer KOBE-KB-60. The driven length turned out to be 35.3 m. After this driving, the load test gave the results plotted in Fig. 13. A maximum vertical load of 5.9 MN was applied and the pile was still far away from failure.

EVALUATION OF LOAD TEST RESULTS

The load-settlement curves obtained are significantly different from those corresponding to the test pile driven into the coralline deposit (floating pile) or resting on the rocky substratum (column pile). In the first case, failure load was easily reached and the second type of test was used to analyze pile behaviour with an axial load double the service load (5.9 MN).

In the first test performed (PC-11), no plug was formed and the failure load obtained was lower than expected. The results of the 5 remaining tests were used to correlate ultimate loads with blowcounts during driving measured at test depths (Fig. 14).

Failure loads were obtained using the criteria of Mazurkiewicz (1) and refusals correspond to a Delmag D-44 hammer. Parameters used to calculate Hiley's formula are given in Table 1.

TABLE 1 - PARAMETERS USED WITH HILEY'S FORMULA
DELMAG D-44

Weight	42.2 KN
Rated Energy	118.0 KN.M
Efficiency	0.85
Coefficient of Restitution	0.40
Anvil Stiffness	350 Kg/cm^3
Quake	0.1"

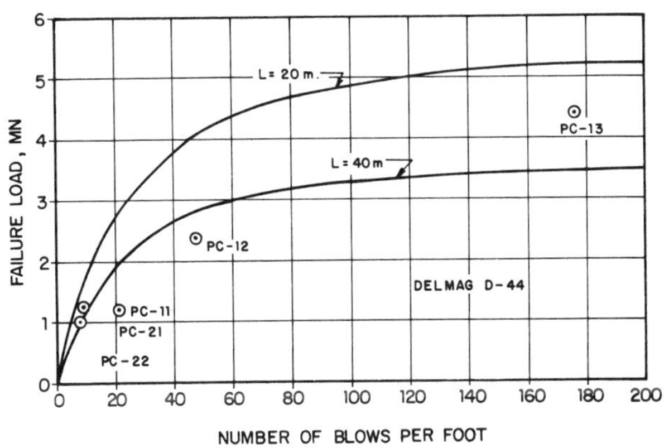

Fig. 14 – COMPARISON OF FAILURE LOADS WITH HILEY'S FORMULAE

In view of the results, during construction of the wharf, the piles were driven until they were embedded in the rocky substratum and, therefore, it was considered necessary to reach refusals lower than 1 mm/blow, measured through a depth of at least 0.3 m. However, the results were also analyzed from the point of view of the behaviour of the piles driven into the granular coralline deposit.

The load tests showed that unit resistance by the point and by the shaft were much lower than would occur in a granular siliceous soil of similar granulometry and density.

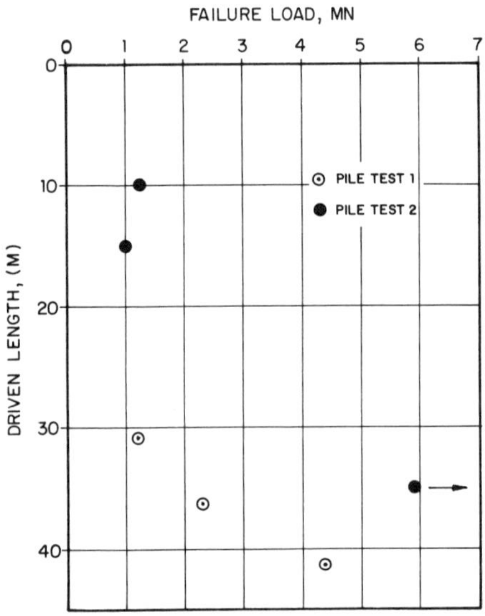

Fig. 15 - FAILURE LOADS VERSUS DEPTH.

The ultimate loads of the six load tests are plotted in Fig. 15. With this data, and assuming a constant value of the unit skin friction according to Datta et al. (2), it is possible to derive the values of the parameters given in Table 2. As a comparison, the values of f_s and q_p have been estimated for a similar pile driven in sand with N = 20 (average value obtained in the coralline deposit), using one of the empirical correlations valid for siliceous sand (3); these results are also shown in Table 2. As may be seen, the differences are significant: the resistant parameters deduced from the load tests are 3 to 4 times lower than those estimated by correlation with the number N. By way of an example, the table also includes the failure loads of a 35 m long pile corresponding to both pairs of values. Similar values of ultimate unit skin friction have been reported elsewhere (4).

TABLE 2

	Deduced from load tests (1)	Estimated by correlation with N (2)	Ratio (2)/(1)
Unit resistance at point, MN/m^2, q_p	2	6	3.0
Unit resistance by shaft, MN/m^2, f_s	14.7	54.0	3.7
35 m shaft failure load, MN	2.1	7.2	3.4

For static loads and low stress levels, these materials have a resistant behaviour similar to that of granular siliceous soils (5). However, the behaviour of granular siliceous and calcareous materials as regards pile driving differs very significantly. The main differences are summarized below (6):

a. Calcareous particles are normally less resistant than siliceous. Therefore, the grain fragmentation phenomena which take place under the high stress levels produced at the point during driving (responsible for the decrease of q_p as the pile diameter increases in siliceous sands) are much more pronounced in calcareous deposits, in which the generalized breaking of grains and collapse of cemented structures drastically increase soil compressibility (7).

   It should be emphasized that this difference in behaviour is not shown in the static or dynamic penetration tests, since the stress levels developed are not very high.

b. Calcareous deposits are normally quite porous, either because they are formed by coral fragments or due to cementing phenomena immediately after their deposition, which occur without significant changes in density. During driving, the calcareous particles are degraded in contact with the shaft and are displaced towards the inside of the porous formation. Consequently, the lateral pressures on the pile are low.

In summary, the lower load capacity of the piles driven in granular calcareous deposits seems to be due to undesirable phenomena which occur during driving.

REFERENCES

1. Fellenius, B.H., "Test Loading of Piles and New Proof Testing Procedure", Jour. Geotech. ASCE, 101 GT9, 1975.
2. Datta, M., Gulathi, S.K. and Rao, G.V., "An Appraisal of the Existing Practice of Determining the Axial Load Capacity of Deep Penetration Piles in Calcareous Sands", Proc. 12th Annual Conf. Offshore Tech., Vol. 4, Houston, 1980.
3. Coyle, H.M. and Castello, R.R., "New Design Correlations for Piles in Sand", Proc. ASCE, J. of Geotech. Engng., GT 7, July, 1981.
4. Angemeer, J., Carlson, E.D., Stroud, S. and Kurzeme, M., "Pile Load Test in Calcareous Soils Conducted in 400 feet of Water from a Semi-Submersible Exploration Rig", OTC 2311, 1975.
5. De La Gándara, J., G. Puyuelo, J., and Velasco, M., "Comportamiento de Pilotes Hincados en un Depósito Granular Calcáreo", Bol. Sociedad Española Mec. Suelo, Madrid, 1982.
6. McClelland, B., "Design of Deep Penetration Piles for Ocean Structures", Proc. ASCE, J. of Geotech. Engng., GT 7, July, 1974.
7. Vesic, A.S., "Experiments with Instrumented Pile Groups in Sand", ASTM Special Technical Pub. N° 44, 1968.

GENERAL EFFECTIVE STRESS METHOD FOR PILES IN CLAY
PART I - THEORY

By Robert C. Kirby[(1)], M. ASCE, Melvin I. Esrig[(2)], M. ASCE, and Benton S. Murphy[(3)]

SUMMARY

The axial capacity of a pile in clay can be divided among the capacity provided by the pile tip and the capacity provided by shear resistance along the pile shaft. The shaft resistance is normally calculated using empirical methods which incorporate one or two preconstruction soil parameters. The most common approach is to use empirical methods based on undrained shear strength. Most geotechnical engineers would agree that the ultimate shear resistance along the pile shaft is controlled by the effective stress at the pile face at failure and the effective stress friction angle. However, the major unknown in the application of an effective stress method is the prediction of the effective stress at the pile face at failure. This stress is influenced by initial effective stress and the changes in effective stress associated with pile installation, consolidation around the pile, and pile loading. The quasi effective stress methods in use make simplifying assumptions about effective stress at the pile face at failure that do not reflect the complex pile installation, set-up and loading process. This paper describes the theoretical aspects of a 6-year program of engineering research in which effective stress is tracked from initial conditions through pile loading to failure. The analytical and/or soil behavior models for each event in the life of a pile are described and the limitations of these models are identified.

---

(1) Associate, Woodward-Clyde Consultants, Clifton, N.J.
(2) Principal, Woodward-Clyde Consultants, Wayne, N.J.
(3) Staff Research Engineer, Amoco Production Company

PURPOSE OF PAPER

Esrig et al (5) described the initial development of a general effective stress method for prediction of axial capacity for displacement piles in clay. The purpose of this paper is to provide an up-to-date summary of the essential elements of the general effective stress method developed during the Amoco Pile Research Program. This summary is necessary because the technical reports generated during the project fill many volumes and because partial results of the research at various stages of development are scattered among a number of technical papers (1, 2, 5, 6, 7, 9, 10, 14, 18, 19, 20, 25). A companion paper (11) considers field and laboratory data used to test the ESM, restates the ESM in a form better suited to practical application, and presents aids for application of the method for preliminary design.

BASIS OF GENERAL EFFECTIVE STRESS METHOD FOR PREDICTION OF AXIAL PILE CAPACITY

The axial capacity of a pile in clay can be divided among the capacity provided by the pile tip and the capacity provided by shear resistance along the pile shaft. Most geotechnical engineers would agree that the shear stress along the pile shaft at failure, τ_f, is related to the radial effective stress on the pile face at failure, $\bar{\sigma}_{rf}$, and the effective stress friction angle of the soil, $\bar{\phi}$. The stress conditions for an element of soil at the pile face are as shown in Fig. 1. Inspection of Fig. 1 indicates that τ_f can be written as

$$\tau_f = \bar{\sigma}_{rf} \sin \bar{\phi} \quad \ldots\ldots\ldots\ldots\ldots\ldots (1)$$

Effective stress friction angle, $\bar{\phi}$, is an engineering property that can be measured in the laboratory. On the other hand, $\bar{\sigma}_{rf}$, is influenced by the state of stress before pile installation, the engineering properties of the soil, the geometry of the pile, the pile installation procedure, and the changes in stress and soil properties associated with pile installation, setup, and loading.

Our efforts at developing the general effective stress method have shown that the mean effective stress, $\bar{p}$, is a more predictable measure of effective stress than radial effective stress. As a result, we have decided

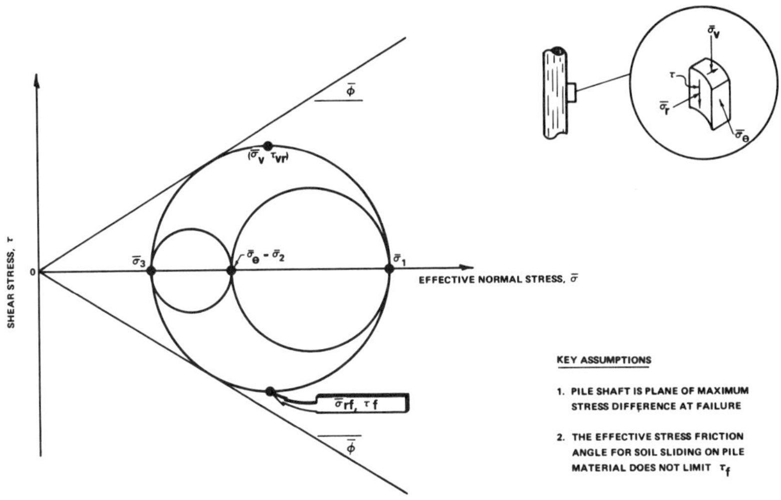

FIGURE 1 MOHR CIRCLE REPRESENTATION OF STRESS SYSTEM AT FAILURE UNDER COMPRESSIVE AXIAL PILE LOAD

to track mean effective stress from initial conditions to pile loading to failure, and then to convert mean effective stress to radial effective stress based on analysis of the stress system at failure. Equation 1 can be rewritten to reflect this approach as follows:

$$\tau_f = \bar{p}_f \frac{\bar{\sigma}_{rf}}{\bar{p}_f} \sin \bar{\phi} \quad \ldots\ldots\ldots\ldots\ldots\ldots \text{(2)}$$

The ratio $\bar{\sigma}_{rf}/\bar{p}_f$ defines the relationship between mean effective stress and radial effective stress at failure and is called the Load Module, LM.

The effective stress friction angle at the pile-soil interface may be lower than the effective stress friction angle of the soil. We can account for the effects on τ_f of a low interface friction angle by introducing an Interface Factor, F_i. Equation 2 can be rewritten to include the Load

Module and the Interface Factor and becomes

$$\tau_f = \bar{p}_f \text{ LM } F_i \sin \bar{\phi} \quad \ldots \quad (3)$$

The major challenge of the general effective stress method for prediction of axial pile capacity is the prediction of $\bar{p}_f$ which is influenced by initial effective stress and the changes in effective stress associated with the major events in the life of a pile. Tracking mean effective stress from initial conditions to pile loading to failure can be viewed as a problem of addition as illustrated in Table 1. The following sections of this paper consider the analytical and/or soil behavior models used to predict the stress states and/or stress changes identified in Table 1. Having predicted $\bar{p}_f$, we then consider methods for estimating the Load Module, LM, and the Interface Factor, F_i.

Event	Stress Change		Cumulative Stress	
Site Investigation	none		$\bar{p}_o$	initial effective stress
Pile Installation	change in effective stress caused by pile installation	$\Delta\bar{p}_{inst}$	$\bar{p}_i$	effective stress after pile installation
Pile Set-up	change in effective stress caused by consolidation around a displacement pile	$\Delta\bar{p}_{cons}$	$\bar{p}_{ac}$	effective stress after consolidation
Pile Loading	change in effective stress caused by pile loading	$\Delta\bar{p}_{load}$	$\bar{p}_f$	effective stress at pile loading to failure

Table 1 Stresses Associated With Major Events In The Life Of A Displacement Pile In Clay

The ultimate capacity at the pile tip contributes little to the overall capacity of relatively long piles in soft marine sediments. As a result, this was not a topic of the research described herein and conventional methods for estimating tip capacity from undrained shear strength can be used.

DESCRIPTION OF PRECONSTRUCTION STATE OF THE SOIL DEPOSIT

One of the first steps toward the prediction of the behavior of soil when subjected to a prescribed loading system or displacement pattern is to define the initial or preconstruction **state** of the soil. The specific parameters used to describe initial conditions differ from one researcher to another but most agree that initial state of stress and stress history are key parameters. The concept of a **critical state** (21) has proven useful in the development of the general effective stress method for prediction of axial pile capacity. As a result, the conventions used in this paper to describe the **state** of the soil follow closely those used by Roscoe et al (21). The following paragraphs review these conventions and show how they are related to conventions more familiar to geotechnical engineers in the United States.

The **state** of a small element of soil (a point) within a soil deposit is defined by the density and the state of stress at the point. The density is conveniently defined by void ratio, e. Definition of the stress system at a point would require six parameters in the most general case. However, the initial stress system on a soil element can be defined by two stress parameters if the horizontal stress is isotropic (i.e. $\bar{\sigma}_2 = \bar{\sigma}_3$) and the shear stresses on horizontal and vertical planes are zero. The most obvious stress parameters for this situation are the vertical and horizontal effective stresses, but experience suggests that stress parameters which are related to average normal effective stress and average shear stress are more useful for analysis. As a result, the stress parameters $\bar{p}$ and q, as defined below, were adopted by Roscoe and his co-workers (21).

$$\bar{p} = \frac{\bar{\sigma}_1 + \bar{\sigma}_2 + \bar{\sigma}_3}{3} \qquad (4)$$

$$q = \bar{\sigma}_1 - \bar{\sigma}_3 \qquad (5)$$

where $\bar{\sigma}_1$, $\bar{\sigma}_2$ and $\bar{\sigma}_3$ are the major, intermediate and minor principal effective stresses, respectively. Note that the definitions of $\bar{p}$ and q differ from the definitions used by Lambe (13).

The stress parameters $\bar{p}$ and q, and the density parameter e, are sufficient to define the initial **state** at a point within a soil deposit. The **state** of the soil could be represented by plotting a point in $\bar{p}$-q-e space (see Fig. 2) but three dimensional plots are impractical for routine use. This problem is overcome by unfolding the three dimensional plot along the $\bar{p}$ axis and plotting the projections of the **state** of the specimen in $\bar{p}$-q space and $\bar{p}$-e space. If the **state** of the specimen changes from that at point 1 to that at point 2, this change can be represented by the **state path** from point 1 to 2 in Fig. 2.

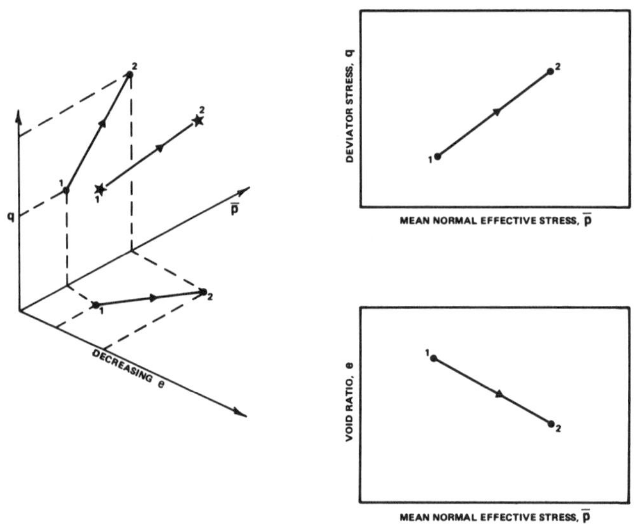

A. POINTS IN $\bar{p}$, q, e SPACE

B. EQUIVALENT REPRESENTATION IN $\bar{p}$ – q AND $\bar{p}$ – e SPACE

FIGURE 2 FORMAT FOR DISPLAYING STATE DATA AND CONCEPT OF STATE PATH

The **state path** of a clay during one dimensional compression and swelling is shown in Fig. 3. The soil is undergoing one-dimensional virgin compression as the **state** changes from point 1 to 2. The slope of the **state path** in $\bar{p}$-q space is constant because the ratio of horizontal to vertical effective stress is constant and equal to K_o for a normally consolidated soil. One-dimensional swelling occurs from point 2 to point 3 and the **state path** in

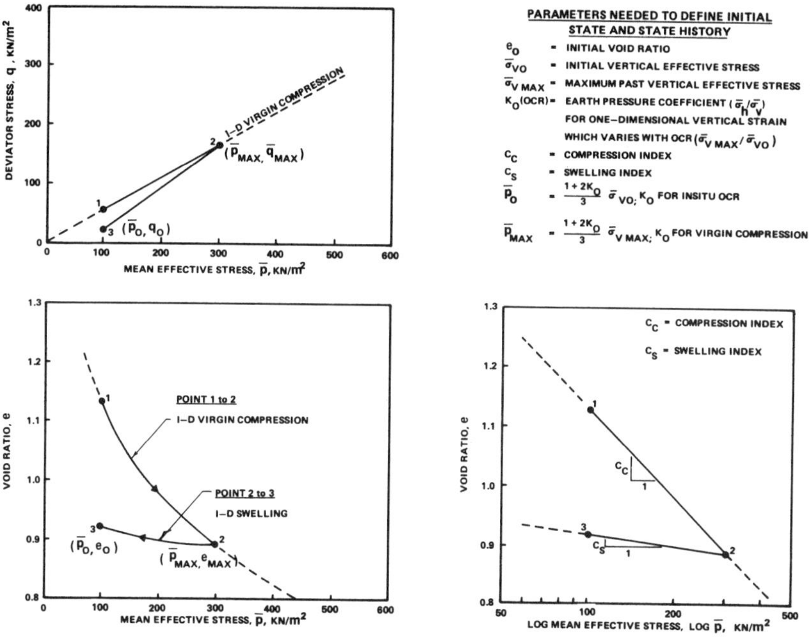

FIGURE 3 TYPICAL BEHAVIOR DURING ONE–DIMENSIONAL (I–D) COMPRESSION AND SWELLING, AND PARAMETERS NEEDED TO DEFINE INITIAL STATE AND STATE HISTORY

$\bar{p}$-q space moves more rapidly toward the $\bar{p}$ axis during swelling as K_o increases. One-dimensional virgin compression and swelling is also shown as e vs log $\bar{p}$ in Fig. 3. The slope of the virgin compression line is the compresson index, C_c, and the slope of the swelling line is the swelling index, C_s. Roscoe et al (21) plot e vs ln $\bar{p}$ and call the slope of the virgin compression line λ and the slope of the swelling line κ.

The information contained in Fig. 3 provides a complete description of the preconstruction **state** and **state history** for an element of soil with current **state** defined by point 3. Construction of Fig. 3 requires a considerable amount of information on soil properties and stress conditions and the necessary parameters are listed in Fig. 3. Our experience indicates that the estimation of **initial state** and **state history** is the most important and most difficult task for application of the general effective stress method for prediction of axial pile capacity. However, this difficulty is shared by all rational approaches to prediction of the performance of earth supported structures. The SHANSEP method described by Ladd and Foott (12) is an example of the successful practical application of a method for prediction of the stability of embankments on soft ground which requires characterization of **initial state** and **state history.**

EFFECTIVE STRESS AROUND A PILE AT THE TIME OF INSTALLATION

Having defined the initial or preconstruction **state** of the deposit, we are now ready to consider the effective stress around a displacement pile at the time of installation. Installation of closed-end cylindrical piles requires the displacement of a volume of soil equal to the volume of the pile. The soil in the path of a displacement pile is brutally displaced by the advancing pile tip and forced to occupy a zone immediately around the pile shaft. The minimum size of this remolded zone can be calculated by matching the volume of the remolded zone to the volume of the pile. For full displacement piles (i.e., closed-end piles) this zone extends from the pile face to a radial offset from the face of the pile of approximately 40 percent of the radius. For example, the thickness of the remolded zone for 12-3/4 inch (320 mm) diameter closed end pipe piles is expected to be about 2-1/2 inches (65 mm). Open-end pipe piles will displace a volume of soil equal to the volume of steel if the pile behaves as a "cookie-cutter". If plugging of open-end pipe piles occurs the volume of displaced soil could approach that of full displacement piles.

The remolding of the soil immediately around a displacement pile in clay occurs at constant volume because the time required for dissipation of the excess pore pressure associated with pile installation is very large compared with the time required for pile installation. Consequently, the effective stresses within the zone of remolded soil around a displacement pile are those associated with remolding at constant volume.

The concept of **critical states** deals explicitly with the **state** of soils when sheared to large strains. Roscoe and his co-workers at the University of Cambridge (21) adopted the concepts of Casagrande (3) and Taylor (24) relative to critical void ratio, expanded the concept to sands and clays, and proposed the following:

Concept of Critical States. The **state** of a soil specimen reaches or approaches a **critical state** during a shear test with monotonic loading; this **critical state** is a unique combination of effective stresses and void ratio at which the specimen can continue to be deformed without further change in these parameters.

The concept of a **critical state** for shear at constant volume of specimens of normally consolidated clay and of overconsolidated clay is illustrated in Fig. 4. The **states** of the normally consolidated clay specimens prior to shear lie along the virgin compression line (points 2 and 2' in Fig. 4). Shear of the normally consolidated specimens at constant volume causes a decrease in mean effective stress and an increase in the ratio $q/\bar{p}$. The mean effective stress and the ratio of $q/\bar{p}$ approach stable values at large strains (points 3 and 3').

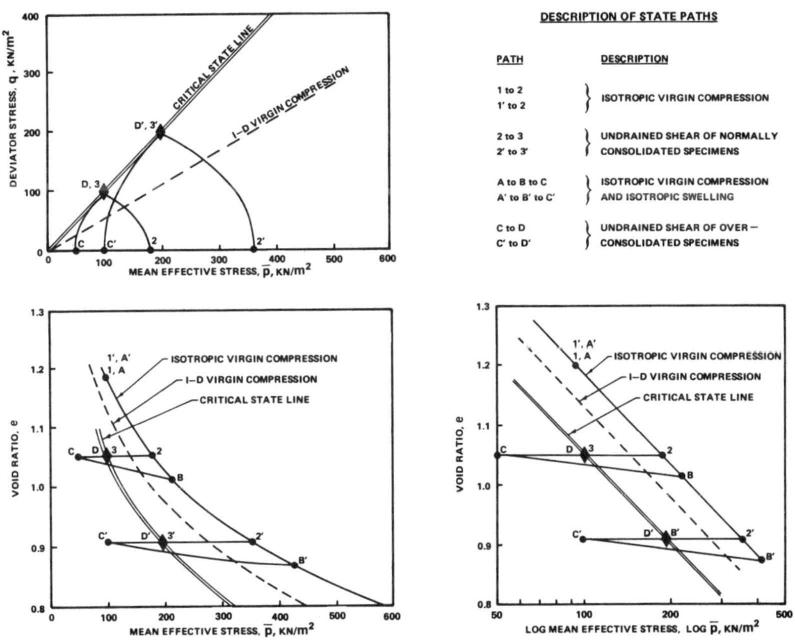

FIGURE 4 ILLUSTRATION OF CONCEPT OF CRITICAL STATES

The initial **states** of overconsolidated specimens are shown in Fig. 4 by points C and C'. The specimens selected for discussion have the same preshear void ratio as the normally consolidated specimens (points 2 and 2') but much lower preshear effective stresses. The overconsolidated specimens were normally consolidated to $\bar{p}_{max}$ in the past (points B and B') and then unloaded and allowed to swell under a mean effective stress equal to $\bar{p}_o$ (points C and C') prior to shear. Shear at constant volume of the overconsolidated specimens is accompanied by an increase in mean effective stress and an increase in the ratio $q/\bar{p}$. At large strains, the mean effective stress and the ratio of $q/\bar{p}$ approach constant values designated as points 3 and 3'.

The end points of the four triaxial compression tests shown in Fig. 4 lie along the line of **critical states.** Experiments show that for a particular soil the ratio of the mean effective stress at the **critical state,** $\bar{p}_{cs}$, to the initial mean effective stress of normally consolidated specimens, $\bar{p}_{nc}$, is approximately constant. As a result, the line of **critical states** is parallel to the virgin compression line in e-log $\bar{p}$ space. The locus of **critical states** in $\bar{p}$-q space has been found to be a straight line through the origin with slope **M.** **M** is related to the effective stress friction angle, $\bar{\phi}$, as follows:

$$\mathbf{M} = \frac{6 \sin \bar{\phi}}{3 - \sin \bar{\phi}} \quad ; \text{ triaxial compression } \bar{\sigma}_2 = \bar{\sigma}_3 \qquad \ldots\ldots\ldots (6)$$

$$\mathbf{M} = \frac{6 \sin \bar{\phi}}{3 + \sin \bar{\phi}} \quad ; \text{ triaxial extension } \bar{\sigma}_2 = \bar{\sigma}_1 \qquad \ldots\ldots\ldots (7)$$

Thus, defining the line of **critical states** for a point within a soil deposit only requires two parameters in addition to those required to define **initial state** and **state history,** i.e., effective stress friction angle, $\bar{\phi}$, and $\bar{p}_{cs}/\bar{p}_{nc}$. Both parameters can be determined from a series of undrained triaxial compression tests with pore pressure measurements. These tests should be conducted at several void ratios to check that $\bar{p}_{cs}/\bar{p}_{nc}$ is a constant value, at least for the interval of void ratio and stress that is of interest.

Having defined the **initial state, state history,** and line of **critical states** at a point within a soil mass, we can predict the mean effective stress of the soil if it were sheared to the critical state by monotonic loading in the triaxial test. However, the remolding caused by pile installation is greater than the remolding in the triaxial test and pile installation cannot be considered monotonic loading. The influence of degree of remolding on mean effective stress at the **critical state** should be small but the influence of cyclic loading is to reduce the mean effective stress at large strains. As a result, the mean effective stress within the zone of remolded soil around a driven pile, $\bar{p}_i$, is expected to be less than or equal to the mean effective stress at the critical state, $\bar{p}_{cs}$, as determined from monitonic loading of laboratory specimens in the triaxial test.

$$\bar{p}_i = \eta\, \bar{p}_{cs};\ \eta \leq 1 \qquad (8)$$

Values of η have been inferred from experiments included within the Amoco Pile Research Program. The limited data now available suggest η to be 0.9 for full displacement jacked piles and 0.6 for full displacement driven piles. Further information on these experiments is contained in a companion paper (11) on testing and application of the general effective stress method for prediction of axial capacity of piles in clay.

TOTAL STRESS AND PORE PRESSURE AROUND A DISPLACEMENT PILE AT THE TIME OF PILE INSTALLATION

The mean effective stress within the remolded zone around a displacement pile at the time of pile installation is independent of the total stress changes during pile installation because the soil is sheared to the **critical state** at constant volume. The mean effective stress within the soil surrounding the pile will increase during consolidation as the excess pore pressures generated by pile installation dissipate. Prediction of the changes in effective stress during consolidation requires a knowledge of the total stresses and pore pressures at the start of consolidation. Thus the total stress, pore pressure and effective stress around a displacement pile at the time of pile installation must be predicted to allow analysis of effective stress changes during consolidation.

Fig. 5 is a schematic representation of a long displacement pile in clay immediately after pile installation. The remolded zone around the pile contains a volume of soil equal to the volume of the pile. The introduction

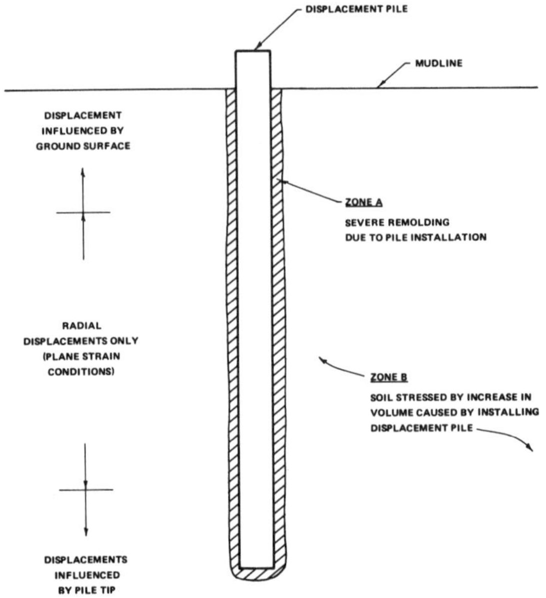

FIGURE 5 SCHEMATIC REPRESENTATION OF ZONES AROUND A DISPLACEMENT PILE IN CLAY IMMEDIATELY AFTER INSTALLATION

of the pile is accommodated by displacements within the soil mass around the pile. These displacements are expected to be upward and outward near the ground surface and downward and outward near the pile tip. The effects of the ground surface and pile tip decrease with distance from these end conditons and, for long piles, there is a large central region in which the net displacements are essentially radial. If we consider a horizontal slice of this central region, and assume that only radial displacements occur, the relatively simple analytical model of undrained plane-strain cylindrical cavity expansion can be used to predict changes in total stress.

Palmer (17) discusses in detail the analysis of plane-strain cylindrical cavity expansion and demonstrates how pressuremeter test results can be interpreted to determine the stress-strain behavior of the soil. Palmer's approach can be extended to allow calculation of the stress changes caused by plane strain cavity expansion in a material that remains at constant volume during shear and obeys any stress-strain relationship assigned by the analyst. The writers (10) have analyzed cavity expansion in materials with stress-strain behaviors typical of soils with a wide range of overconsolidation ratios using an extension of Palmer's method. However, it is our opinion that a simple elastic-plastic stress-strain model is sufficiently accurate for practical application because the predicted stresses are relatively insensitive to the details of the shape of the stress-strain curve. Fig. 6 illustrates the behavior of a typical soil element around an expanding cavity. Fig. 7 presents the equations for calculating stress changes and introduces the concept of Displacement Ratio, DR, which allows consideration of small displacement piles and full displacement piles.

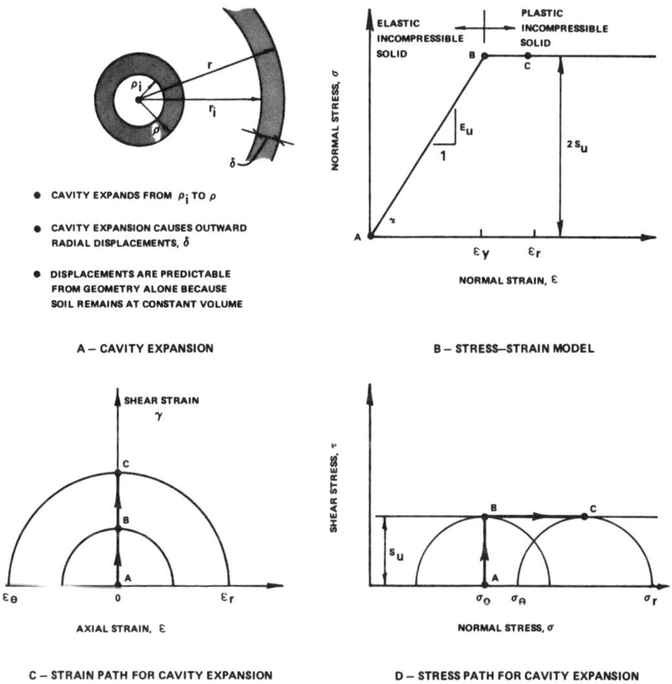

FIGURE 6 STRESS – STRAIN BEHAVIOR, STRAIN PATHS, AND STRESS PATHS FOR TYPICAL ELEMENT OF SOIL IN VICINITY OF EXPANDING CYLINDRICAL CAVITY

Fig. 7 indicates that a boundary can be identified between material that is elastic and material that has yielded. The material within the yielded region is confined horizontally by the elastic material beyond a radial offset R and vertically by the fixed smooth horizontal plates of the analytical model. The procedure to locate this boundary is simply to express strain

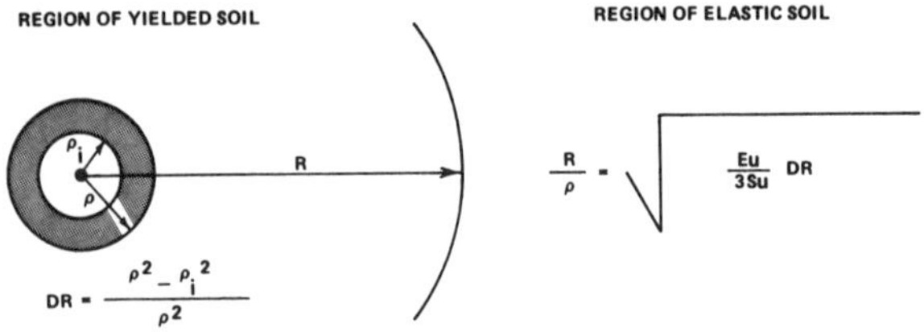

STRESSES

YIELDED REGION	ELASTIC REGION
$\Delta\sigma_r = Su\,(1+2 \ln R/r)$	$\Delta\sigma_r = Su\,(R^2/r^2)$
$\Delta\sigma_\Theta = \Delta\sigma_r - 2Su$	$\Delta\sigma_\Theta = -\Delta\sigma_r$
$\Delta\sigma_v = \Delta\sigma_r - Su$	$\Delta\sigma_v = 0$
$\Delta p = Su\,(2 \ln R/r)$	$\Delta p = 0$

FIGURE 7 STRESS CHANGES AROUND EXPANDING CYLINDRICAL CAVITY WITHIN AN INCOMPRESSIBLE ELASTIC–PLASTIC MATERIAL

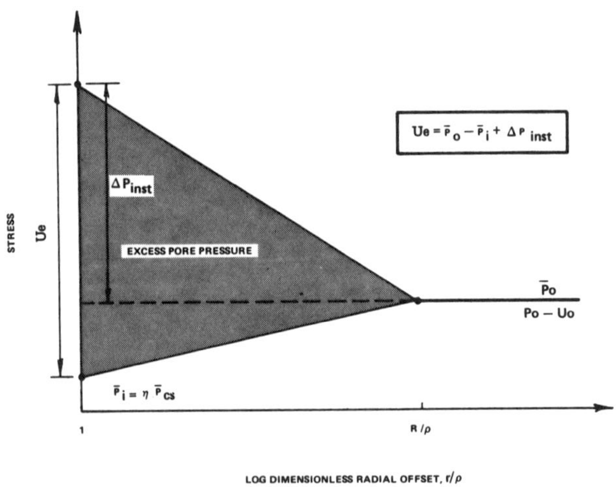

FIGURE 8 APPROXIMATE PROCEDURE FOR CALCULATION OF STRESSES AROUND A DISPLACEMENT PILE IN CLAY

in terms of radial offset and to solve for the radial offset at which a strain large enough to cause yield, ε_y, is reached. The yield strain can be shown to be $(E_u/S_u)^{-1}$. As a result, E_u/S_u appears in the definition of the radial offset of the elastic-plastic boundary, R, and in the various expressions for stress within the yielded region.

Fig. 8 illustrates a procedure for estimating the changes in mean effective stress around a displacement pile, as modeled by an expanded cavity. The mean effective stress at the face of the pile is $\bar{p}_i$ (see Eq. 8) and the effective stress at the elastic-plastic boundary is $\bar{p}_o$. The variation of $\bar{p}$ from $\bar{p}_i$ to $\bar{p}_o$ is assumed log-linear. The variation of mean total stress is log-linear (see Fig. 7) and the calculated excess pore pressure is also log-linear. The excess pore pressure at the pile face can be calculated as follows:

$$U_e = \Delta p_{inst} - \Delta \bar{p}_{inst} \quad \text{(9.a)}$$

$$U_e = 2\, S_u \ln \left[\frac{E_u}{3S_u}\right]^{\frac{1}{2}} - \left(\bar{p}_i - \bar{p}_o\right) \quad \text{(9.b)}$$

A significant result of the analysis is that plane-strain cavity expansion causes a relatively large increase in total stress and pore pressure. For normally consolidated soils, the maximum change in vertical total stress is about equal to the initial vertical total stress. For overconsolidated soils, the change in vertical total stress is greater than the initial vertical total stress. The increase in vertical total stress is a reaction to cavity expansion provided in the plane strain model by the fixed smooth horizontal plates. The increase in vertical total stress is maximum at the pile face and decreases log-linearly to zero at the boundary between the elastic and yielded regions.

The mechanism for development of large changes in vertical total stress in the field, and the depth at which the plane strain cavity expansion model become applicable, are important considerations. Fig. 9 illustrates conceptually the situation in the field. Material which has yielded during pile installation is surrounded by material that has remained elastic. The radius of the yielded region is equal to the pile radius at the ground surface and increases gradually with depth until it reaches its maximum value at a depth where surface effects are no longer felt. The gradual increase in radius of the yielded region is consistent with the concept that vertical displacements are maximum near the ground surface and decrease gradually with depth. Vertical displacements reduce the shear strains in the horizontal plane and thus significantly reduce the size of the yielded region. The material that has not been strained to the yield point serves as an elastic vessel to contain the yielded soil.

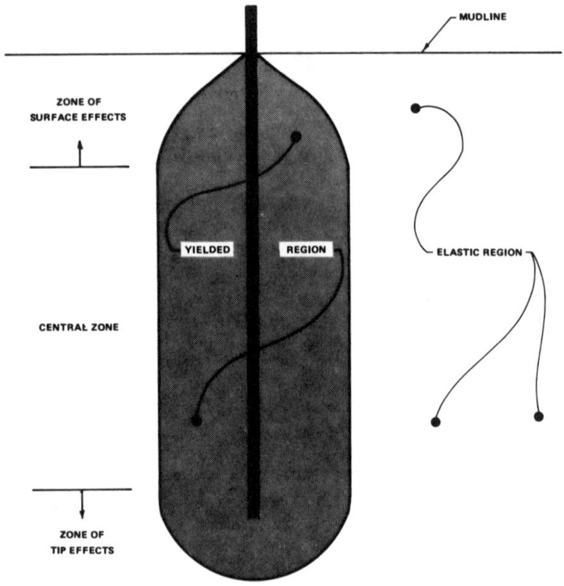

FIGURE 9 GEOMETRY OF YIELDED REGION AROUND A FULL DISPLACEMENT PILE IN CLAY

The increase in vertical total stress produced by plane-strain cavity expansion is provided by the "roof" of the elastic vessel containing the yielded region. The minimum thickness of the "roof" of the elastic vessel can be calculated from limiting equilibrium analysis. Fig. 10 shows the free body being considered. A one radian segment of material is loaded at the bottom by the change in vertical total stress caused by cavity expansion. This change is vertical total stress is accomodated by the development of shear stresses as shown in Fig. 10. These shear stresses act over the thickness of the roof containing the yielded zone. The minimum thickness of the elastic roof, and thus the minimum depth of surface effects, can be

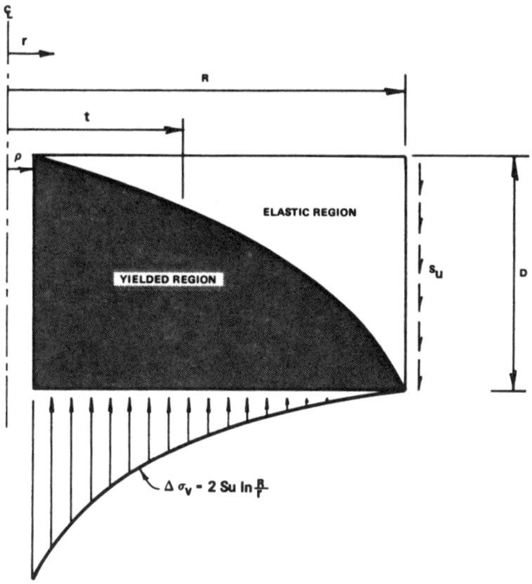

LIMITING EQUILIBRIUM EQUATION FOR ONE RADIAN SEGMENT

$$\int_{r=\rho}^{r=t} \Delta\sigma_v \, r dr = \frac{s_u \, t \, D}{FS}$$

ANALYSES USING THIS EQUATION INDICATE

minimum FS for $\frac{t}{R} \approx 0.6$

minimum depth for FS = 1, $\frac{D}{R} \approx 0.6$

FIGURE 10 ESTIMATE OF MINIMUM DEPTH OF SURFACE EFFECTS FOR CAVITY EXPANSION

calculated if it is assumed that the shear stresses are equal to the shear strength of the material. Such calculations show that the minimum thickness of the elastic roof is approximately 60 percent of the radius of the yielded region. The actual depth of surface effects will be larger than that predicted from this limiting equilibrium approach. However, even if significant surface effects extend to twice the depth predicted by limiting equilibrium analysis, these surface effects would still be limited to a depth of 10 to 12 pile diameters.

CONSOLIDATION AROUND A DISPLACEMENT PILE

The expanding cavity model suggests that the pore pressures generated during pile installation vary rapidly from a maximum value at the pile face to zero at the boundary between the yielded region and elastic region (see Fig. 8). Pore water flows radially away from the pile during consolidation and the volume of the soil within the yielded region decreases. This decrease in volume allows inward displacement of the boundary between the yielded region and the elastic region and causes a reduction in radial total stress as the elastic region is unloaded. Thus, consolidation around a driven pile is a particularly challenging problem because the pore pressure changes during consolidation are associated with both an increase in effective stress and a decrease in total stress.

Randolph and Wroth (19) developed a closed form solution for consolidation around a displacement pile assuming soil behaves as a compressible two-phase elastic material and that displacements and flow are only in the radial direction. Leifer et al (14) report the expansion of this solution to consider the variation of the properties of the two-phase elastic material with radial offset.

Miller et al (15) developed a computer solution for consolidation around a driven pile in which the soil was modeled using the modified Cam-Clay soil model described by Roscoe and Burland (22). Randolph et al (20) developed a similar solution to this problem. There is little doubt that the modified Cam-Clay model is likely to produce more accurate insights to **state** changes during consolidation than an elastic soil model but there is no reason to expect that the modified Cam-Clay model accurately predicts the behavior of the soil in the remolded zone around a displacement pile. A review of the effects of sample disturbance on the behavior of soils in the laboratory is presented below and helps to define the limitations of the modified Cam-Clay model.

Fig. 11 serves to illustrate the typical behavior of specimens of clay during one-dimensional vertical compression, as predicted using the modified Cam-Clay soil model. The **initial state** of the specimen prior to sampling is shown by Point A. Sampling produces an isotropic state of stress (Point B) and swelling causes a reduction in mean effective stress from Point B to Point C. Behavior during one-dimensional reloading of the soil can be

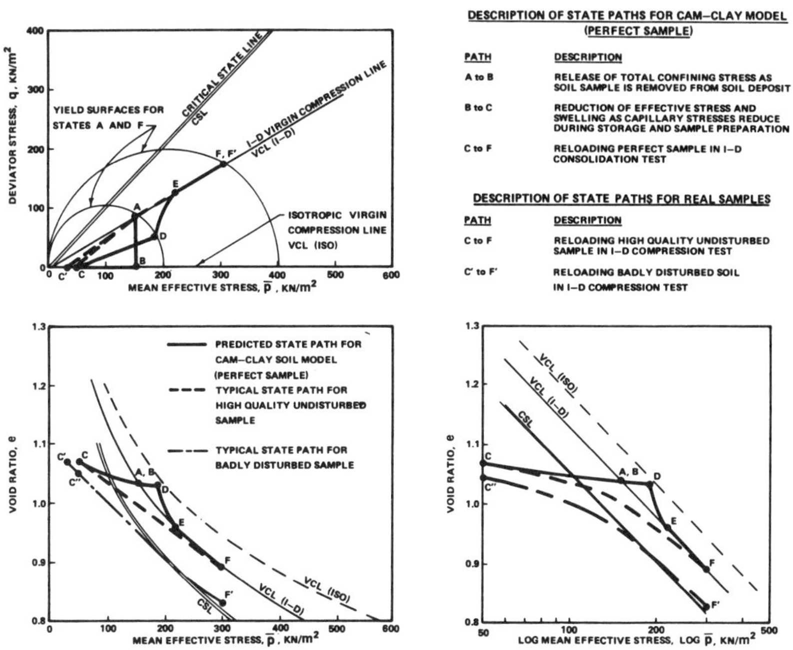

FIGURE 11 CONSOLIDATION BEHAVIOR OF UNDISTURBED SAMPLE OF CLAY REALITY VS. PREDICTED BEHAVIOR USING MODIFIED CAM–CLAY SOIL BEHAVIOR MODEL

predicted using the modified Cam-Clay model. The soil behaves as an elastic material from point C to D where the **state path** reaches the yield surface defined by the presampling **state** of the material. Thereafter, the material yields during further one-dimensional compression and the **state path** approaches the one-dimensional virgin compression line in $\bar{p}$-q and e-$\bar{p}$ space.

Typical behavior of a high qualilty undisturbed soil specimen is also shown in Fig. 11. The reloading curve in e- log $\bar{p}$ space is rounded compared to that predicted by modified Cam-Clay and meets the virgin compression line representing the behavior of the in-situ undisturbed soil at about point F, which is a stress level about twice the level of stress at Point A. A typical **state path** for reloading behavior of a badly disturbed specimen is also shown in Fig. 11. The initial effective stress after severe sample disturbance (remolding) is less than the initial effective stress for the high quality undisturbed sample (point C' vs point C). The badly disturbed soil specimen is more compressible during reloading than was the high quality undisturbed specimen and the compression curve in e-$\bar{p}$ space approaches the virgin compression curve for the in-situ material very slowly. It was assumed in constructing Fig. 11 that the compression line for the badly disturbed specimen meets the virgin compression line at about 40 percent of the initial void ratio. This assumption is consistent with observations on the effects of remolding and sample disturbance reported by Schmertmann (23). Inspection of Fig. 11 suggests that the Cam-Clay model does not provide a good prediction of the behavior of soils during initial one-dimensional loading in the laboratory because of the effects of sample disturbance. However, the Cam-Clay model does provide a prediction of soil behavior for the perfect sample (no sample disburbance) which serves as a reference and can be modified to account for sample disturbance. Of course, the deficiencies of the Cam-Clay model for modeling real soil behavior are not limited to sample disturbance. However, the point of this discussion is that the effects of sample disturbance not captured by the Cam-Clay model can lead to serious errors in predicting the behavior of badly disturbed soil specimens.

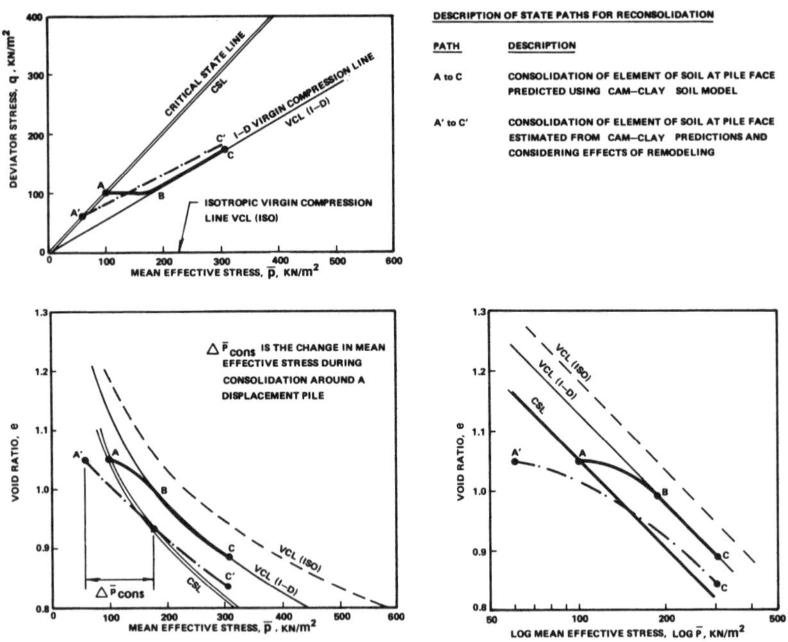

FIGURE 12 CONSOLIDATION BEHAVIOR OF SOIL AROUND A DRIVEN PILE REALITY VS. PREDICTED BEHAVIOR USING MODIFIED CAM–CLAY SOIL BEHAVIOR MODEL

Fig. 12 illustrates the behavior of an element of soil adjacent to a displacement pile during consolidation. This soil element lies within the zone of severe remolding (Zone A) shown in Fig. 5 and is undergoing one-dimensional radial compression during consolidation. The difference between Fig. 11 and Fig. 12 is that Fig. 11 considers consolidation of soils in the conventional one-dimensional laboratory consolidation test whereas Fig. 12 considers one-dimensional radial consolidation of the soil next to a driven pile. Consider first the behavior predicted using the modified Cam-Clay model. One-dimensional compression of the material causes rapid adjustments in stress ratio as the stresses seek those appropriate for one-dimensional compression of a plastic material (Path A to B). Thereafter, one-dimensional compression follows the virgin compression line in e-$\bar{p}$ and p-$\bar{q}$ space. The behavior predicted by the modified Cam-Clay soil model does not reflect the severe remolding caused by pile installation. The general effects of severe remolding by pile installation should be similar to the effects of severe sample disturbance. These effects were illustrated in Fig. 11 for laboratory one-dimensional consolidation tests. The estimated effects of remolding on the **state path** during consolidation for the soil around a driven pile are illustrated in Fig. 12. The precise **state path** cannot be defined because the effects of sample disturbance are difficult to quantify. The **state path** illustrated in Fig. 12 is based on the assumption that the virgin compression line for the undisturbed soil is reached at a void ratio equal to about 40 percent of the initial void ratio.

Inspection of the **state paths** in Fig. 12 shows that one effect of remolding is to shift the **state path** in e-$\bar{p}$ space to the left of that predicted by the Cam-Clay model. The **state path** for the soil within the remolded zone crosses the projection of the **critical state line** at a mean effective stress of about 170 kN/m^2 for the example in Fig. 12 and the value of $\bar{p}$ is within 10 percent of the value of $\bar{p}_{cs}$ within the stress range of 120 to 270 kN/m^2. The closeness of the state path to the critical state line is significant to stress changes during pile loading that are discussed later.

Our discussion of stress changes during reconsolidation has focused on the **state path** for one-dimensional radial compression of the soil within the remolded zone around a displacement pile. Having investigated the general characteristics of the **state path**, we need to predict the magnitude of the change in effective stress during reconsolidation. The magnitude of the increase in mean effective stress during consolidation is equal to the excess pore pressures at the start of consolidation, minus the change (reduction) in mean total stress during consolidation. The initial excess pore pressures are as defined by analysis of cavity expansion (see Fig. 8). The changes in mean total stress are related mainly to the average Bulk Modulus of the soil within the yielded region that is experiencing volumetric strains during consolidation and the Shear Modulus (or Young's Modulus for shear at constant volume) of the soil within the elastic region that is undergoing shear at constant volume during consolidation. Fig. 13 shows how the change in mean effective stress during consolidation, $\Delta \bar{p}_{cons}$, is related to excess pore pressure at the pile face, U_e, and the ratio of average Bulk Modulus to Young's Modulus for undrained loading, $\bar{B}_{avg}/E_u$. This relationship was generated using the elastic solution developed by Randolph and Wroth (19), as generalized by Leifer et al (14), and was found to be essentially independent of Poisson's Ratio and the size of the yielded region.

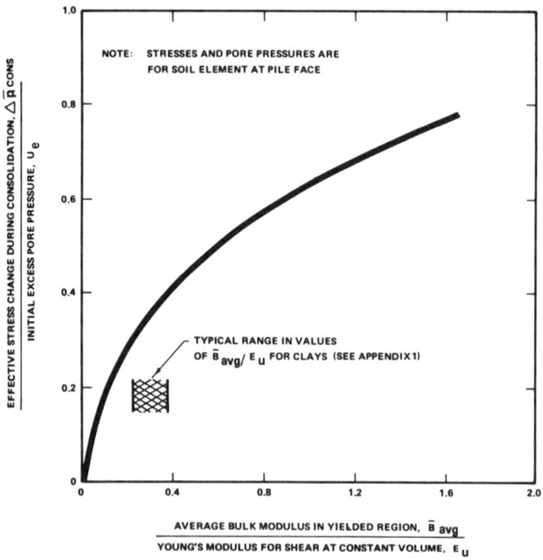

FIGURE 13 INFLUENCE OF MODULI ON EFFECTIVE STRESS CHANGES DURING CONSOLIDATION

Predictions of $\Delta \bar{p}_{cons}/U_e$ using Fig. 13 must be based on values of $\bar{B}_{avg}/E_u$ consistent with observed soil behavior - not on values of $\bar{B}_{avg}/E_u$ predicted using elastic theory or simplified soil models such as modified Cam-Clay. Appendix I contains calculations that suggest typical values of $\bar{B}_{avg}/E_u$ range from about 0.20 to 0.35 for soils with a wide range of stress histories and overconsolidation ratios. For this range in $\bar{B}_{avg}/E_u$, Fig. 13 indicates that $\Delta \bar{p}_{cons}$ ranges from 30 to 40 percent of U_e, the excess pore pressure generated during pile installation.

We now have enough information to estimate the final state of stress after consolidation. The consolidation path was shown in Fig. 12. The change in mean effective stress during consolidation for the soil element at the face of the pile is 30 to 40 percent of the initial excess pore pressure at the pile face. This pore pressure can be predicted using Equation 9. The following input parameters for Equation 9 can be used for the example shown in Fig. 12:

$$\bar{p}_o = 145 \text{ kN/m}^2$$

$$\bar{p}_i = 60 \text{ kN/m}^2$$

$$S_u = 50 \text{ kN/m}^2$$

Substituting these values in Equation 9, and assuming $E_u/S_u = 400$ (see Appendix I) yields a calculated value of U_e equal to 330 kN/m^2. The change in mean effective stress during consolidation, $\Delta \bar{p}_{cons}$, is estimated as 35 percent of U_e or 115 kN/m^2. The effective stress at the pile face after consolidation ($\bar{p}_i + \Delta \bar{p}_{cons}$) is calculated to be 175 kN/m^2. Thus, the predicted mean effective stress after consolidation (175 kN/m^2) is approximately 20 percent higher than the initial mean effective stress (145 kN/m^2) for this example.

The change in mean effective stress during consolidation is shown in Fig. 12. The final **state** in e-$\bar{p}$ space happens to fall at the approximate location where the recompression line crosses the critical state line. This is a result of happenstance but it is clear that the final **state** lies very close to the projection of the **critical state line** in e-$\bar{p}$ space for the range of stress increases expected during consolidation.

PILE LOADING

Having tracked mean effective stress from initial effective stress, $\bar{p}_o$, through pile installation and consolidation (see Table 1), we now have an estimate of the mean effective stress after consolidation, $\bar{p}_{ac}$. The calculation of mean effective stress at the pile face at failure, $\bar{p}_f$ is simple in principle. The **state** of the soil changes from $\bar{p}_{ac}$ to $\bar{p}_f$ as the soil is sheared at constant volume by pile loading. Large shear strains are reached at the pile face at failure and the mean effective stress at failure corresponds to the mean effective stress at the **critical state.** The **state** of the soil after consolidation was shown to be in the vicinity of the line of critical states in e-$\bar{p}$ space (see Fig. 12). As a result, shear of the soil at the pile face by pile loading causes only a small change in mean effective stress. This presumes that the line of **critical states** shown in Fig. 12 describes the behavior of soil at large strains after experiencing pile installation and reconsolidation. Laboratory data presented in a companion paper (11) suggest this is probably a correct assumption.

The total stress changes during pile loading are not relevant to pile capacity for loading rates large enough to preclude drainage. However, knowledge of total stress changes is necessary to evaluate field measurement of pore pressures generated by pile loading and to estimate the ultimate pile capacity for drained and undrained loading. Esrig et al (7) report the results of finite element computations of stresses around a compressible pile supported in an elastic medium. These analyses suggest that the increase in mean total stress caused by pile loading for a soil element at the pile face is generally less than 10 percent of the applied shear stress for the same element of soil. The limit of the applied shear stress is roughly related to the undrained shear strength. Thus, the change in mean total stress caused by pile loading to failure is expected to be less than 10% of the undrained shear strength.

If a pile is loaded so quickly that drainage cannot occur, and if the change in effective stress during loading is essentially zero as predicted by the analysis described in this paper, the pore pressure generated by pile loading would be equal to the change in mean total stress, i.e., less than 10 percent of the undrained shear strength. The writers (10) have presented data which support this prediction and additional data is included in a companion paper (11).

If a pile is loaded so slowly that drainage could occur, the increase in mean effective stress would equal the increase in mean total stress which is expected to be less than 10 percent of the undrained shear strength. This increase in mean effective stress is less than 5 percent of the mean effective stress for undrained pile loading and the corresponding maximum increase in pile capacity would be less than 5 percent.* Thus, the expected difference in capacity for drained vs. undrained loading of piles should not be significant. It is interesting to note that field observations consistent with this conclusion might be interpreted to indicate that slow and rapid pile loading both occur under drained conditions, when the correct interpretation, in our opinion, is that both drained and undrained loading of piles occur at constant volume.

STRESS SYSTEM FOR PILE LOADING TO FAILURE

Equation 3 is the basic equation for prediction of shear resistance along the pile shaft at failure, τ_f, and is repeated below.

$$\tau_f = \bar{p}_f \, LM \, F_i \sin \bar{\phi} \qquad (3)$$

We have tracked the mean effective stresses shown in Table 1 from the initial state of stress, $\bar{p}_o$, to the stress at the pile face at failure, $\bar{p}_f$. Analysis of the stress system for pile loading to failure allows us to estimate the Load Module and the Interface Factor. Load Module, LM, defines the relationship between $\bar{p}_f$ and radial effective stress at the pile face at failure, $\bar{\sigma}_{rf}$. The Interface Factor, F_i, is a correction factor to account for the reduction in τ_f associated with an effective stress friction angle for soil sliding on the pile material, $\bar{\phi}_{ss}$, which is less than the effective stress friction angle for the soil, $\bar{\phi}$.

---

* A mean total stress change of $0.1S_u$ corresponds to $0.05\bar{p}$ if we assume $S_u/\bar{p}_f$ is 0.5. The actual value of $S_u/\bar{p}_f$ would probably be less than 0.5.

Load Module

Fig. 14 illustrates the stress changes during pile loading using Mohr Circles to represent the state of stress. The stress conditions after consolidation for a soil element at the pile face are illustrated in Part A.

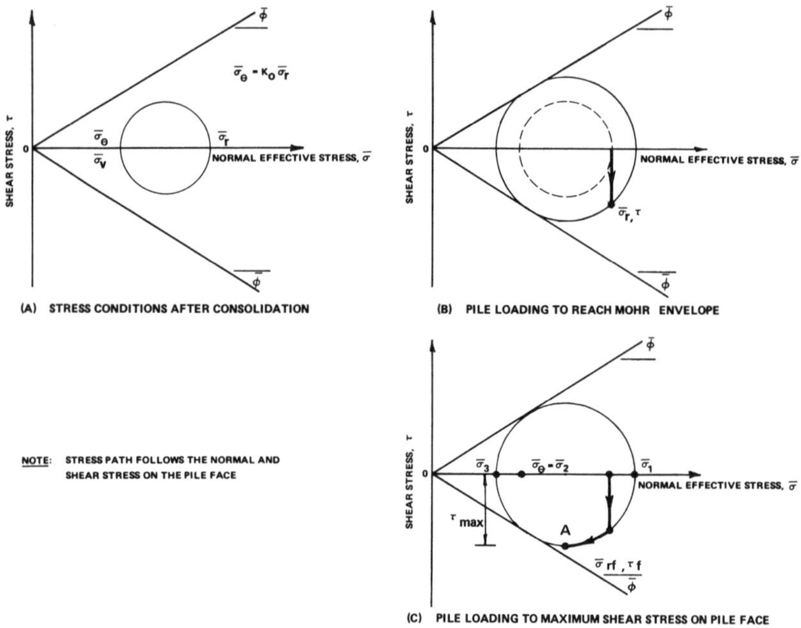

FIGURE 14 – MOHR CIRCLE REPRESENTATION OF EFFECTIVE STRESS CHANGES DURING PILE LOADING

The radial effective stress, $\bar{\sigma}_r$, is the major principal stress, and the vertical and circumferential stresses, $\bar{\sigma}_v$ and $\bar{\sigma}_\theta$, are the intermediate and minor principal stresses. The element of soil at the pile face was subjected to one dimensional radial strains during consolidation. This strain path causes the vertical and circumferential stresses to approach similar values during consolidation and leads to stress ratios, $\bar{\sigma}_v/\bar{\sigma}_r$ and $\bar{\sigma}_\theta/\bar{\sigma}_r$, approximately equal to the earth pressure coefficient for one-dimensional loading, K_o.

Axial loads applied to the pile are resisted by vertical shear stresses on the pile face. However, the effective stress state around a pile during pile loading is complicated by changes in normal effective stresses and reorientation of principal planes in a soil which has experienced the trauma of pile installation and reconsolidation. Nevertheless, it is the writers' opinion that reasonable assumptions can be made and these assumptions lead to stress changes that are also reasonable.

A key assumption is that there is no change in mean effective stress during pile loading. The analysis of stress changes during pile installation, reconsolidation, and loading developed in this paper support this assumption. Data on pore pressures developed during pile loading (10) are generally consistent with this assumption.

We also need to make assumptions about the failure law and the stress system on the failure plane. We have assumed the stress system at failure can be defined by a Mohr Envelope and that the pile face is a plane of maximum stress difference (i.e., maximum shear stress) at failure. The use of a Mohr Envelope is a generally accepted assumption but the assumption that the failure plane is a plane of maxium stress difference $(\bar{\sigma}_1-\bar{\sigma}_3)_{max}$, rather than a plane of maximum obliquity $(\bar{\sigma}_1/\bar{\sigma}_3)_{max}$, is controversial.

The Mohr circle in Fig. 14C represents the stress conditions at failure for an element of soil next to the pile. The normal and shear stresses on the pile face are shown by point A in Fig. 14C. The Mohr-circle and stress state on the pile face are based on the assumptions discussed above, namely, $\Delta\bar{p}$ during loading is zero, Mohr Envelope is valid, and the pile face is plane of maximum stress difference. In addition, it was assumed that the intermediate principal stress $(\bar{\sigma}_2 = \bar{\sigma}_\theta)$ does not change during pile loading.

The path of the stress changes from part A to part C of Fig. 14 is influenced by changes in normal stress and/or rotation of principal planes. We have assumed in constructing Fig. 14B that pile loading initially causes no change in radial effective stress. Thus, the stress path for the pile face is a vertical line in Fig. 14B. When the Mohr Circle reaches the Mohr Envelope, the stress conditions on the pile face will not allow sliding to occur and continued application of shear stresses as pile loading continues is accompanied by decreases in radial effective stress as the stress conditions on the pile face move toward values that will allow sliding to occur.

The stress path in Fig. 14C is considered to provide a reasonable approximation to the actual stress path. The "corner" where the vertical path meets the Mohr circle is a consequence of an abrupt change in assumptions about $\bar{\sigma}_r$ and a smoother transition would be more realistic.

Another implication of the stress paths shown in Fig. 14C is that the stress conditions on the vertical plane (and other planes as rotation of principal stresses occurs) satisfy the Mohr failure law before the stress conditions at the pile face do. It is our opinion that this is not a concern to the model because pile plunging will not occur until the pile face becomes a failure plane.

Fig. 14 indicates that the maximum shear resistance at the pile face can be calculated using Equation 1 and that $\bar{\sigma}_{rf}$ is equal to $(\bar{\sigma}_{1f} + \bar{\sigma}_{3f})/2$. Since $(\bar{\sigma}_1 + \bar{\sigma}_3)/2$ does not change during pile loading, we can calculate $\bar{\sigma}_{rf}$ directly from the stress conditions after consolidation as shown below.

$$\bar{\sigma}_{rf} = \frac{\bar{\sigma}_1 + \bar{\sigma}_3}{2} = \frac{\bar{\sigma}_r (1 + K_o)}{2} \quad \ldots\ldots\ldots\ldots\ldots (10)$$

Similarly, since $\bar{p}$ does not change during pile loading, we can calculate $\bar{p}_f$ from the stress conditions after consolidation as shown below.

$$\bar{p}_f = \bar{p}_{ac} = \frac{\bar{\sigma}_r (1 + 2 K_o)}{3} \quad \ldots\ldots\ldots\ldots\ldots (11)$$

The ratio $\bar{\sigma}_{rf}/\bar{p}_f$, the Load Module, can be calculated by combining Eqs. 10 and 11, as follows:

$$LM = \frac{\bar{\sigma}_{rf}}{\bar{p}_f} = \frac{3 (1 + K_o)}{2 (1 + 2 K_o)} \quad \ldots\ldots\ldots\ldots\ldots\ldots\ldots (12)$$

K_o can be related to friction angle by the empirical expression proposed by Jaky (8).

$$K_o = 1 - \sin \bar{\phi} \quad \ldots\ldots\ldots\ldots\ldots\ldots\ldots\ldots (13)$$

Substituting Equation 13 into Equation 12 yields values of LM that vary from 1.075 for $\bar{\phi}$ of 20° to 1.15 for $\bar{\phi}$ of 35°.

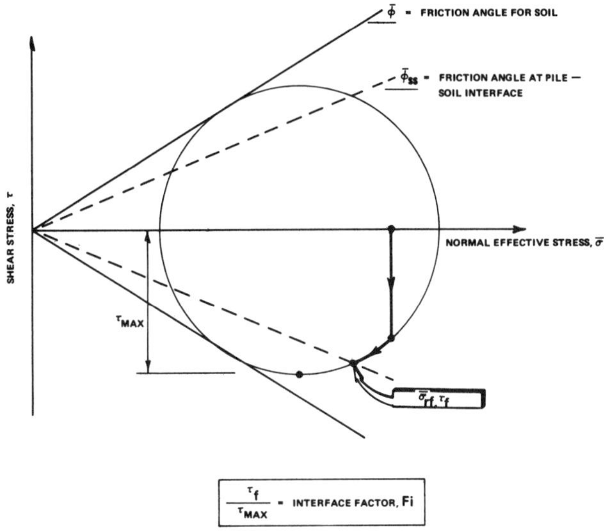

FIGURE 15 INFLUENCE OF INTERFACE FRICTION ON SHEAR RESISTANCE DURING PILE LOADING

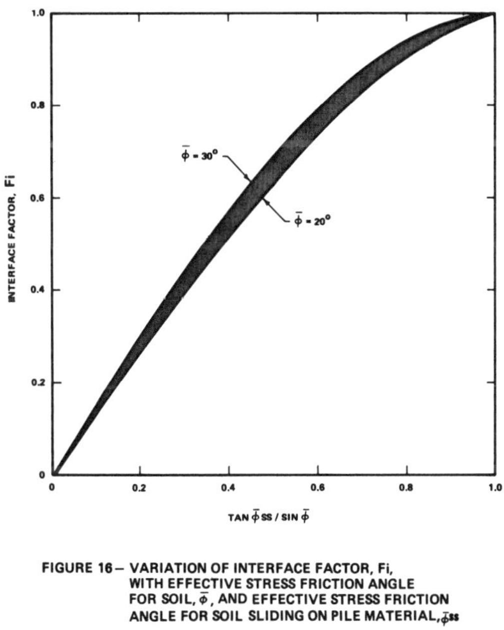

FIGURE 16 – VARIATION OF INTERFACE FACTOR, Fi, WITH EFFECTIVE STRESS FRICTION ANGLE FOR SOIL, $\bar{\phi}$, AND EFFECTIVE STRESS FRICTION ANGLE FOR SOIL SLIDING ON PILE MATERIAL, $\bar{\phi}_{ss}$

Influence of Reduced Friction Angle at Pile Face

The friction angle for soil sliding on the pile material, $\bar{\phi}_{ss}$, may be less than the friction angle for the intact soil, $\bar{\phi}$. If this is the case, the interface friction angle will limit the shear resistance that can be mobilized at the pile face, as illustrated in Fig. 15. The changes in normal and shear stresses at the pile face from Part C of Fig. 14 are reproduced in Fig. 15. The increase in shear stress during rotation of principal planes is limited because the friction angle at the pile-soil interface allows slip to occur at the pile face.

Analyses of Fig. 15 show that the shear stress that can be developed at the pile face is not controlled by interface friction when tan $\bar{\phi}_{ss}$ is equal to or greater than sin $\bar{\phi}$. On the other hand, the shear stress that can be developed on the pile face would be reduced to zero for tan $\bar{\phi}_{ss}$ equal to zero. The effects of interface friction can be accounted for using an Interface Factor, F_i, which varies with tan $\bar{\phi}_{ss}$/sin $\bar{\phi}$ and with $\bar{\phi}$. The variation of F_i with tan $\bar{\phi}_{ss}$/sin $\bar{\phi}$ for $\bar{\phi} = 20^o$ and $\bar{\phi} = 30^o$ is shown in Fig. 16.

SUMMARY AND CONCLUDING REMARKS

This paper started with a basic expression for shear resistance at the pile face at failure, Equation 3, and went on to describe the analytical and soil behavior models used to estimate the parameters included in Equation 3. In general, our present analytical capabilities are in excess of our abilities to provide the necessary input parameters. As a result, the emphasis in the near future should be on experimental work, guided by our understanding of the theoretical aspects of the problem. A companion paper (11) tests the general effective stress method against field and laboratory data and presents the method in a form suitable for practical application. A summary of the status of development of the analytical and soil behavior models follows.

Initial State and State History

Characterization of **initial state** and **state history** is essential for application of any rational method for prediction of soil behavior in terms of effective stress. This can be a very difficult task but can be achieved using available technology.

Stress Changes Due to Pile Installation

The concept of a **critical state** is used to predict effective stresses in the remolded zone around a driven pile. The effective stresses are believed to be less than those at the **critical state** and the correction factors to account for this are empirical. More field data will be required to better define these correction factors.

The total stress changes due to pile installation are predicted using the plane-strain cavity expansion model. This model ignores many details of the pile installation process including the gradual advancement of the pile tip, but the model is responsive to the displacement patterns after pile installation. Extensions to consider surface and tip effects are possible but the writers do not feel these extensions are urgently needed. The main problem with the analysis is selection of undrained shear strength and normalized undrained modulus, E_u/S_u. The undrained shear strength appropriate for cavity expansion is the more elusive of these two parameters.

Stress Changes During Consolidation

Solutions for stress changes during consolidation have been developed and suggest that the **state paths** during consolidation are strongly influenced by the constitutive laws assumed to represent the material behavior. All the constitutive laws applied to this problem neglect the influence on soil behavior of severe remolding around the pile. Moreover, the simple constitutive laws (elastic theory, Cam-Clay) imply relationships between drained and undrained behavior that differ from empirical observations. Nevertheless, these analyses, when modified to account for sample disturbance and more accurate relationships between drained and undrained modulus, yield predictions of **state** after consolidation that are very close to the line of critical states. This is considered an essential result of any analytical model because of the data which suggest only small changes in effective stress during pile loading(10).

Improvements in our understanding of soil behavior during consolidation will require further experimental work in the laboratory and the field.

Additional analytical work would probably not be productive until such experimental data are available.

Stress Changes During Pile Loading

The analysis of stress changes during consolidation provides the starting **state** for pile loading. The effective stress changes during pile loading are governed by this starting **state** and the mean effective stress at the **critical state.** The analysis presented herein suggests that these changes in effective stress are small and can be assumed as zero for practical purposes. A consequence of essentially zero change in effective stress during pile loading is that the pore pressure generated by pile loading is equal to the change in mean total stress. The change in mean total stress during pile loading is estimated to be less than 10 percent of the undrained shear strength and field measurements of pore pressure during pile loading generally support this. Thus, any analytical work on this problem must predict little or no change in effective stress during pile loading. The stress changes during pile loading presented herein satisfy this requirement but do involve assumptions regarding reorientation of principal planes and values of the intermediate principal stress. However, these assumptions do not have serious implications on pile capacity.

Load Module

The relationship between mean effective stress at failure and radial effective stress at failure is defined by the Load Module. The controversial elements of this analysis are the value of intermediate principal stress at failure and whether the pile face is a plane of maximum stress difference or maximum obliquity. Neither of these questions leads to large differences in the numerical value of the Load Module and further analysis does not appear to be necessary.

Interface Factor

The analysis of the effects of reduced friction angle at the pile face are summarized in terms of an Interface Factor. Further analysis is not considered necessary. However, more data are required on interface friction angle, particularly as it compares to friction angle for intact soil or residual friction angle.

ACKNOWLEDGEMENT

The development of a general effective stress method for prediction of axial capacity for displacement piles in clay began in 1971. R. G. Bea, then of Shell Oil Company, was concerned that methods for prediction of axial pile capacity might not be reliable for the unprecedented loads from fixed offshore platforms being planned at the time. He expressed this concern to M. I. Esrig of Woodward-Clyde Consultants and they agreed that a new look at methods for prediction of axial pile capacity using effective stress concepts, rather than undrained shear strength, might be productive. By 1974, Bea's idea that a new approach for prediction of pile capacity might be useful had developed to the point that he persuaded B. S. Murphy of Amoco Production Company and T. W. Miller of Exxon to consider funding a pilot study by Woodward-Clyde Consultants. This pilot study was completed by M. I. Esrig and R. C. Kirby in September 1975 and was funded by Shell Oil Company and Amoco Production Company, with Exxon deciding not to participate.

The results of the pilot study were encouraging. J. A. Focht of McClelland Engineers reviewed the results and concluded that the methodology had promise. B. S. Murphy of Amoco agreed to organize a follow-on study and seek industry-wide support for a cooperatively funded project. The Amoco Cooperative Axial Pile Project (ACAPP) received the financial support of eleven oil companies. C. P. Wroth and M. F. Randolph from the University of Cambridge were retained by Woodward-Clyde Consultants to assist with portions of the theoretical work. The ACAPP reports were issued in December 1976.

Interest in the general effective stress method within the oil industry continued to grow. Exxon retained McClelland Engineers and the University of Cambridge to work on the problem of consolidation around a displacement pile under separate research contracts. B. S. Murphy of Amoco decided to organize a third cooperative project, the Effective Stress Axial Capacity Cooperative (ESACC), and retained L. M. Kraft and J. A. Focht of McClelland

Engineers to critique the results of the ACAPP project. Amoco was faced with the task of developing a project that would receive wide support and make use of the talents of the growing number of consultants working on the general effective stress approach to axial pile capacity. The emphasis of the ESACC project was on laboratory and field experiments to test the analytical methods that had been developed. The research team selected by Amoco for this project included Woodward-Clyde Consultants, the University of Cambridge, and McClelland Engineers. R. E. Olson of the University of Texas at Austin was named the Technical Advisor. His role was to evaluate the results of the ESACC project and to report his findings to the project participants. The ESACC project was sponsored by Brown and Root, Inc., the Federal Highway Administration, The Building Research Station (England) the Norwegian Geotechnical Institute, and thirteen oil companies. Work began in May 1978 and final reports were published in July 1980.

The financial support of all the participating organizations made this research possible. In addition, the interest, enthusiasm and constructive criticism of engineers from the sponsoring organizations, and other consultants involved with the project, contributed significantly to the technical quality and balance of this research. K. A. Blenkarn of Amoco Production Company provided strong support to B. S. Murphy and the concept of cooperatively sponsored engineering research on pile foundations. This research would not have been completed without Dr. Blenkarn's support.

APPENDIX I - ESTIMATED VALUES OF $\bar{B}_{avg}/E_u$ FOR CONSOLIDATION AROUND A DISPLACEMENT PILE

Young's Modulus for clays during undrained shear can be measured in the laboratory, backcalculated from field measurements on full scale earth structures, or measured using in-situ testing devices such as the pressuremeter. A common practice is to normalize undrained modulus, E_u, to undrained shear strength, S_u. Values of E_u are difficult to accurately measure in the laboratory because they are influenced by sample disturbance and the details of the laboratory procedures for consolidation and shear. D'Appolonia et al (4) suggest E_u/S_u values of about 1200 for predicting initial settlements under surface loadings at high factors of safety based on the review of measured performance of large scale field tests.

Other researchers have reported much lower values (100 to 300), possibly because they were backcalculated at lower factors of safety. Isotropically consolidated undrained triaxial compression tests reported by D'Appolonia et al (4) present values of E_u/S_u for clays calculated at a shear stress equal to 2/3 the undrained shear strength of the specimens. The E_u/S_u values decrease with increasing overconsolidation ratio (OCR). The following expressions provide a good fit to their data for values of OCR of 2 or more:

clays of low plasticity $$\frac{Eu}{Su} = \frac{800}{OCR} \quad \text{(A1.a)}$$

clays of high plasticity $$\frac{Eu}{Su} = \frac{400}{OCR} \quad \text{(A1.b)}$$

Laboratory tests simulating one-dimensional consolidation with zero horizontal strain followed by plane-strain shear at constant volume in the horizontal plane have been conducted as part of our pile research by D. M. Wood of the University of Cambridge in the True Triaxial Test device. Partial results of these tests are reported by Kirby and Esrig (10). Wood's tests suggest the stress strain behavior for undrained shear in the horizontal plane is similar to that measured in isotropically consolidated undrained triaxial compression tests. Thus, as a first approximation, we suggest using the values of E_u/S_u contained in Equations A1.a and A1.b for cavity expansion and subsequent consolidation.

The bulk modulus, $\bar{B}$, for small stress changes, is related to the compression ratio CR, and the mean effective stress $\bar{p}$ as follows:

$$\frac{\bar{B}}{\bar{p}_o} = \frac{2.3}{CR} \quad \text{(A2)}$$

CR is defined as the volumetric strain per log cycle of stress change and is related to compression index, C_c, and void ratio, e_o, as follows:

$$CR = \frac{C_c}{1 + e_o} \quad \text{(A3)}$$

Typical values of CR are 0.15 for normally consolidated clays of low plasticity and 0.3 for normally consolidated clays of high plasticity. CR values for recompression are 4 to 8 times lower than those for virgin compression. The recompression value should apply away from the pile while a value approaching that for virgin compression should apply for the remolded soil. If the average is used, and if we assume CR for reloading to be 6 times smaller than CR for virgin loading, we find:

$$\text{clays of low plasticity} \quad \frac{\bar{B}_{avg}}{\bar{p}_o} = 54 \qquad \text{(A4.a)}$$

$$\text{clays of high plasticity} \quad \frac{\bar{B}_{avg}}{\bar{p}_o} = 27 \qquad \text{(A4.b)}$$

We now have values of E_u/S_u as a function of OCR and values of $\bar{B}_{avg}/\bar{p}_o$ for clays of low and high plasticity. We are most interested in the ratio of $\bar{B}_{avg}/E_u$ for consolidation analysis. We can determine this ratio if we can estimate $S_u/\bar{p}_o$. $S_u/\bar{p}_o$ varies with overconsolidation ratio and with plasticity index. A value of $S_u/\bar{p}_o$ equal to 0.3 for normally consolidated soils is a rough first approximation for most clays. The increase in $S_u/\bar{p}_o$ with OCR can be approximated as:

$$\frac{S_u}{\bar{p}_o} = \left(\frac{S_u}{\bar{p}_o}\right)_{nc} OCR^{0.8} \qquad \text{(A5)}$$

Combining equations A1 and A5 and substituting $(S_u / \bar{p}_o)_{nc} = 0.3$, yields:

$$\text{clays of low plasticity} \quad E_u/\bar{p}_o = 240\ OCR^{-0.2} \qquad \text{(A6.a)}$$

$$\text{clays of high plasticity} \quad E_u/\bar{p}_o = 120\ OCR^{-0.2} \qquad \text{(A6.b)}$$

Typical values of the ratio of $\bar{B}_{avg}/E_u$ can be calculated by combining Eqs. A4 and A6.

$$\frac{\bar{B}_{avg}}{E_u} = 0.22\ OCR^{0.2} \quad \text{for clays of low and high plasticity} \qquad \text{(A7)}$$

The value of $\bar{B}_{avg}/E_u$ ranges approximately from 0.22 to 0.35 for values of OCR from 1 to 8.

APPENDIX II - REFERENCES

1. Carter, J.P., Randolph, M.F., and Wroth, C.P., "Stress and Pore Pressure Changes in Clay During and After the Expansion of a Cylindrical Cavity," International Journal on Numerical and Analytical Methods in Geomechanics, Vol. 3, No. 4, 1979, pp. 305-322.

2. Carter, J.P, Randolph, M.F., and Wroth, C.P., "Some Aspects of the Performance of Open and Closed-End Piles," Proceedings of Conference on Numerical Methods in Offshore Piling, Institution of Civil Engineers, London, 1979, pp. 165-170.

3. Casagrande, A., "Characteristics of Cohesionless Soils Affecting the Stability of Slopes and Earth Fills," Journal of the Boston Society of Civil Engineers, 1936, pp. 257-276.

4. D'Appolonia, D.J., Poulos, H.G., and Ladd, C.C., "Initial Settlement of Structures on Clay," Journal of the Soil Mechanics and Foundation Division, ASCE, Vol. 97, No. SM10, 1971, pp. 1359-1377.

5. Esrig, M.I., Kirby, R.C., Bea, R.G., and Murphy, B.S., "Initial Development of a General Effective Stress Method for the Prediction of Axial Capacity for Driven Piles in Clay," Proceedings of Ninth Offshore Technology Conference, Houston, Texas, Vol. 3, 1977, pp. 495-506.

6. Esrig, M.I., and Kirby, R.C., "Soil Capacity for Supporting Deep Foundation Members in Clay," Behavior of Deep Foundations, ASTM STP 670, R. Lundgren, Editor, American Society for Testing and Materials, 1979, pp. 27-63.

7. Esrig, M.I., Kirby, R.C., and Murphy, B.S., "Advances in General Effective Stress Method for the Prediction of Axial Capacity for Driven Piles in Clay," Proceedings of Eleventh Offshore Technology Conference, Houston, Texas, Vol. 1, 1979, pp. 437-448.

8. Jaky, J., "The Coefficient of Earth Pressure at Rest," Journal of the Society of Hungarian Architects and Engineers, pp. 355-358.

9. Kirby, R.C., and Wroth, C.P., "Application of Critical State Soil Mechanics to the Prediction of Axial Capacity for Driven Piles in Clay," Proceedings of Ninth Offshore Technology Conference, Houston, Texas, Vol. 3, 1977, pp. 483-494.

10. Kirby, R.C., and Esrig, M.I., "Further Development of a General Effective Stress Method for Prediction of Axial Capacity for Driven Piles in Clay," Proceedings of Conference on Recent Developments in the Design and Construction of Piles, Institution of Civil Engineers, London, 1979, pp. 335-344.

11. Kirby, R.C., Esrig, M.I., and Murphy, B.S., "General Effective Stress Method for Piles in Clay, Part II - Application," in preparation, January 1983.

12. Ladd, C.C., and Foott, R., "New Design Procedure for Stability of Soft Clays," Journal of the Geotechnical Engineering Division, ASCE, Vol. 100, No. GT7, 1974, pp. 763-786.

13. Lambe, T.W., "The Stress Path Method," Journal of the Soil Mechanics and Foundation Division, ASCE, Vol. 93, No. SM6, 1967, pp. 309-331.

14. Leifer, S.A., Kirby, R.C., and Esrig, M.I., "Effect of Radial Variation of Material Properties on Stress Changes due to Consolidation Around a Driven Pile," Proceedings of Conference on Numerical Methods in Offshore Piling, Institution of Civil Engineers, London, 1979, pp. 157-163.

15. Miller, T.W., Murff, J.D., and Kraft, L.M., "Critical State Soil Mechanics Model of Soil Consolidation Around a Driven Pile," Proceedings of Tenth Offshore Technology Conference, Vol. 4, 1978, pp. 2237-2242.

16. Olson, R.E., "Comments of Technical Advisor on ESACC Work," Amoco Production Company, July, 1980.

17. Palmer, A.C., "Undrained Plane-Strain Expansion of a Cylindrical Cavity in Clay: A Simple Interpretation of the Pressuremeter Test," Geotechnique, Vol. 22, No. 3, 1972, pp. 451-457.

18. Randolph, M.F., Steenfelt, J.S., and Wroth, C.P., "The Effect of Pile Type on Design Parameters for Driven Piles," Proceedings of Seventh European Conference on Soil Mechanics and Foundation Engineering, Vol. 2, 1979, pp. 107-114.

19. Randolph, M.F., and Wroth, C.P., "An Analytical Solution for the Consolidation Around a Driven Pile," International Journal on Numerical and Analytical Methods in Geomechanics, Vol. 3, No. 3, 1979, pp. 217-229.

20. Randolph, M.F., Carter, J.P., and Wroth, C.P., "Driven Piles in Clay - The Effects of Installation and Subsequent Consolidation," Geotechnique, Vol. 29, No. 4, 1979, pp. 361-393.

21. Roscoe, K.H., Schofield, A.N., and Wroth, C.P., "On the Yielding of Soils," Geotechnique, Vol. 8, 1958, pp. 22-53.

22. Roscoe, K.H., and Burland, J.B., "On the Generalized Stress-Strain Behavior of "Wet" Clay," Engineering Plasticity, Ed. J. Heyman, F.A. Lecke, Cambridge, 1968.

23. Schmertmann, J.H., "The Undisturbed Consolidation Behavior of Clay," Transaction, ASCE, Vol. 120, 1955, pp. 1201-1233.

24. Taylor, D.W., Fundamentals of Soil Mechanics, Wiley, 1948.

25. Wroth, C.P., Carter, J.P., and Randolph, M.F., "Stress Changes Around a Pile Driven into Cohesive Soil," Proceedings of Conference on Recent Developments in the Design and Construction of Piles, Institution of Civil Engineers, London, 1979, pp. 345-354.

APPENDIX III - NOTATION

$\bar{B}$	=	effective stress Bulk Modulus
$\bar{B}_{avg}$	=	average effective stress Bulk Modulus within the yielded region
CR	=	compression ratio, $\varepsilon_{vol}/\Delta \log \bar{p}$
CSL	=	critical state line
C_c	=	compression index for virgin compression, $\Delta e/\Delta \log \bar{p}$
C_s	=	swelling index, $\Delta e/\Delta \log \bar{p}$
DR	=	displacement ratio
ESM	=	effective stress method for prediction of axial capacity for displacement piles in clay
E_u	=	Young's Modulus for undrained loading
e	=	void ratio
F_i	=	Interface Factor; a correction factor which accounts for the effects on τ_f of the effective stress friction angle at the pile-soil interface
G_u	=	Shear Modulus for undrained loading
K_o	=	earth pressure coefficient for one-dimensional compression
LM	=	Load Module; defines the relationship between mean effective stress and radial effective stress for pile failure under axial load
OCR	=	overconsolidation ratio, $\bar{\sigma}_{vmax}/\bar{\sigma}_{vo}$
$\bar{p}$	=	mean effective stress

$\bar{p}_{ac}$	=	mean effective stress after consolidation
$\bar{p}_{cs}$	=	mean effective stress at the critical state
$\bar{p}_f$	=	mean effective stress on the pile shaft for pile failure under axial load
$\bar{p}_i$	=	mean effective stress after pile installation
$\bar{p}_{nc}$	=	mean effective stress for normally consolidated clay
$\bar{p}_o$	=	initial mean effective stress
q	=	deviator stress, $\sigma_1 - \sigma_3$
r	=	radial offset
R	=	radial offset from centerline of displacement pile to the boundary between the elastic region and the yielded region
S_u	=	undrained shear strength
t	=	radial offset at outside limit of integration for limiting equilibrium analysis
U_e	=	excess pore pressure at the pile face at the time of pile installation
VCL	=	virgin compression line
Δ	=	displacement of elastic-plastic boundary
Δp	=	change in mean total stress
$\Delta \bar{p}_{cons}$	=	change in mean effective stress caused by consolidation around a driven pile
$\Delta \bar{p}_{inst}$	=	change in mean effective stress caused by pile installation
$\Delta \bar{p}_{load}$	=	change in mean effective stress caused by pile loading
$\Delta \sigma_r$	=	change in radial total stress
$\Delta \sigma_v$	=	change in vertical total stress
$\Delta \sigma_\theta$	=	change in circumferential total stress
ε_{vol}	=	volumetric strain
η	=	corrective factor which relates $\bar{p}_i$ to $\bar{p}_{cs}$

κ	=	swelling index, $\Delta e / \Delta \ln \bar{p}$
λ	=	compression index for virgin compression, $\Delta e / \Delta \ln \bar{p}$
ρ	=	pile radius
$\bar{\sigma}_h$	=	horizontal effective stress
$\bar{\sigma}_r$	=	radial effective stress
$\bar{\sigma}_{rf}$	=	radial effective stress on the pile shaft for pile failure under axial load
$\bar{\sigma}_v$	=	vertical effective stress
$\bar{\sigma}_{vmax}$	=	maximum past vertical effective stress
$\bar{\sigma}_{vo}$	=	initial vertical effective stress
$\bar{\sigma}_1$	=	major principal effective stress
$\bar{\sigma}_2$	=	intermediate principal effective stress
$\bar{\sigma}_3$	=	minor principal effective stress
$\bar{\sigma}_\theta$	=	circumferential effective stress
τ_f	=	shear stress on the pile shaft for pile failure under axial load
τ_{max}	=	maximum shear stress
$\bar{\phi}$	=	effective stress friction angle for soil
$\bar{\phi}_{ss}$	=	effective stress friction angle for soil sliding on pile material

PROCEDURES FOR ANALYSIS OF LATERALLY LOADED PILE GROUPS IN SOFT CLAY

by Dewaine Bogard, A.M.[1], and Hudson Matlock, F.ASCE[2]

INTRODUCTION

A reevaluation of the results of a series of static and cyclic lateral load tests on circular pile groups at Harvey, Louisiana (Ref 6) has revealed that the pile groups exhibited very consistent trends of behavior, as compared with the results of companion single-pile tests. A back-analysis of the results of the experiments using the BMCOL 76 computer program (Ref 2) has led to the development of procedures to predict the behavior of similar pile groups in soft clays.

Criteria for the design of laterally loaded single piles in soft clay have been recommended earlier (Ref 7), and have been adopted by the American Petroleum Institute (Ref 1). The recommended criteria produce nonlinear curves of soil resistance p (force per unit length of pile) as a function of pile deflection y. The curves are then used to analyze the bending of the pile. Static p-y curves may be developed for short-time monotonic loading. Cyclic p-y curves are quasi-static representations of lower-bound peak resistances used for design for conditions involving progressively-increased cyclic loadings such as may occur during the buildup of an offshore storm.

The procedures which are discussed herein have been developed for the formulation of similar nonlinear curves for piles within groups. They follow a pattern consistent with the earlier criteria, with the values of soil resistance p and pile deflection y being modified to include effects of group interaction on both the lateral soil resistance and the pile group deflections.

CURRENT METHODS OF PILE GROUP ANALYSIS

Methods for elastic analyses of laterally-loaded piles and pile groups were developed by Poulos (Refs 8 and 9). The methods were developed to predict the pile-head response at a surface support. The variations in deflection and bending moment along the piles were not considered. The piles must be of constant cross-section, and the pile-head restraints must be either fully-fixed (no rotation) or fully free (no bending moment). The soil must be assumed to be elastic, and have constant and uniform properties with depth.

Focht and Koch (Ref 4) recognized that such assumptions are not appropriate for offshore pile group foundations. A method for pile group analysis was then developed which combined the Poulos approach for

[1]Ertec, Inc., 3535 Briarpark Dr., Houston, Texas 77042
[2]Ertec, Inc., 3777 Long Beach Blvd., Long Beach, California 90807

predicting the pile-head response with a p-y approach for determining the bending moments and deflections along the piles. Although the method proposed by Focht and Koch represents a significant improvement over the elastic analyses, it is also limited in applicability by the need to assume psuedo-elastic soil properties for use in determining the "y-multipliers" for modifying the single-pile p-y curves.

The Focht and Koch procedures were applied to the Harvey, Louisiana, pile group tests by Focht (Ref 5). Comparisons were made between measured and computed pile-head response only; the variation in bending moment along the piles was not considered. Reasonable agreement was reported between computed and measured response of the single pile under static loading; however, the static load behavior of the pile groups was considerably more nonlinear than that predicted. Poor agreement was reported for the cyclic load cases, for both the single-pile and pile group tests.

Among the conclusions reported by Focht is the observation that, for the pile groups tested, the assumption of linear pile group interaction factors did not appear valid.

The approach to pile group analyses discussed herein is similar to that discussed by Focht and Koch, except that the effects of pile group interaction on pile deflections are not assumed to be linear; rather, the procedures are based on the assumption that such effects vary nonlinearly with pile deflection.

BASIC CONCEPTS OF PILE GROUP BEHAVIOR

A laterally loaded pile develops resistance to deflection as a nonlinear function of the relative displacement between the pile and the soil. The increases in lateral pressure created by the pile must be absorbed by the surrounding soil, creating zones of increased shear and normal stress near the pile, which decay rapidly in magnitude with radial distance. For piles within groups, the zones of stress overlap, forming larger zones of stress in the soil surrounding the pile group. Integrating the lateral strains which arise from the superposition of the individual-pile stress bulbs would yield an increase in pile group deflection required to develop an equivalent level of lateral resistance against the individual piles, as compared to isolated piles.

In order to develop a method for the analysis of pile groups, it is first necessary to obtain reasonable estimates of the increases in pile deflection which result from the overlapping of the stresses in the soil around the group. If the magnitude of the additional deflections can be reasonably well estimated, the analysis of pile groups can be performed using the conventional methods of analysis developed for single piles, provided that the structural restraints are identical for each pile in the group, and therefore all piles in the group respond equally to the applied load.

Patterns of stress and deformation around two pile groups are shown in concept in Fig 1. For pile groups with relatively wide spacing, such as in Fig 1a, the development of zones of plastic flow very near the individual piles limits the degree of overlapping of stress and deformation, resulting in only minor increases in lateral deflection. For pile groups with close spacing, such as that shown in Fig 1b, the zones of plastic flow begin to overlap and develop around the pile group as a whole, significantly increasing the deflections of the pile group.

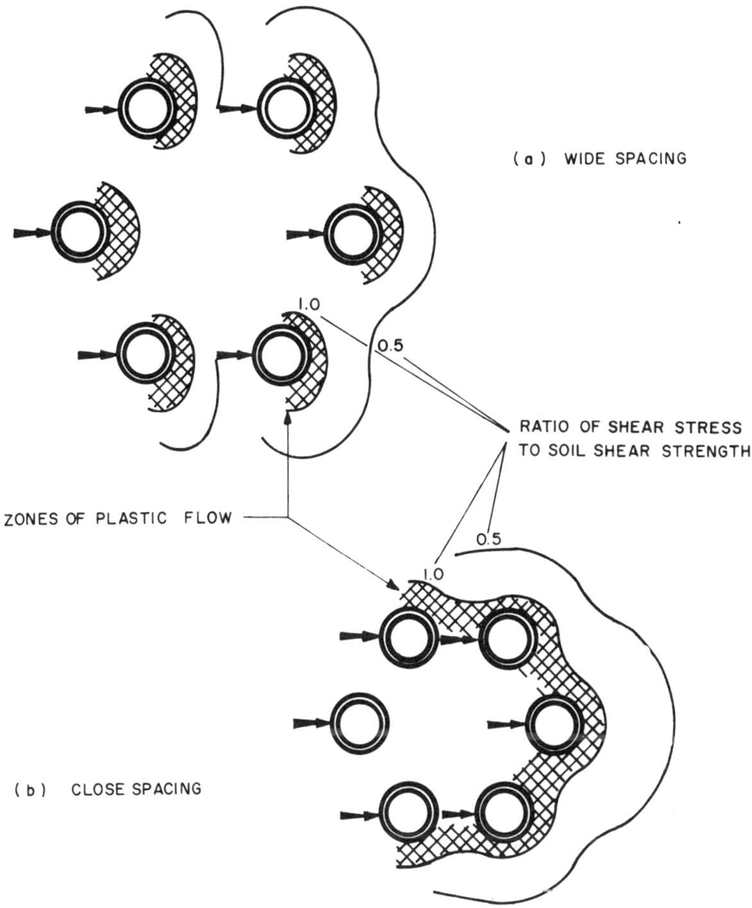

Fig 1. Schematic Illustration of Pile Group under Lateral Loading

Fig 1 suggests a key concept: However complex may be the patterns of stress and deformation which develop in the soil, there is some portion of deflection that arises from deformations in the soil around the group as a whole, and another portion of deflection that is due to localized stress and deformation near the individual piles. For the case of pile groups in soft clay soils, it should be expected that the portion of deflection from each source will increase nonlinearly with increases in lateral resistance.

HYPOTHESIS FOR PILE GROUP ACTION

In order to develop a method by which the magnitudes of the additional deflections may be estimated, the behavior of two geometrically-similar pile groups will be examined.

First, consider a pile group consisting of a number of piles placed at a spacing of one diameter between centers, that is, with adjacent piles being almost in contact. Such a pile group can be expected to act as a single unit, with patterns of behavior associated with the diameter of the group, rather than with the diameters of the individual piles within the group. The effective bending stiffness of the group is simply the sum of the bending stiffnesses of the individual piles. The lateral resistance of the soil, and the corresponding deflections, will be equal to those for an imaginary pile having a diameter equal to the outside dimension of the pile group.

The response of each pile within the group can be evaluated on a per-pile basis by developing a p-y relationship for each pile which consists of the values of soil pressure p for the imaginary pile divided by the number of piles in the group, with corresponding values of deflection y being those developed for the imaginary pile. Due to the extremely close spacing of the group, the contribution to deflection y from local deformations in the soil immediately adjacent to each pile will be negligible. For closely-spaced pile groups, the behavior will thus be dominated by the portion of deflection arising from global deformation of the soil mass, with little influence from the relative deflections of individual piles within the mass.

Next, consider a second pile group having the same outside dimensions, but with a smaller number of piles arranged at some wider spacing. The bending stiffness of the group will again be the sum of the bending stiffnesses of the individual piles. As compared with the first group at any given value of deflection, the lateral resistance of the soil against each pile will increase, yet the total resistance against the group will decrease, due to the smaller number of piles in the group.

In evaluating the behavior of the individual piles within the second group on a per-pile basis, the lateral resistance p at any pile deflection y may be estimated by distributing the lateral resistance p for the imaginary pile equally among all the piles in the group. The ultimate lateral resistance against each pile, however, cannot exceed that which would be developed if each pile were independent of the group.

Corresponding values of deflection y developed for the imaginary pile will not be appropriate for the piles in this group, since the degree of overlapping of the individual-pile stress bulbs is less. For pile groups with non-contiguous pile spacing, the additional pile deflections which result should decrease in linear proportion to increases in pile spacing. The values of deflection from the imaginary pile which are based on a center spacing of one pile diameter may thus be divided by the actual center spacing of the piles in order to obtain the component of deflection due to soil mass deformation.

The component of deflection due to localized deformation in the soil immediately adjacent to the individual piles will not be affected by the deformations in the soil surrounding the pile group, and may be estimated using current single-pile theory.

The total deflection y for a pile within the group at any value of soil pressure p will then be the sum of deflection from the two components; that due to global deformation of the soil mass surrounding the pile group, and that due to local deformations concentrated in the soil very near the individual piles within the group. In terms of p-y curve construction, this concept is shown in Fig. 2. Curve A is a conventional single-pile p-y curve. Curve B shows the component of pile deflection arising from soil mass deformation, as calculated from the modified p-y curve for the imaginary pile.

As shown in the conceptual construction of the group-pile p-y curve C, the maximum resistance is limited to the lower peak of the two curves. At each value of soil resistance p, the deflections from Curves A and B are simply added, up to the value of maximum resistance. Thereafter, the deflections increase only in accordance with Curve B. Any rebound along Curve A is neglected; the single-pile deflection component is assumed to remain at Point A.

The relative positions of the two components of pile deflection are dependent upon the number and spacing of the piles in the group. For widely-spaced pile groups, Curve A will represent the component of deflection from the imaginary pile and Curve B the single-pile p-y curve. The p-y construction described above is not affected; for pile deflections beyond that required to achieve the peak resistance, Curve C merely follows the single-pile p-y curve rather than the curve developed for soil mass deformation.

The nonlinear p-y curves thus described will reasonably accommodate the extremes in behavior expected for pile groups placed at widely different center spacings. At close spacings, the lateral resistance per pile from the imaginary pile will be small, and the deflections large, so that the deflections from the individual pile will not contribute significantly to the total deflection; as the spacing increases, the resistance per pile will increase at such a rate so that the contribution to deflection from the imaginary pile will rapidly diminish, leaving only the component of deflection from the individual pile. For intermediate spacings, there will be a gradual transition from one extreme of behavior to the other, as the portion of deflection from each source reflects the changes in the number and spacing of piles within the group.

DEVELOPMENT OF NONLINEAR RESISTANCE-DEFLECTION CURVES FOR STATIC LOADING

The procedures for developing nonlinear resistance-deflection (p-y) curves for individual piles within groups may be summarized as follows:

1. Using the procedures recommended in Refs 1 and 7 for single piles, develop the variation in ultimate resistance with depth

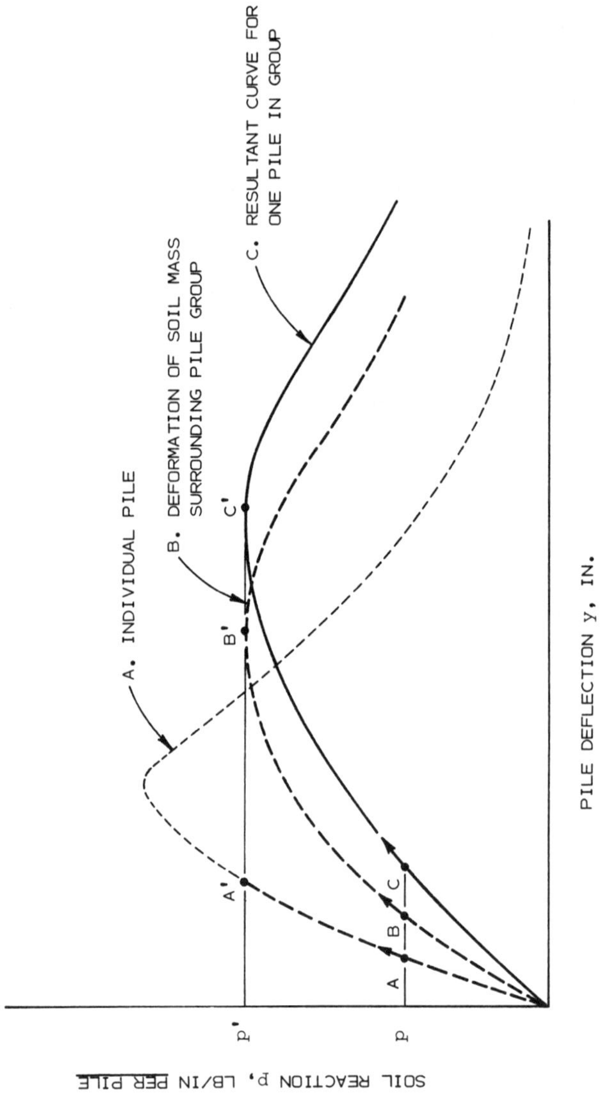

Fig 2. Conceptual Construction of P-Y Curves for Piles Within Groups

for the individual piles within the group and for an imaginary pile having a diameter equal to the gross diameter of the pile group.

2. Develop nonlinear curves of soil resistance p and pile deflection y for the imaginary pile. Distribute the values of soil resistance p equally among the piles in the group. Corresponding values of deflection y are obtained by dividing those developed for the imaginary pile by the center spacing of the piles, expressed in pile diameters.

3. At each value of soil pressure p from Step 2, determine the component of deflection from the individual piles, again using procedures recommended in Refs 1 and 7.

4. The nonlinear curves for a pile within the group are then completed by adding the two components of deflection at each value of soil pressure.

The limiting, or ultimate, lateral resistance is taken as the smaller of the two values; that computed on a per-pile basis from the imaginary pile, or that computed for the individual piles. For closely-spaced pile groups, those from the imaginary pile will control. For widely-spaced pile groups, those from the individual piles will be the smaller.

The nonlinear curves may then be used to analyze the pile group for bending on a per-pile basis using conventional single-member solutions. The behavior of the pile group may also be examined by replacing the bending stiffness of the single pile with the sum of the bending stiffnesses of all the piles in the group, and multiplying the values of soil resistance per unit length of pile by the number of piles in the group, provided that each pile has identical conditions of head restraint.

For those cases where the piles are not arranged in a symmetric pattern, or where the structural connectivity differs among the piles in the group, the response of the pile group and the distribution of shears and bending moments among the individual piles within the group may be determined using more sophisticated techniques, which properly account for effects of structural connectivity and load directionality on the relative deflections and rotations of the individual piles (Ref 3).

EFFECTS OF CYCLIC LOADING ON PILE GROUP BEHAVIOR

The results of the cyclic lateral load tests at Harvey, La. (Ref 6) indicated that the initiation of cyclic degradation in resistance occurred at a value of pile deflection related to the individual pile diameters, rather than the diameter of the pile groups. Near the surface, the losses in resistance were associated with vertical heave and with the separation of pile and soil, or gapping. At depths where the soil was sufficiently confined so that the soil movement was horizontal, the losses in resistance were the equivalent of those normally associated with remolding. The losses in resistance at depth are thought to be initiated by the enforced reversal of plastic strain and slip in the soil. In either case, the effects of cyclic degradation are most severe in the soil immediately adjacent to the pile, with the severity of the effects decreasing rapidly with radial distance.

If the piles in a group are separated by sufficient distance, the local heave and gap formation will be confined to the soil very near the individual piles. As the gaps form around the individual piles, the lateral resistance is significantly reduced, reducing the stress transferred to the surrounding soil; thus little group action will be observed. At depth, the effects of group action will be the same as those observed during static loading at equivalent values of lateral resistance. The effects of cyclic loading on a widely-spaced pile group will thus be a reduction in the observed group effects as compared with the static behavior as the soil resistance decreases during cyclic loading.

If the piles in the group are closely spaced, the surface gaps initiated around the individual piles will overlap and join, forming a continuous gap around the entire pile group. At depth, the zones of plastic shear and remolding will also overlap, creating a zone of remolded and weakened clay around the perimeter of the group. Closely-spaced pile groups will thus exhibit more severe effects of cyclic loading than will isolated piles, especially in the increases in pile group deflection required to sustain an equal value of shear per pile, as compared to an equivalent static loading.

MODIFICATION OF STATIC CURVES FOR CYCLIC LOADING

The psuedo-depth defined as x_r in the current single-pile criteria (Ref 7) is related to the depth required to maintain full contact between the pile and soil during cyclic loading. The term x_r is also used to estimate the increase with depth of the cyclic minimum lateral resistance at large deflections.

The term x_r may be used as an aid in approximating the degree of pile group interaction by reducing the component of deflection from the imaginary pile (which represents the global soil deformation) by the ratio of the depth x to the term x_r computed at the same depth for the individual pile. The component of deflection from the individual pile (which represents the relative deflection within the soil mass) will remain unchanged. The result will be a gradual transition in the predicted behavior with depth, from no pile group interaction at the surface to full interaction at and below the depth x_r.

At the surface, complete loss of soil resistance is assumed to occur at a pile deflection of $15y_c$, as suggested in Ref 7. As the depth increases, the value of resistance at the deflection $15y_c$ is also assumed to increase in the same manner, being the ratio of x/x_r multiplied by the peak cyclic resistance.

For deflections greater than that required to achieve the peak cyclic resistance, the p-y curve for a group pile is assumed to be parallel with that developed for the single pile; that is, the component of deflection from the imaginary pile at the peak cyclic resistance is added to the value $15y_c$ for the single pile. In effect, this construction assumes that no elastic recovery of deformation occurs in the soil mass during continued cyclic loading, although the quasi-static form of the cyclic p-y curve indicates a decrease in lateral soil resistance with increases in cyclic pile deflection.

COMPARISON OF PREDICTED AND MEASURED STATIC PILE GROUP BEHAVIOR

In order to demonstrate the procedures described above, the methods will be used to formulate nonlinear p-y curves for the pile groups tested at Harvey, Louisiana, and reported in Ref 6. The curves will be used to analyze the pile groups for bending on a per-pile basis using the computer program BMCOL 76. The analyses will be performed using the deflections measured at the two support points as boundary conditions. The computed values of shear and bending moment will then be compared to those measured during the static and cyclic pile group tests.

A complete description of the experimental apparatus and procedures used in the Harvey pile tests appears in Ref 6, and will not be repeated here. The experiments were performed on restrained-head single piles and circular groups of 5 and 10 piles each. The piles were nominal 6-inch diameter standard pipes (6-5/8-inch actual diameter) arranged with centers in a circular pattern 35 inches in diameter (41-5/8-inch outside diameter). The pipe piles were driven to a final penetration of 40 feet below the bottom of an excavated pit.

Total lateral loads applied to the groups were measured by specially-constructed load cells. In addition, the loads applied to selected individual piles were measured. The bending moments along the piles were measured by insertable units designed to reflect the actual curvature in the pipe piles, and yielded an output that was proportional to the bending moment. All the load-sensing units were carefully calibrated prior to and during the test program.

The shear strength profile selected for use in developing the nonlinear curves was obtained using an in-situ vane-shear device. The shear strength profile is shown in Fig 3. The value at the mudline was taken as 1.5 psi, increasing to 3.0 psi at a depth of 1 foot, thereafter increasing with depth at a rate of 0.077 psi per foot.

The results of the laboratory triaxial compression tests on intact samples were used to determine the values of ε_{50} (strain at half the maximum compressive stress difference) for developing the shapes of the p-y curves. The values of ε_{50} generally increased with depth, being 0.008 near the surface, and increasing to 0.02 at a depth of 11.5 feet, below which depth the value was assumed to remain constant.

The variation in ultimate lateral resistance with depth for the single pile and for the group piles is shown in Fig 4. A value of 0.25 was used for the J coefficient, in order to obtain the best fit with the results of the static load tests on the single pile. The values are given for the pile groups on a per-pile basis, that is, the ultimate resistance for the 41-5/8-inch diameter imaginary pile has already been divided by the number of piles in the group.

It is of interest to compare the magnitudes of lateral resistance calculated for the two pile groups, as compared to the single pile. The 10-pile group shows an ultimate lateral resistance per pile of only half that of the single pile; the 5-pile group exceeded the predicted single-pile capacity near the surface, and at depths greater than those shown in Fig 4. The 5-pile group is thus at a spacing where group action begins to dominate the behavior under static loading; the behavior of the 10-pile group will clearly be dominated by group effects.

The behavior predicted for the single pile under static loading is compared with the experimental data in Figs 5, 6, and 7. The pile-head response, shown in Figs 5 and 6, shows reasonable agreement among

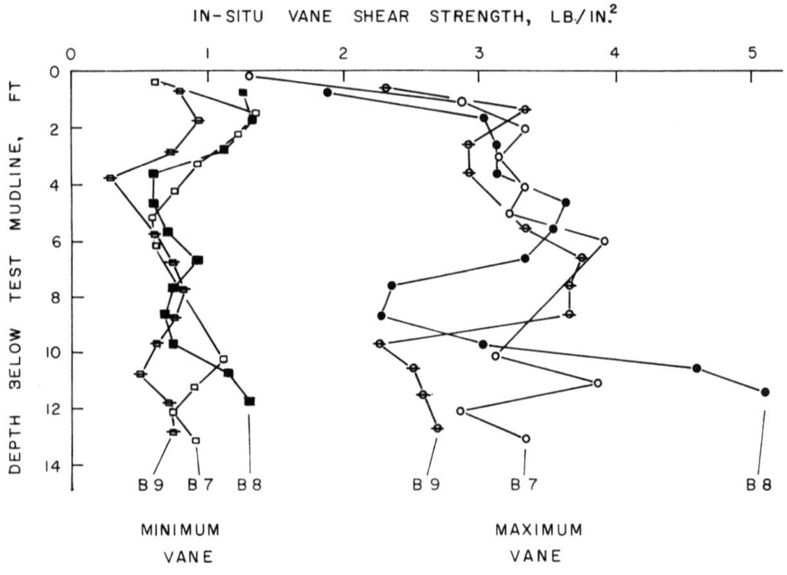

Fig 3. In Situ Vane Shear Strength Profile at Harvey, Louisiana Test Site

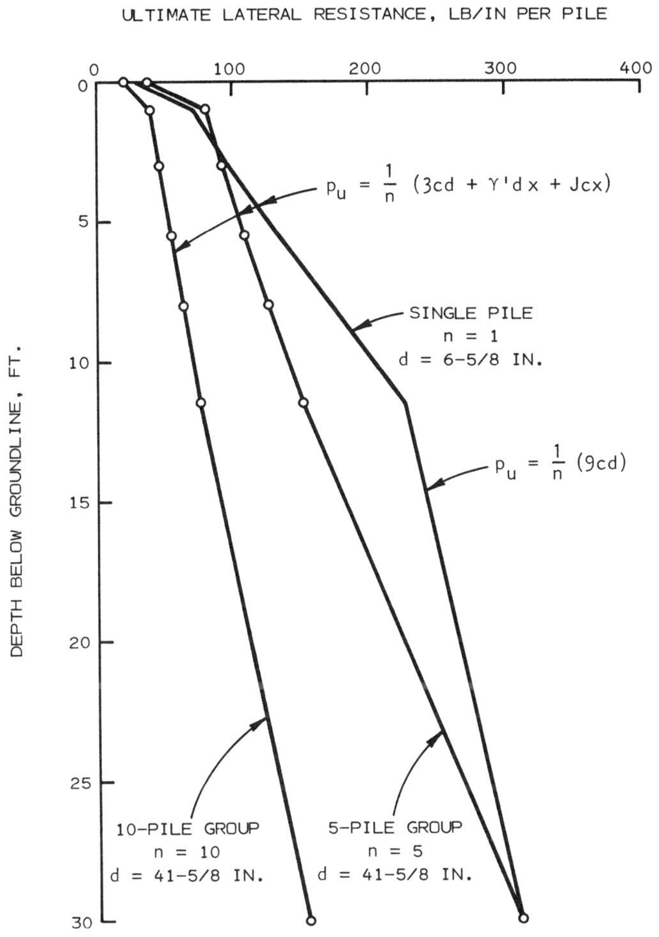

Fig 4. Variation in Ultimate Lateral Resistance with Depth

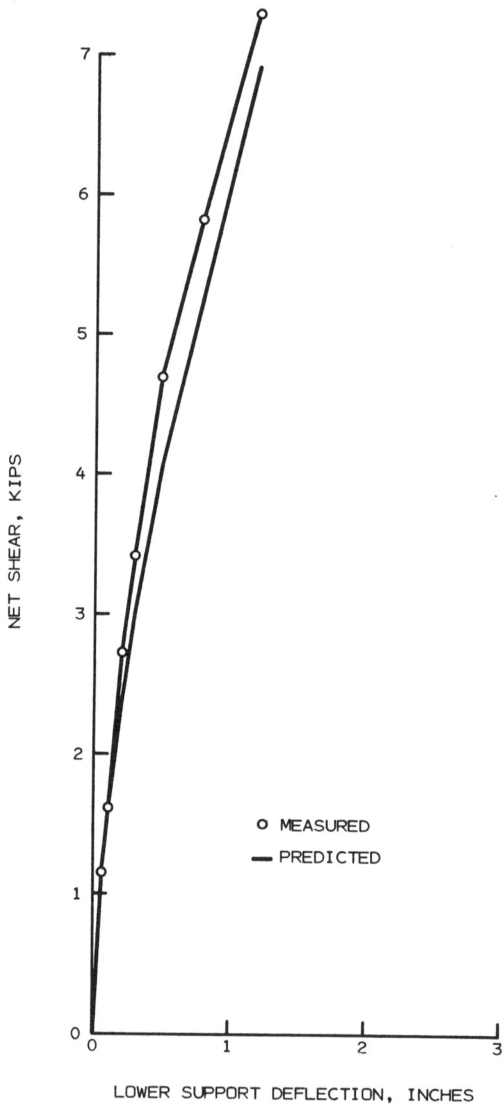

Fig 5. Development of Lateral Load with Static Deflection of Single Pile

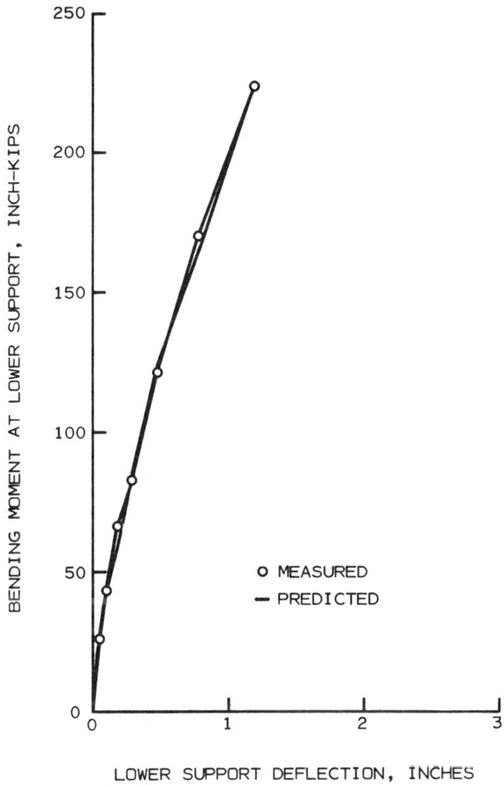

Fig 6. Development of Bending Moment with Static Deflection of Single Pile

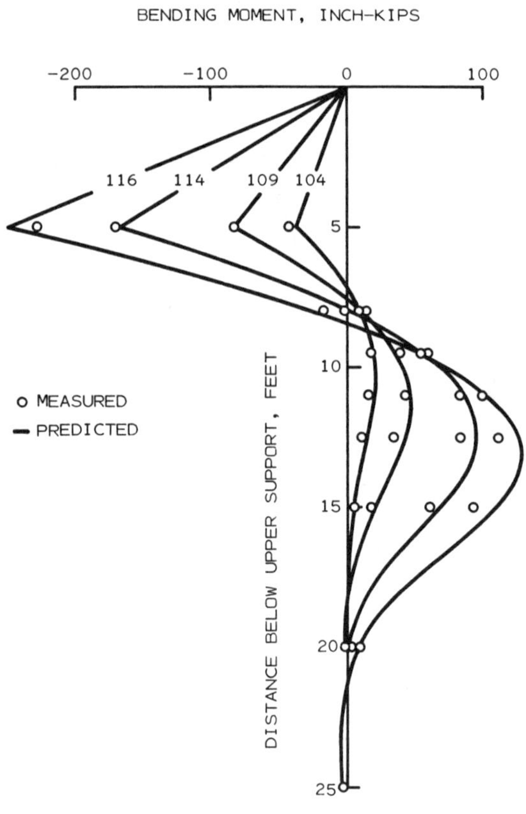

Fig 7. Bending Moment Along Single Pile During Static Tests

predicted and measured values of lateral shear and bending moment with increasing pile deflection. The predicted and measured variations in bending moment along the pile, shown in Fig 7, also agree reasonably well. This agreement indicates that both the magnitude and distribution of lateral resistance with depth was satisfactory, and that the shapes of the resistance-deflection curves were also reasonably close to those developed during the experiments.

The nonlinear curves developed for the static loading of the 5-pile group are shown in Fig 8. Included in Fig 8 are the curves developed for the single pile, the imaginary pile (as modified to account for spacing), and the group pile, which consists of the sum of the deflections from the other two curves at each selected value of lateral resistance. At this particular depth, 3 feet below the mudline, the component of deflection from the imaginary pile was approximately twice that from the single pile; along the full length of the piles, the deflections from the imaginary pile ranged from 1 to 6 times those of the single pile. It should be noted that the values of deflection for the group pile are affected by the relative values of ultimate resistance for the individual piles and the imaginary pile in addition to the differences in the diameters; thus, there is no single "y-multiplier" or "stretch-factor" which can be applied along the full length of the piles.

The predicted behavior of the 5-pile group is compared to that measured in the experiments in Figs 9, 10, and 11. The behavior predicted agrees very closely with that measured, with the predicted shears and bending moments falling within the range of measured values, and being reasonably near the average values. Bending moment diagrams from only four of the tests are shown in Fig 11. Similar agreement was noted throughout the range of loading; the intermediate tests are omitted from the figure for clarity of presentation.

The agreement among the predicted and measured values of shear and bending moment at the lower support and among the bending moments measured along the piles indicates that the methods used to develop the support curves reproduced the distribution of lateral resistance with depth for the pile group reasonably well over the entire range of pile deflection.

The nonlinear curves developed for static loading of the 10-pile group are shown in Fig 12. Included are the nonlinear curves for the single pile, the imaginary pile (as adjusted for spacing), and the composite curve for a group pile. As seen in the figure, the group effects completely dominate, with the single-pile contribution being only a small part of the total deflection. For this particular pile group, the contribution to deflection from the single pile varied from 1 to 6 percent of the total deflection, and could thus have been omitted. The single-pile deflection was included, however, both for reasons of consistency, and to aid in the later construction of the nonlinear curves to be used for comparisons with the results of the cyclic load tests.

The behavior predicted for the 10-pile group is compared in Figs 13, 14, and 15 with that measured during the experiments. Again, the predicted behavior agrees reasonably well with that measured. The values of predicted shear given in Fig 13 lie along the lower bound of those measured. A better fit could have been obtained by adjusting the soil parameters within the range of reported values; this was not done so that a more valid comparison of the results could be performed.

Better agreement was obtained between the predicted and measured values of bending moment, as shown in Figs 14 and 15. The predicted values

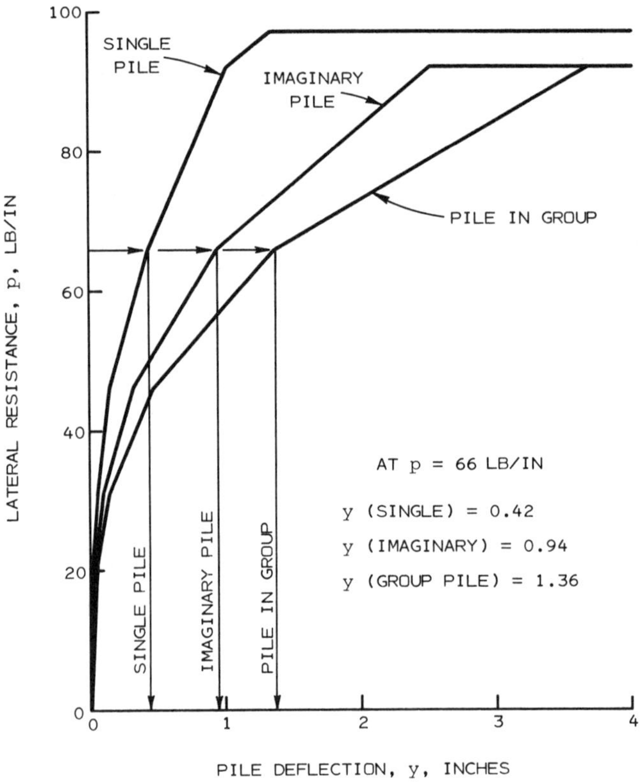

Fig 8. Development of Static p-y Curve for the 5 - Pile Group

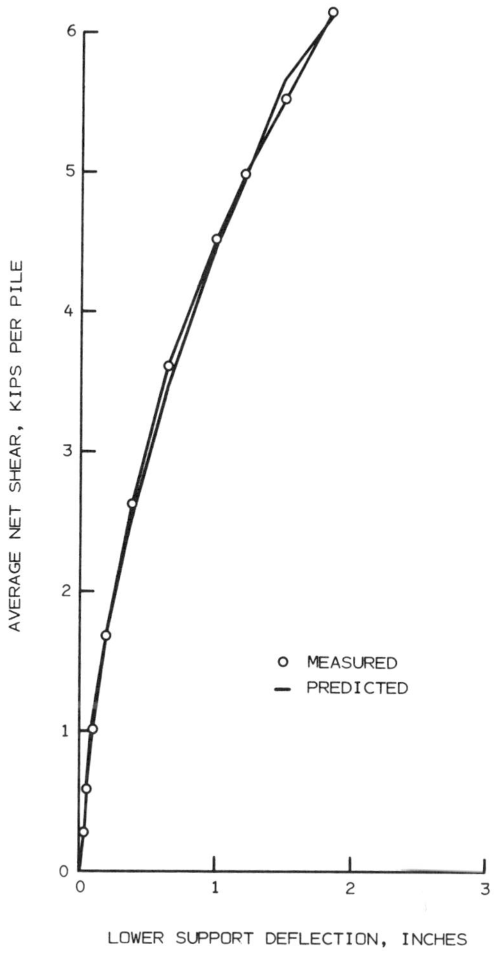

Fig 9. Development of Lateral Load with Static Deflection of 5 - Pile Group

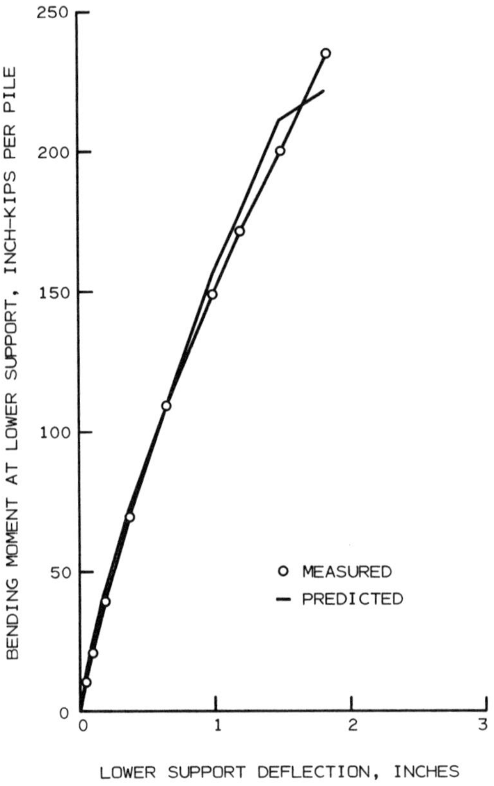

Fig 10. Development of Bending Moment with Static Deflection of 5 - Pile Group

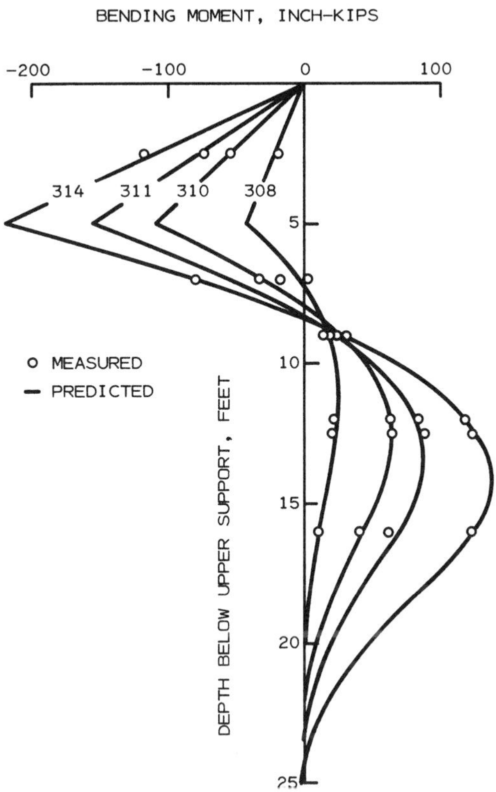

Fig 11. Bending Moment Along Piles During Static Tests of the 5 - Pile Group

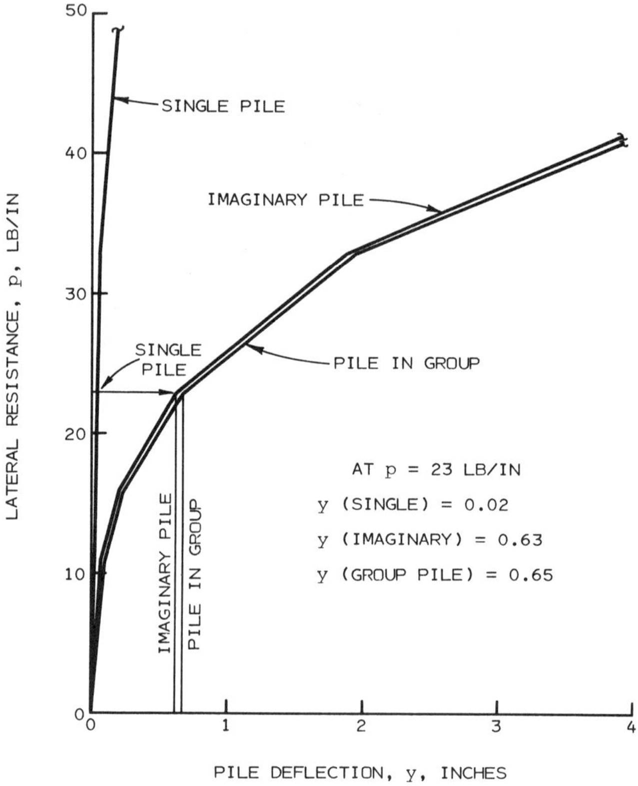

Fig 12. Development of Static p-y Curve for the 10 -Pile Group

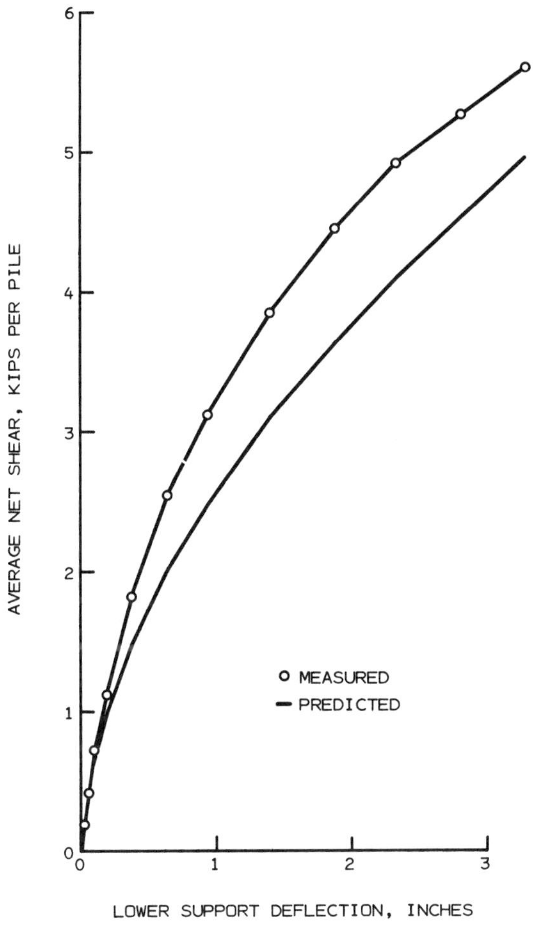

Fig 13. Development of Lateral Load with Static Deflection of 10 - Pile Group

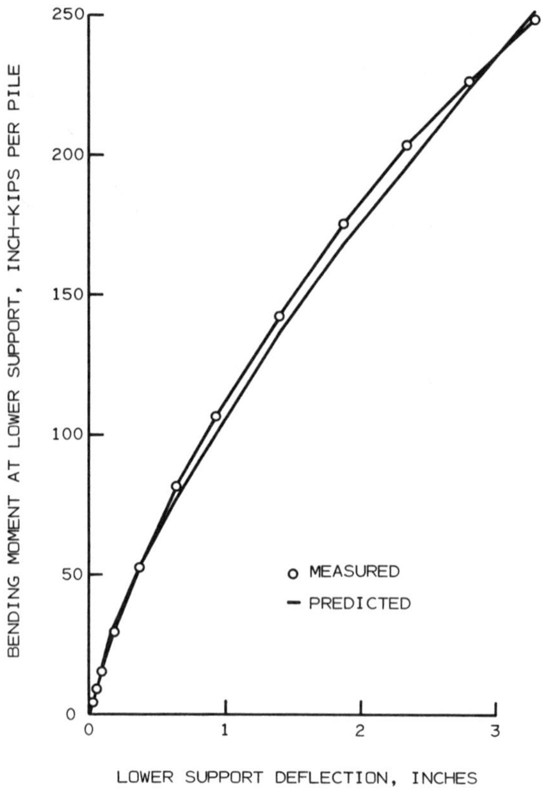

Fig 14. Development of Bending Moment with Static Deflection of 10 - Pile Group

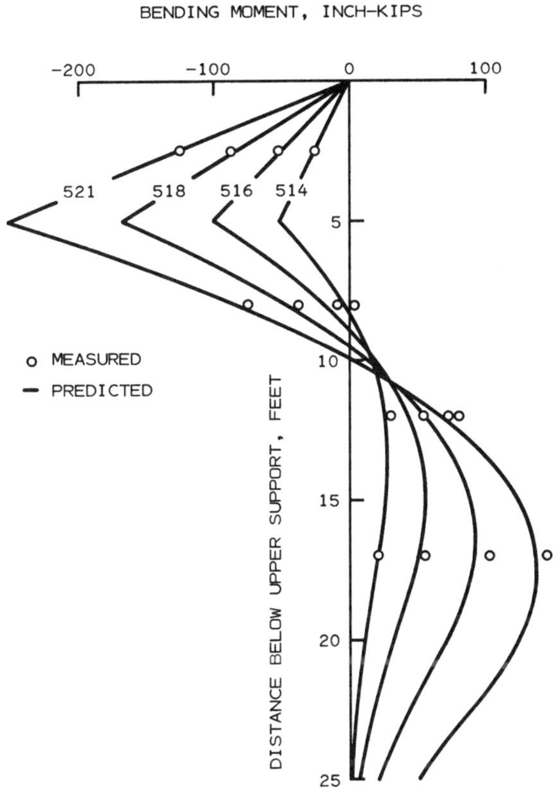

Fig 15. Bending Moment Along Piles During Static Tests of 10 - Pile Group

are near the average values shown, with most of the predicted values being within the range of measured values. Again, only four of the bending moment diagrams are given, for clarity of presentation. Similar agreement was obtained for the intermediate load cases not shown.

COMPARISON OF PREDICTED AND MEASURED CYCLIC PILE GROUP BEHAVIOR

The behavior predicted for the single pile during cyclic loading is compared in Figs 16, 17, and 18 to that measured during the experiments. In order to achieve the agreement shown, the peak cyclic resistance was taken as 0.34 times the ultimate static resistance, rather than the value of 0.72 suggested in Ref 7. As can be seen in Fig 3, this ratio is approximately that of the ratios of the residual to the maximum vane shear strengths from the in-situ vane shear tests. With this exception, the procedures followed in developing the support curves were the same as those given in Ref 7.

The nonlinear curves developed for the 5-pile group at the 36-inch depth are shown in Fig 19. At this depth, the ratio of the depth x to the computed depth x_r was 0.28, so that the contribution to deflection from the imaginary pile was only 28 percent of that shown in Fig 8 for static loading. As shown in the figure, the contribution to deflection from the imaginary pile has been reduced to a value smaller than that from the single pile. The predicted decrease in the added deflection will reduce considerably the computed group effects; as noted in Ref 6, the piles in this group did exhibit patterns of response similar to the single pile, except at the larger deflections.

The predicted behavior of the 5-pile group under cyclic loading is compared to that measured in Figs 20, 21, and 22. The agreement shown in the figures among both shear and bending moment at the lower support, and among the bending moments along the piles, indicate that the method of constructing the cyclic p-y curves gave satisfactory agreement.

The nonlinear curves developed for the 10-pile group at the 36-inch depth are shown in Fig 23. The contribution to deflection from the imaginary pile is only 28 percent of that shown in Fig 12, yet the curves indicate that the behavior will again be dominated by group effects, as was the case for the static loading.

The predicted behavior for the 10-pile group under cyclic loading is compared to that measured in Figs 24, 25, and 26. Although the agreement between the predicted and measured response at the lower support, shown in Figs 24 and 25, is not as good as that for previous tests, it is considered satisfactory. Better agreement is shown in Fig 26, which shows the predicted and measured bending moments developed along the piles.

During the cyclic loading of the 10-pile group, the gaps around the individual piles joined, forming a continuous gap around the pile group as a whole. At the end of the tests at large deflections, the gap was found to extend several feet into the soil. The result was the creation of a highly nonlinear and inelastic system, which is extremely difficult to model. During the cyclic tests of the 5-pile group, the gaps around the individual piles did not overlap, so that the piles tended to revert to solitary-pile behavior, with group effects only becoming apparent at large deflections.

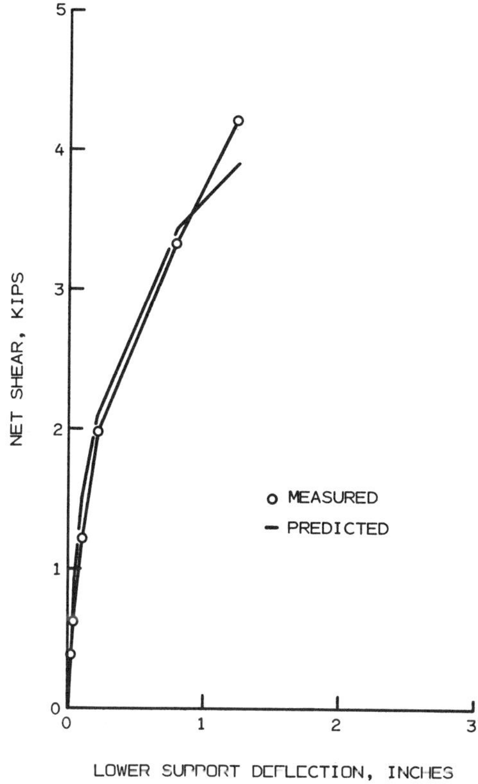

Fig 16. Development of Lateral Load with Cyclic Deflection of Single Pile

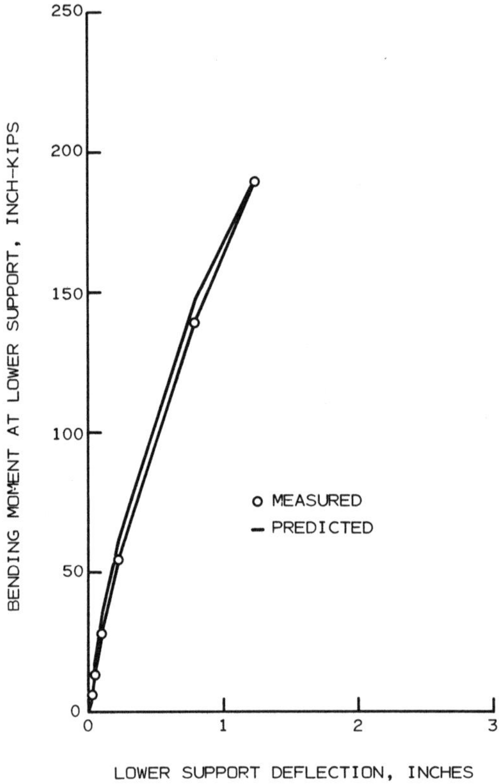

Fig 17. Development of Bending Moment with Cyclic Deflection of Single Pile

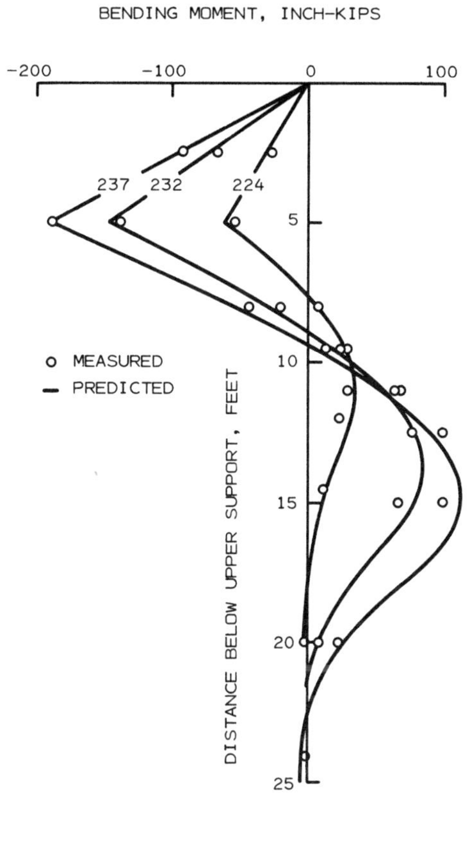

Fig 18. Bending Moment Along Single Pile During Cyclic Tests

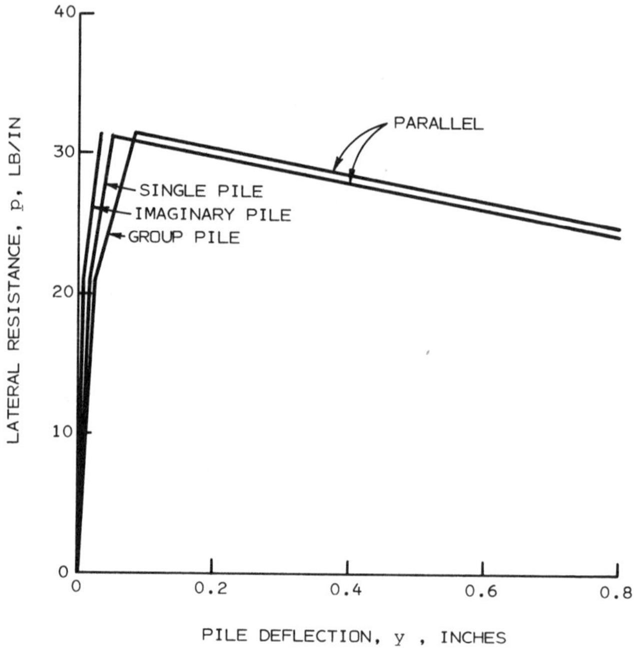

Fig 19. Development of Cyclic p-y Curve for the 5 - Pile Group

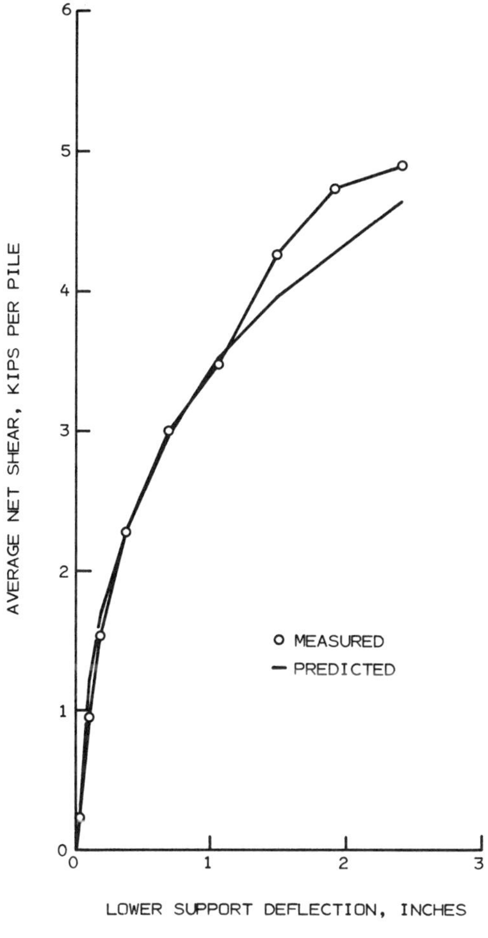

Fig 20. Development of Lateral Load with Cyclic Deflection of 5 - Pile Group

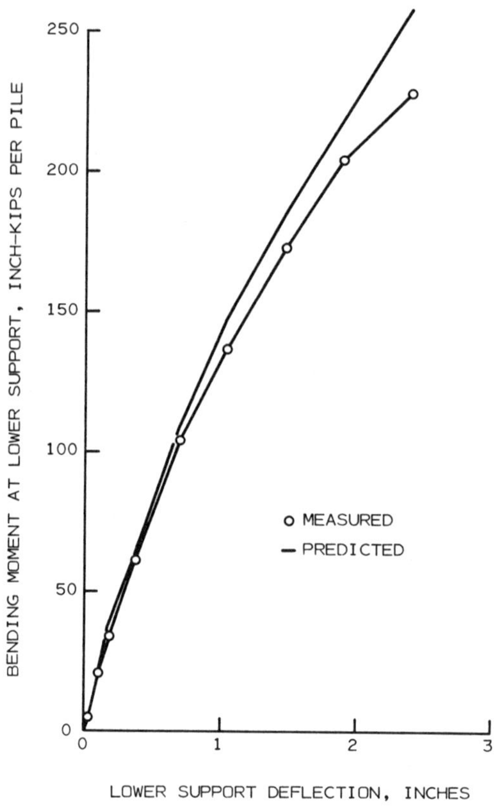

Fig 21. Development of Bending Moment with Cyclic Deflection of 5 - Pile Group

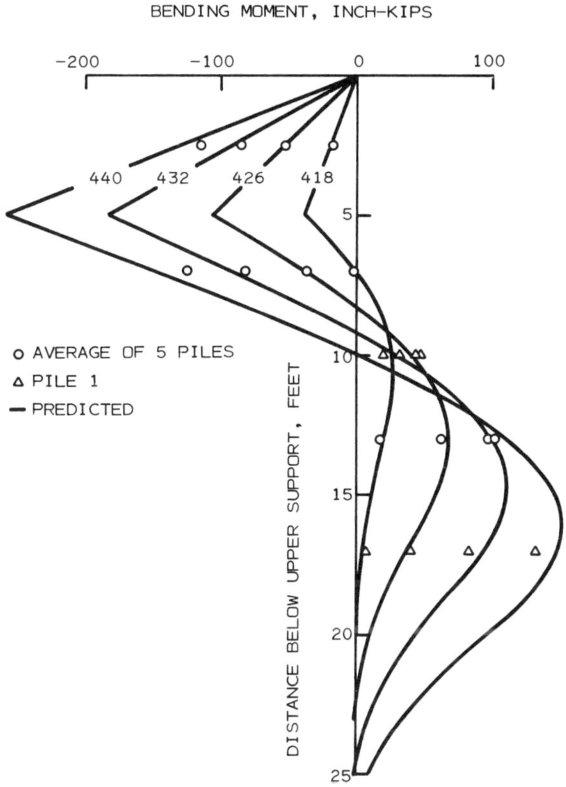

Fig 22. Bending Moment Along Piles During Cyclic Tests of 5 - Pile Group

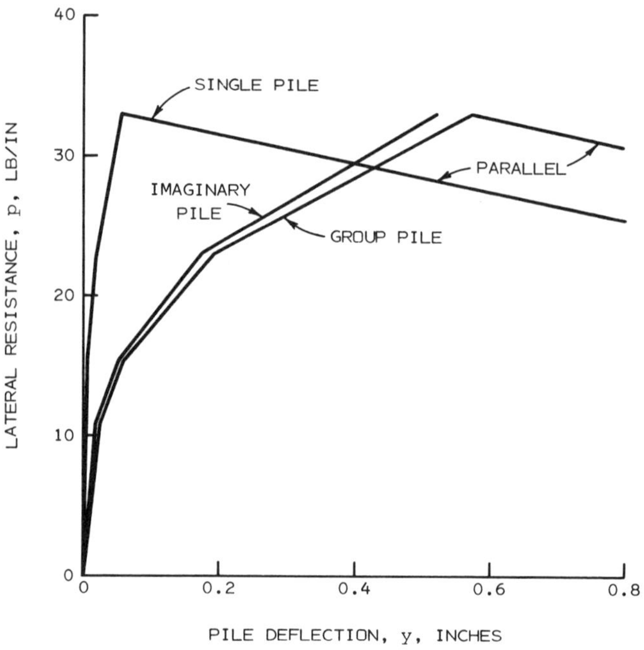

Fig 23. Development of Cyclic p-y Curve for 10 - Pile Group

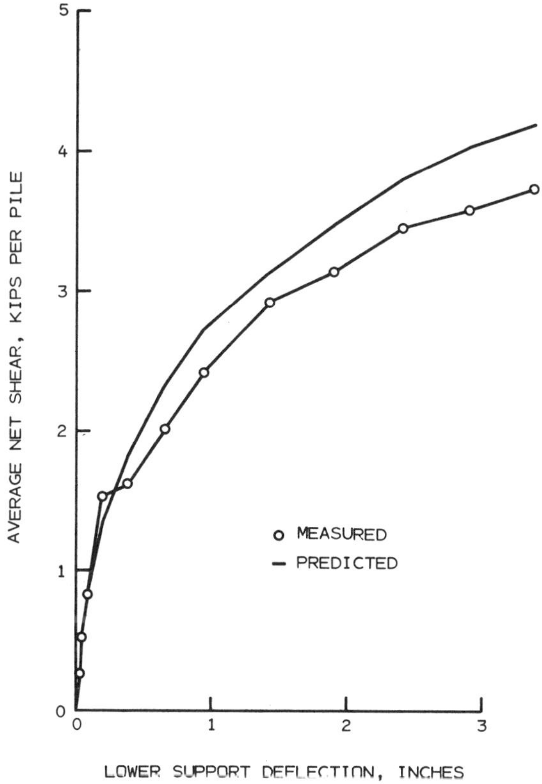

Fig 24. Development of Lateral Load with Cyclic Deflection of 10 - Pile Group

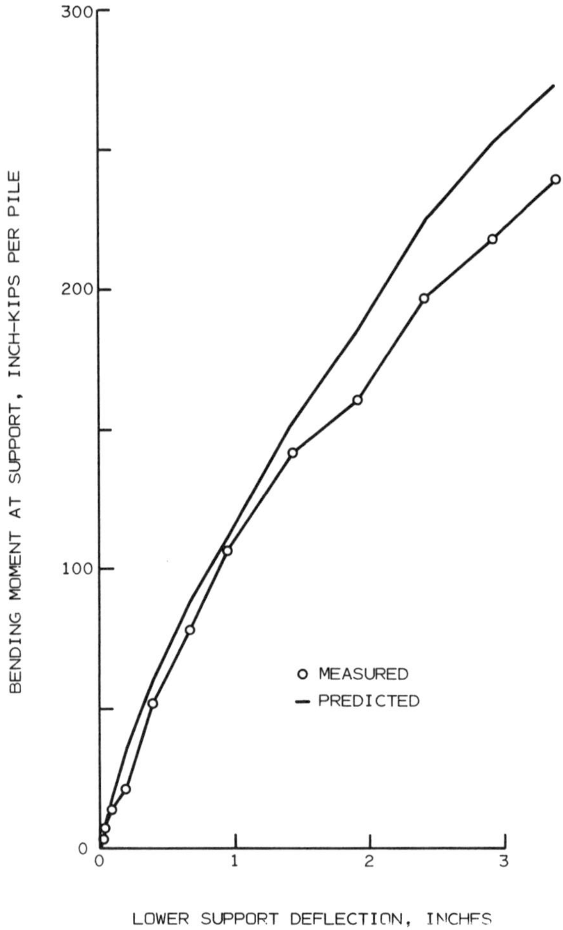

Fig 25. Development of Bending Moment with Cyclic Deflection of 10 - Pile Group

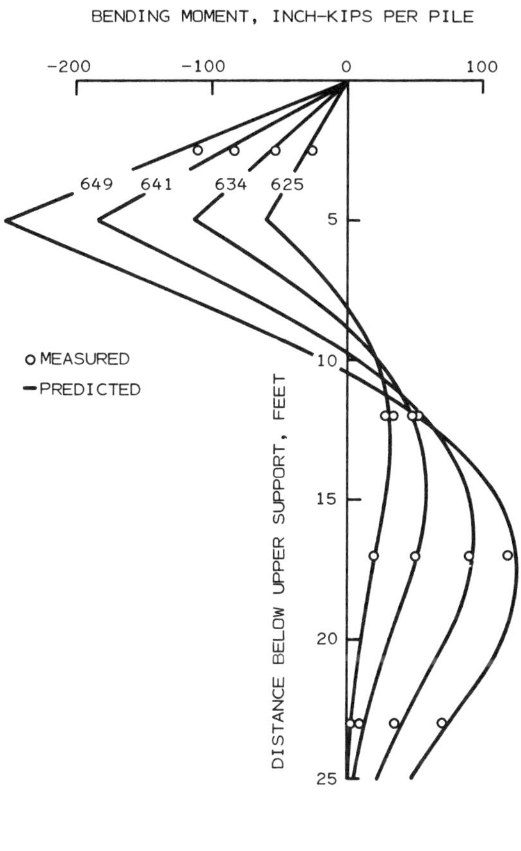

Fig 26. Bending Moment Along Piles During Cyclic Tests of 10 - Pile Group

CONCLUSIONS AND RECOMMENDATIONS

The agreement between the predicted and measured behavior of the pile groups indicates that the procedures recommended for the development of nonlinear resistance-deflection curves reproduced both the magnitude and distribution of lateral resistance with depth reasonably well over a wide range of pile deflections, for both static and cyclic loading of the pile groups tested. The agreement indicates that the use of two independently-formulated nonlinear relationships for the components of deflection, from group-pile interaction and from individual-pile deflection within the field of global soil deformation, is a reasonable and valid approach to the problem.

The procedures are not represented as a rigorously correct solution of a complex problem; they merely represent a convenient method by which the effects of the superposition of lateral stresses in the soil surrounding pile groups may be estimated sufficiently well to perform analyses of the group within commonly-accepted tolerances for geotechnical design. As such, the recommended procedures have been shown to work satisfactorily for the pile group tests performed in Harvey, La.

The procedures were developed for analysis of pile groups having structural restraints and loading conditions which result in approximately equal pile-head deflections and rotations among the piles in the group. For such groups, the variations in shear and bending moment among the individual piles are small, and can be disregarded in the structural design of the piles. Such groups can be analysed as a single unit or on a per-pile basis using conventional single-member solutions, provided that the predominant mode of displacement is lateral translation. If the mode of displacement of the group is rotational, either by torsion or tilting (rocking), the stress patterns which are created in the surrounding soil are significantly different; thus, pile group effects predicted by use of the procedures will likely be in error.

The predominant source of variations in shear, bending moment, and axial load among the piles in a group is the variation in structural connectivity, with the additional influence of load directionality becoming important in the case of nonsymmetrical pile groups or structural configurations. For such cases, the pile group cannot be treated as a unit, but must be analysed using techniques such as those described in Ref 3, in which the piles are solved independently and simultaneously, with the interconnectivity between the piles and the structure being maintained throughout the iterative solution process. For such groups, the p-y curves to be used for the individual piles may be developed using the methods described earlier, provided that the mode of displacement of the group is primarily lateral translation.

REFERENCES

1. API Recommended Practice for Planning, Designing, and Constructing Fixed Offshore Platforms. American Petroleum Institute, Washington, D.C., Tenth Edition, March, 1979.

2. Bogard, Dewaine, and Hudson Matlock, "A Computer Program for the Analysis of Beam-Columns Under Static Axial and Lateral Loads," Proceedings, Ninth Annual Offshore Technology Conference, OTC 2953, Houston, Texas, May, 1977.

3. Bryant, Larry M., and Hudson Matlock, "Three-Dimensional Analysis of Framed Structures with Nonlinear Pile Foundations," Ninth Annual Offshore Technology Conference, OTC 2955, Houston, Texas, May, 1977.

4. Focht, J.A., Jr., and K.J. Koch, "Rational Analysis of the Lateral Performance of Offshore Pile Groups," Proceedings, 1973 Offshore Technology Conference, OTC 1896, Houston, Texas, May, 1973.

5. Focht, J.A., III, "A Comparison Between Measured and Calculated Pile Group Behavior Under Lateral Loading," M.S. Report, The University of Texas at Austin, March, 1982.

6. Matlock, Hudson, W.B. Ingram, A.E. Kelley, and D. Bogard, "Field Tests of the Lateral Load Behavior of Pile Groups in Soft Clay," Proceedings, Twelfth Annual Offshore Technology Conference, OTC 3871, Houston, Texas, May, 1980.

7. Matlock, Hudson, "Correlation for Design of Laterally Loaded Piles in Soft Clay," 1970 Offshore Technology Conference, OTC 1204, Houston, Texas, May, 1970.

8. Poulos, Harry G., "Behavior of Laterally Loaded Piles: I - Single Piles," Journal of the Soil Mechanics and Foundation Division, ASCE, Vol. 97, No.SMS, May, 1971.

9. Poulos, Harry G., "Behavior of Laterally Loaded Piles: II - Pile Groups," Journal of the Soil Mechanics and Foundation Division, ASCE, Vol. 97, No. SMS, May, 1971.

DEVELOPMENT OF P-Y CURVES FOR LAYERED SOILS

By Michael Georgiadis[1], A.M. ASCE

INTRODUCTION

The most usual method currently utilized in the analysis of laterally loaded offshore piles is based on a modified Winkler spring hypothesis and involves the solution of the conventional elastic beam theory differential equation. The soil surrounding the pile is represented by a set of non-linear springs which are assumed to act independently. A more elaborate model which considers shear coupling between these springs has been presented by Georgiadis and Butterfield (5). The most important element in predicting the pile response is the determination of the appropriate spring characteristics, i.e. the evaluation of the soil resistance versus pile deflection relationships, known as P-Y curves. These curves are commonly determined using the strength-deformation properties of the soil as obtained from standard laboratory tests.

Criteria for constructing P-Y curves for both static and cyclic loading have been presented by Matlock (8), Reese et al. (11), Welch and Reese (13) and Reese et al. (10) for various types of soil conditions. These criteria were based on the results of lateral load tests on instrumented piles, which were performed in rather homogeneous soils.

Information about the lateral behavior of piles in layered soil profiles is very limited. Some analytical studies have been performed by Davisson and Gill (2), Khadilkar et al. (7), Naik and Peyrot (9) and Dordi (3) for two-layer soils to define the influence of the pile length, the thickness of the upper layer and the ratio of the stiffness of the upper layer to the stiffness of the lower layer, on the pile response. However, these analyses are based on simplified assumptions and do not consider the non-linearity of the soil, which is one of the main advantages of the P-Y approach. They are limited to predicting either the ultimate load that creates soil failure, without taking into account bending of the pile, or the response of a beam in elastic layers of known moduli of subgrade reaction.

A research program has been carried out at the University of Texas, which involved two sets of tests. The first consisted of small scale laboratory tests on a 25 millimeter diameter pile embedded in a layered system of stiff clay overlying soft clay (1). A field test was conducted by Hargrove (6) at the Lake Austin area with a 152 millimeter diameter pile and a layered soil consisting of medium dense sand overlain by a stiff clay layer. Apart from these two sets of experiments no other published experience on the behavior of laterally loaded piles has been encountered by the author. In addition the literature does not include any special criteria for the construction of P-Y curves in layered soils.

---

[1] Project Engineer, D'Appolonia S.A., Brussels, Belgium

An approach usually adopted to determine the P-Y curve at a depth H_i in a layered soil profile is to use the P-Y criteria developed for homogeneous soils, assuming either that all the soil above H_i has the same strength properties as those at H_i, or an average undrained shear strength of the material above H_i. These assumptions however, can either underestimate or overestimate the pile response to lateral loading, depending on whether the upper layer is stronger or weaker than the lower layer respectively.

This paper presents a new approach for developing P-Y curves for layered soil profiles. This approach involves the determination of an equivalent depth for the top of each layer as a function of the actual depth, the overburden pressure and the strength properties of the overlying layers. Depending on the properties of the overlying soil, the equivalent depth can be either smaller or larger than the actual depth. In order to demonstrate the equivalent depth method, two examples of computing the lateral pile behavior with the new P-Y approach are presented, one for a strong material overlying a weaker material (Austin pile test) and the other for a weak material overlying a strong material. The results are compared to the experimental data for the first case and to the pile response computed from the conventional homogeneous-soil P-Y analysis for both cases.

2.0 METHODOLOGY OF THE NEW APPROACH

As mentioned in the introduction the key parameter of the new approach in the development of P-Y curves for a layered soil profile is the determination of the equivalent depths of all the layers existing below the upper layer. This determination is presented schematically in Figure 1 for a typical layered soil profile which consists of an upper soft layer overlying a stiff layer and another soft layer.

The P-Y curves of the first layer are determined according to the standard criteria for homogeneous soils. To compute the P-Y curves of the second layer, the equivalent depth h_2 of the top of this layer has to be previously determined. In order to do this the force, F_1, required to induce soil failure of a pile segment embedded to the bottom of the upper layer is computed. This is achieved by performing an integration of the ultimate resistance, p_u, of the P-Y curves, over the thickness, H_1, of the first layer. If, for example, the first layer consists of soft clay (8), the force F_1 is given by the following equation:

$$F_1 = \int_0^{H_1} p_u \cdot dH = \int_0^{H_1} N_p \cdot S_u \cdot D \cdot dH \qquad (1)$$

$$N_p = 3 + \frac{\sigma_H}{S_u} + J \cdot \frac{H}{D} \quad \text{or} \quad N_p = 9 \qquad (2)$$

where S_u is the undrained shear strength of the soil,
D is the pile diameter
N_p is a non-dimensional ultimate resistance coefficient.
H is the depth below groundline
σ_H is the overburden pressure at H and
J is an empirical coefficient.

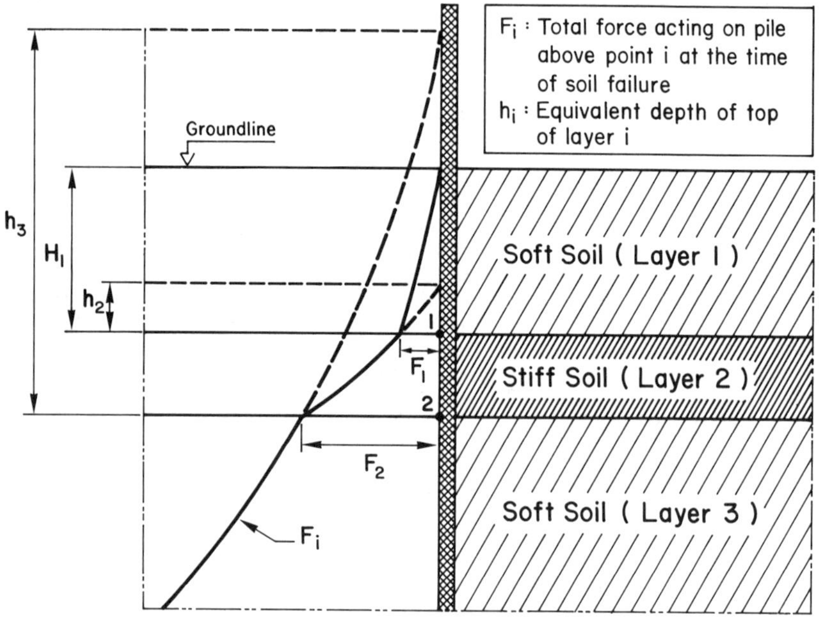

Fig 1 Typical Dermination of Equivalent Depths in a Layered Soil Profile

After the force F_1 has been computed, the embedment depth, h_2, of the same pile in a material having the properties of the second layer is calculated so that the force required to cause failure is equal to F_1. This depth is the equivalent depth of the top of the second layer (i.e. the equivalent thickness of layer 2 soil which provides the same lateral resistance with the actual thickness of layer 1 soil) and is obtained from the solution of the following equation:

$$F_1 = \int_0^{h_2} p_u \cdot dh \qquad (3)$$

where p_u is the ultimate soil resistance of the P-Y curves given by one of the homogeneous soil P-Y criteria; p_u is a function of the equivalent depth, the actual overburden pressure and the strength properties of the second layer.

When the equivalent depth of the top of the second layer has been determined the P-Y curves of this layer can be computed using the conventional P-Y criteria. It is shown in Figure 1 that because of the stiffer nature of the second layer, the equivalent depth, h_2, is significantly smaller than the thickness of the first layer.

In order to develop P-Y curves for the third layer which is overlain by a stiffer layer, the same methodology has to be followed.

The first step is to compute the force F_2 which is required to cause failure using the P-Y curves already computed for the first and second layers. After that the equivalent depth h_3 is calculated and P-Y curves can be developed in the third layer using the strength and deformability properties of that layer and considering that it starts at a depth equal to h_3 below groundline. Figure 1 illustrates that because of the high stiffness of the second layer, the equivalent depth, h_3, of the third layer is larger than the thickness of the two top layers.

The same procedure can be applied in the determination of P-Y curves in a soil profile having any number of layers. It can also be applied in a soil profile whose properties vary with depth by dividing it into a large number of thin layers which have constant soil properties.

It should be noted from Figure 1 that in the case of the two extreme soil conditions, i.e. extremely soft soil over very stiff soil or vice versa, the following conclusions could be drawn:

- If, for example, the upper layer is a very soft clay and the lower layer is a dense sand, the equivalent depth h_2 will practically be zero even if the thickness of the top layer is large. This means that the only contribution the first layer has on the P-Y curves of the second layer is to increase the overburden pressure. Wedge failure criteria will be applied (resulting in lower ultimate resistance) because the layer starts essentially at zero depth. On the other hand, if the homogeneous soil P-Y criteria for layered soil were applied, deep failure would have to be considered (if the thickness of the first layer is large) as it is assumed that the strength properties of the second layer also apply to the first layer, resulting in an overestimation of the ultimate resistance.

- If the upper layer is a dense sand and the lower layer is a soft clay, the equivalent depth h_2 will be very large and therefore deep failure mode will have to be applied to the second layer because of the high strength of the first layer. This would not occur if homogeneous soil P-Y criteria were applied, in which case wedge-type failure would be considered.

The new approach for determining P-Y curves for layered soils has been incorporated in a modified version of the COM624 computer program developed at the University of Texas (12). This program was used with and without the new P-Y criteria, in the applications described in the following section.

APPLICATIONS

In order to check the results of the new method for developing P-Y curves and to compare them with the results of the homogeneous soil P-Y approach for layered soil, two example problems were considered. One involved the Austin field pile test (2) and the other the lateral loading of a full size offshore pile embedded in an idealized layered soil profile of soft clay overlying a stiff clay. The results obtained from the field test and from the two sets of predictions (with and without the new P-Y criteria) are compared in terms of groundline deflections and maximum bending moments.

Example 1: Austin Pile Test

This test involved the cyclic lateral loading of a 152 millimeter diameter free headed pile. The properties of the soil profile used and the pile characteristics, as presented by Hargrove (6), are shown in Figure 2.

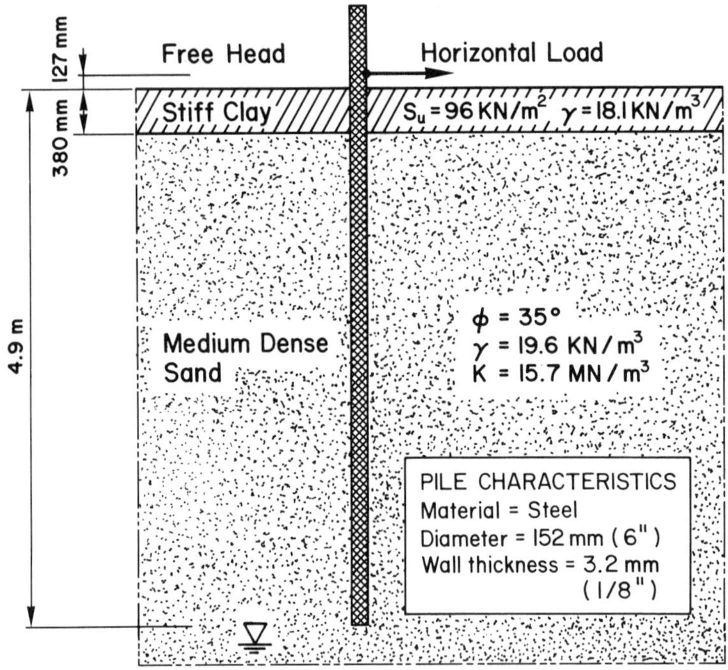

Fig 2: Example 1 - Austin Pile Test: Soil Profile and Pile Characteristics

The soil profile consists of a 380 millimeter thick stiff clay layer overlying a medium dense sand. The properties of the clay were obtained from in-situ small-sized vane tests and some trixial tests. The properties of the sand were determined from the results of dynamic penetration tests and direct shear tests. Strain gauges were installed inside the pile in order to determine the bending moments developed during loading. Apart from the bending moments, the pile deflection and inclination were measured at groundline. The pile was loaded and unloaded ten times up to 26 kiloNewtons. The groundline deflections and the maximum bending moments measured during the first and the tenth loading cycles are presented in Figures 3 and 4 respectively.

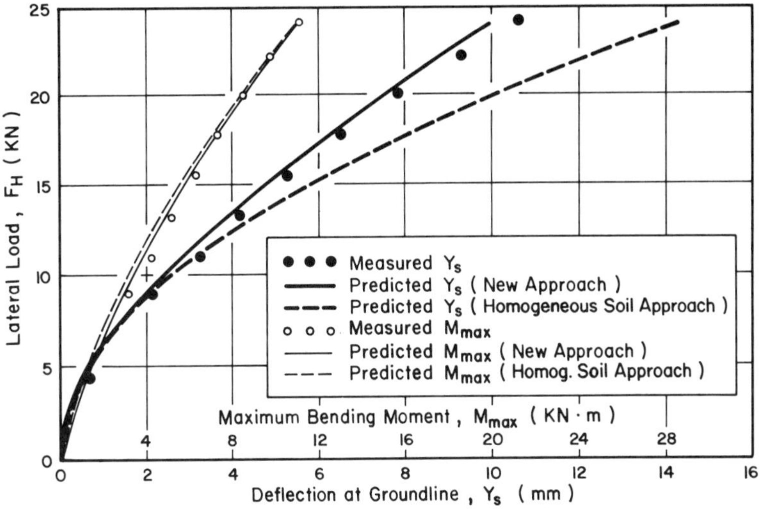

Fig 3: Computed and Measured Groundline Deflections and Maximum Bending Moments, Example 1 - Austin Pile Test, First Load Cycle

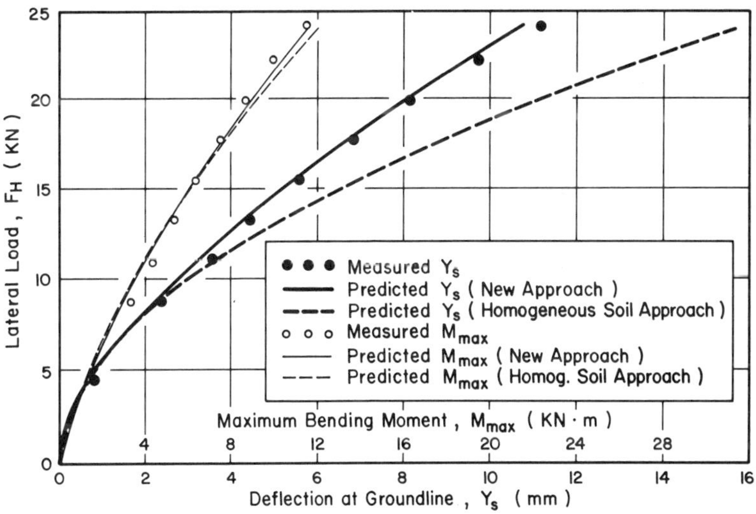

Fig 4: Computed and Measured Groundline Deflections and Maximum Bending Moments, Example 1 - Austin Pile Test, Tenth Load Cycle

Since the ground water table was below the tip of the pile, the pile response was predicted using the stiff clay above water table P-Y criteria developed by Welch and Reese (13) and the sand P-Y criteria developed by Reese et al. (10). Two sets of analyses were performed using the COM624 (12) computer program. The first analysis was conducted utilizing the program in its original form in which the homogeneous soil approach is adopted to determine the P-Y curves of the second layer. The second analysis was performed using a modified version of the program, in which the new P-Y approach for layered soils was utilized. The results of these two analyses are also presented in Figures 3 and 4 indicating that the predictions of the new method are in excellent agreement with the experimental results in terms of both maximum bending moments and deflections at groundline and for both static and cyclic loading conditions. The agreement is good for all load levels and this supports the validity of the new approach. The pile deflections at groundline computed from the homogeneous soil P-Y approach for layered soil are larger than the ones measured in the field because, as mentioned in the previous section, the stiffer nature of the upper clay layer was not taken into account in the evaluation of the P-Y curves of the underlying sand layer.

Example 2: Full Size Offshore Pile

As expected, the results of the previous example indicated that the homogeneous soil P-Y approach for layered soil gives conservative results when the upper layer is stronger than the lower layer. In order to investigate the difference between the results of the new and the homogeneous soil approaches when the upper layer is weaker than the lower layer, a two meter diameter offshore pile was analyzed for cyclic loading. The soil profile considered, is shown in Figure 5 and consists of a ten meter thick layer of very soft clay which is overlying a layer of very stiff clay. The soil is submerged and therefore the P-Y criteria developed by Matlock (8) for clays were adopted. The pile is assumed to be fixed headed and consequently the maximum bending moments occur at groundline.

The results of the two analyses performed using the original and the modified versions of the COM624 computer program are presented in Figure 6 in terms of maximum bending moments and groundline deflections. It is noted that in this case the use of the homogeneous soil P-Y approach for layered soil underestimates the bending moments by about ten percent and the groundline deflections by thirty percent as compared with the equivalent depth method. These significantly large differences occur because of the very soft consistency of the first layer, which results in a very small equivalent depth for the top of the second layer. Because of this small equivalent depth, deep failure in the stiff clay of the second layer starts at a greater depth when the new P-Y approach is applied instead of the homogeneous soil method.

It should also be noted that the large difference between the groundline deflections computed with the two methods, would result in additional differences between the maximum bending moments if the pile was a member of a pile group and the group response was computed according to the Y-modifier method. This method uses the single pile displacement to compute the displacement of the group and then to determine the bending moments of the piles in the group (Focht and Koch) (4).

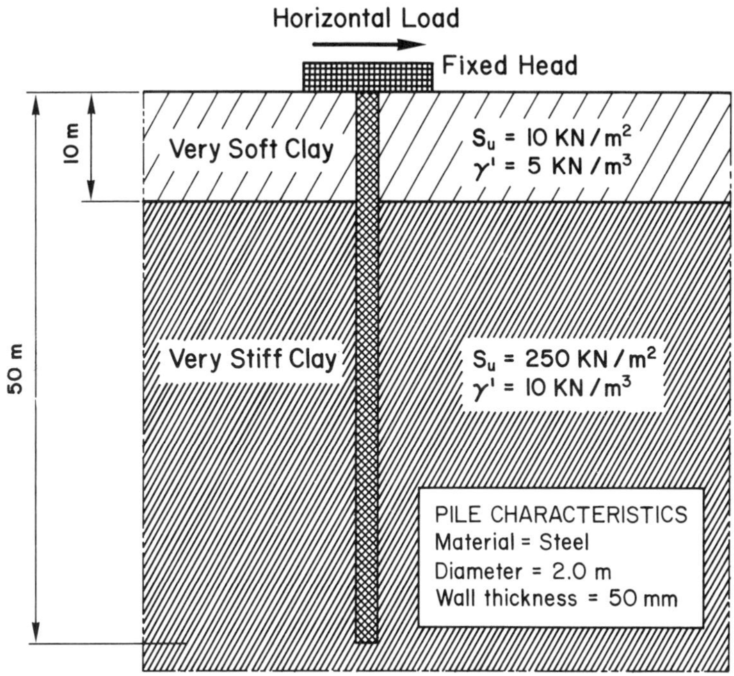

Fig 5: Example 2 - Soil Profile and Pile Characteristics

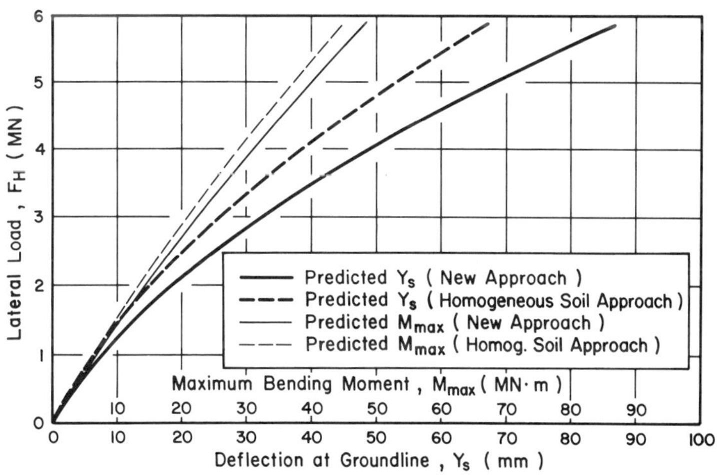

Fig 6: Computed Groundline Deflections and Maximum Bending Moments
Example 2 - Cyclic Loading

CONCLUSIONS

A new approach has been developed for determining P-Y curves in layered soils, which allows the P-Y curves in a specific layer to be dependent on both the properties of that layer and the overburden produced by different soil types above. The soil layering is taken into account by computing equivalent depths for each of the underlying layers. These equivalent depths are functions of the actual depths, the overburden pressure and the strength properties of the overlying layers. The new approach for layered soils has been incorporated in a modified version of the computer program COM624 developed at the University of Texas for computing lateral pile response.

Lateral pile response predicted using the new approach and the homogeneous soil P-Y approach for layered soil were compared to field test results obtained from the literature. Excellent agreement was found between the field test results and those predicted by the new method in terms of both maximum bending moment and groundline deflections for both static and cyclic loading conditions. The pile response computed by the homogeneous soil P-Y approach used by some people, which assumes homogeneous soil properties throughout the entire depth, was found to either overestimate or underestimate the actual pile capacity with respect to the equivalent depth method and the field test results, depending on whether the upper layer is softer or stiffer than the underlying layer, respectively.

ACKNOWLEDGEMENT

The author would like to thank Professor L.C. Reese for providing the data of the field test performed at Austin by the University of Texas, which enabled the comparison with the results of the new approach to be conducted.

REFERENCES

1. Allen, J. D., and Reese, L. C., "Small Scale Tests for the Determination of P-Y Curves in Layered Soils," Proceedings of the Twelfth Annual Offshore Technology Conference, Paper OTC 3747, May 1980, pp. 109-116.

2. Davisson, M. T. and Gill, H. L., "Laterally Loaded Piles in a Layered Soil System," Journal of the Soil Mechanics and Foundations Division, ASCE, Vol. 89, No. SM3, Paper 3509, May 1963, pp. 63-94.

3. Dordi, C. M., "Horizontally Loaded Piles in Layered Soils," Proceedings of Specialty Session 10, The Effect of Horizontal Loads on Piles due to Surcharge of Seismic Effects, Ninth International Conference on Soil Mechanics and Foundation Engineering, Tokyo, July 1977, pp. 65-70.

4. Focht, J. A., and Koch, K. J., "Rational Analysis of the Lateral Performance of Offshore Pile Groups," Proceedings of the Fifth Offshore Technology Conference," Paper OTC 1896, May 1973, pp. 701-708.

5. Georgiadis, M. and Butterfield, R., "Laterally Loaded Pile Behavior," Journal of the Geotechnical Engineering Division, ASCE, Vol. 108, No. GT1, Paper 16828, January 1982, pp. 155-165.

6. Hargrove, J. Q., "Field Test of a Pile Under Lateral Loading in a Layered Soil Profile," Geotechnical Engineering Thesis GT81-6, Civil Engineering Department, University of Texas at Austin, May 1981.

7. Khadilkar, B. S., Chandrasekaran, V. S., and Rizvi, I. A., "Analysis of Laterally Loaded Piles in Two-Layered Soils," Proceedings, Eight International Conference on Soil Mechanics and Foundation Engineering, Moscow, 1973, Vol. 21, pp. 155-158.

8. Matlock, H., "Correlations for Design of Laterally Loaded Piles in Soft Clays," Proceedings, Second Annual Offshore Technology Conference, Paper OTC 1204, April 1970, pp. 578-588.

9. Naik, T. R., and Peyrot, A., "Analysis and Design of Laterally Loaded Piles and Caissons in a Layered Soil System," Methods of Structural Analysis, Proceedings of the National Structural Engineering Conference, ASCE, Madison, Wisconsin, August 1976, Vol. II, pp. 589-606.

10. Reese, L. C., Cox, W. R., and Koop, F. D., "Analysis of Laterally Loaded Piles in Sand," Proceedings, Sixth Annual Offshore Technology Conference, Paper OTC 2080, May 1974, pp. 473-483.

11. Reese, L. C., Cox, W. R., and Koop, F. D., "Field Testing and Analysis of Laterally Loaded Piles in Stiff Clay," Proceedings, Seventh Annual Offshore Technology Conference, Paper OTC 2312, May 1975, pp. 671-690.

12. Reese, L. C., and Sullivan, W. R., "Documentation of Computer Program COM624, Analysis of Stresses and Deflections for Laterally Loaded Piles, Including Generation of P-Y Curves," Geotechnical Engineering Center, Department of Civil Engineering, University of Texas at Austin, 1980.

13. Welch, R. C., and Reese, L. C., "Lateral Loading of Deep Foundations in Stiff Clay," Journal of the Geotechnical Engineering Division, ASCE, Vol. 101, No. GT7, Paper 11456, July 1975, pp. 633-649.

LATERAL PILE ANALYSIS IN WEAK CARBONATE ROCKS

By A.F. Abbs, BSc MICE[1]

INTRODUCTION

This paper presents an analytical procedure for lateral pile analysis in weak carbonate rocks. In recent years there has been extensive development of piled offshore structures in weak carbonate rocks which are subjected to lateral loads. However, performance and methods of analysis for these materials are at best poorly understood. The method outlined in this paper for lateral pile analysis was developed from work on offshore platforms in the Middle East, although data from many sources were used in evaluation of procedures.

The primary objective was to develop a method of lateral pile analysis which was practical and simple to perform, whilst retaining a realistic model of what was believed to be occurring. The purpose of this was to be able to better understand the behaviour of these materials during lateral pile loading. The method had to make use of data from conventional site investigations so that it could be adopted without far-reaching changes in field techniques.

Moreover, it was known that investigations by various companies over a period of several years had produced a wide range of estimated lateral pile performance. Estimates have been based on various methods, including both simple elastic formulae and p-y methods which were developed from different references such as stiff clay, sand, soft clay or even pressuremeter (Refs. 2,4,5,6). The effect of this variety of techniques was found to introduce differences in calculated behaviour at least as great as that due to variations in the material strength from one location to the next.

The materials in question cover a variety of rocks with grain sizes ranging from clay up to gravel, with unconfined strengths ranging from about 0.5 MN/m^2 up to 5 MN/m^2. The materials are often porous and are usually variable with dry densities in the range 1.3 to 1.9 MN/m^2. Previous experience with driven piles in these materials has shown that low unit skin friction values are often encountered owing to particle crushing and high porosities. It was believed that similar phenomena may occur during lateral loading.

FIELD INVESTIGATION TECHNIQUES

Typical field investigation techniques consist of conventional rotary coring, with a fairly large diameter core being taken (76 millimetres being typical). Recently methods of logging the materials have become fairly standardised following the method of Clark & Walker

---

[1] Senior Engineer, Dames & Moore, London, UK

(Ref. 1). This gives a detailed geological description of the rock which unfortunately does not necessarily relate to its mechanical properties. Therefore it has now become common to generate a simple mechanical log comprising a profile of unconfined strength with depth. This is done by performing a large number of unconfined tests (with strain measurement) at the site, usually on board ship, with additional data points being provided by point load test indices which are converted to unconfined strength. (The ratio of unconfined compressive strength to point load test index is generally in the range 3 to 4 for these materials, significantly less than published figures for conventional rock).

Using these data and other laboratory test data a design profile for each boring (or group of borings) is developed. Figure 1 shows an example of such a profile.

These design profiles require a large degree of judgement, because of the scatter of data generally found. Figure 1 is typical with a slightly conservative "mean" line being selected. The effects of taking "upper bounds" "mean values" or "lower bounds" are discussed later in this paper.

PROPOSED METHOD OF ANALYSIS FOR LATERAL CAPACITY

As there was no published or generally accepted method of analysis suited to the conditions in question, it was decided to make use of existing methods of lateral pile analysis to provide an estimate of the lateral load/deflection response. It was decided to continue use of the p-y method.

Results of a large number of unconfined compression tests showed that most of the very weak and weak rocks encountered brittle failure at less than 1 per cent strain. For low strain conditions the behaviour of these materials is controlled by the inter-particle cementation and by the deformability of the rock mass (intact rock resistance). At larger strains, the interparticle bonding breaks down and it may be expected that the stress/ strain behaviour will principally be controlled by the frictional properties of the material (residual resistance). Therefore a combined 'hybrid' approach was proposed to take account of the brittle behaviour of the weak rocks, as indicated on Figure 2.

The methods used to generate the curves and their application for pile design are discussed below.

Intact Rock Resistance Curves

These curves are calculated on the basis of the unconfined compressive strength (assumed to be equal to twice the shear strength) and the elasticity modulus of each stratum, using the calculation procedure for stiff clay (static case) proposed by Reese et al. (Ref 5). The elasticity modulus for the stratum as a whole is obtained by factoring the laboratory moduli in a manner adapted from Hobbs (Ref.3), based on the Rock Quality Designation of the rock core.

These Intact Rock Resistance curves are used to calculate response

SIMPLIFIED LOG OF BORING	UNCONFINED COMPRESSIVE STRENGTH q_f MN/m²	UNCONFINED MODULUS INTACT ROCK MN/m²	ROCK MASS MODULUS MN/m²	ROCK MASS STRAIN % AT 50% σ_f	DRY DENSITY Mg/m³	RESIDUAL FRICTION ANGLE φ°
CALCARENITE	8·0	1350	810	0·49	1·9	30
CALCARENITE	1·8	300	150	0·60	1·6	30
CALCARENITE & CORAL	0·9	150	135	0·33	1·4	30
CALCARENITE	4·0	680	610	0·33	1·7	30
CALCARENITE	2·2	370	370	0·30	1·6	30
CALCARENITE	5·4	910	910	0·30	1·8	30
CALCILUTITE	0·4	70	70	0·29	1·2	28
CALCARENITE	2·6	440	440	0·30	1·6	30
CALCARENITE & CALCISILITE	3·6	610	425	0·42	1·7	28
CALCISILTITE & CALCILUTITE	0·2	30	6	1·67	1·2	28

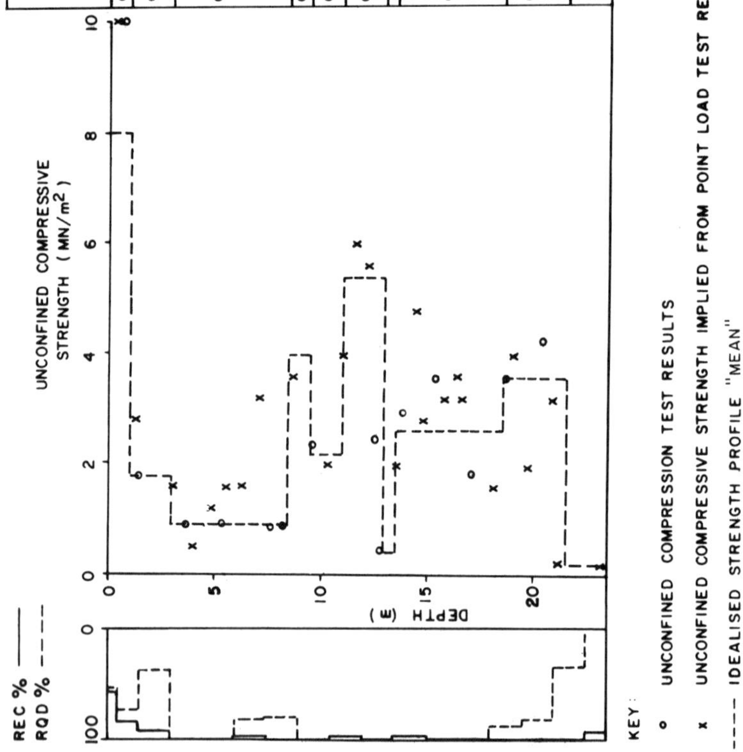

Figure 1. Typical Idealised Design Parameters

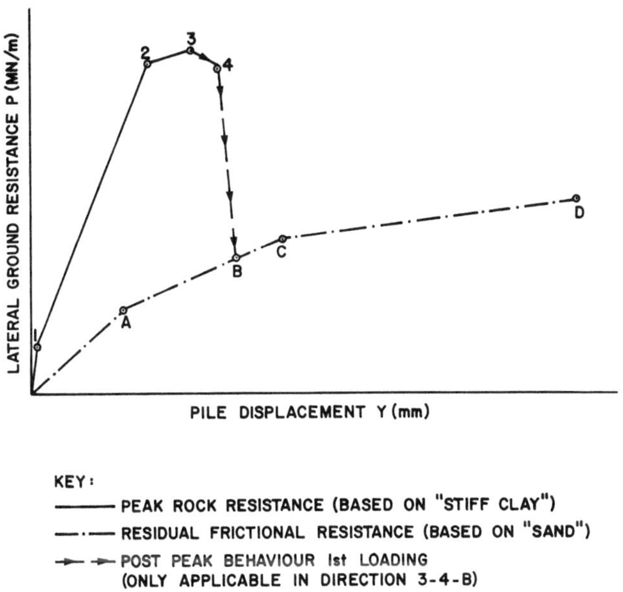

Figure 2. Make-up of Hybrid p-y Curves

to initial loading and for subsequent load cycles provided that calculated deflection does not exceed the deflection at the point corresponding to the peak stress. This assumes that the behaviour is essentially elastic in this range. Once this point is passed, inter-particle bonding is assumed to be destroyed and the resistance is assumed to be represented by the "residual frictional resistance" curve. The slope of the drop from peak resistance to the residual frictional resistance curve is arbitrary. On first loading the slope would probably be more shallow than shown but as the structures are subject to cyclic loading the actual slope is not critical. On subsequent load cycles the residual frictional curve is assumed to apply. The fall from the peak to the residual curve is therefore assumed to occur during an additional 10% displacement after the point of peak resistance. This slope is convenient from a computational point of view and represents the most dramatic fall that is anticipated from the more brittle materials encountered in the field.

Rock Residual Frictional Resistance Curve

The frictional resistance curves are used for rock strata in which calculated displacement has exceeded the peak value on the intact resistance curve. The frictional resistance curves should be calculated using the procedures for sands (cyclic case) published by Reese et al. (Ref.6). The friction angles on Figure 1 are residual values based on previous experience and test results. (Carbonate materials frequently exhibit high peak friction angles near or above 40° but after very few cycles of shear the values frequently fall to a residual value of between 24° and 33°. This reduction in friction angle is largely due to particle crushing which is dependent upon normal stress level and occurs much more in carbonate rocks than in siliceous materials).

Figure 3 presents typical examples of these combined curves together with the assumed strength profile.

DISCUSSION OF RESULTS

This section discusses results from the proposed method based on data from a typical project.

The assumed simplified profiles are presented on Figure 4 which show upper and lower bounds selected from the scatter of data points found.

The results of the analysis are presented on Figure 5 and the following points are evident from this example analysis.

The upper bound profile results in only very small deflections, even at the maximum shear force of 2.0 MN. (The maximum bending moments were also found to be small, less than 0.75 MNm at the maximum load).

However the lower bound profile shows the onset of failure at a shear of about 1.0 MN which is below likely values for the maximum design shear for offshore structures. The mean profile shows onset of

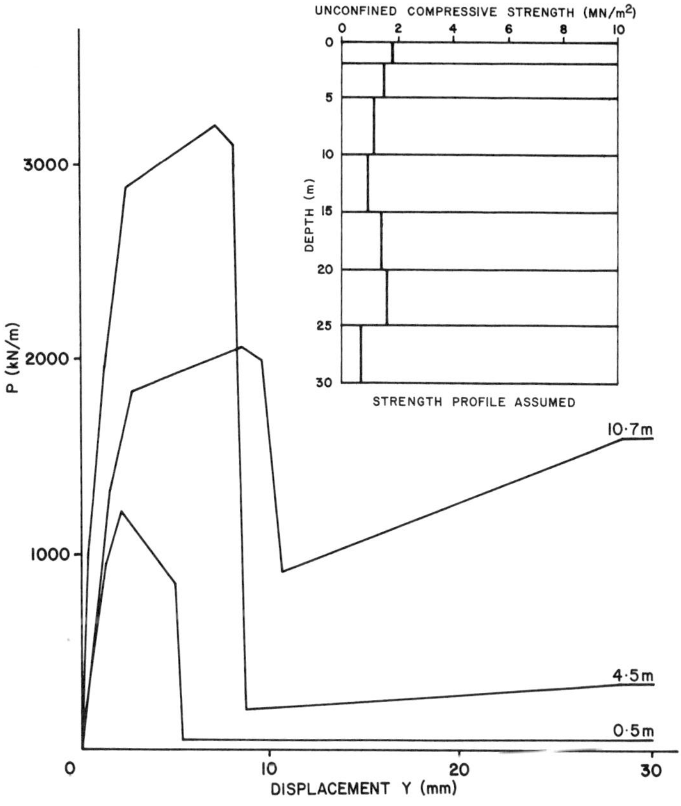

Figure 3 Typical p-y Curves for Brittle Carbonate Rock

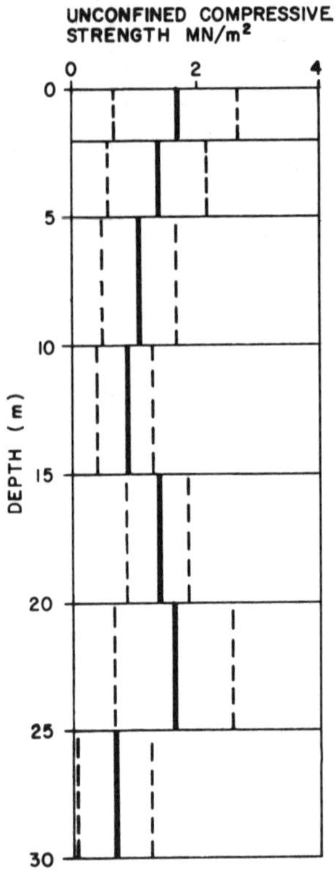

MEAN σ_f MN/m²	LOWER BOUND MN/m²	UPPER BOUND MN/m²
1·70	0·70	2·70
1·40	0·60	2·20
1·10	0·50	1·70
0·90	0·50	1·30
1·40	0·90	1·90
1·65	0·70	2·60
0·70	0·10	1·30

Figure 4. Idealised Strength Profiles

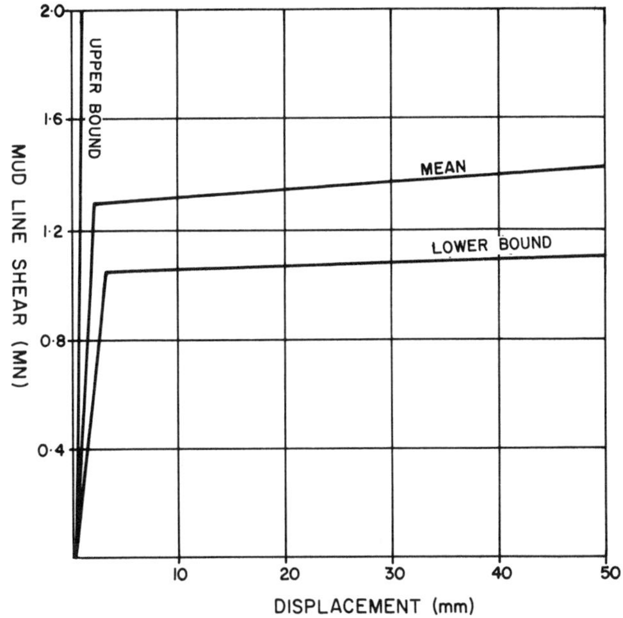

Figure 5. Pileload Displacement Behaviour with Different Assumptions of Strength Profile

design shear for offshore structures. The mean profile shows onset of failure at about 1.3 MN, which is also low.

As can be seen, the load deflection curve is highly non-linear, this is due to the sudden drop of the p-y curves, especially at shallow depth, from the peak rock curve to the residual rock curve. The effect of this drop is that once failure is reached at the surface the rock almost completely loses strength and the pile effectively becomes cantilevered from the point where intact rock resistance still exists. Because of the dramatic nature of the loss of lateral resistance, a "wave front" of rock failure moves down the pile at a constant applied load to about 7 metres depth. This unstable behaviour has dramatic effects on the bending moments and stresses in the pile, such that total pile failure could result. Figure 6 shows displacement and bending moments calculated for the lower bound profile. It is readily seen that a small increase in load produces a disproportionate increase in bending moment and deflection.

This analysis demonstrates that it is important to design piles so that the risk of failure is eliminated; failure of the rock does not just result in gradually increasing movement, it can result in gross over-stress of the pile. Moreover the effects of rock failure have significant repercussions on the axial capacity of the pile. As an illustration, for the mean design strength profile in question the ultimate axial pile capacity was calculated to be 15 MN for a 16 metre long grouted pile. If rock failure to a depth of 5 metres was caused through excessive lateral loading, this ultimate capacity would be reduced to about 9 MN. This is because after crushing, the rock grout bond is destroyed and pile capacity in the failed zone is derived from simple friction.

With all the above in mind it should be remembered that the probability of the "lower bound" profile existing is small. Consequently it is recommended to use the mean profile as this is considered more realistic, but a suitable safety factor should be incorporated to allow for uncertainties in the analysis.

It is considered that at the present time the introduction of more sophisticated design procedures than currently proposed would be unjustified because of the limitations of the data available, and the uncertainties in the analysis.

PILE TESTS

Regrettably there are virtually no lateral pile tests performed in these materials which have reached failure. This is not really surprising in view of the costs involved on such tests offshore. A recent paper by Schirmer and Fujioka (Ref.7) showed very brittle non-linear response for a short pile in partially cemented coral in Guam. This results lends support to the approach outlined in this paper.

Recent work applying the method discussed above in weak calcarenite gave excellent predictions of performance in the elastic range (for example 5 millimetres predicted, compared to 4 millimetres measured, at 0.6 MN horizontal force). Unfortunately these tests were not taken to

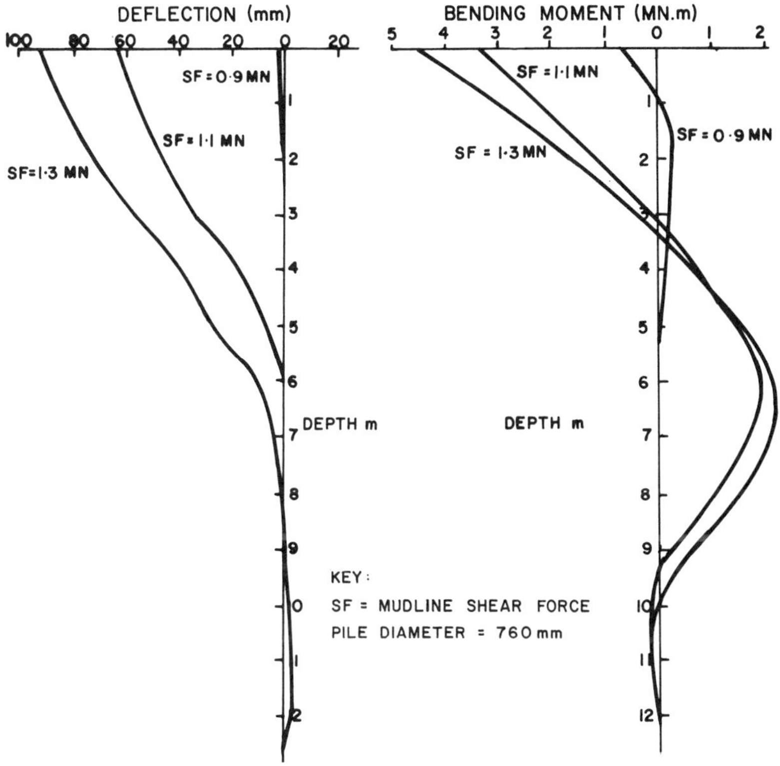

Figure 6 Pile Displacement and Bending Moments for Lower Bound Strengths

failure so that the results are only of passing interest. The need for field tests in these materials is pressing, especially considering the brittle collapse behaviour predicted, and the many uncertainties involved in the method of analysis.

CONCLUSIONS

A method for studying lateral pile performance in weak carbonate rocks has been proposed.

Although the recommended design methods are relatively simple, a degree of judgement is required from the designer. The methods of lateral analysis have been adapted from methods established for completely different non-carbonate soils. These methods have been adapted on the basis of judgement with very little actual observed behaviour of similar prototype foundations. Until large scale pile load test data is available there can be no check on the validity of the design methods.

REFERENCES

1. Clarke A.R. & Walker BF (1977): "A proposed scheme for the classification and nomenclature for use in the engineering description of Middle East sedimentary rocks", Geotechnique 27,1.

2. Gambin M. (1979): "Calculation of Foundations subjected to Horizontal Forces using Pressuremeter Data", Sols-Soils No 30/31.

3. Hobbs N.B. (1974): "Factors affecting the prediction of settlement of structures on rock, with particular reference to Chalk and Trias", Proc. Conf. on Settlement of Structures, Cambridge, UK.

4. Matlock H. (1970): "Correlations for Design of laterally loaded piles in Soft Clay", OTC 1204.

5. Reese L.C. & Case W.R. (1975): "Field Testing and Analysis of Laterally loaded Piles in Stiff Clay", OTC 2312.

6. Reese L.C. et al. (1974): "Analysis of Laterally Loaded Piles in Sand", OTC 2080.

7. Schirmer H. & Fujioka M. (1982): "Moment Foundation Design in tropical/ Residual Soils", Proc Conf. Engineering and Construction in Tropical and Residual Soils, Honolulu.

SITE SPECIFIC DESIGN OF LATERALLY LOADED PILES

by S. Kay, H.J. Kolk and W.R. van Hooydonk[1]

INTRODUCTION

The conventional approach for assessing the response of single piles under lateral loading is to use horizontal load-displacement (p-y) criteria. These are based on semi-empirical correlations between pile load test data and basic soil properties, which have been established for a limited number of soil types. Considerable engineering judgement is required when extrapolating these criteria to pile foundations in soils with vastly different characteristics.

The response of pile groups subjected to lateral loading is normally analysed using a combination of the above p-y criteria and elastic solutions. Significant simplifications are made in this analysis, which generally leads to unrealistically large deflections.

Considerable costs can be associated with the design and construction of offshore jackets and pile foundations so that these structures are able to resist large horizontal loads and overturning moments.

The authors have developed alternative design methods, using site specific in-situ load-deformation data, which are believed to yield a more accurate assessment of the lateral load response of both single piles and pile groups. The proposed methods for single piles are presented in the first part of this paper, whereas the proposed method for pile groups is presented in the second part.

PART I - SINGLE PILES

Conventional design practice

The current design practice for assessing the load-deformation behaviour of offshore piles is to use the beam-column method. In this analysis, the pile is discretised as either a series of finite difference stations or as a series of beam finite elements. The soil support is of the Winkler type, consisting of a series of independent discrete or continuously distributed lateral springs. The resistance-displacement relationships are the non-linear "p-y" curves.

[1] Computing Manager, Offshore Projects Manager and Special Projects Manager, Fugro B.V., Leidschendam, The Netherlands.

Provided satisfactory p-y curves can be specified, the results of deflection and moment are reliable (23). The computer time involved is very small, thus allowing parametric studies to be cheaply carried out. If no site specific testing is available, the p-y curves have to be synthesized from "standard" curves such as those proposed by API or DNV (1, 5).

The API rules are commonly used to derive three sets of curves which depend upon soil type, i.e. soft clay, stiff clay and sand.

The curves for soft clay are based primarily on lateral load tests on a 12.75 inch diameter pile at the Sabine River in Texas (10).

The p-y criteria for stiff clay (19) were developed using lateral load test data from a site near Austin. The overconsolidated stiff clay was of marine origin.

More recently, Sullivan et al (22) have proposed unified p-y criteria for both clay types, which are based upon test results from the former sites and test sites at Manor and San Francisco Bay. All the clay criteria have been generalised by expressing the p-y curves in terms of pile diameter, undrained shear strength, ε_{50} (= the axial strain in a triaxial compression test, at half the maximum deviator stress), effective unit weight, depth and empirical coefficients. Engineering judgement has to be employed when selecting any of these p-y criteria for a given offshore site by comparing the various soil types.

The p-y curves for sand (18) are based on lateral load tests on piles at Mustang Island (4). These curves are expressed in a generalised format which includes pile diameter, effective angle of internal friction, empirical coefficients and the modulus of subgrade reaction. A great difficulty is the selection of the latter value when developing p-y criteria for a given offshore site.

Considerable engineering judgement is required when p-y criteria need to be developed for clays and sands which differ markedly from the materials tested in the above lateral pile load tests. It is even more difficult to develop p-y criteria for completely different soils, such as soft rock, silts and carbonate materials. The latter problem especially has prompted the authors' company to develop the in-situ testing techniques and associated analytical methods described in the following sections.

Site specific p-y method

The authors company has developed in-situ lateral pile load test equipment in order to investigate the soil response during static and cyclic loading at a proposed platform site. This equipment consists of a remote controlled seabed rig which enables lateral loading of two test piles (Fig. 1). This rig can be deployed from any North Sea type soil investigation vessel or other vessels with adequate handling facilities. The test piles have an outer diameter of 14 inches and an embedment length of either 3.5 or 6.0 m (L/D = 10 and 17 respectively). The piles are driven into the seabed using free fall hammers which are wireline operated from the vessel. Various combinations of either static or cyclic horizontal loading and overturning moment can be applied by using two hydraulic jacks at

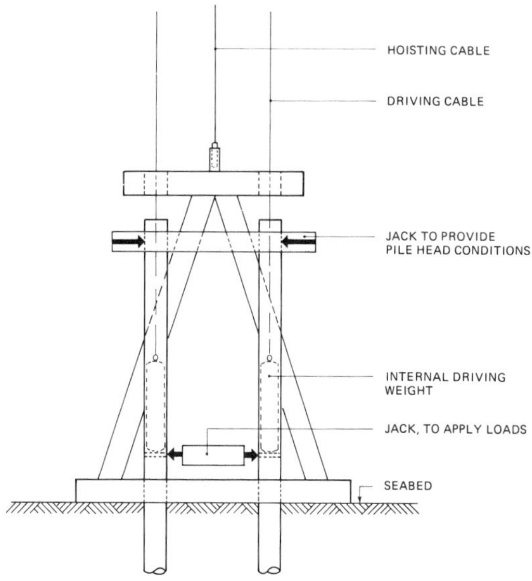

Fig. 1 - Lateral load test frame

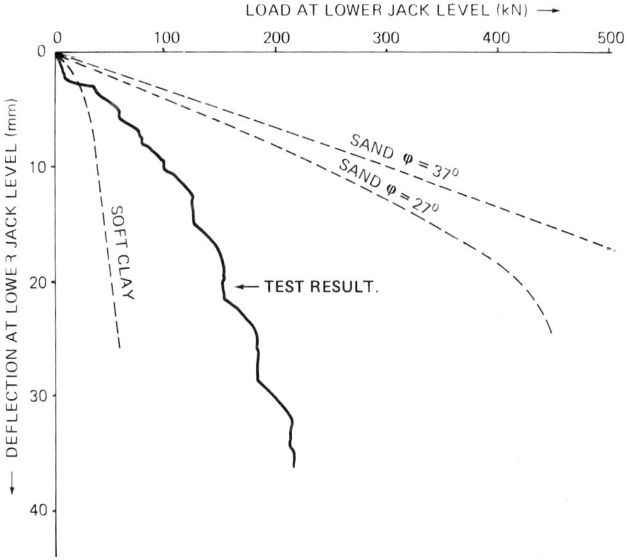

Fig. 2 - Typical lateral load test result

different levels above the seafloor. The loads and deflections at both jack levels are measured and recorded on chart plotters and tape on board the vessel. For further information on the equipment, reference should be made to a paper by King et al (8).

A typical test result is shown in Fig. 2. The results can be used in two ways, i.e. either by selecting standard type p-y criteria which compare best with the recorded data, or by developing new p-y criteria using the load-deflection results.

The first approach is illustrated in Fig. 2. It can be observed that for this specific site (calcareous silt) the soil response is between a normally consolidated clay (c_u/p_o = 0.3) and a loose sand (ϕ = 27°).

A detailed description of the analytical methods used in the second approach is given by Reese and Cox (Ref. 17). A non-dimensionalised p-y curve obtained by applying this method is presented in Fig. 3.

The p-y curves obtained from the test piles by either of the former approaches can be extrapolated to the prototype foundation piles and greater depths by non-dimensionalising the p-y curves using basic soil properties and accounting for diameter size. This method is basically the same as employed for generation of standard p-y curves in conventional soils.

Finite element-pressuremeter approach

An alternative approach for analysing laterally loaded offshore piles is to use a finite element method. Thus, the soil is considered as a solid bonded to the surface of the pile. The very nature of the continuum approach is physically more satisfactory than a p-y or Winkler type approach, where the soil is modelled as a series of uncoupled lateral springs. In addition, as will be shown in Part II, the finite element method offers distinct advantages over other currently used methods for analysing pile groups.

The finite element method uses real soil properties such as shear modulus, Poisson's ratio, and soil strength properties (undrained shear strength or effective angle of internal friction). The most critical parameter in this analysis is the shear modulus.

The authors recommend the establishment of this value using in-situ pressuremeter tests. The pressuremeter used by the authors company until now is the "Stressprobe". This is a push-in pressuremeter that has been developed by the Building Research Establishment, and is especially suitable for use offshore.

The pressuremeter, in a non-inflated condition, has an outer diameter of 78 mm and a length of 695 mm. It can be pushed into the soil at the bottom of a borehole by means of the hydraulic ram system developed by the author's company (Fig. 4). The pressuremeter is provided with a specially designed shoe. The design of the shoe is such that virtually all the soil enters the hollow stem of the pressuremeter during insertion. The disturbance at the circumference of the pressuremeter is thus minimised. The membrane is inflated using a down-hole pressure developer. This basically consists of a ram, activated by an electric motor, which develops the oil pressure required for the inflation of the membrane. A detailed

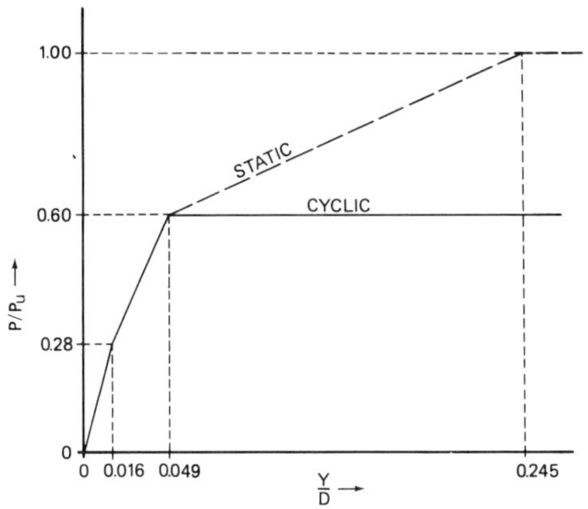

WHERE:

P = LATERAL SOIL RESISTANCE FOR DEFLECTION y [F/L]

P_u = ULT. STATIC SOIL RESISTANCE [F/L]

y = LATERAL PILE DEFLECTION [L]

D = PILE DIAMETER [L]

Fig. 3 - Typical p-y curve result

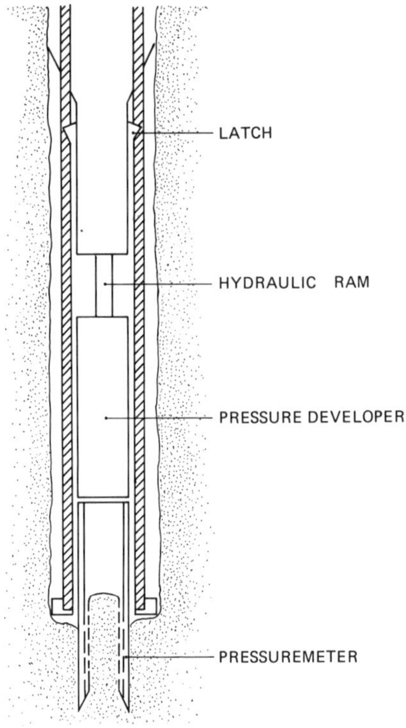

Fig. 4 - Pressuremeter test setup

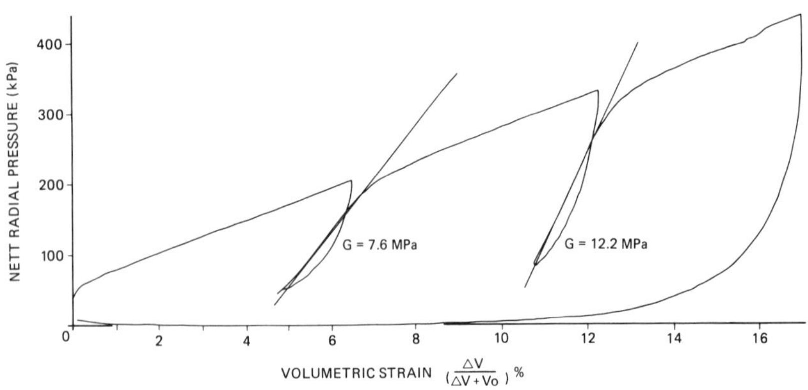

Fig. 5 - Typical pressuremeter test result

description of the stressprobe is given in a paper by Henderson et al (7). A pressuremeter test preferred by the authors consists of performing two deflate-inflate cycles prior to reaching the maximum volumetric strain of about 20 per cent. A typical pressuremeter test result is presented in Fig. 5. The slope of the deflate-inflate loops is an indication of the in-situ shear modulus (7). By performing a series of pressuremeter tests at various elevations in a borehole, one can obtain a profile of in-situ shear modulus versus depth.

The pressuremeter tests can also be used to establish values for in-situ strength. Alternatively one can use cone penetrometer and/or laboratory strength tests to estimate the in-situ strength properties. Poisson's ratio can either be estimated or it can be established from carefully performed triaxial tests on undisturbed samples.

Axisymmetric finite element program for non-axisymmetric loading

The axisymmetric continuum is idealised as an assemblage of finite elements interconnected at their nodal circles. The displacements are represented by three components u, v and w in the radial, vertical and tangential directions (Fig. 6a). The normal polynomial expressions are used to express the variation of displacement within the r-z plane and a Fourier series is used in the θ direction. Since the material and geometric properties remain constant in this direction, the problem can then be simplified to a quasi two-dimensional one (26).

The first application of the above concepts was made by Wilson (24) and further use for pile analysis has been reported by Randolph, Suen, and Kuhlemeyer (14, 15, 21, 9). A simple Fortran program listing is given by Smith (20). Meissner and Winnicki have incorporated elasto-plastic algorithms (11, 25).

For the elastic solution of a laterally loaded pile, only the first harmonic needs to be solved. A choice exists between using loads and displacements which are symmetric (about the $\theta = 0$ axis) or anti-symmetric (symmetric about the $\theta = \pi/2$ axis) (Fig. 6b).

For other loading studies, using the zero harmonic to apply a twist about the z-axis, only the anti-symmetric option can be used. Thus, for computational convenience, this was used for the laterally loaded pile. Hence, the three displacement and load components vary as follows:

$$u = u^* \sin\theta \qquad R = R^* \sin\theta \qquad (r)$$
$$v = v^* \sin\theta \qquad Z = Z^* \sin\theta \qquad (z)$$
$$w = w^* \cos\theta \qquad T = T^* \cos\theta \qquad (\theta)$$

where u^*, v^* and w^* represent the nodal displacement amplitudes, and R^*, Z^* and T^* represent the amplitudes of the nodal circle forces. They have dimensions of force per unit radian.

A possible limitation is the sinusoidal assumption made for the variation of the forces and displacements in the θ-direction. However, both Randolph and Suen have performed comparison checks against the Poulos integral equation solutions and found satisfactory agreement (14, 21).

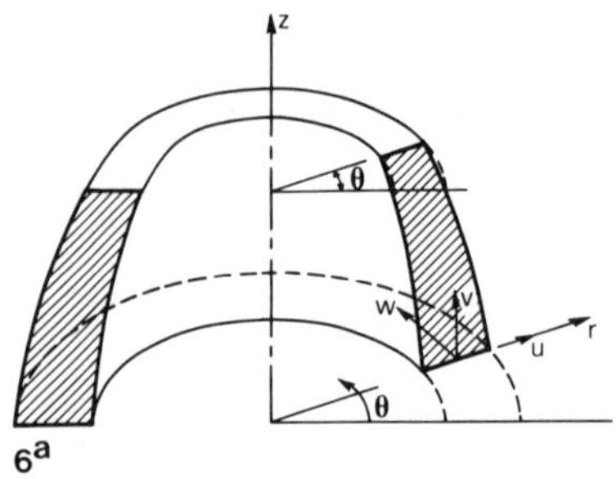

6a

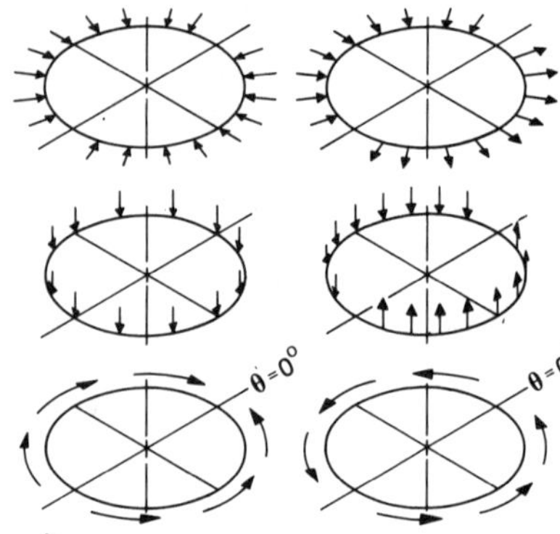

6b (a) (b)

Fig. 6a - An axisymmetric solid
Coordinates and displacements

Fig. 6b - Symmetric (a) and anti-symmetric (b)
load and displacement components

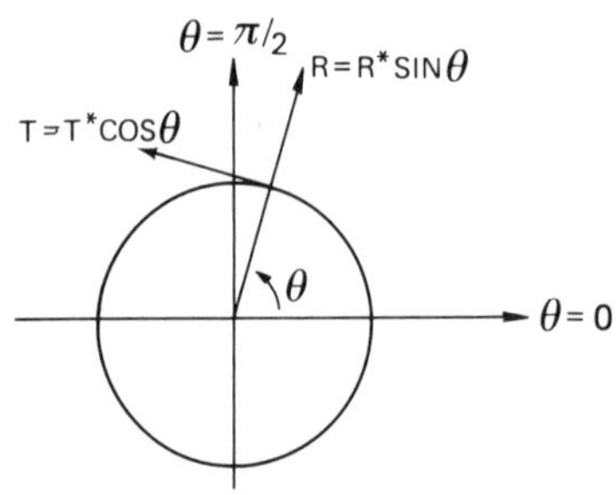

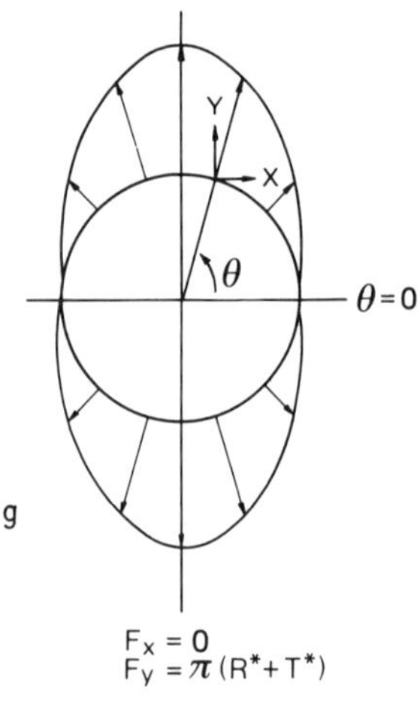

Fig. 7a - Lateral force loading

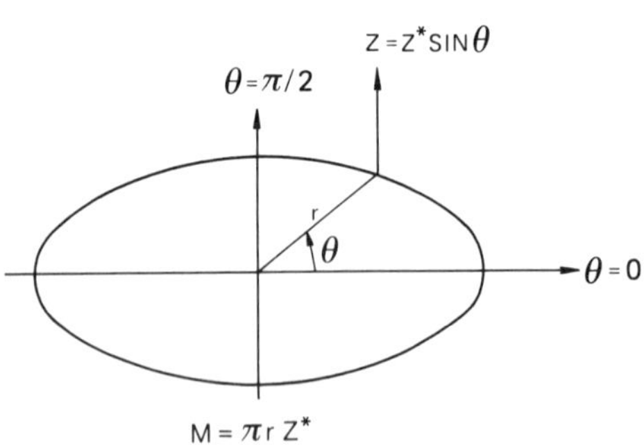

Fig. 7b - Overturning moment loading

The equivalent lateral load is applied by the radial and tangential forces R* and T*. From Fig. 7a, their components along the cartesian x, y axes are given at any angle θ by:

$$F_x = R \cos\theta - T \sin\theta$$
$$= (R^* - T^*) \sin\theta \cos\theta$$

$$F_y = R \sin\theta + T \cos\theta$$
$$= R^* \sin^2\theta + T^* \cos^2\theta$$

The total forces are obtained by integrating around the nodal circle from θ = 0 to 2π, and it can be shown that:

$$F_x = 0$$
$$F_y = \pi\,(R^* + T^*)$$

Thus, the pile displaces along the y (θ = π/2) axis under the equivalent lateral load applied in the same direction.

F_y can be applied via any suitable combination of the R* and T* terms. The R* term alone was used to present the reported results, since studies using various proportions of R* and T* indicated insignificant variations in the single pile stiffness.

Proposed design procedure for single piles

The above-mentioned elastic finite element program is used as the basic tool for the design. The elastic method of solution may be justified by the fact that most of the soil remains at a relatively low level of shear stress and that only the near-surface soils become plastic. Simple techniques to incorporate non-linearity are dealt with in the following paragraphs.

The steps in the proposed method are as follows:

a. Perform a laterally loaded fixed-head pile analysis using the axisymmetric finite element program with non-symmetric loads.

b. Develop a Winkler beam-column type model to obtain agreement with the finite element model. This is done by extracting the effective "p-y" curves at the pile/soil interface model from the element output. The radial displacements are known and the amplitudes of the reactive forces are obtained from multiplying the element stiffness matrix by the element nodal displacement amplitudes. The equivalent lateral force F_y can be found from the magnitudes of R* and T*. The lateral forces can be converted into pressure values and a best fit curve (or line) is drawn through this data. Since this is an elastic analysis, the p-y curves have a constant slope k (= modulus of subgrade reaction).

Plasticity incorporation

At the time of writing, a linear elastic-perfectly plastic finite element program is being prepared, using multiple harmonic analysis (11, 25).

Two approaches are meanwhile proposed to incorporate plasticity effects.

The first approach is to examine the magnitudes of the principal stresses and strains obtained from the output of the elastic finite element analysis, and to soften the Young's modulus E appropriately, and to repeat the analysis until satisfactory convergence has occurred. This approach may be likened to the "secant modulus" approach used in non-linear finite element work, and can be seen to involve both more finite element work and the derivation of modified Winkler (beam-column) models.

The second approach is to accept that the initial slope of the p-y curves (the modulus of subgrade reaction) represents the elastic solution, and to apply a limiting pressure value P_{ult}, above which the slope of the p-y curve is zero. This involves only modifying the p-y data obtained from the equivalent Winkler (beam-column) model.

Illustrative case

The approach outlined above has been applied to analyse an offshore pile.

The pile is 1.83 m OD and has a 60 mm wall thickness to 10 m below pile head, and 50 mm thereafter. The soils at the site consist of normally consolidated calcareous sands and silts. Within the critical pile length, l_c[2], bands of cemented sands and silts occur. Their thickness is so small that they could be ignored in this simple case study.

The pressuremeter test results indicated that the soil model could be approximated by the formula G = 0.35 z (MPa, m), where G is the shear modulus and z the depth from mudline. A Poisson's ratio ν of 0.30 was used to derive the equivalent Young's modulus-depth relationship of E = 0.91 z (MPa, m).

The finite element mesh used consisted of 96 8-noded isoparametric elements (Fig. 8). The mesh has a radial and vertical extent of twice the effective pile length. This is less than that suggested by Kuhlemeyer, since it was found that further increases gave no significant solution improvement. The run-time of the program was approximately 4 minutes on a VAX 11/780 computer system.

Figures 9 and 10 show the resulting displacement patterns. From the deformed mesh plot, it can be seen that a fixed-headed analysis has been performed.

Fig. 11 shows the modulus of subgrade reaction values (k) back-figured from the element analysis. The solid line is the equivalent Winkler (beam-column) model. The scatter of the data points is a consequence of the "de-lumping" algorithm employed and of the stepped E-modulus profile for the soil. It is thought that the negative k-values at depths below 20 m result from the fact that both the computed displacements and pressures are very small. Small inaccuracies due to computer roundoff have a marked effect on the results. In any case, these negative k-values are at depths at or below the l_c value, where k-value variations have little or no effect upon the deflected pile shape.

[2] l_c, the critical pile length, is the pile length beyond which the pile head deflection is not influenced by further pile length increases.

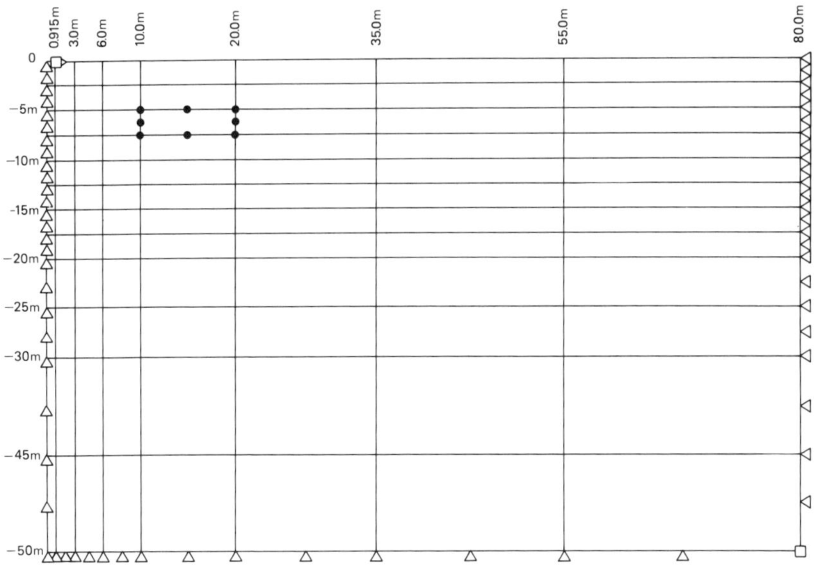

Fig. 8 - Finite element mesh for laterally loaded pile

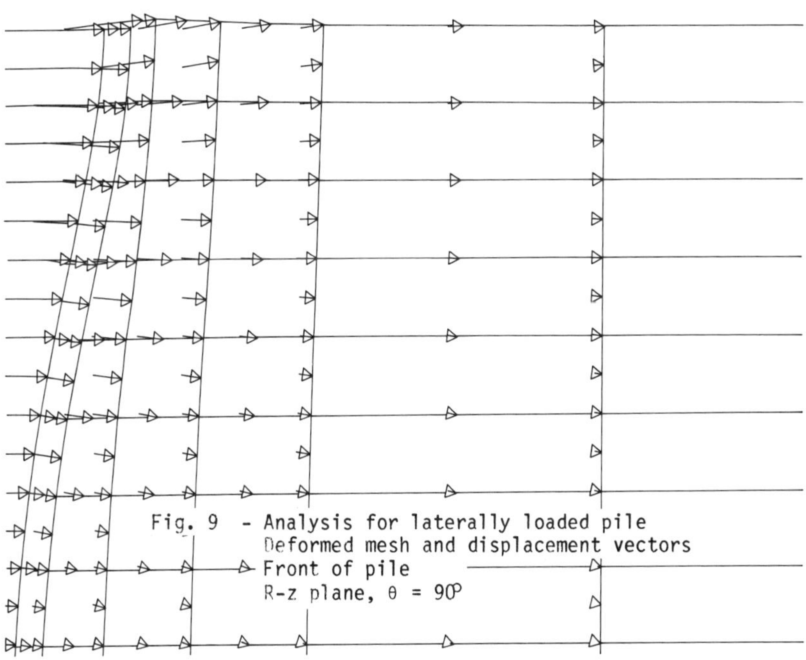

Fig. 9 - Analysis for laterally loaded pile
Deformed mesh and displacement vectors
Front of pile
R-z plane, θ = 90°

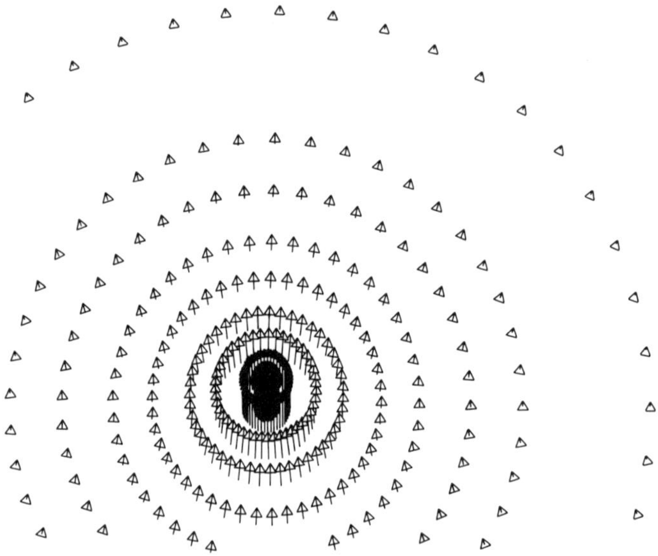

Fig. 10 - Analysis for laterally loaded pile
Displacement vectors R-θ plane, z= 0.0 m

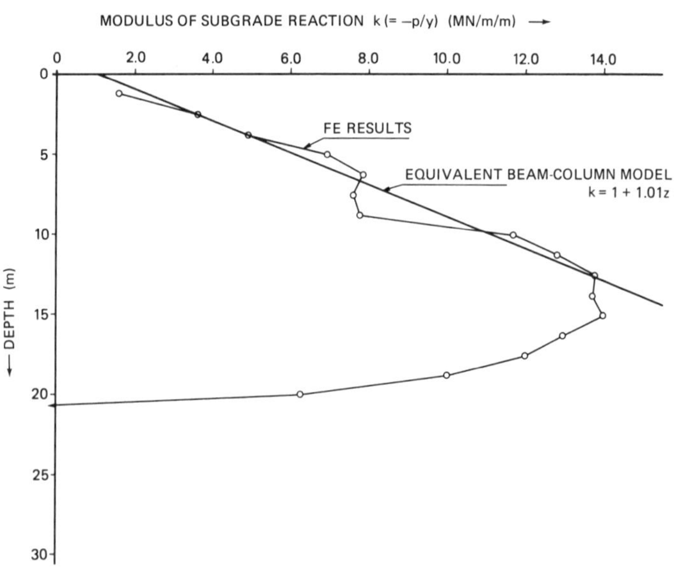

Fig. 11 - Comparison of finite element and
equivalent beam-column k-values

Fig. 12 shows that the proposed equivalent Winkler model reproduces the finite element results to within engineering accuracy, particularly within the upper levels of the pile.

Since the soil modulus was assumed to vary linearly with depth, the single pile results may be compared with the solutions obtained by using the equations suggested by Randolph (15):

Table 1 - Single pile behaviour (elastic)

Method	Equivalent Solid Pile Section Modulus ($MN\text{-}m^2$)	Single Pile Head Lateral Stiffness (MN/m)
FEM	40752 (0 - 10 m) 40261 (10 m +)	69.9
Randolph	40261 (0 m +)	75.9

PART II - PILE GROUPS

Poulos/p-y approach for pile groups

This approach, suggested by Focht and Koch (6), is usually adopted in current offshore pile foundation design. It is based on an assessment of the non-linear single pile behaviour, to which is added a summation of linear elastic interaction factors, accounting for the effect of other piles in the group. Thus, according to this method, the deflections at mudline are:

$$y_g = y_s + \Delta y$$

or

$$y_k = \rho_{EP} \cdot H_k + \rho_E \sum_{\substack{j=1 \\ j \neq k}}^{n} \alpha_{kj} \cdot H_j \qquad (1)$$

The steps in the Poulos/p-y method are briefly as follows:

a. ρ_{EP} is derived from beam-column analyses, using the non-linear "p-y" curves discussed in Part I.

b. ρ_E and the interaction factors α_{kj} are determined from elastic half space theory.

For a uniform value of soil Young's modulus, Poulos (12, 13) has presented charts from which ρ_E and α_{kj} may be determined.

For a linearly varying soil modulus, the finite element based normalised solutions given by Randolph (15) may be employed. Algebraic expressions are also presented for the constant modulus case.

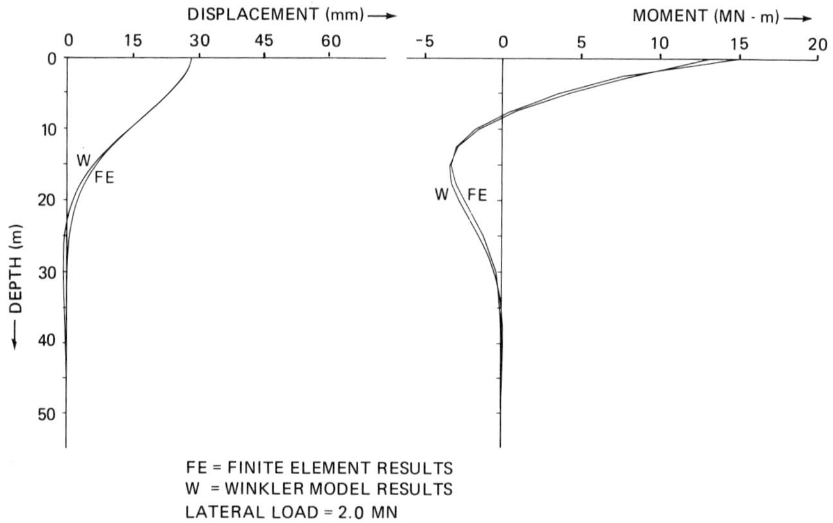

Fig. 12 - Comparison of pile displacements and bending moments Y-modifier = 1.0

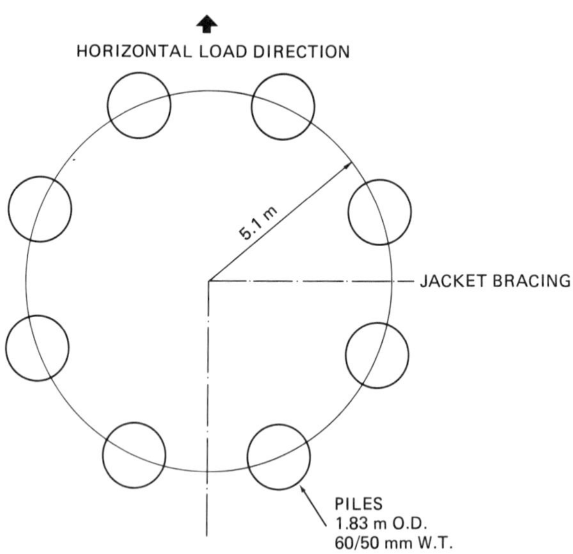

Fig. 13 - Pile group layout

Both approaches require a knowledge of the value of l_c, the critical length of the pile. l_c is usually shorter than the actual pile length, and can be directly evaluated from equations given by Randolph (15). Alternatively, l_c is sometimes taken from the results of Step a, as the level at which the first zero deflection is reached in the beam-column analysis.

c. For equal pile displacements at mudline, $y_k = y_g$, the Poulos p-y relationship gives n equations with n + 1 unknowns: $H_1, \ldots, H_n$ and y_g.
Because the individual pile loads are not necessarily equal, the problem is solved by assuming various total group deflections y_g and inverting the set of n equations to find the unknown H_j-values. Since the term ρ_{EP} is non-linear, an iterative technique is required. From the individual pile loads, the group lateral load, and hence the lateral group-displacement curve, may be determined.

d. The displacement and moment distribution in an individual pile is derived from beam-column analyses using p-y data, "softened" to an extent that for the given load the mudline deflection equals y_g. The "softening" is done by multiplying the y-values by a coefficient larger than 1 (y-modifier).

Limitations of Poulos/p-y approach

The limitations stem from the assumptions made in the p-y (Winkler) and Poulos (continuum) approaches. Inaccuracies generally arise due to simplifications made in the derivation of the ρ_{EP}, ρ_E and α terms.

The results for ρ_{EP} are strongly influenced by the k-values implied in the p-y data. If no site-specific data are available, then "standard" p-y curves have to be used. As mentioned previously, these have been determined for only a limited number of soil types from a limited number of tests.

The ρ_E and α terms are influenced by the selection of a single E-value (Poulos), or G and m* values (Randolph) for the soil. This can be particularly difficult for sites with layers of soft and stiff soils. In addition, no allowance can be made for scour, pile stickup or varying pile wall thicknesses within the critical length l_c.

The terms ρ_E and ρ_{EP} should have identical values in the elastic range. Since they have been derived from two different models, discrepancies usually occur. Thus, varying lateral stiffnesses can be derived, depending upon the soil constants assumed and the source of ρ_E.

Both the solutions of Poulos and Randolph define α as the ratio of the amount of soil displacement due to adjacent pile loading and the single pile displacement. That is, the additional soil displacements are assumed equal to the pile displacements. This would occur only if the pile was perfectly flexible and therefore able to follow the moving soil displacement exactly. If the pile was infinitely stiff, then the soil would move around the pile shaft. No additional movement of the pile would be caused, resulting in a zero interaction factor. Since offshore piles can be classed as neither perfectly flexible nor perfectly stiff, the α factors so derived may be conservative and thus over-predict the amount of additional displacement.

Finally, the y-modifier approach essentially softens the p-y curves at all depths. The resulting computed moment and deflection curves will be realistic only at mudline, where the match is applied. Below mudline, the deflected shape, and hence the bending moments, may be over-estimated.

Proposed design procedure for pile groups

The finite element output, used in Part I to obtain the single pile deflected shape, is re-analysed to determine the additional displacements within the soil mass surrounding the loaded pile.

No plasticity effects are incorporated into the elastic solutions (this assumption is also made in the Poulos/p-y approach). In fact, the additional displacements would generally decrease.

Steps a and b in the proposed method are given in the Section "Proposed design procedure for single piles". The remaining steps are:

c. Extract the profiles of lateral soil displacement versus depth from the results of Step a. This has to be done for the positions within the soil mass around the loaded pile corresponding to the remaining pile positions within the group. These displacements have to be the components in the appropriate load direction. Due to the differing departure angles ψ (and spacings s), the procedure has to be repeated for all the other loaded pile positions within the group. If this was to be done manually, advantage could be taken at this stage of symmetry of the pile group. However, the task is easily performed by a digital computer. For each loaded pile j, the soil movement profile Δy_{kj} is thus accumulated. This is an analogous procedure to Step b in the Poulos/p-y method.

d. Perform an additional beam-column run in which the p-y curves found in Step b are offset by the Δy_{kj} values found in Step c. This pile-in-moving soil technique, which takes account of the relative pile/soil rigidity and pile fixity, is commonly used on land for piles adjacent to slopes (3) and for pile design in mudslide areas offshore (2).

e. If required, obtain the pseudo interaction factors by comparing the results of Steps d and b (or a).

f. Obtain equivalent y-modifiers to simulate the mudline displacements and moment response. The beam-column data so derived (k-values, y-modifiers) can be used by structural engineers. It should be appreciated that, at other depths, equivalence will not be satisfied, due to incompatibility between the Winkler and continuum response models.

g. If required, ensure complete compatibility by repeating Step a and applying the cumulative nodal forces (equivalent to the cumulative soil displacement profiles found in Step c) simultaneously with the original pile head load. The resulting output will be the group pile deflection, from which the bending moments can be calculated. The mudline displacements and moments will be identical to those from Step f.

h. The above procedure gives the lateral load-displacement at mudline behaviour for a single pile in the group. If the geometry of the group is such that widely differing individual pile loads are anticipated, then follow a similar procedure to Step c in the Poulos/p-y method in order to obtain the lateral group load-displacement curve.

Illustrative case

The above approach has been applied to analyse an offshore pile group, the layout of which is shown in Fig. 13.

The 8 piles are equally spaced around a 5.1 m radius pitch circle. The pile dimensions are all similar to each other and identical to those used in Part I.

By taking the components of the surface soil displacements (Fig. 10) in the direction of loading, the perfectly flexible interaction factors can be derived. A typical result is shown in contour form (Fig. 14). As expected, most interaction occurs in front of the loaded pile. This can also be seen from Fig. 10 where, for a given radius, the displacements in front of the pile are larger than those at right angles to the pile.

Applying the sum of the surface soil displacements to the original equivalent beam-column model, the group displacement at mudline can be obtained. Fig. 15 shows the displacement and moment behaviour of a group pile at the design y-modifier level of 4.07.

The group pile results may be again compared with the solutions obtained by using the Randolph equations.

Table 2 - Group pile behaviour (elastic)

Method	Average Group Pile Head Lateral Stiffness (MN/m)
FEM + perfectly flexible pile interaction factors	23.9
FEM + pile-in-moving soil analysis	29.5
Randolph	29.6

A broader spread of answers for group stiffness than for single pile stiffness was obtained. It appears that the Randolph equations underestimate the perfectly flexible pile interaction factors.
By using the deflected soil profiles, a 25% increase in group stiffness resulted.

In passing it is noted that for the actual design study

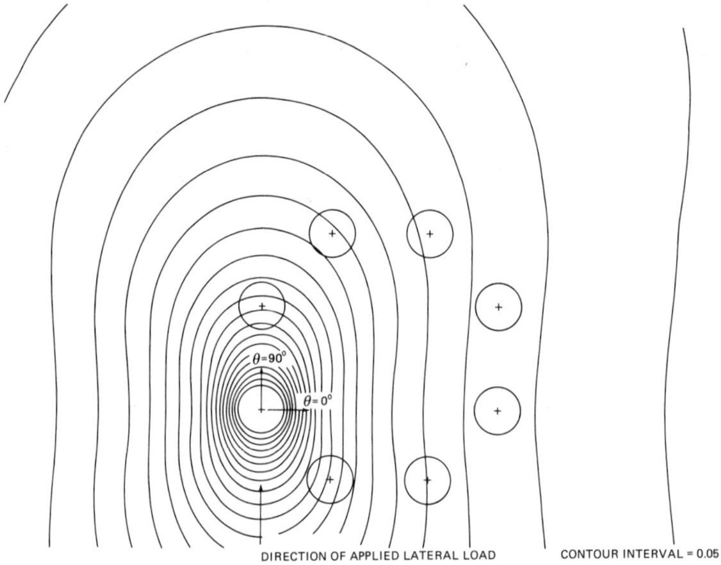

Fig. 14 - Analysis for laterally loaded pile flexible interaction factor contours R-θ plane, z≐ 0.0 m

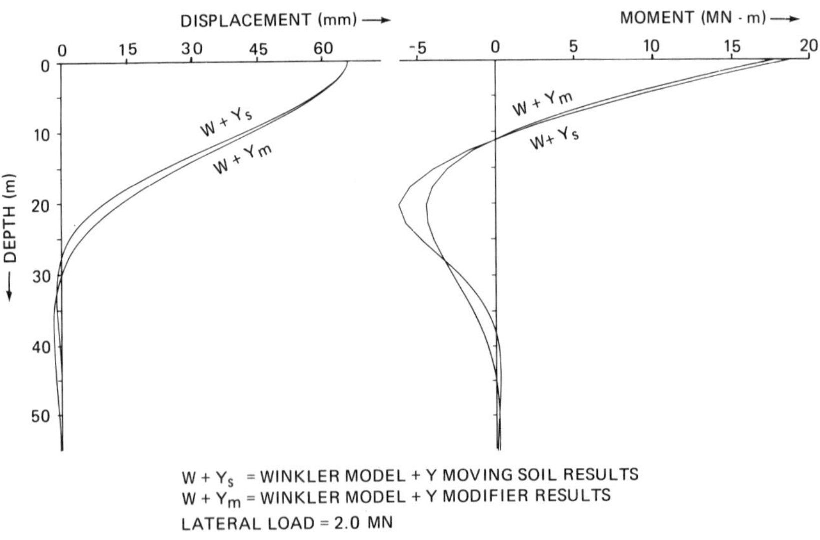

Fig. 15 - Comparison of pile desplacements and bending moments Y-modifier = 4.07

plasticity was incorporated using the first approximate method discussed previously. This resulted in an increase in pile head deflections and bending moments of less than 10 percent. In addition, in-situ lateral pile load tests indicated that cyclic loading effects were not applicable for the lateral load levels of interest.

Furthermore, the study also considered the effects of scour. This was done by using very small E-values for the soil mesh above scour level and reducing the E-values below scour level to account for the effects of overburden removal.

Limitations of the finite element approach

The proposed approach, although it is considered to be an improvement on the Poulos/p-y method, cannot be considered to represent the perfect solution to the problem of a laterally loaded pile group.

The effects of pile installation, causing soil disturbance and/or gapping adjacent to the pile shaft, have not been considered. Although the finite element method is capable of incorporating these effects, there is uncertainty as to what new values should be inserted, unless the finite element analysis results are calibrated against the in-situ lateral pile load test data.

In the derivation of the soil displacement profiles at the remaining pile positions, the second pile is assumed to be non-existent. Although the reduction in actual displacement has been taken into account by means of the pile-in-moving soil technique, the procedure should ideally be checked by analysing two piles, one of which is loaded, by means of a 3-D finite element analysis.

The incorporation of soil plasticity into the existing 2-D program will further increase the computer run-time and storage requirements. Such a program would still be only capable of reproducing the failure zone around a single pile. In practice, for closely spaced piles, the highly stressed zones will overlap one another and this effect cannot be handled by an axisymmetric program. Thus, in the long term, attention must be turned to a more rigorous 3-D solution.

Despite these limitations, it is believed that the 2-D FE analysis achieves a more satisfactory design, both geotechnically and mathematically, provided reasonably conservative "real" soil properties are selected and sensitivity analyses are performed.

In this respect, a companion paper is planned in which it is proposed to report the back-analysis of lateral load test data from a site where pressuremeter testing has been performed.

CONCLUSIONS

The most significant advantage is that, unlike the Poulos/p-y approach, both the single pile and the pile group behaviour can be obtained from the same finite element analysis. That is, no errors due to incompatibility between the Winkler and continuum analyses can occur. Furthermore, the generality of the finite element method is that pile and soil non-homogeneity and non-linearity effects can be readily incorporated. The group design procedure outlined has been shown to be capable of by-passing the intermediate step of interaction factor determination. By using a "pile-in-moving soil" approach, which takes the pile rigidity into consideration, lower pseudo α factors can be calculated resulting in an increase in lateral group stiffness.

Also, the same procedure, generally employing the same finite element model and program, can be used to obtain other pile group stiffness matrix terms (16), for example due to overturning moment, twist and axial load.

REFERENCES

1. AMERICAN PETROLEUM INSTITUTE (1982). "Recommended Practice for Planning, Designing and Constructing Fixed Offshore Platforms - API RP 2A".

2. Bogard, D. and Matlock, H. (1977). "A Computer Program for the Analysis of Beam-Columns under Static Axial and Lateral Loads". 9th Offshore Technology Conference, Houston, Texas, Vol. 3, Paper 2953.

3. CIAD (1972) "Door Grond Horizontaal Belaste Palen". Eindrapport CIAD-Projectgroep, Zoetermeer, The Netherlands.

4. Cox, W.R., Reese, L.C. and Grubbs, B.R. (1974). "Field Testing of Laterally Loaded Piles in Sand". 6th Offshore Technology Conference, Houston, Texas, Vol. 2, Paper 2079.

5. DET NORSKE VERITAS (1977). "Rules for the Design, Construction and Inspection of Offshore Structures". Appendix F: Foundations, pp. F13-F17.

6. Focht, J.A. and Koch, K.J. (1973). "Rational Analysis of the Lateral Performance of Offshore Pile Groups". 5th Offshore Technology Conference, Houston, Texas, Vol. 2, Paper 1896.

7. Henderson, G., Smith, P.P.K. and St. John, H.D. (1979). "The Development of the Push-in Pressuremeter for Offshore Site Investigation". Society for Underwater Technology Conference, London.

8. King, R.W., Hooydonk, W.R. van, Kolk, H.J. and Windle, D. (1980). "Geotechnical Investigations of Calcareous Soils on the North West Shelf, Australia". 12th Offshore Technology Conference, Houston, Texas, Vol. 2, Paper 3772.

9. Kuhlemeyer, R.L. (1979).
"Static and Dynamic Laterally Loaded Floating Piles".
A.S.C.E., Vol. 105, No. GT2, pp. 289-304.

10. Matlock, H. (1970).
"Correlations for Design of Laterally Loaded Piles in Soft Clay".
2nd Offshore Technology Conference, Houston, Texas, Vol. 1, Paper 1204.

11. Meissner, H.E. (1976).
"Laterally Loaded Pile in Cohesionless Soil".
Numerical Methods in Geomechanics, ASCE, Vol. 3, pp. 1353-1365.

12. Poulos, H.G. (1971).
"Behaviour of Laterally Loaded Piles: I - Single Piles".
A.S.C.E., Vol. 98, No. SM4, pp. 711-731.

13. Poulos, H.G. (1971).
"Behaviour of Laterally Loaded Piles: II - Pile Groups".
A.S.C.E., Vol. 98, No. SM4, pp. 733-751.

14. Randolph, M.F. (1977).
"A Theoretical Study of the Performance of Piles".
Ph.D. Thesis, University of Cambridge.

15. Randolph, M.F. (1981).
"The Response of Flexible Piles to Lateral loading".
Geotechnique, Vol. 31, No. 2, pp. 247-259.

16. Randolph, M.F. and Poulos, H.G. (1982).
"Estimating the Flexibility of Offshore Pile Groups".
2nd. Int. Conf. Numerical Methods in Offshore Piling, Austin, Texas, Preprint.

17. Reese, L.C. and Cox, W.R. (1969).
"Soil Behaviour from Analysis of Tests of Uninstrumented Piles Under Lateral Loading".
Performance of Deep Foundations, ASTM STP444, pp. 160-176.

18. Reese, L.C., Cox, W.R. and Koop, F.D. (1974).
"Analysis of Laterally Loaded Piles in Sand".
6th Offshore Technology Conference, Houston, Texas, Vol. 2, Paper 2080.

19. Reese, L.C., Cox, W.R. and Koop, F.D. (1975).
"Field Testing and Analysis of Laterally Loaded Piles in Stiff Clay".
7th Offshore Technolgy Conference, Houston, Texas, Vol. 2, Paper 2312.

20. Smith, I.M. (1981).
"Programming the Finite Element Method, with Application to Geomechanics".
John Wiley and Sons Ltd, London, pp. 120-128.

21. Suen, P.C. (1978).
"Axisymmetric Structure under Non-axisymmetric Loads".
B.Sc. Thesis, University of Manchester.

22. Sullivan, W.R., Reese, L.C. and Fenske, C.W. (1980).
"Unified Method for Analysis of Laterally Loaded Piles in Clay".
1st Int. Conf. Numerical Methods in Offshore Piling, ICE, London, pp. 135-146.

23. Williams, D.J. (1980).
Discussion, 1st Int. Conf. Numerical Methods in Offshore Piling, Institution of Civil Engineers, London, p. 200.

24. Wilson, E.L. (1965).
"Structural Analysis of Axisymmetric Solids".
Journal American Institute of Aeronautics and Astronautics, Vol. 3, pp. 2269-2274.

25. Winnicki, L.A. and Zinkeiwicz, O.C. (1979).
"Plastic (or Visco-plastic) Behaviour of Axisymmetric Bodies Subjected to Non-symmetric Loading - Semi-analytical Finite Element Solution".
Int. Jnl. Num. Meth. Eng., Vol. 4, pp. 1399-1412.

26. Zinkeiwicz, O.C. (1971).
"The Finite Element Method in Engineering Science".
McGraw Hill Book Co. Inc., New York, pp.254-273.

LIST OF SYMBOLS

c_u	soil undrained shear strength
D	pile diameter
E	soil Young's modulus
F_x, F_y	resultant applied forces along x, y axes
G	soil shear modulus
H_j, H_k	lateral forces applied to piles j and k
j	loaded pile number
k	soil modulus of subgrade reaction, also pile number under consideration
L	pile length
l_c	critical pile length
m	rate of change of soil shear modulus G with depth
m*	$= m\ (1 + 3\nu/4)$
n	number of piles in group
p	lateral soil force per unit length of pile
p_o	total vertical soil pressure
r	radial distance
R,R*	radial force, radial force amplitude
T,T*	tangential force, tangential force amplitude
u,u*	radial displacement, radial displacement amplitude
v,v*	vertical displacement, vertical displacement amplitude
w,w*	tangential displacement, tangential displacement amplitude
y	lateral pile displacement
y_g, y_k	group mudline displacement of pile k
Y_s	single displacement of pile k
z	vertical distance
Z,Z*	vertical force, vertical force amplitude

α_{kj}	lateral interaction factor, additional displacement of pile k due to load on pile j
ϕ	soil friction angle
ν	soil Poisson's ratio
θ	tangential angle
ρ_E	elastic single pile displacement per unit load
ρ_{Ep}	elasto-plastic single pile displacement per unit load
Δy	additional displacement due to interaction
Δy_{kj}	soil displacement of pile k due to load on pile j

DYNAMIC TESTING AND ANALYSIS OF GAROUPA PGP-1 PILES

By Rune Dahlberg,[1] and Sergio Matos[2]

1. INTRODUCTION

With the purpose of verifying the bearing capacity of the foundation piles for the Garoupa platform offshore southeast Brazil and improving the design basis for future pile foundations at the Campos Basin, PETROBRAS decided to instrument the driving of some of the 36 foundation piles for the Garoupa jacket. The pile monitoring which was performed by Det norske VERITAS covered both initial driving and redriving and provided useful information regarding pile capacity and hammer performance. The "set up" effects were of particular interest since the observed blowcount during initial driving was considerably lower than expected from driveability studies.

This paper presents information about the Garoupa jacket, soil conditions at the Garoupa site, pile capacity predictions, objectives and scope of dynamic pile testing, instrumentation techniques and analysis of dynamic data. The overall pile monitoring results provided useful data which could be used for verification of:

- hammer performance as expressed through the measured energy transmitted to the pile

- static and dynamic resistance forces along the pile and their distribution along the pile through CAPWAP analysis

- set up effects as derived from CAPWAP predicted pile capacities during driving and redriving.

2. PLATFORM DATA

The Garoupa platform PGP-1, which is a drilling and production platform, is the first deep water platform installed on the Brazilian Continental Shelf. The water depth at the platform location is 120 m and the weight of the jacket is 80 MN. At launching the weight including four preinstalled piles, flotation tanks, risers and anodes amounted to about 100 MN. The total weight of the platform as installed is 130 MN with a design deck load of 160 MN. The jacket dimensions are 75 x 55 m at base level and 55 x 25 m at top level.

The eight leg steel jacket has four skirt piles around each leg plus two additional skirt piles along each of the two longer sides,

---

[1] Principal Surveyor, Det norske VERITAS, Høvik, Norway
[2] Civil Engineer, Petrobras, Rio de Janeiro, Brazil

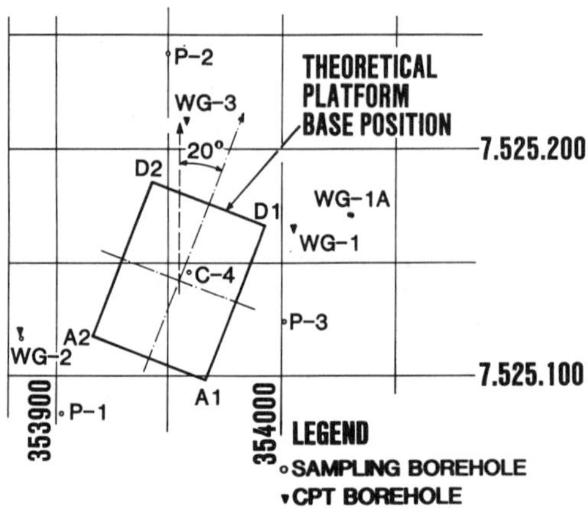

Fig. 1 Site investigation program

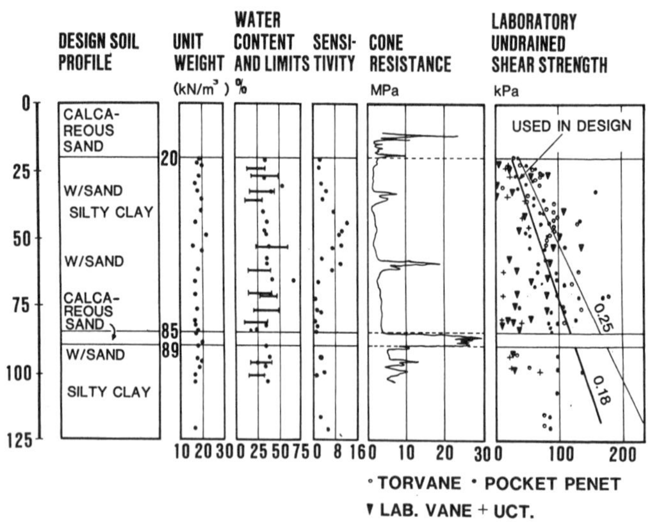

Fig. 2 Soil conditions

thus in total 36 piles. The wall thickness of the piles is 32 mm and the piles were designed for a penetration of 85 m.

The jacket was designed by Interconsult S.p.A. Milan for installation over a preinstalled submarine template through which 9 wells had been partially drilled by a mobile drilling rig to allow for early production. A consortium between Micoperi S.p.A. Milan and Montreal Engenharia S.A. Rio de Janeiro had the responsibility for construction and installation of the platform.

3. SOIL CONDITIONS

For assessment of the soil conditions at the selected site, a comprehensive program of geotechnical investigations was undertaken, including soil sampling and downhole cone penetration testing. Fig. 1 illustrates the geotechnical site investigation program. The design soil profile with representative laboratory test results and CPT results are shown in Fig. 2.

The top 20 m consists of fine to medium calcareous sand with shell fragments, partially cemented. Grain size analysis showed percentages of gravel size particles (> 2 mm) as high as 80% to 90%, due to cementation. For the same reason, unit weight values as low as 15 KN/m^3 were encountered. A mineralogical analysis of some samples of this sand gave the following average results: $CaCO_3$ 89.3%; $MgCO_3$ 7.5%; SiO_2 0.7% and organic material 1.6%. The particle shape was angular and very irregular, which would lead to considerable interlocking effects during shear. In fact, from direct shear and static triaxial CIU testing, the undrained angle of shearing resistance was found to vary widely from 39° to 59°. Cone penetration testing gave cone resistances of up to about 23 MPa in this layer and friction ratios in the order of 0.20% to 0.75%.

From 20 m down to 85 m, a layer of silty clay with fine sand and sparse coarser sand lenses occurs. Fine sand percentage is in the range of 15% to 25% in weight, whereas silt and clay size particle contribution is in the order of 50% and 25%, respectively. No significant changes were noticed in the unit weight nor in the limits and water content, throughout this depth. A sensitivity increase was detected from about 30 to 55 m depth, reaching values over 8 (which might explain the easy penetration of the piles). Below about 60 m, sensitivity decreased to the limits of moderately sensitive soils; however, the results of laboratory testing at these depths indicated a possible disturbance effect that might have reduced the sensitivity values. Cone penetration testing showed a trend of linear increase of soil resistance with depth, as expected for normally consolidated soils. Laboratory test results are widely scattered, possibly due to strong disturbance effects.

At about 85 m sloping slightly downwards in the SW-NE direction, a highly cemented calcareous sand layer with 3 to 4 m thickness is encountered with a cone resistance of about 25 MPa. Below that depth, interbedded layers of silty clay and sand exist.

4. PILE CAPACITY PREDICTIONS

Based on the geotechnical conditions at the site, the calcareous sand layer at 85 m was selected as a bearing layer for the 1219 mm (48 in) diameter piles. Due to the scattering of the laboratory results, pile capacity predictions were made using a Su/σ_{vo}' ratio of 0.18 for the silty clay throughout the depth of interest, interpreted as a lower bound from the CPT results. An average submerged unit weight of 8.5 kN/m^3 was assumed for the soil. For the upper calcareous layer, the angle of shearing resistance was conservatively set equal to 30 degrees. A limiting unit skin friction was assumed equal to 50 kPa. For the bearing layer, a limiting tip resistance was assessed equal to 4.5 MPa.

Based on the λ-method (1) a single pile capacity of 22.9 MN was predicted assuming about 1.5 m embedment of the pile tip in the bearing layer. A 23 per cent reduction in shaft capacity for group piles was allowed for in the design. This gave the following predicted factors of safety, see Table 1, for the actual loading conditions.

Table 1. Factor of safety for single piles and pile groups

Member	Pile capacity MN	Loading condition Operating MN	Storm MN	Factor of safety Operating	Storm
Single pile	22.9	10.5	13.5	2.18	1.70
Pile group	74.4	36.9	47.3	2.02	1.57

5. JUSTIFICATION FOR DYNAMIC PILE MEASUREMENTS

Current methods used for predicting the capacity of large offshore piles are correlated against pile load test data which do not cover the high capacities and great pile penetrations encountered with offshore piles. The range of soils conditions covered by the available pile test data is also very limited. Together with other uncertainties which have to be considered and coped with at the design stage, the safety of the pile foundation may very well turn out to be either on the conservative or on the non-conservative side of the intended safety level.

Dynamic measurements on high capacity piles will have the objectives of verifying the capacity of the piles tested and providing data for updating current design methods, which will be of benefit also for the design of future pile foundations in similar soils.

For the Campos area and elsewhere the presence of calcareous soils leads to great uncertainty in the pile foundation design due to the complex load transfer mechanism in such soils. Recommendations for the design of piles in such soils have been given (e.g. 2,3). Common to recommendations proposed so far is the fact that they are all based on very limited experience without backup of full scale pile load test data. Local experience may then be improved by taking dynamic measurements during driving as a substitute for static pile load tests. In view of those facts and the awareness of the importance of the foundation contribution to the platform integrity, PETROBRAS decided to have the driving of some of the piles instrumented. Det norske VERITAS was contracted with for providing both pile driving monitoring and analysis. For the analysis the CAPWAP method (Case Pile Wave Analysis Program) was selected due to the long experience and good results achieved with this method so far.

For the Campos area the Garoupa data now represent an important reference. However, PETROBRAS has already carried out dynamic pile measurements on other platforms in the same area.

6. INSTRUMENTATION AND ANALYSIS PROCEDURES

The techniques used were developed in the 1960's and reference is made to (4).

The instrumentation consists of pairs of accelerometers and force transducers which are attached by bolting to the pile head or as in this case to the follower head. After preamplification all signals are transmitted via cable to the registration system on the crane barge. At this point after averaging (and integration of acceleration) the force and velocity time histories are displayed on oscilloscope for control of hammer performance and functioning of the system. All data are stored on tape for further analysis. Processing and analysis can be done in real time to provide a blow by blow record of maximum energy delivered to the pile, maxima of force, velocity, acceleration and maximum displacement of pile top as well as the so-called Case Method capacity at time of driving.

Further analysis is done after driving either on board the crane barge or in the office in order to check or supplement the initial results. For long offshore piles with their irregular cross-sectional area the Case Method prediction, which is based on a closed form solution and assumption of an uniform pile, does not provide a realistic pile capacity. In the further analysis certain representative "good" blows will therefore be selected for CAPWAP analysis which uses a discrete model of the pile with springs and dashpots representing the soil resistance forces along the pile, see Fig. 3. In this method the measured acceleration at the pile top is used as a boundary condition and a wave equation type analysis is performed. The magnitude and distribution of static and dynamic driving resistance forces are assumed and varied until a best possible agreement is obtained between measured and computed pile top force. When an acceptable pile top force match has been obtained, the corresponding soil resistance distribution is taken as the most likely set of soil strength parameters that produce the measured data. Further details on this method are found in (4). Due to the soil disturbance caused by pile driving, the CAPWAP predicted capacity derived from data taken during normal driving, will underestimate the long term capacity of the pile. A more realistic capacity will be obtained if the same analysis is carried out on data obtained in a redriving situation, i.e. after a delay which allows excess pore pressures to dissipate in the soil. The amount of regain in pile capacity due to this "set-up" effect and the time required for the pile to regain its full capacity vary from soil to soil.

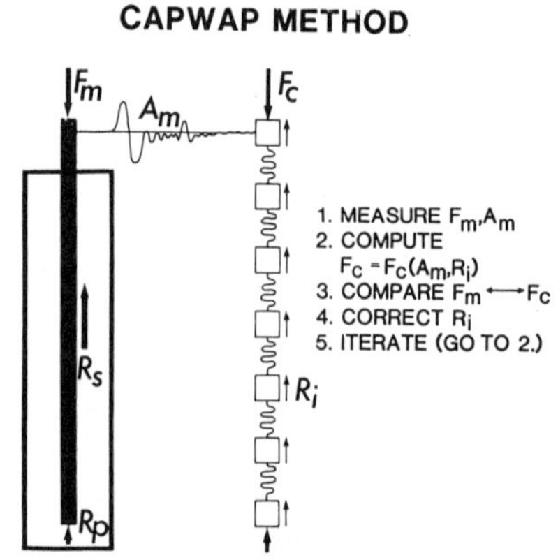

Fig. 3 Principle of CAPWAP method

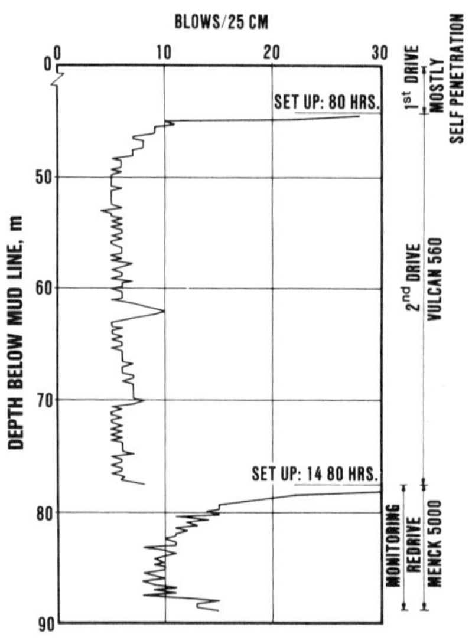

Fig. 4 Typical blowcount vs depth record

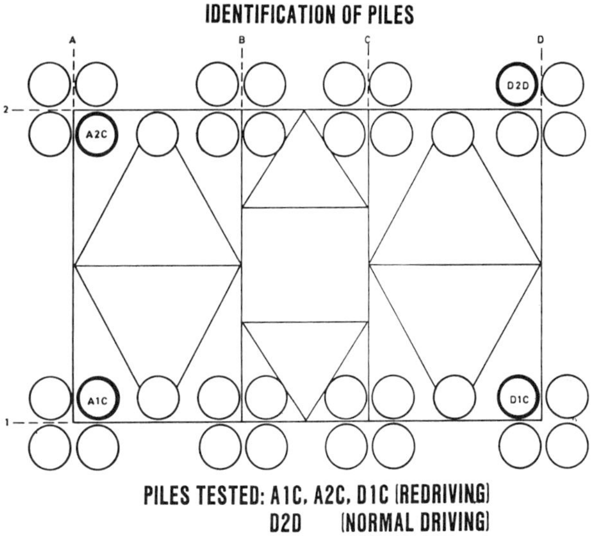

Fig. 5 Identification of piles

7. SCOPE AND SEQUENCE OF PILE MONITORING

The Garoupa jacket was launched on 1st May 1980 after a few days standby due to high swell and strong current. Setting of the jacket above the template went successfully and smoothly. Unforeseen installation problems, however, mainly caused by the presence of a very hard layer just a few meters below mudline and a very loose layer underneath delayed the pile installation. Much caution had to be used during the initial phase of pile driving since the piles penetrated by their own weight to 40-50 m depth after having passed through the harder upper layer, see Fig. 4. Further driving to target penetration was then accomplished with a much lower blowcount than expected which led to some doubt regarding the long term pile capacity. At this point PETROBRAS decided to redrive some piles after a certain set up period.

The preinstalled corner piles A1/C, A2/C and D1/C shown in Fig. 5 were selected for monitored redriving and pile D2/D was monitored during normal driving for comparison. The driving history of these piles is shown in Table 2.

Table 2. Driving history of monitored piles

Pile no.	1st hrs.	Drive m	2nd hrs.	Drive m	3rd hrs.	Drive m	4th hrs.	Drive m
A1/C	-	44.25	80	77.75	1480	88.85	-	-
A2/C	-	46.00	52	88.25	1325	88.50	-	-
D1/C	-	47.50	10	68.20	122	82.25	1475	93.00
D2/D	-	53.50	22	88.0				

As seen from Table 2 the redrive took part after a set-up period of 1325 to 1480 hrs. or about two months. At that time piles A1/C and and D1/C had their tips in the silty clay layer about 10 m and 5 m respectively above the bearing layer while pile A2/C had been driven about one metre into this layer. At redriving pile A2/C was driven about 0.25 m further into the bearing layer while pile A1/C was driven to about the same depth and pile D1/C through the bearing layer with the objective to check its thickness if further penetration would be deemed necessary for the other piles.

In order to minimize the delay caused by the pile monitoring two followers were instrumented while other activities went on. Since the initial blows are the most interesting in a redriving situation, it was agreed to use a sufficiently heavy hammer which was capable of delivering high energy blows from the very start. The reason is that the relative displacement imposed between pile and soil has to be large enough throughout the length of the pile, in order to mobilize simultaneously all resistance forces; if not, the pile capacity will be underestimated. For this reason the hammer, before being moved onto the instrumented piles, was also "warmed up" by driving another, non-instrumented, pile. This explains the rather constant hammer efficiency measured throughout the redriving, see Fig. 6. The efficiency is defined as the energy transmitted to the pile divided by the manufacturers rated energy for the hammer. The energy is derived as the integral of force times velocity over time. For the Menck MRBS 5000 hammer used during redriving, the measured efficiency was 50 to 60 per cent while the Vulcan 560 used on pile D2/D had a measured efficiency of only 40 to 50 per cent since it was malfunctioning at that occasion.

8. RESULTS OF CAPWAP ANALYSIS

The CAPWAP analysis was carried out jointly between Det norske VERITAS and Goble & Associates Inc., in Cleveland. Since 1981 VERITAS operate their own processing and analysis system which is a new generation of the system used for analysis of the Garoupa data.

The static soil resistance distribution reached through the CAPWAP analysis shows three distinct layers of interest along the pile shaft. Firstly, a high resistance layer at about 6 to 10 m depth which gave the large driving resistance, secondly a zone between 10 and 40 to 50 m which shows a very low resistance and thirdly the lower part down to 88 m which gives the major contribution to the pile capacity. This is in good agreement with the overall observations during pile driving,

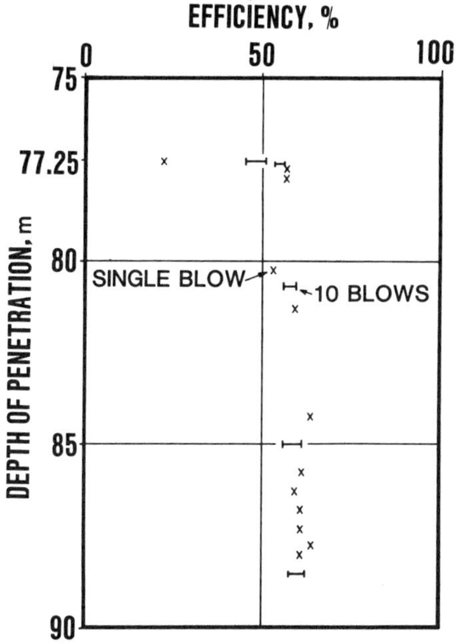

Fig. 6 Measured hammer efficiency (Menck MRBS 5000)

see Fig. 4.

In total 9 blows were selected for CAPWAP analysis. Table 3 contains a summary of the results.

Table 3. Summary of CAPWAP analysis results

Pile no.	Blow no.	Depth of pen.	CAPWAP resistance Skin Total MN	Skin Top 9m MN	Tip MN	Blows/25 cm Measured	Blows/25 cm Computed	Computed tip displ. Max. mm	Computed tip displ. Final mm
A1/C	2	77.2	21.6	6.9	1.0	1000	1170	1.5x)	0.3
	57	77.3	18.0	1.8	1.3	250	165	6.9	1.5
	750	88.6	6.2	0.2	0.9	13	15	19.0	16.5
A2/C	2	88.2	24.8	6.2	1.9	> 1000	2250	1.0x)	0.0
	10	88.2	23.2	2.9	0.8	500	1600	1.0x)	0.3
	152	88.4	17.2	1.7	1.2	90	120	3.6	2.0
D1/C	2	82.2	17.5	4.4	0.7	> 1000	4800	2.0	0.0
	803	87.7	11.7	0.4	0.9	32	40	9.9	6.4
D2/D	1054	88.1	6.0	0.4	0.6	23	26	13.7	8.6

x) All resistance not mobilized since computed maximum tip displacement is less than the quake (= 2 mm).

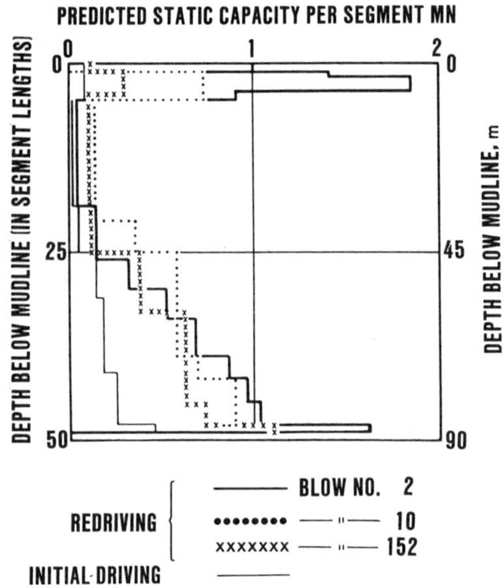

Fig. 7 Result of CAPWAP analysis

Table 3 and Fig. 7 show the dramatic decrease in pile capacity from the initial blows during redriving until full degradation of the soil resistance. Of particular interest is the tremendous decrease in resistance within the upper 9 metres. It is obvious that it is not advisable to count on the peak resistance of the upper layers but rather the residual value, since there will be significant relative cyclic displacements between the pile wall and the soil in this region during a storm in addition to lateral displacements of the pile. Another piece of information which is contained in Table 3 is the computed pile toe displacement for the blows analysed. For pile Al/C, blow No. 2, the maximum displacement during the wave pass was 1.5 mm with a final permanent displacement of 0.3 mm. Since the soil displacement required to mobilize the ultimate resistance (quake) was assumed to be 2 mm in the computations, the full soil resistance was not mobilized for this blow. The same result was obtained for two other blows, see Table 3. The comparison between measured and computed blowcount is a further quality check on the final "match" between computed and measured pile top force.

The discussion above indicates that the resistance in the upper zone might have started to degrade before the resistance of the deeper soil layers had been mobilized. This will give a CAPWAP capacity which is somewhat less than the ultimate capacity. For the Garoupa piles this will, however, have only minor influence, see section 9.

Damping can be introduced in the CAPWAP analysis either as Case damping or Smith damping. Case skin damping is defined as the sum of all viscous skin damping factors (J) divided by the pile's impedance, EA/c, and similar for the toe damping. Smith skin damping is defined as the sum of all viscous skin damping factors (J) divided by the static skin resistance while the Smith toe damping is obtained by dividing the toe viscous damping factor by the static toe resistance. For the actual CAPWAP analyses the Smith skin damping used varied between 0.5 and 0.7 s/m and the Smith toe damping between 0.4 and 0.6 s/m.

9. ASSESSMENT OF LONGTERM STATIC PILE CAPACITY

There is a basic difference between the static pile capacity predicted by the CAPWAP method and the long term static pile capacity. The difference lies in the failure modes. During a dynamic loading as imposed by a hammer blow, the steel tube punches unplugged through the soil (although some local plugging can occur in very hard or dense soils). In this case the driving resistance is the sum of outer and inner skin resistance plus the toe resistance of the steel section. During static (or quasi-static) loading of the pile in the long term a plugged behaviour will often be a more adequate assumption. In this case the static pile capacity will be the sum of outer skin resistance and the toe resistance of the full cross-sectional area of the pile. The conversion between these two types of failure modes is worth much attention.

The Garoupa piles are assumed to penetrate without plugging during driving. A two metre driving shoe with a wall thickness about 20 per cent greater than the pile itself will probably lead to much less inner than outer unit skin friction. Let us assume a ratio between inner and outer unit skin friction of 0.3 for the silty clay layers between 20 and 87 m depth during redriving.

With the assumption of an average submerged unit weight of the soil of 8.5 kN/m^3 the mobilized outer unit skin friction (c_a) in the silty clay layer has been plotted as (c_a/σ_{vo}') where σ_{vo}' is the effective overburden pressure, see Fig. 8. It is interesting to note the low skin friction above 40 to 50 m depth which is in good agreement with the observed behaviour of the pile. From 50 to 60 m the major static resistance is obtained. In the same figure is shown the residual outer skin friction acting during normal driving which is only about 25 per cent of that during redriving. The ratio between inner and outer unit skin friction during normal driving has been assumed equal to 0.5.

In Fig. 9 is also shown the gradual degradation of unit skin friction in the cemented calcareous sand layer close to mudline. The conservative assumption of equal inner and outer skin friction has been made. In the long term no more than 0.7 MN of capacity is considered achievable from the upper 20 m. This corresponds to an average unit skin friction of 9 kPa in this layer.

The contribution to long term pile capacity from the toe assuming plugged conditions is estimated to a maximum of 5 MN corresponding to an average unit end resistance of 4.4 MPa at 1.5 m penetration into the very dense calcareous sand layer (cone resistance in the range 25 to 30 MPa).

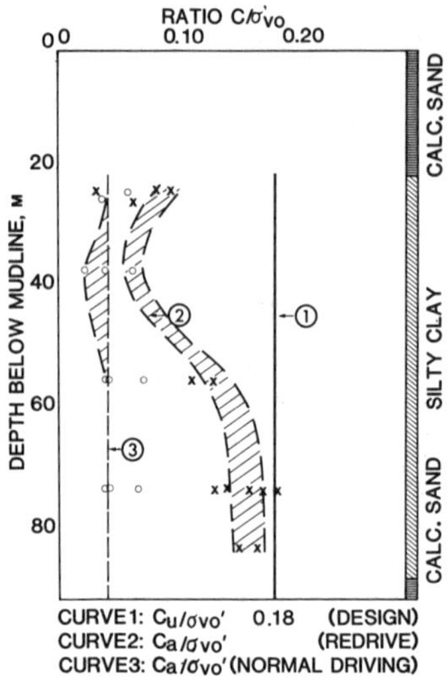

Fig. 8 Interpretation of CAPWAP results

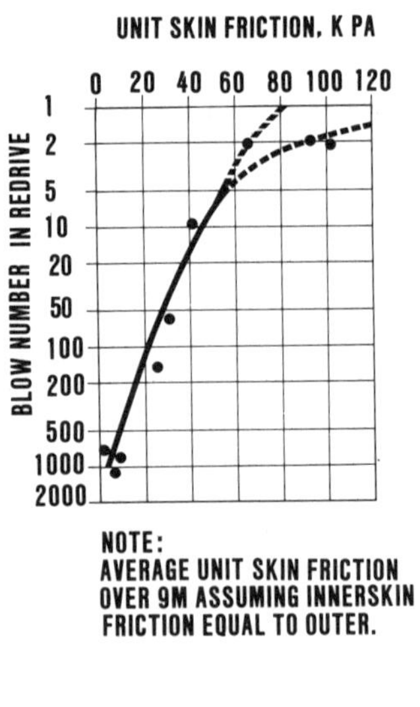

Fig. 9 Interpretation of CAPWAP results

Taking the skin resistance for the silty clay layers directly from the (c_a/σ_{vo}') ratios in Fig. 8, the above assumptions lead to the following long term capacity for a single pile penetrated to 88 m depth below mudline, see Table 4.

Table 4. Single pile capacity evaluated from CAPWAP results

Depth interval m	Contribution to capacity MN	Accumulated capacity MN
Skin resistance		
0 - 20	0.7	0.7
20 - 40	1.0	1.7
40 - 60	3.5	5.2
60 - 88	10.7	15.9
Toe resistance	5.0	20.9

It is reasonable to expect some further increase in skin resistance from 20 to 45 m (see Fig. 8), say 1 MN which corresponds to a (c_a/σ_{vo}') ratio of 0.08 which is still a rather low degree of mobilization. It should be borne in mind that the cone penetrometer tests carried out by Wimpey Laboratories give no reason to doubt that the undrained shear strength increases linearly with depth.

Based on the above interpretation of available data and the CAPWAP results a single pile capacity of 21.9 MN is obtained for 88 m pile penetration. This capacity, which is in exellent agreement with the prediced capacity based on the λ-method, gives the following "verified" factors of safety, see Table 5.

Table 5. "Verified" factors of safety

Member	Pile capacity MN	Loading condition Operating MN	 Storm MN	Factor of safety Operating	 Storm
Single pile	21.9	10.5	13.5	2.08	1.62
Pile group x)	74.1	36.9	47.3	2.00	1.57

x) 20 per cent reduction in single pile shaft resistance assumed to account for group effect.

10. THE FUTURE ROLE OF OFFSHORE DYNAMIC PILE TESTING

Dynamic pile testing during driving seems to be a practical and economical means of testing offshore piles to failure for verification of design assumptions. So far, this kind of testing has been limited to pile installations where the hammer is operating above water. Techniques are, however, available also for underwater dynamic measurements where the signals may be transmitted without use of cables. A promising method for data transmission without cables is hydroacoustic data transmission based on a battery driven hydroacoustic system modulated by digitized measurement data.

The pile monitoring results are of vital interest already during pile installation. Such immediate information is made available on site by use of the so-called Pile Driving Analyzer (PDA) which provides real time data for control of hammer performance, pile integrity and pile capacity at the time of driving. It may often be a further advantage to carry out the CAPWAP analysis while the pile installation is still underway since the Case method capacity may be questionable for long offshore piles with irregular cross section. A portable system for processing and CAPWAP analysis, now available, can easily be brought to an on-shore base at convenient distance from the offshore site, thereby cutting down the time between measurements and final conclusions.

11. CONCLUSIONS

For the installation of the Garoupa platform a program was established to instrument the driving of some piles with the purpose of verifying the bearing capacity. Being the first structure in the Campos area and having the piles installed through and resting on calcareous soils, the Garoupa platform presented an opportunity to improve local and general knowledge about the behaviour of calcareous soils.

Due to an easier penetration than expected, it was decided to redrive some piles after a certain set up period. Three piles were redriven after about two months.

On site analysis of the data from the instrumentation was performed using the so-called Case Method to provide a preliminary assessment of the static pile capacity at the time of driving. On shore a reanalysis was done of the static pile capacity using the more reliable CAPWAP method. This provided a basis for assessment of the long-term static pile capacity and the set-up effects.

For both driving and redriving situations, a non plugged pile was assumed since the downward propagating shock wave is unlikely to accelerate simultaneously both the steel pile and the soil plug. The difference between the static capacity after full set up predicted by the CAPWAP method and the long term static capacity can be explained mainly by different failure modes during driving and static loading. The regain of static pile capacity between the time of initial driving and redriving is a measure of the set up effect. In the present case the ratio of static capacity at the time of redriving to that during initial driving was 2.5 to 3 for the silty clay layers and almost ten for the shallow calcareous sand layer.

The large increase in skin friction in the calcareous sand layer is attributed to cementation of the soil to the pile wall. However, few blows after initiating the redrive a tremendous decrease in skin friction was noted. For this reason a contribution of only 0.7 MN was considered achievable for the upper 20 m. With 16.2 MN shaft resistance in the underlying silty clay, and a plugged tip contribution of 5 MN, the long term capacity is estimated to be 21.9 MN, giving the safety factors shown in Table 5. The CAPWAP predicted shaft resistance is only 1 MN less than that predicted at the design stage based on the λ-method which gives support to the continued use of this method for design of piles for the Campos Basin.

The monitoring work, especially in the redrive, proved to be essential in this case as an acceptance basis for the piles, mainly because observed blow count was considerably lower than expected from driveability studies. In the authors opinion the instrumentation of pile driving should be established as part of the pile installation acceptance criteria for piled foundations in new areas to be developed or in areas where the subsoil conditions are different from common practice in piling. This will also improve the data base for design of future pile foundations in similar soils.

12. ACKNOWLEDGEMENTS

The authors wish to express their gratitude to their organizations for assistance in writing this paper, and especially to PETROBRAS, for permitting its publication.

13. REFERENCES

1. Vijayvergiya, V. A. and Focht, J. A. Jr., "A New Way to Predict Capacity of Piles in Clays", Offshore Technology Conference, Vol. 2, May 1972, pp. 865-874.

2. Datta, M. Gulhati, S. K. and Rao, G. V., "An Appraisal of the Existing Practice of Determining the Axial Load Capacity of Deep Penetration Piles in Calcareous Sands", Ofshore Technology Conference, May 1980, Paper No. OTC 3867.

3. Beringen, F. L., Kolk, H. J. and Windle, D., "Cone Penetration and Laboratory Testing in Marine Calcareous Sediments", ASTM Symposium on Calcareous Soils of D18 Committee, January 1981.

4. Rausche, F., Goble, G. G. and Moses, F., "A New Testing Procedure for Axial Pile Strength", Offshore Technology Conference, May 1971, Paper No. 1481.

ANALYSIS OF OFFSHORE PILE DRIVING - A REVIEW

by G. G. Goble[1] (M.ASCE)

INTRODUCTION

The cost of offshore construction operations requires that reliable estimates be made of the time required for pile installation. The only means available to solve this problem is by analysis using "wave equation" computer programs. This tool also provides a means of estimating pile capacity. Since static load tests on offshore piles are very difficult, time consuming and expensive dynamic analysis has considerable practical usefulness.

In the past few years dynamic measurements during pile driving have been made with increasing frequency. These measurements are analyzed to evaluate both driving system performance and static pile capacity. This paper will discuss the general problem of the analysis of offshore pile driving and will point out those areas where questions arise or problems exist.

Two different programs provide the basis of most of the current computer codes used in the analysis of offshore pile driving problems. The first is the program developed over several years time at Texas A & M University (8,14) and second a program called WEAP developed by Goble and Associates under contract with the Federal Highway Administration (3). These programs have been modified to meet particular requirements so they appear in several proprietary codes. Both programs use a similar soil resistance definition based on an elastoplastic spring with a linear dashpot as suggested by E.A.L. Smith (16). The most important difference between the two programs is the more detailed thermodynamic treatment of diesel hammers contained in WEAP.

Dynamic measurements have become increasingly common in recent years. Measurements have been made that only sampled the first relative maximum of the pile top force and then compared it with families of wave equation analyses. Complete force time records have also been recorded and then used to evaluate driving system parameters and soil resistances. Finally measurements of force and acceleration are used in the analysis of a model that does not include the hammer. Delivered energy can be calculated as can the soil resistance parameters.

CHARACTERISTICS OF OFFSHORE PILE DRIVING

To understand the charateristics of offshore pile driving consider the measurements in Figures 1 and 2. The pile systems involved in

[1] Chairman, Dept. of Civil, Environmental and Architectural Engineering, University of Colorado, Boulder, CO 80309

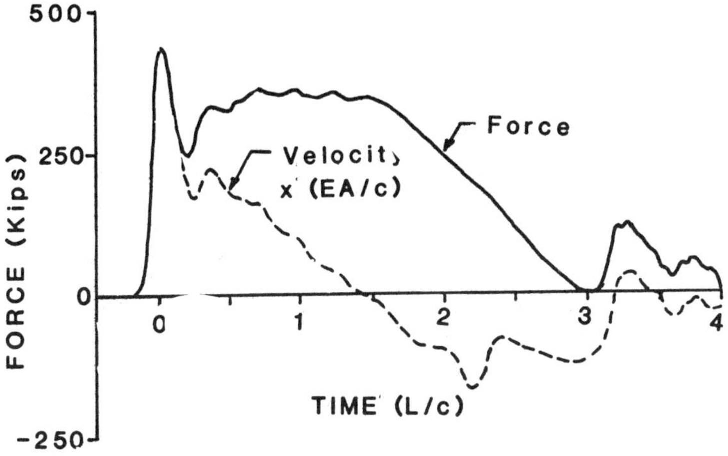

Typical Land Pile Force, Velocity-Time Measurement With Significant Shaft Resistance

Figure 1

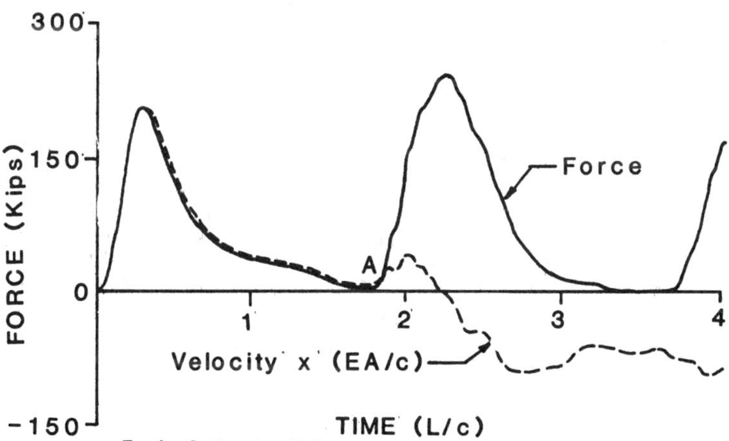

Typical force, Velocity-Time Measurement with Offshore Character

Figure 2

these tests represented typical situations for land (Fig. 1) and offshore piles (Fig. 2). In the land case the presence of side friction forces near the pile top induces reflected compression forces that cause the force and velocity records to begin to deviate from each other. Clearly the impact event is mixed with the soil reflections.

The record from the offshore pile, on the other hand, exhibits the force velocity proportionality, indicating no skin resistance, until almost all of the hammer energy has been transferred to the pile. Both force and velocity are nearly zero at point A in Fig. 2. In this case it should be possible to extract reliable driving system parameters from the first part of the measurements while in the land example such is not the case. There is a pitfall, illustrated by the measurement given in Figure 2, in using only the force record to attempt to determine soil resistances. Note that at the point where the force curve increases in value near the 2L/c time the velocity also increases. (L is the pile length and c the velocity of wave propatation). Thus, the beginning of this increase is not due to reflected forces from soil resistance but rather from an input force. This particular input force arises from the impact of the hammer assembly on the pile top. Both force and velocity records must be available to correctly analyze the measurements.

It should be pointed out that the input and reflected forces in pile driving are commonly separated in the offshore case primarily because all offshore hammers are "light". In land applications the ratio of ram weight to pile weight will typically be on the order of two for steel piles. Offshore ram pile weight ratios will commonly be less than 1/8. Thus, in the offshore case the hammer input force will decay much more rapidly relative to the time of pile tip reflection than is normal for land piles.

MEASUREMENTS AND THEIR EVALUATION

Measurements of force during pile driving have been made in the offshore environment for some time. Steel piles are used exclusively the availability of a reliable modulus of elasticity makes the use of resistance strain gages quite natural. Since the impact strains are large it is quite easy to obtain accurate measurements using bridge circuits. Perhaps, the use of semiconductor gages should be considered since they show a much greater sensitivity and probably could be used with simpler instrumentation systems. However, they offer other difficulties in use so probably resistance gages will continue to be dominant.

Motion measurements have not been as frequently performed. Due to the availability of a simple reference system, acceleration measurements are the most attractive. Although measurements of this type have been made extensively for almost 20 years (5) the view was expressed at the first meeting of this conference in 1979 that they were very difficult if not impossible.

In commonly used analyses it is desirable that frequency content in the signal up to about 2 khz be retained. Higher frequency motions are usually of no importance. Therefore, a mechanical filter is desirable to remove the high frequency content. A successful solution is to use piezoelectric accelerometers having quite high natural frequencies. These devices are mounted to the pile with a mounting system that does not transmit the frequencies in the range of the natural frequency of the accelerometer. In this way nonlinear accelerometer outputs are avoided. Electronic filtering of the output signal then provides a satisfactory measurement.

Seismic devices for the direct measurement of velocity can also be considered. Such devices become very complex due to the need for a control system to prepare the transducer for the next hammer blow. There is probably little advantage in such a device.

An important reason for making both force and velocity measurements (integrated from acceleration) is the increased reliability that comes from checking the two measurements against each other. Force and velocity are related by

$$F(t) = \frac{EA}{c} v(t) \quad (1)$$

where F is the force in the stress wave, E is the material modulus, A is the pile cross sectional area and v is the velocity of particle motion in the stress wave. Since F and v are obtained independently, if they show this proportionality as in Figures 1 and 2 it gives strong confidence in the entire measurement system. This proportionality holds only during wave propagation in one direction. When reflections arrive at the measurement point the force and velocity will no longer be proportional.

ANALYSIS FOR DRIVING SYSTEM PERFORMANCE

The most popular measure of driving system performance is energy. Due to the historic use of energy deriving from dynamic formula it is natural to place importance on it. If force and velocity measurements are available then the energy passing through the point of measurement can be calculated from the relationship

$$E(t) = \int_0^t F(t)\, v(t)\, dt \quad (2)$$

The maximum value of this quantity has been given the name ENTHRU (11). Since motion measurements are not always made the force velocity proportionality from Eq. 1 is sometimes noted and energy is then calculated from

$$E(t) = \frac{c}{EA} \int_0^t F^2(t)\, dt \quad (3)$$

This relationship is correct so long as force and velocity are proportional. For offshore piles this requirement is frequently satisfied as shown in Figure 2. However, a limit should be placed on the length of time over which the integration is carried. In no case should the integration be allowed to continue more than a time of 2L/c. As shaft resistance is located further up the pile the ENTHRU will be overestimated. The problem is illustrated in Figure 1. Furthermore, offshore piles commonly have a nonuniform cross section and so the force and velocity will lose proportionality.

A more interesting approach to the analysis of driving system performance is to determine the parameters used in wave equation analysis from measurements made at the pile top. The commonly required parameters for air/stream hammers are capblock stiffness, coefficient of restitution, helmet weight and ram impact velocity. It would seem reasonable that helmet weight should be known since its measurement is quite a direct operation. Unfortunately, this may not always be true. However, it is possible to obtain these quantities with a reasonable level of accuracy. It is a commonly used procedure using force measurements only together with wave equation analysis. Determination and collection of data on capblock parameters and ram impact velocities should be particularly useful for improving the reliability of wave equation analysis. Data of this type should be assembled to obtain mean values and dispersion about the mean.

It should be noted that measurement devices are available for measuring ram impact velocity directly and this information is now being regularly measured. One approach is to make direct measurements using transducer devices attached to the ram (14). Another approach is to calculate the impact velocity indirectly from measurement at the pile top using impulse-momentum procedures (10).

ANALYSIS FOR SOIL PERFORMANCE

In recent time the CAPWAP (Case Pile Wave Analysis Program) method (13) has been used with increasing frequency for evaluating static capacity and soil performance. This procedure models the pile and soil in the manner of Smith. The hammer is not included in the model. One of the two measured quantities, usually velocity, is treated as an input and the soil resistance parameters are adjusted to give the best possible agreement between the calculated and measured values of the other quantity. In its original version this analysis used an optimization procedure to arrive at the "best" agreement. Experience showed that usually some human interaction could improve the agreement at the end of the automated analysis.

As the method began to be used on long offshore piles and when a minicomputer was implemented into the analysis, an interactive approach was used. Thus, an operator makes the modifications in the soil resistance based on an examination of plots of calculated and measured quantities.

A somewhat different procedure was proposed by Dolvin and Poskitt (2). They took advantage of the separation of the impact portion of the record from the soil reflection and used only one of the measurements, force, in their analysis. In the first phase of the analysis values are calculated for the driving system parameters. In the second phase of the analysis the total hammer pile system is modeled including the hammer using the previously calculated driving system parameters and the measured forces are used in a sequential procedure to determine soil parameters. The Dolvin and Poskitt method uses an automated search procedure for calculating soil parameters. The measured acceleration is used in an independent computation to "check" the results. This approach suffers from the fact that the hammer input is not always separable from the soil response. In addition stray friction forces, sometimes present between the pile and the jacket, might be mistaken for driving system input effects.

The CAPWAP method emphasizes the evaluation of the elasto-plastic ultimate resistance and the damping constant. It is also possible to evaluate quake values and in some cases quake is found to vary quite widely from the commonly used 0.1 inches (1). The analysis of pile plugging has also been done with some success (4).

DRIVEABILITY ANALYSES

One of the primary reasons for doing wave equation analysis is to analyze the effectiveness of a proposed driving system prior to going to the field. It is appropriate to discuss the difficulties surrounding this analysis.

Ian Smith (17) pointed out that for the long piles characteristic of the offshore environment there can be difficulties with the integration procedure. His parameter studies indicated perhaps blow count (penetration) was not seriously influenced by analysis methods that in fact gave quite erroneous force values. Integration methods are available that will provide a correct results of wave propagation (12). More attention should be given to this topic. However, care must be exercised in modifying the integration procedure. Currently used programs have been extensively "calibrated" by experience. Drastic changes in the analysis may cause the loss of this calibration.

Ian Smith also showed advantages in using a three dimensional soil model (really two dimensional due to symmetry). Such an approach is attractive since soil constitutive properties can be used as input in the analysis and a link is established between the subsurface investigation information and the pile driving analysis soil parameters. Considerable concern has been expressed about "radiation" damping and its absence from the traditional E.A.L. Smith soil model.

The problems with the use of a finite element model are as follows:

1. The model still contains an interface element that must be crudely described.
2. Constitutive properties are not really that easily obtained from commonly available subsurface investigation information.
3. Much larger running times will be required in the analysis.

Still the approach has the tremendous advantage of rationality and its use will probably expand.

Holloway (9) showed the importance of residual stresses in wave equation analysis. His primary concern was to show that residual stresses can substantially affect apparent soil resistance distribution. They also have a strong influence on the bearing graph particularly in the high blow count region. Based on the results of Holloway's work and on Reference (12) it is clear that all driveability analyses must include residual stress analysis capabilities.

CONCLUSIONS

Wave Equation analysis presents the only means currently available to estimate offshore pile driveability. Pile axial capacity is also estimated. When these techniques are combined with dynamic measurments, the capacity estimations are substantially improved and driving system performance can also be evaluated.

Much could be done to improve these analyses.

1. Dynamic analysis integration procedures should be studied to select improved procedures and determine the consequence of changing procedure on the results due to "calibration" of input data in current programs.

2. Dynamic analysis incorporating a finite element soil model should be investigated and correlated with field measurements.

3. Measured driving system parameters must be collected to determine mean values for particular equipment together with dispersion. This information would make it possible to estimate the reliability of a Wave Equation analysis.

4. Wave Equation analysis must include a residual stress analysis.

Measurements combined with CAPWAP analysis are beginning to be used to evaluate pile capacity. this apporach is particularly useful if combined with restrike measurements and a careful soils analysis.

REFERENCES

1. Authier, J and Fellenius, B.H., "Quake Values Determined from Dynamic Measurements," Proceedings of the International Seminar on the Application of Stress-Waves Theory on Piles, Stockholm, June 1980, Edited by H. Bredenberg, A.A. Balkema, Rotterdam, 1981.
2. Dolvin J. and Poskitt, T.J., "An Optimization Method for Pile Driving Analysis," Proceedings, Second International Conference on Numerical Methods in Offshore Piling, Austin, Texas, April, 1982.
3. Goble, G.G. and Rausche, F., "Wave Equation Analysis of Pile Driving - WEAP Program," Volumes I, II, and III, FHWA Report No. FHWA-IP-76-14.1, Goble and Associates, 4423 Emery Industrial Parkway, Warrensville Heights, Ohio, 44128, 1976, Updated 1980.
4. Goble, G.G. and Rausche, F., "Pile Driveability Predictions by CAPWAP," Proceedings of the Conference on Numerical Methods in Offshore Piling, May 1979, Institution of Civil Engineers, London.
5. Goble, G.G., Scanlan, R.H. and Tomko, J.J., "Dynamic Studies on the Bearing Capacity of Piles," Vol. I and II, Case Institute of Technology, 1967.
6. Hery, P., "Residual Stress Analysis in WEAP," Masters Thesis, Dept. of Civil, Environmental and Architectural Engineering, University of Colorado, Boulder, May 1983.
7. Hirsch, T.J., Carr, L., Lowry, L.L., "Pile Driving Analysis - Wave Equation User Manual TTI Program," Volume I, II and III, FHWA Report No. FHWA-IP-76-13.1, Texas Transportation Institute, Texas A. & M. University, College Station, Texas 77840, April 1976.
8. Hirsch, T.J., Koehler, A.M. and Sutton, V.J.R, "Selection of Pile Driving Equipment and Field Evaluation of Pile Bearing Capacity during Driving for the North Sea Forties Field," Proceedings, Seventh Annual Offshore Technology Conference, Paper No. 2247, May 1975.
9. Holloway, D.M., Clough, G.W. and Vesic, A.S., "The Effect of Residual Driving Stresses on Pile Performance Under Axial Loads," OTC 3306, May 1978.

10. Likins, G.E., "Evaluating the Performance of Pile Driving Hammers," Informal Presentation, Pile Driving Analyzer, Fourth Annual User's Day, Amsterdam, May 1982.
11. Michigan State Highway Commission, "A Performance Investigation of Pile Driving Hammers and Piles," Final Report, Lansing, Michigan, March 1965.
12. Rausche, F., "Comparison of Accuracy of Integration Techniques in Wave Equation Analysis, "Informal Presentation, Pile Driving Analyzer, Fourth Annual User's Day, Amsterdam, May 1982.
13. Rausche, F., Moses, F. and Goble, G.G., "Soil Resistance Predictions from Pile Dynamics," Journal of the Soil Mechanics and Foundations Division, ASCE, Vol. 98, No. SM9, Proc. Paper 9220, September 1972.
14. Samsom, C.H., Hirsch, T.J. and Lowry, L.L., "Computer Study of the Dynamic Behavior of Piling," Journal of the Structural Division, ASCE, Paper No. 3608, ST4, August 1963.
15. Schaap and Ims, "Fugro Impact Velocity Measurement System," Informal Presentation, Pile Driving Analyzer Fourth Annual User's Day, Amsterdam, May 1982.
16. Smith, E.A.L., "Pile Driving Analysis by Wave Equation," Journal of Soil Mechanics and Foundations, ASCE, No. 86, August 1960.
17. Smith, I.M. and Chow, Y.K., "Three Dimensional Analysis of Pile Driveability," Proceedings, Second International Conference on Numerical Methods in Offshore Piling, Austin, Texas, April, 1982.

INSTRUMENTATION AND PERFORMANCE OBSERVATIONS OFFSHORE

by

Elmo DiBiagio* and Kaare Høeg*, M.ASCE

INTRODUCTION

This paper illustrates how geotechnical and structural instrumentation and full-scale monitoring programs can be used to advantage for offshore engineering projects and practice. This use of instrumentation is in a way an extension of the "observational method", a procedure which foundation engineers traditionally rely on in the persuit of their work. Instrumentation provides the engineer with observational capabilities that greatly surpass his own powers of observation, and when properly used has an almost unlimited potential for increasing our ability to cope with our engineering tasks.

Offshore engineering spans over many disciplines and covers a vast array of activities; thus, it is not surprising that offshore instrumentation applications are also varied in scope and purpose.

Instrumentation is, for instance, playing an increasingly important role in improving the effectiveness and quality of offshore site investigations. Improved in-situ testing devices have been made possible through the use of modern instrumentation methods and components. Self-contained instrument packages set on the sea floor make it possible to obtain records of important design parameters such as earthquake motions, variations in wave pressure on the sea floor, in-situ pore pressures, and movements in submarine slopes at locations far from shore.

Monitoring programs are used to install and evaluate the performance of offshore structures, and instrumentation of this kind provides a basis for improved design and contributes significantly to advances in the state of the art in offshore engineering.

Equally important are those cases where instrumentation is needed for special purposes; to aid in the solution of an unexpected problem that has arisen, for research, or to test a new construction procedure.

* Norwegian Geotechnical Institute, Oslo, Norway

MOTIVATION AND NEED FOR OFFSHORE INSTRUMENTATION

The motivation for instrumenting offshore structures varies from case to case depending on the type of structure, previous experience, uncertainties in design, size, environmental conditions and in some cases operational requirements. Experience shows that the amount of instrumentation included in monitoring programs for offshore structures is not necessarily in direct proportion to the size and complexity of the project. This point is illustrated vividly in Fig. 1 for two structures for which remarkably different amounts of instrumentation have been proposed.

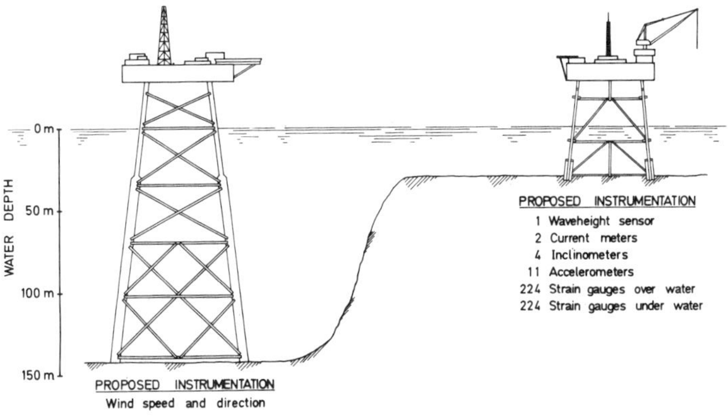

Fig. 1. Proposed instrumentation for two offshore structures.

A number of the more obvious reasons for instrumenting fixed offshore structures are pointed out below.

Instrumentation for installation of structures

For most fixed structures some instrumentation will be needed to monitor the placement and installation of the structure on site. Equipment for precise positioning of the structures or for monitoring tensions in tow and mooring lines are examples. A good illustration of a structure that will require exceptionally accurate positioning and control during installation is Phillips Petroleum Co.'s Maureen platform (8) which is scheduled for installation in the North Sea in 1983. The 120,800 ton, 263 m high steel gravity base structure will be towed to the Maureen Field and positioned to within 15 cm of predrilled wells under 96 m of water.

For jack-up rigs improvements in safety as well as speed of installation can be obtained by monitoring leg loads and penetration while the rig is being jacked up or down (1).

Instrumentation and monitoring programs assist in bringing a gravity structure from a floating position onto the sea floor without damaging the structure or disturbing the foundation soil (4). A number of specialized instrumentation systems provide the installation crew with information needed to assess and fully control the installation of the structure.

Dynamic measurements and the use of pile driving analyses assist in the drivability predictions for piles and provides installation control. Furthermore, measured dynamic pile penetration resistance during the last stage of driving provides an indication of the axial pile capacity.

Structures that are fabricated in parts and assembled offshore may require extensive instrumentation in order to mate the parts properly. This was the case for the three sections of the Cognac platform which was assembled in 312 m water depth (11).

The installation and use of instrumentation of the kind described above is just a small part of a complicated construction procedure that must be undertaken before the structure can be put to its intended use; thus, there will generally be little objection from the contractors or platform owners to providing instrumentation for this purpose.

Instrumentation for operation of the structure

Once a structure is installed and put into service, instrumentation may be a useful aid for proper operation of the facility. Environmental data, for example, is needed for air traffic control and planning ship movements. Strain gauges may be used as a means of deck weight control. Measurements of pitch and roll, hawser pull and forces in loading hoses may be needed for safe operation of offshore loading buoys (9). Likewise, measurements of deck displacements and inclination of drilling risers, or movements of pipeline and conductors may have direct bearing on operational matters.

On concrete production platforms temperature measurements may be needed for operation of cooling systems used to control the temperature of the concrete around embedded well conductors or in the walls of storage cells.

Many gravity base structures are equipped with under-base drainage systems which are used to improve the stability of the structures (7). Optimum operation of the drains is assured if flow rates and pore water pressure in the foundation soil are monitored.

Instrumentation that can be shown to be useful from the operational standpoint will normally be given high priority by the platform operator.

Instrumentation for performance monitoring

Three fundamental reasons for monitoring the performance of any structure, whether onshore or offshore, are: (a) to confirm that the behaviour of the structure is in accordance with predictions, (b) to be more knowledgeable of any discrepancy in performance that is found and thereby be in a better position to take corrective action, if this is possible, and (c) to improve design procedures and practice for future structures.

Monitoring programs are particularly important in offshore engineering because of uncertainties in the prediction of environmental loads, in the complex design analyses, and in foundation design parameters. The latter point is a consequence of the general difficulty in carrying out in-situ tests and in obtaining good quality undisturbed soil samples from offshore sites. Examples of load-related measurements are measurement of hydrodynamic pressure or forces in selected members or nodes of a steel structure, and measurement of cyclic stress levels needed for fatigue analyses. Examples of foundation-related measurements for a gravity structure are measurement of settlement, pore water pressure and cyclic displacements at mudline, Fig. 2. For a piled structure measurement of reaction load on pile groups and load transfer along the piles are of basic interest.

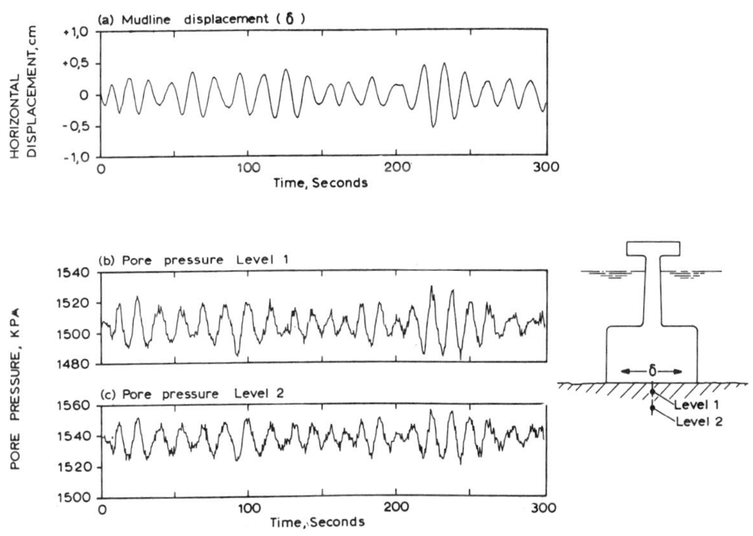

Fig. 2. Measured cyclic variations in pore pressure and horizontal displacment at mudline for a North Sea gravity structure.

Performance observations do not always require instrumentation, of course. For example, monitoring programs may include periodic visual inspections by divers or remote controlled vehicles to detect flaws and damage or to check for scour around structures.

The owner does not usually regard performance monitoring instrumentation with the same sense of urgency that is associated with instrumentation needed for operational reasons or to monitor installation of structures. The reason is that it is difficult to establish a direct link between the purpose for which the instrumentation is installed and the primary function of the structure - to produce hydrocarbons.

Instrumentation for certification and recertification

One important factor which influences decisions to provide performance monitoring programs or not is the attitude of regulatory agencies and classification companies. If they require field measurement data for proof of satisfactory performance for certification or recertification of a structure, instrumentation will be provided. On the other hand if none is required, very often none will be provided.

Instrumentation for trouble shooting

In the real engineering world things do not always work out as they are supposed to do in spite of the best of plans. When something does go wrong or an unexpected event occurs, instrumentation can be helpful to assess damage, for planning remedial action or as an aid in avoiding a reoccurrence of the trouble. There are a number of instances where instrumentation has been used for this purpose offshore, for instance after a ship collision; however, one generally hears little about these applications.

Instrumentation for research and development

There is almost universal agreement on the usefulness of instrumentation for carrying out research on full-scale structures; however, the dividing line between research instrumentation and routine or control instrumentation is not always easily established. Whether a proposed instrumentation program for an offshore structure constitutes research or not is a question that will usually be debated, sometimes quite energetically. Nevertheless, programs have been and are being implemented on offshore structures, and these are contributing significantly to advances in the state of the art of offshore engineering. Projects of this kind range from small ones involving only a few instruments and little cost to massive Industry sponsored projects such as the Brent B Research Project (12), Fig. 3, and the foundation monitoring project for the Magnus structure (3) which is mentioned in Section 3.

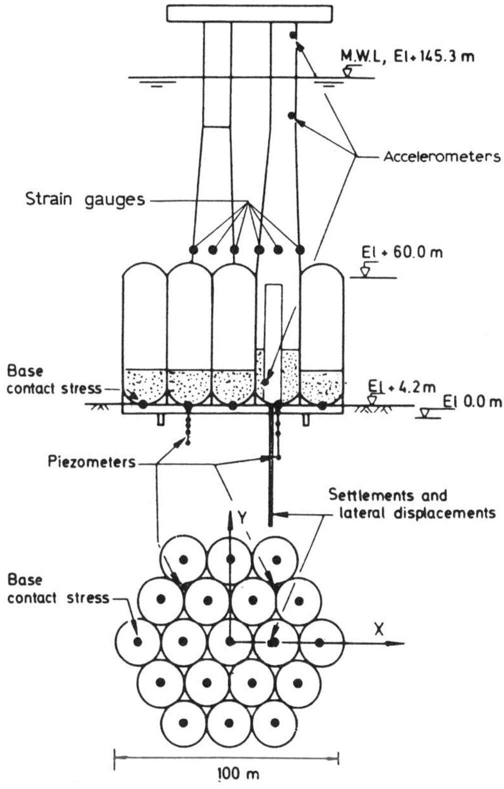

Fig. 3. Geometry of the Shell/Esso Brent B structure and location of sensors monitored during the two-year research project.

EXAMPLES OF OFFSHORE INSTRUMENTATION APPLICATIONS

In order to illustrate offshore instrumentation concepts and procedures, three typical applications will be presented; a gravity base structure, a steel piled structure and an offshore loading buoy.

Instrumentation for gravity structures

This example illustrates a case where instrumentation is essential for proper installation of an offshore structure as well as to document the long-term behaviour of the structure and its foundation.

Mobil's Statfjord B gravity platform shown in Fig. 4, the largest structure built to date, has a base area of 18,000 sq. m and a net weight of 800,000 tons. The deck alone is 8 stories high and weighs 43,000 tons.

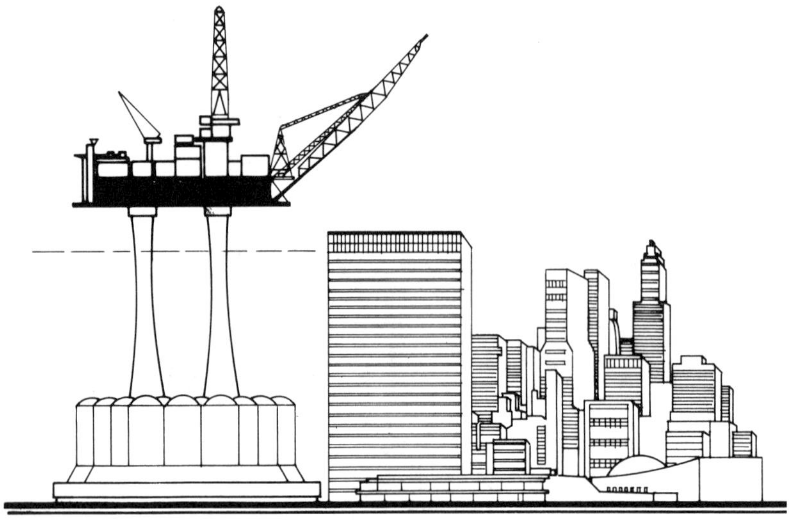

Fig. 4. Statfjord B gravity structure compared to United Nations building and Manhattan sky line.

Gravity structures are built in protected waters near shore, and then towed to the offshore location and set on the sea floor in the course of a few days. From the foundation point of view the load transfer is almost instantaneous. To bring a gravity structure from a floating position safely on to the sea floor without overstressing the structure or damaging the foundation soil should not be done without the aid of instrumentation. Consequently these structures are equipped with a variety of geotechnical and structural instrumentation systems specifically intended to aid in the physical installation of the structure on site (4). Other systems are installed for long-term performance monitoring (2,4,5).

Generally speaking, a comparatively larger number of instruments and instrumentation systems are used to monitor installation than to document long-term performance. As many as 10 or 12 different systems may be used to monitor the installation phase, whereas only 4 or 5 may be required to observe the key parameters needed for performance monitoring programs. Usually, however, the long-term monitoring systems are more extensive in that some of the instruments used primarily for the installation phase are also included in the long-term program as these are already installed on the structure and can be monitored at little extra cost.

Table I summarizes the number and types of instruments that were used for the installation of Statfjord B.

TABLE I

SUMMARY OF INSTRUMENTATION FOR MONITORING INSTALLATION
MOBIL - STATFJORD B CONDEEP PLATFORM

Measurement	Number of instruments	Type
Wave height	1	Waverider buoy
Wind speed and direction	1	Anemometer
Clear tow path	3	Sidescan sonar
Under keel clearance	4	Echosounder
Draught	2	Pressure sensor
Touchdown	4+[4]	Pressure sensor
Penetration depth	5	Special device
Water pressure under base	10+[10]	Pressure sensor
Inclination	2	Servo-inclinometer
Ballast water level	43	Pressure sensor
Short-term-settlement	1+[1]	Pressure sensor
TOTAL	76+[15]	

Note:

Numbers in brackets [ ] denote redundant instruments.

The relative location of and different types of instruments incorporated in the performance monitoring program for Statfjord B are shown in Fig. 5. These are summarized in Table II below.

The strain gauges at the top and bottom of one tower are used to determine the forces and moments that are caused by environmental loads. The strain gauges on the inclined braces around the base are used to obtain information regarding cyclic forces on the caisson during storm periods. The accelerometers are used to measure accelerations which are processed to find resonant frequency of vibration, mode shapes and cyclic displacements of the deck and base. Graduated scales for visual observations of scour are mounted at 6 locations on the periphery of the base.

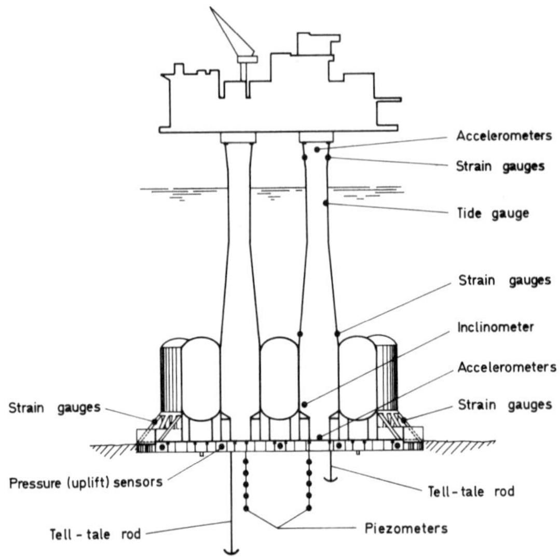

Fig. 5. Instrumentation for monitoring the long-term performance of Statfjord B.

TABLE II

SUMMARY OF PERFORMANCE MONITORING INSTRUMENTATION
MOBIL - STATFJORD B CONDEEP PLATFORM

Measurement	Number of instruments	Type
Wind speed and direction	1	Anemometer
Inclination	2	Servo inclinometer
Long-term settlement	2	Tell tale rod
Water pressure under base	10+[10]	Pressure sensor
Pore water pressure	10+[10]	Pressure sensor
Tidal variations	1+[1]	Pressure sensor
Strain top/bottom tower	16+[16]	Strain gauges
Strain in inclined braces	14+[14]	Strain gauges
Linear acceleration	4	Servo-accelerometer
Angular acceleration	5	Servo-accelerometer
TOTAL	65+[51]	

Note:
Numbers in brackets [ ] denote redundant instruments.
Wave data is obtained from a neighbouring platform.

Instrumentation for a piled steel structure

Some instrumentation is needed to monitor positioning, ballasting and up-ending operations during the installation of steel jackets and templets. Furthermore, instrumentation is used to perform dynamic measurements during pile installation, as pointed out in Section 2.

The instrumentation systems that have been useful for monitoring the performance of steel structures can be grouped into two main categories: those which are used to obtain information related to applied loads and stresses in the structure, and those whose function is to monitor the displacements and motions caused by the applied loads. The types of measurements that are of general interest are illustrated in Fig. 6.

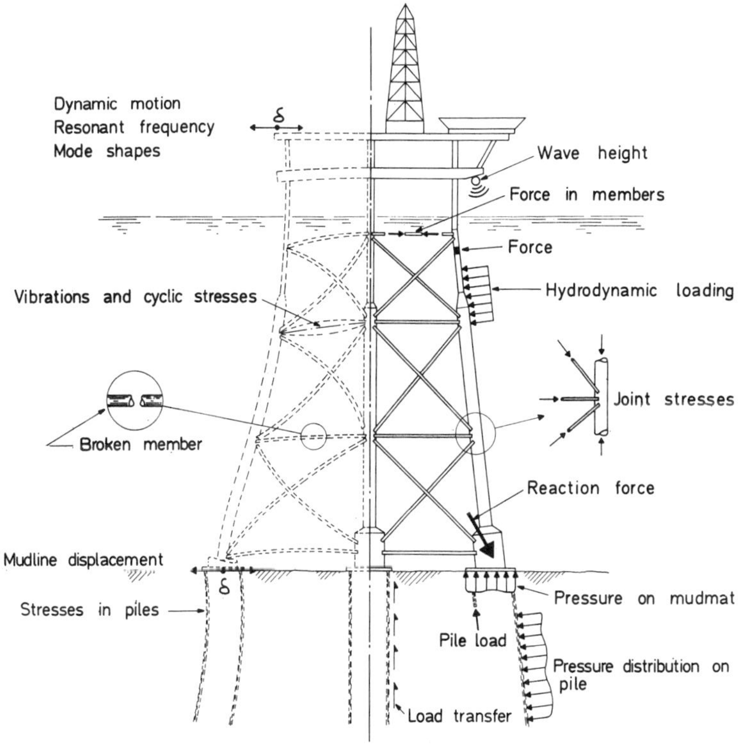

Fig. 6. Types of field measurement of interest for steel piled structures.

Wave heights can be monitored by means of waverider buoys or with wave staffs and radar or infra-red devices mounted on the structure. Measurement of hydrodynamic pressure on the structure by means of pressure sensors is seldom done because of the difficulty and high costs involved. Hydrodynamic loads on an isolated part of the structure are occasionally monitored, however, by means of force collars that measure the total resultant force on an isolated element of the structure. Strain gauges are commonly used to monitor the total loads in individual members or nodes. An alternate method of obtaining information regarding wave loading is to install a dummy member, known as a wave boom, which is specially instrumented with strain gauges, load cells or pressure transducers.

Field measurements needed to evaluate the fatigue life of the structure can be obtained by strain-gauging selected members and recording the number of stress cycles that exceed specified threshold values for various sea states. This information together with statistical wave data can then be used to estimate the total number of stress cycles that will occur during the life of the structure. These measurements will generally be made on members for which the design or anticipated stresses are known with confidence. The field measurements from this member can then be used to estimate cyclic stress levels for other more complex members.

A number of instrumentation techniques are being used to detect flaws or defects such as cracked or severed members in steel structures. One method is based on vibration measurements wherein changes in natural periods of vibration and mode shapes are used to detect changes in integrity of the structure (10). Another technique used for this purpose is acoustic emission (6).

Probably the most difficult type of measurement program to carry out on a structure of this kind is one intended to monitor the performance of driven foundation piles. Unfortunately, because of this, our design analyses for offshore piles contain significant uncertainties. To illustrate the complexity of this type of monitoring program the various types of instruments and measurements that have been included in the foundation monitoring program for one leg of the Magnus platform (3) are summarized below. Naturally, a pile monitoring program does not have to include measurement of all the parameters listed below to provide useful information. For the Magnus project:

- one leg is thoroughly instrumented with strain gauges to determine the time history of the load acting on the pile group resulting from environmental forces,

- earth pressure transducers are fitted to the mud mat to determine the proportions of the load carried by the mud mat and pile group respectively,

- individual piles are fitted with strain gauges to measure the distribution of load within the pile group,

- strain gauges are installed at several cross sections along the length of one pile to determine load transfer with depth,

- pressure transducers are installed at several levels on one pile to measure pore water pressure and earth pressure distribution, and

- settlement gauges and accelerometers are installed to monitor the short- and long-term vertical settlement and movements of one leg.

Unfortunately it will not be possible to carry out the whole measurement program described above. During up-ending of the Magnus platform several of the main instrumented piles became displaced in their guides and were lost. However, new piles were instrumented, but because of time limitations it was not possible to reinstall the same amount of instrumentation. The accident with the Magnus piles is a somber reminder of the unexpected difficulties and disappointments that can occur in instrumentation work. Hopefully this will not discourage similar monitoring programs in the future.

Instrumentation for an offshore loading buoy

This example illustrates a case where instrumentation is used predominately for operational reasons and no geotechnical instrumentation is

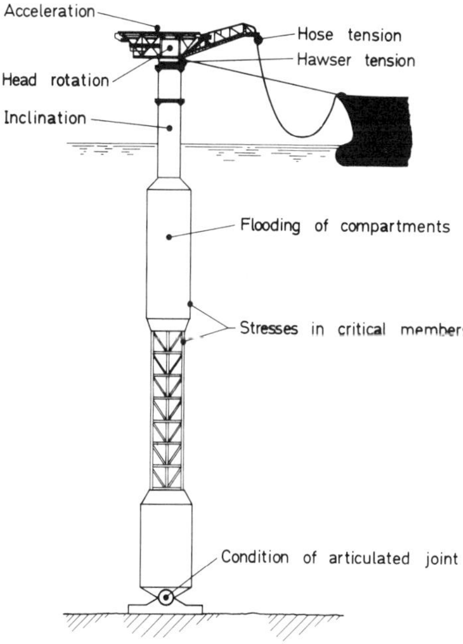

Fig. 7. Measurement program for an articulated loading buoy.

included. Fig. 7 illustrates the principal features of an articulated loading buoy of the type installed in the North Sea. The various types of instrumentation that can be useful on structures of this kind are indicated on the figure. It should be pointed out, however, that the example given is not modelled after any one particular structure. It has been presented only to illustrate commonly used instrumentation.

Most of the instrumentation indicated in the figure have obviously useful operational functions. Measurements of wind speed, forces in the loading hose and hawser tension are needed for safe operation of the facility. Furthermore, measurements of pitch and roll and measurements of the absolute orientation of the rotating top part of the buoy are also valuable for operational reasons. The need to monitor acceleration of the helideck has been proposed on some structures as a requirement for safe operation of helicopters, for example to confirm that aircraft can land without overstressing landing gear.

The need for specialized instrumentation to verify the integrity and operational status of the swivel joint at the base can be justified on several grounds. Similar arguments can be stated for measurements of stresses in critical parts of the buoy. It is a fact that instrumentation of this kind has been used partly as a preventative measure to avoid a reoccurrence of operational problems that have been experienced on loading buoys, though not necessarily on articulated buoys however.

COMPARISON BETWEEN ONSHORE AND OFFSHORE INSTRUMENTATION

It may be informative to point out some of the principal differences between onshore and offshore instrumentation projects.

Operating conditions

Construction of offshore structures is a complicated procedure which generally takes place over a long period of time. In some cases the structures are fabricated in sections at one or more locations and then subsequently assembled at an offshore site. At the offshore site it is not feasible to install foundation monitoring instruments beforehand as can be done on most onshore projects. Because of the size and complexity of these structures, installation of instruments is difficult and certainly far more expensive than for conventional structures onshore. Cable installation and handling procedures on many structures are extremely difficult and risky.

Instrument requirements

Instruments that are to be used offshore must operate under conditions that are generally far more severe than those encountered on most conventional projects. The principal difference - and the most worrisome one - is that the instruments must operate in a seawater environment, often under very high ambient pressure. In addition, instruments must often operate over long periods of time and in many cases service or replacement is impossible. The need for redundancy is evident.

The operating conditions for offshore instruments limit significantly the types of materials that can be used and put stringent demands on waterproofing and sealing techniques. Long cable runs, often in a noisy electrical environment, influence the selection of the preferred type of output signal from the instruments. Galvanic isolation of individual measurement loops in many instances is advisable and necessary to prevent one defective instrument from influencing unfavorably the performance of others. Furthermore, instruments that are installed in hazardous areas must comply with stringent electrical certification codes.

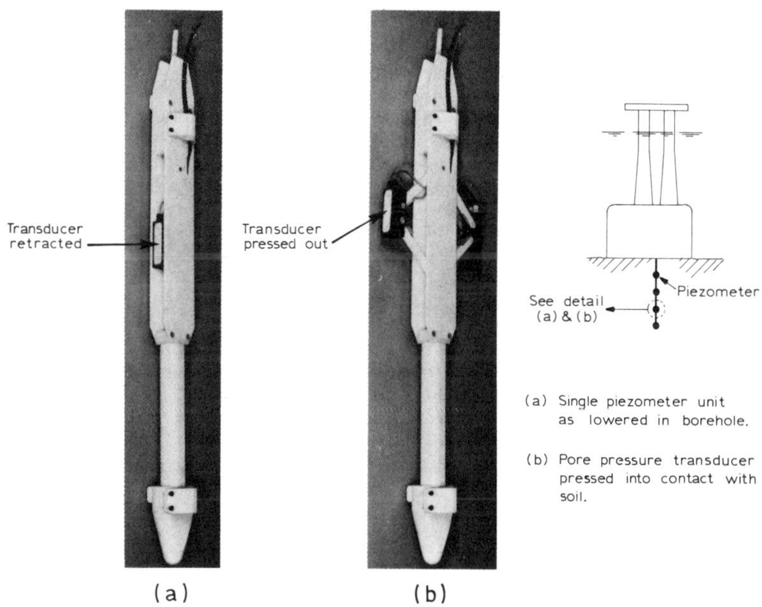

Fig. 8. Photograph of a special offshore piezometer unit. Several such units are linked together to form a string of instruments which is installed in a single borehole.

Very few standard instruments will meet all the requirements mentioned above; thus, specially designed instruments are commonplace in offshore applications. See Fig. 8 for example. This of course contributes to the high cost of implementing these instrumentation projects. Most instruments used for geotechnical and structural monitoring projects are based on resistance strain gauges, vibrating-wire sensors, linear differential transformers, and force-balance servo devices such as inclinometers and accelerometers. The advantages of having instruments with frequency output signals and the need for long-term stability are clearly manifested in the widespread use of vibrating-wire type instruments.

Data acquisition systems

For onshore structures the time at which the maximum loading occurs is usually known reasonably well; thus, monitoring programs and recording equipment can be designed with this in mind. Marine structures are dimensioned on the basis of a statistical design storm, and one does not know when the design storm, or in fact any major storm, will occur. This is similar to the earthquake instrumentation and monitoring for onshore structures. Consequently the data acquisition system must be designed to assure that the most critical data will be recorded. For this reason offshore recording systems must record continuously or be triggered automatically when specified threshold values, such as wave height or acceleration level, are detected. Automatic recording systems are necessary in some cases because structures are unmanned or simply because platform operating personnel do not have time.

For many conventional instrumentation projects it may be adequate to take readings on a daily, weekly or monthly basis. However, because much of the offshore data is of a dynamic nature the data acquisition systems must scan and record data at very high rates, for example up to 10 times per second for accelerometers used to monitor dynamic motions.

The principal types of data acquisition systems currently used offshore are computer based systems with recording on digital magnetic tape and PCM (pulse code modulation) systems with recordings in digital form on analogue tape. Not only do these recording systems have to cope automatically with large volumes of data, they must do this in the offshore environment which includes poor quality mains supplies and numerous power failures. The performance record for data acquisition systems in the offshore environment is not always good.

Data analysis

Analysis of offshore performance data is by far more complicated and time consuming than for conventional instrumentation projects. Many individuals have underestimated the differences in costs and time required to process and report offshore data. This is partly because of the large volumes of data involved and partly because of special processing and analysis techniques that are needed for some of the data, for example to determine displacements from measured accelerations, Fig. 9.

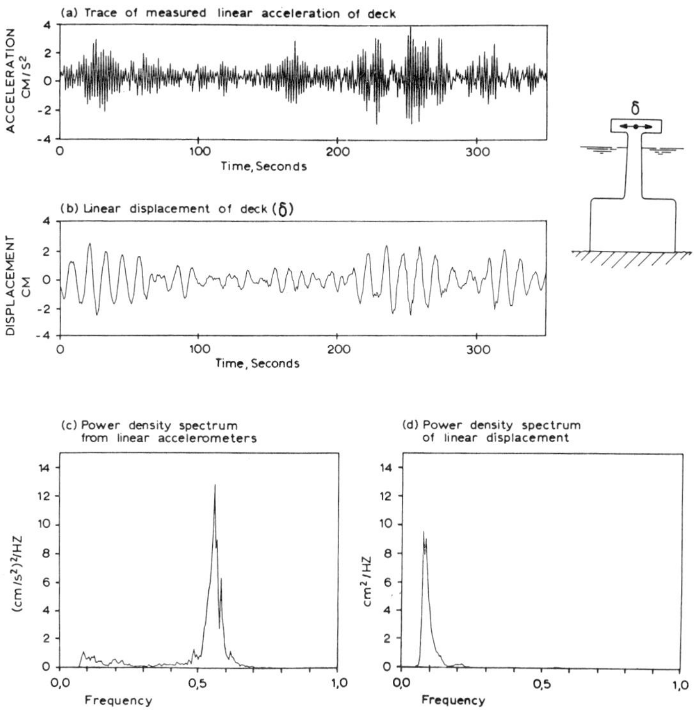

Fig. 9. Traces of measured accelerations and samples of results obtained from analysis of accelerometer data for a North Sea gravity structure.

In order to reduce the amount of work involved in searching through and processing large amounts of data onshore, some data acquisition systems are programed to perform data processing tasks offshore, such as reduction of data to statistical parameters and generation of statistical reports offshore, when the computer is not engaged in other activities. This concept was used for the data acquisition system used on the Brent B Research Project (12) and it is currently being used on several other structures.

PERFORMANCE OF INSTRUMENTS IN THE OFFSHORE ENVIRONMENT

On most offshore projects, instrumentation work has to be integrated into a complicated construction sequence, and the systems have to be installed in bits and pieces simultaneously with welding, burning, concreting and other construction activities potentially hazardous to delicate instrumentation system components. In the case of gravity base structures a large number of instruments have to be mounted during the early stages of construction, and many of the sensors and a major portion of the platform cabling are submerged in salt water under relatively high pressure for perhaps two or three years before the structure is installed and the instruments are put into use for the first time. Because of the adverse working conditions it is to be expected that instruments will fail and some certainly do. For this reason the key factor to be considered when planning an offshore instrumentation project is to provide sufficient redundancy so that some instrument failure will not ruin the project.

When an instrumentation system does not operate satisfactorily or fails completely, it is not always easy to ascertain the cause of the malfunction, because instruments are often unaccessible or cannot be recovered for study. Because transducers are vulnerable to damage there is a tendency to assume that defective transducers and system failures are synonymous. This may be the case, of course, but a transducer that appears to be defective may simply have gone off-scale and thereby cease to operate through no fault of its own. Furthermore, a system failure may just as well be a result of a fault in one of the other components in the system, such as submerged cables and junction boxes, or the result of damage caused by an external event or accident, often undetected.

In spite of the preceeding statements, experience shows that with the use of modern instrumentation technology it is possible to successfully implement complex instrumentation programs in the offshore environment. In instrumentation work, probably more so than in many other fields of applied technology, improvements in the state of the art and increased knowledge as to how to avoid pitfalls are derived primarily from past experience. Thus, it is natural to expect that the overall reliability of the instrumentation systems on the more recently instrumented offshore structures should be better than for earlier ones. Accordingly, any appraisal of instrument reliability has to be qualified as to how and when the project was carried out.

The above point is clearly manifested by the authors' experience at the Norwegian Geotechnical Institute, (NGI). The first geotechnical and structural instruments installed by NGI on concrete gravity structures took place in early 1973. When the first structure was installed on site in mid-1975, of a total of 102 instruments approximately 10 percent of the instruments did not operate satisfactorily. Since then, for each successive structure the performance record for the instrumentation systems has in general improved steadily. Some 5 years and 2000 instruments later in August 1981 when the last gravity structure was installed (Statfjord B, Fig. 4), the performance record was 100 percent even though some instruments which were not accessible for service or repair had been installed 44 months earlier.

In broad terms it can be stated that the instruments used in the long-term monitoring programs for the gravity structures have performed well. A general observation is that the vast majority of the instruments that were operating properly when the structures were installed are still operating.

Even though there have been some instrument failures, this has not adversely affected the monitoring programs, because all the critical data has been obtained through redundancy in the form of duplicate instruments and independent back up systems.

EXAMPLES OF INCREASED KNOWLEDGE THROUGH INSTRUMENTATION

The role that monitoring programs have played in confirming and improving our understandings of the installation and behaviour of gravity base structures in the North Sea illustrates the value that instrumentation can have in offshore engineering.

The introduction of the gravity base platform concept in the early 1970's brought with it a multitude of novel design problems. There was little precedence on which to base the design of these new structures. Almost everything was new - structural concepts, materials, foundation concepts, methods of construction and installation, and the severe environmental conditions that are characteristic of the North Sea.

The situation is significantly different today. Within a decade many of the original uncertainties associated with the design of the gravity base structures are no longer points of concern. Much of this evolution can be credited to the extensive monitoring programs that have been carried out on these structures. Some of the principal results and findings of the instrumentation programs are listed below.

- It has proven possible to install gravity structures safely without causing damage to the structure or the foundation soil. Fourteen gravity structures have been installed successfuly without mishap or unexpected complications. In all cases information derived from instrumentation has been used as a basis for decision making by those responsible for installation of the structures.

- The penetration resistance of the foundation skirts under the platform base may be predicted based on cone penetration tests during the site investigation. This has been confirmed by field observations during the installation of the structures.

- Measured long-term settlements and rate of consolidation have been compared with theoretical predictions. Theory and practice agree reasonably well for magnitudes but the measured field rate is faster than theoretically predicted.

- The combination of creep settlements due to sustained loads and settlements due to cyclic loading, amounts to approximately 3 to 18 mm/year. This field information is very important as the uncertainties associated with the related theoretical analyses are very great.

- Long-term differential settlements have been small and of no practical significance.

- Foundation pore water pressure build-up and dissipation during and after storm periods has been confirmed. As predicted the magnitude of the observed excess pore pressure has been small in the stiff North Sea deposits.

- Redistribution of base contact stresses with time and due to cyclic loading appears to be of a minor magnitude.

- Long-term horizontal displacements due to preferential direction of environmental loading appear to be small and of no practical significance.

- Dynamic motion of those structures for which a thorough analysis of the recorded accelerometer signals has been carried out, confirm that the actual behaviour of the structures and foundation is in general agreement with predictions made using current design methods. This applies to the magnitude of cyclic motions as well as the natural periods of the system. Measurements confirm that the resonant frequencies lie above the frequency band where the bulk of the wave energy occurs.

- Strain gauge instrumentation in the concrete has been used to confirm acceptable stress levels and to determine creep rates for concrete subjected to cyclic loading.

- Strain gauge measurements in the towers of several structures have provided a basis for determining environmental loading on the towers.

- Only one instance of significant scour has been observed around the base of a gravity structure. In this case the scour was local and concentrated at one corner of the square-shaped base.

The profession has learned a great deal about the behaviour of gravity platforms from field performance observations during a relatively short period of time. This increased know-how makes us better equipped to proceed with the design of new structures in much deeper water.

CONCLUSIONS AND FUTURE TRENDS

Instrumentation is essential to many of the engineering tasks to be performed in the offshore environment. Performance monitoring may for instance be required during platform installation, as control during operation, and to improve design procedures and practice for future structures.

Reliability estimates may be updated based on field observations, and this evaluated experience makes it posssible to proceed to more difficult and challenging offshore undertakings. Performance monitoring is the only way to check the results from model tests and theoretical analyses, and herein lies the promise for real improvements in the state of the art.

The knowledge gained, during a relatively short period of time, about the behaviour of gravity platforms subjected to severe storms, clearly demonstrates the merits of field instrumentation programs. The recorded performance observations will assist significantly when gravity structures are to be designed for offshore areas with much deeper water and much softer foundation soils than those encountered at the locations of present gravity structures.

The problems facing offshore engineers and instrumentation engineers in the future will be no less challenging and exciting than they are today:

- There should be an increased interest in testing offshore piles. Despite the long history and extensive use of pile supported platforms, the profession still does not have any significant measurements of pile behaviour under a structure during storm conditions. With increasing water depths, the pile foundations will represent a larger percentage of the total platform costs, and more emphasis will be placed on pile installation techniques, construction monitoring and control.

- The dynamic and fatigue behaviour of deepwater platforms needs to be studied with the aid of field instrumentation. Field measurements of soil-structure interaction and damping will be required, as fixed piled and multi-base gravity structures are designed for more than 300 m of water.

- The increased interest in jack-up rigs and the trend to use these rigs as more permanent structures, should lead to an increased use of instrumentation for monitoring installation and performance as well as removal. The need for this is evident in view of the relatively poor safety record for jack-up rigs in the past.

- Instrumentation will be used to investigate mudslides and their effects. Progress in this subject may ultimately require offshore field tests with blasting to trigger local slides to evaluate seafloor stability, run-out distance of sliding masses, and forces imposed by the mudflow on platform foundations and pipelines.

- Site investigations will continue to benefit from improved instrumentation technology. Instruments will increase the reliability of in-situ testing procedures as well as sampling equipment and techniques. Field measurements of in-situ pore water pressures in "underconsolidated" deposits are needed.

It is to be expected that our ability to implement instrumentation projects in the future will be significantly improved because of technological advances in other fields. Improved alloys and manufacturing techniques will provide better instruments for use in the offshore environment. Developments in computer technology will improve the reliability of data acquisition systems and simplify data processing tasks. In some cases the use of satelite communication links to transmit data to shore will help to reduce the turn around time needed to report on the behaviour of offshore structures.

REFERENCES

1. Ansquer, P. F. and Antalovsky, S. A. (1982). "Rig jacking on soft soil foundations: Improvements in safety and speed by monitoring leg loads". Offshore Technology Conference, 14. Houston, Texas 1982. Proceedings, Vol. 4, pp. 357-362.

2. Bourgeois, D. and Leroy, J. P. (1976) "Geophysical and mechanical platform instrumentation". Conference (on) Design and Construction of Offshore Structures". Institution of Civil Engineers, London, 1976. Proceedings, pp. 9-15.

3. Civil Engineering, London. "Magnus - a pioneering venture". Civil Engineering. November 1982.

4. DiBiagio, E., Myrvoll, F. and Hansen, S. B., (1976): "Instrumentation of gravity platforms for performance observations". International Conference on Behaviour of Off-shore Structures, 1. BOSS '76. Trondheim 1976. Proceedings, Vol. 1. pp. 516-27

5. DiBiagio, E., Myrvoll, F. and Hansen, S. B., (1979): "How successful have performance monitoring programmes been for gravity base structures?". International conference on Behaviour of Off-shore Structures, 2. BOSS '79. London 1979. Proceedings, Vol. 3, pp. 709-727.

6. Dumousseau, P. F., Laffont, P. and Thebault, M. (1979) "Experimental study of acoustic emission monitoring of crack propagation in offshore steel tubular joint". Offshore Technology Conference, 11. Houston, Texas 1979. Proceedings, Vol. 1, pp. 593-596.

7. Eide, O., Andresen, A., Jonsrud, R. and Andenaes, E. (1982) "Reduction of pore water pressure beneath concrete gravity platforms". International Conference on Behaviour of Off-shore Structures, 3. BOSS '82. Cambridge, Mass 1982. Proceedings, Vol. 2. pp. 373-382.

8. Engineering News Record (1982). "Platform 'first' edges ahead". Engineering News Record, November 1982, p. 32.

9. Hays, D., McSwiggan, M. and Vilain, R. (1979) "Operation of an articulated oil loading column at the Beryl field in the North Sea". Offshore Technology Conference, 11. Houston, Texas 1979. Proceedings, Vol. 3, pp. 1805-1815.

10. Kenley, R. M., and Dodds, C. J. (1980) "West Sole WE platform: Detection of damage by structural response measurements". Offshore Technology Conference, 12. Houston, Texas 1980. Proceedings, Vol. 4, pp. 111-118.

11. Simpson, W. F.(1979) "Instrumentation system for the Cognac platform". Offshore Technology Conference, 11. Houston, Texas 1979. Proceedings, Vol. 2, pp. 1245-1251.

12. Society for Underwater Technology, London. "Shell Brent "B" Instrumentation Project". Society for Underwater Technology Seminar, London, 1979. Proceedings. 116 p.

SUBJECT INDEX

Page numbers refer to first page of paper

AUTHOR INDEX

Page number refers to first page of paper